Delaunay Mesh Generation

CHAPMAN & HALL/CRC COMPUTER and INFORMATION SCIENCE SERIES

Series Editor: Sartaj Sahni

PUBLISHED TITLES

ADVERSARIAL REASONING: COMPUTATIONAL APPROACHES TO READING THE OPPONENT'S MIND
Alexander Kott and William M. McEneaney

DELAUNAY MESH GENERATION
Siu-Wing Cheng, Tamal Krishna Dey, and Jonathan Richard Shewchuk

DISTRIBUTED SENSOR NETWORKS, SECOND EDITION
S. Sitharama Iyengar and Richard R. Brooks

DISTRIBUTED SYSTEMS: AN ALGORITHMIC APPROACH
Sukumar Ghosh

ENERGY-AWARE MEMORY MANAGEMENT FOR EMBEDDED MULTIMEDIA SYSTEMS: A COMPUTER-AIDED DESIGN APPROACH
Florin Balasa and Dhiraj K. Pradhan

ENERGY EFFICIENT HARDWARE-SOFTWARE CO-SYNTHESIS USING RECONFIGURABLE HARDWARE
Jingzhao Ou and Viktor K. Prasanna

FUNDAMENTALS OF NATURAL COMPUTING: BASIC CONCEPTS, ALGORITHMS, AND APPLICATIONS
Leandro Nunes de Castro

HANDBOOK OF ALGORITHMS FOR WIRELESS NETWORKING AND MOBILE COMPUTING
Azzedine Boukerche

HANDBOOK OF APPROXIMATION ALGORITHMS AND METAHEURISTICS
Teofilo F. Gonzalez

HANDBOOK OF BIOINSPIRED ALGORITHMS AND APPLICATIONS
Stephan Olariu and Albert Y. Zomaya

HANDBOOK OF COMPUTATIONAL MOLECULAR BIOLOGY
Srinivas Aluru

HANDBOOK OF DATA STRUCTURES AND APPLICATIONS
Dinesh P. Mehta and Sartaj Sahni

HANDBOOK OF DYNAMIC SYSTEM MODELING
Paul A. Fishwick

HANDBOOK OF ENERGY-AWARE AND GREEN COMPUTING
Ishfaq Ahmad and Sanjay Ranka

HANDBOOK OF PARALLEL COMPUTING: MODELS, ALGORITHMS AND APPLICATIONS
Sanguthevar Rajasekaran and John Reif

HANDBOOK OF REAL-TIME AND EMBEDDED SYSTEMS
Insup Lee, Joseph Y-T. Leung, and Sang H. Son

HANDBOOK OF SCHEDULING: ALGORITHMS, MODELS, AND PERFORMANCE ANALYSIS
Joseph Y.-T. Leung

HIGH PERFORMANCE COMPUTING IN REMOTE SENSING
Antonio J. Plaza and Chein-I Chang

INTRODUCTION TO NETWORK SECURITY
Douglas Jacobson

LOCATION-BASED INFORMATION SYSTEMS: DEVELOPING REAL-TIME TRACKING APPLICATIONS
Miguel A. Labrador, Alfredo J. Pérez, and Pedro M. Wightman

METHODS IN ALGORITHMIC ANALYSIS
Vladimir A. Dobrushkin

PERFORMANCE ANALYSIS OF QUEUING AND COMPUTER NETWORKS
G. R. Dattatreya

THE PRACTICAL HANDBOOK OF INTERNET COMPUTING
Munindar P. Singh

SCALABLE AND SECURE INTERNET SERVICES AND ARCHITECTURE
Cheng-Zhong Xu

SOFTWARE APPLICATION DEVELOPMENT: A VISUAL C++®, MFC, AND STL TUTORIAL
Bud Fox, Zhang Wenzu, and Tan May Ling

SPECULATIVE EXECUTION IN HIGH PERFORMANCE COMPUTER ARCHITECTURES
David Kaeli and Pen-Chung Yew

VEHICULAR NETWORKS: FROM THEORY TO PRACTICE
Stephan Olariu and Michele C. Weigle

Delaunay Mesh Generation

Siu-Wing Cheng
Tamal Krishna Dey
Jonathan Richard Shewchuk

CRC Press is an imprint of the
Taylor & Francis Group, an informa business
A CHAPMAN & HALL BOOK

CRC Press
Taylor & Francis Group
6000 Broken Sound Parkway NW, Suite 300
Boca Raton, FL 33487-2742

Printed in the United States of America on acid-free paper
Version Date: 2012920

International Standard Book Number: 978-1-58488-730-0 (Hardback)

Library of Congress Cataloging-in-Publication Data

Cheng, Siu-Wing.
Delaunay mesh generation / Siu-Wing Cheng, Tamal K. Dey, Jonathan Shewchuk.
p. cm. -- (Chapman & Hall/CRC computer and information science series)
Includes bibliographical references and index.
ISBN 978-1-58488-730-0 (hardback)
1. Nets (Mathematics) 2. Numerical grid generation (Numerical analysis) 3. Triangulation. I. Dey, Tamal K. (Tamal Krishna), 1964- II. Shewchuk, Jonathan. III. Title.

QA611.3.C43 2012
514'.223--dc23 2012031208

Visit the Taylor & Francis Web site at
http://www.taylorandfrancis.com

and the CRC Press Web site at
http://www.crcpress.com

Contents

Preface

The study of algorithms for generating unstructured meshes of triangles and tetrahedra began with mechanical and aeronautical engineers who decompose physical domains into grids for the finite element and finite volume methods. Very soon, these engineers were joined by surveyors and specialists in geographical information systems, who use meshes called "triangulated irregular networks" to interpolate altitude fields on terrains, in the service of mapmaking, contouring, and visualization of topography. More recently, triangular meshes of surfaces have become prevalent as geometric models in computer graphics. These three groups of customers are the most numerous, but hardly the only ones; triangulations are found in most applications of multivariate interpolation.

Unfortunately, it is fiendishly hard to implement a reliable mesh generator. The demands on a mesh are heavy: it must conform to the geometric domain being modeled; it must contain triangles or tetrahedra of the correct shapes and sizes; it may have to grade from very short edges to very long ones over a short distance. These requirements are sometimes contradictory or even impossible. Most mesh generators are fragile, and sometimes fail when presented with a difficult domain, such as an object with many sharp angles or strangely curved boundaries.

One of the most exciting developments in computational geometry during the last several decades is the development of *provably good* mesh generation algorithms that offer guarantees on the quality of the meshes they produce. These algorithms make it easier to trust in the reliability of meshing software in unanticipated circumstances. Most mesh generators fall into one of three classes: advancing front mesh generators, which pave a domain with triangles or tetrahedra, laying down one at a time; meshers that decompose a domain by laying a grid, quadtree, or octree over it; and Delaunay mesh generators, which maintain a geometric structure called the *Delaunay triangulation* that has remarkable mathematical properties. To date, there are no provably good advancing front methods, and Delaunay meshers have proven to be more powerful and versatile than grid and octree algorithms, especially in their ability to cope with complicated domain boundaries.

In the past two decades, researchers have made progress in answering many intricate questions involving mesh generation: Can a mesher work for all input domains, including those with curved boundaries and sharp edges? If not, when and where must it make compromises? How accurately can a mesh composed of linear triangles or tetrahedra approximate the shape and topology of a curved domain? What guarantees can we make about the shapes and sizes of those triangles or tetrahedra? As a community, we now have algorithms that can tackle complex geometric domains ranging from polyhedra with internal boundaries to smooth surfaces to volumes bounded by piecewise smooth surfaces. And these algorithms come with guarantees.

This book is about algorithms for generating provably good Delaunay meshes, with an emphasis on algorithms that work well in practice. The guarantees they offer can include well-shaped triangles and tetrahedra, a reasonably small number of those triangles and tetrahedra, edges that are not unnecessarily short, topologically correct representations of curved domains, and geometrically accurate approximations of curved domains. As a foundation for these algorithms, the book also studies the combinatorial properties of Delaunay triangulations and their relatives, algorithms for constructing them, and their geometric and topological fidelity as approximations of smooth surfaces. After setting out the basic ideas of Delaunay mesh generation algorithms, we lavish attention on several particularly challenging problems: meshing domains with small angles; eliminating hard-to-remove "sliver" tetrahedra; and generating meshes that correctly match the topology and approximate the geometry of domains with smooth, curved surfaces or surface patches.

We have designed this book for two audiences: researchers, especially graduate students, and engineers who design and program mesh generation software. Algorithms that offer guarantees on mesh quality are difficult to design, so we emphasize rigorous mathematical foundations for proving that these guarantees hold and the core theoretical results upon which researchers can build even better algorithms in the future. However, one of the glories of provably good mesh generation is the demonstrated fact that many of its algorithms work wonderfully well in practice. We have included advice on how to implement them effectively. Although we promote a rigorous theoretical analysis of these methods, we have structured the book so readers can learn the algorithms without reading the proofs.

An important feature of this book is that it begins with a primer on Delaunay triangulations and constrained Delaunay triangulations in two and three dimensions, and some of the most practical algorithms for constructing and updating them. Delaunay triangulations are central to computational geometry and have found hundreds, probably thousands, of applications. Later chapters also cover Voronoi diagrams, weighted Voronoi diagrams, weighted Delaunay triangulations, restricted Voronoi diagrams, and restricted Delaunay triangulations. The last is a generalization of Delaunay triangulations that permits us to mesh surfaces in a rigorous, reliable way. We believe that this book is the first publication to combine so much information about these geometric structures in one place, and the first to give so much attention to modern algorithms.

The book can be divided into three parts of nearly equal length. The first part introduces meshes and the problem of mesh generation, defines Delaunay triangulations and describes their properties, and studies algorithms for their construction. The second part gives algorithms for generating high-quality meshes of polygonal and polyhedral domains. The third part uses restricted Delaunay triangulations to extend the algorithms to curved surfaces and domains whose boundaries are composed of curved ridges and patches.

The first chapter begins by describing the goals of mesh generation and telling a brief history of research in the field. Then it formally defines triangulations as simplicial complexes, and it defines the domains that those triangulations triangulate as other types of complexes. Chapters 2–5 cover Delaunay triangulations, constrained Delaunay triangulations, and algorithms for constructing and updating them in two and three dimensions. Chapter 2 introduces Delaunay triangulations of sets of points in the plane, their properties, and the geometric criteria that they optimize. It also introduces *piecewise linear complexes* (PLCs) as geometric structures for modeling polygonal domains; and triangulations of PLCs, particularly *constrained Delaunay triangulations* (CDTs), which generalize Delaunay triangulations to enforce the presence of specified edges. Chapter 3 presents algorithms for constructing Delaunay triangulations and CDTs, specifically, the incremental insertion and gift-wrapping algorithms. Chapter 4 extends Delaunay triangulations to higher dimensions and reviews geometric criteria that Delaunay triangulations of all dimensions optimize, some of which govern the accuracy of piecewise linear interpolation over triangles and tetrahedra. Chapter 5 reprises the incremental insertion and gift-wrapping algorithms for constructing Delaunay triangulations and CDTs in three dimensions.

Chapter 6 kicks off the middle third of the book with a discussion of *Delaunay refinement algorithms* for generating provably good triangular meshes of PLCs in the plane. Chapter 7 is an interlude in which we return to studying geometric complexes, including Voronoi diagrams, weighted Voronoi diagrams, and weighted Delaunay triangulations, which arm us with additional power to mesh polyhedral domains with small angles, eliminate some particularly troublesome tetrahedra of poor quality known as *slivers*, and handle curved surfaces.

Chapters 8–11 study algorithms for constructing tetrahedral meshes of polyhedral domains represented by three-dimensional PLCs. Chapter 8 presents a straightforward extension of the two-dimensional Delaunay refinement algorithm to three-dimensional domains with no acute angles. Chapter 9 describes an algorithm, new with this book, that meshes PLCs with small angles by constructing a *weighted* Delaunay triangulation. Chapters 10 and 11 describe a *sliver exudation* technique for removing slivers from a Delaunay mesh, thereby providing a mathematical guarantee on the quality of the tetrahedra. Although this guarantee is weak, the algorithm's success in practice exceeds what the theory promises. In both of these chapters, we have substantially improved the results in comparison with the previously published versions.

The final third of the book is devoted to meshing curved surfaces. A piecewise linear mesh cannot exactly conform to a curved surface, so we develop tools in approximation theory and topology to help guarantee the fidelity of a mesh to an underlying curved surface.

Chapter 12 covers topological spaces, homeomorphisms, isotopies, manifolds, and properties of point samples on manifolds. Chapter 13 introduces *restricted Voronoi diagrams*, whose Voronoi cells lie on a manifold, and their dual complexes, *restricted Delaunay triangulations*. We study conditions under which a restricted Delaunay triangulation is a topologically correct and geometrically close representation of a manifold. Chapter 14 describes mesh generation algorithms for curved surfaces and for the volumes they enclose; the meshes are restricted Delaunay triangulations. Chapter 15 makes the difficult jump from smooth surfaces to piecewise smooth surfaces, represented by a very general input domain called a *piecewise smooth complex* (PSC). PSCs bring with them all the difficulties that arise with polyhedral domains, such as enforcing boundary conformity and handling small domain angles, and all the difficulties that arise with smooth surfaces, such as guaranteeing topological correctness and small approximation errors. The algorithms described in the last two chapters and their analyses are considerably improved since their original publication.

At the end of each chapter, we provide historical and bibliographical notes and citations to acknowledge the pioneers who introduced the ideas in each chapter and to reference related ideas and publications. We include exercises, some of which we have assigned in graduate courses on mesh generation or computational geometry. We also use exercises as a way to include many interesting topics and improvements that we did not have enough room to discuss in detail, and theorems we did not have room to prove.

This book would have been impossible without help and advice. We thank our students who implemented versions of many of the algorithms presented in this book and generated pictures of the meshes they produced. Tathagata Ray meshed polyhedra with acute angles and remeshed polygonal surfaces. Joshua Levine meshed piecewise smooth complexes. Kuiyu Li helped to generate some of the figures. The Computational Geometry Algorithms Library (CGAL) project offered us a wonderful platform on which many of our implementations were carried out.

Joshua Levine read the manuscript at an early stage of the book and gave some valuable suggestions. Andrew Slatton read the last chapter carefully and pointed out some deficiencies in an early draft. For conversations that improved the writing in this book, we thank Nina Amenta, Marshall Bern, Jean-Daniel Boissonnat, L. Paul Chew, Herbert Edelsbrunner, Lori Freitag, Omar Ghattas, Anand Kulkarni, François Labelle, Gary Miller, Scott Mitchell, James O'Brien, David O'Hallaron, Edgar Ramos, Jim Ruppert, and Dafna Talmor. We also thank the researchers cited in the bibliography. Skype made possible our intercontinental conversations while writing this book.

We are grateful for the funding provided by the National Science Foundation, the Research Grant Council of Hong Kong, the University of California Lab Fees Research Program, the Alfred P. Sloan Foundation, and the Okawa Foundation that supported not only this book, but also much of the research that made it possible. We also wish to acknowledge the support of our departments: the Department of Computer Science and Engineering at the Hong Kong University of Science and Technology, the Department of Computer Science and Engineering at The Ohio State University, and the Department of Electrical Engineering and Computer Sciences at the University of California, Berkeley.

We are indebted to our families for encouraging us in our intellectual pursuits and giving us unfailing support throughout the years it took to write this book. Siu-Wing gives his thanks and love to Garmen Szeto and Nicole Cheng. Nicole is as old as this book project. The joint endeavour of book writing and child raising would not be as joyous without Garmen's immense energy and loving care. Tamal cannot thank enough his wife Kajari for taking care of many of the family chores which created the space for him to devote time to the book. Soumi and Sounak, their children, kept asking about the book. Their curiosity and enthusiasm helped Tamal remain engaged with the book for six years. Gopal Dey and Hasi Dey, his parents, were a constant inspiration even in their absence since it is they who implanted the love for books in Tamal. Jonathan wishes to acknowledge the cafés in which he wrote his part: the 61C Café in Pittsburgh, Masse's Pastries and Far Leaves Teas in Berkeley, and Leland Tea Company and Samovar Tea Lounge in San Francisco. But his favorite is still Chez Bill and Lynne Shewchuk in Cranbrook, British Columbia, Canada, the world's warmest writing retreat and maker of the best baked macaroni and cheese. His love and blessings rest upon this house.

Siu-Wing Cheng
Tamal Krishna Dey
Jonathan Richard Shewchuk
21 May 2012
Hong Kong, Columbus, and Cranbrook

Chapter 1

Introduction

One of the central tools of scientific computing is the fifty-year old *finite element method*—a numerical method for approximating solutions to partial differential equations. The finite element method and its cousins, the finite volume method and the boundary element method, simulate physical phenomena including fluid flow, heat transfer, mechanical deformation, and electromagnetic wave propagation. They are applied heavily in industry and science for diverse purposes—evaluating pumping strategies for petroleum extraction, modeling the fabrication and operation of transistors and integrated circuits, optimizing the aerodynamics of aircraft and car bodies, and studying phenomena from quantum mechanics to earthquakes to black holes.

The aerospace engineer Joe F. Thompson, who commanded a multi-institutional mesh generation effort called the National Grid Project, wrote in 1992 that

> An essential element of the numerical solution of partial differential equations (PDEs) on general regions is the construction of a grid (mesh) on which to represent the equations in finite form. ... [A]t present it can take orders of magnitude more man-hours to construct the grid than it does to perform and analyze the PDE solution on the grid. This is especially true now that PDE codes of wide applicability are becoming available, and grid generation has been cited repeatedly as being a major pacing item. The PDE codes now available typically require much less esoteric expertise of the knowledgeable user than do the grid generation codes.

Two decades later, meshes are still a recurring bottleneck. The *automatic mesh generation problem* is to divide a physical domain with a complicated geometry—say, an automobile engine, a human's blood vessels, or the air around an airplane—into small, simple pieces called *elements*, such as triangles or rectangles (for two-dimensional geometries) or tetrahedra or rectangular prisms (for three-dimensional geometries), as illustrated in Figure 1.1. Millions or billions of elements may be needed.

A mesh must satisfy nearly contradictory requirements: it must conform to the shape of the object or simulation domain; its elements may be neither too large nor too numerous; it may have to grade from small to large elements over a relatively short distance; and it must be composed of elements that are of the right shapes and sizes. "The right

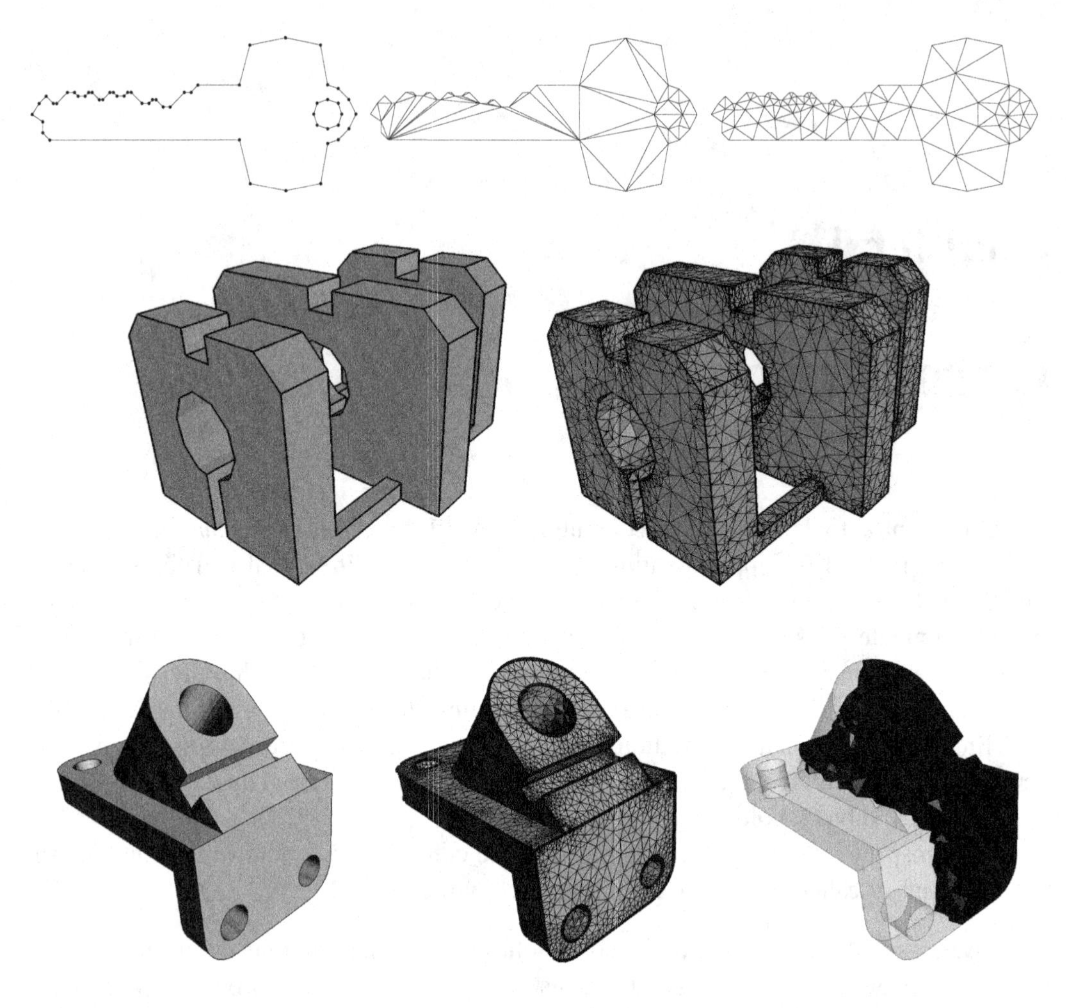

Figure 1.1: Finite element meshes of a polygonal, a polyhedral, and a curved domain. One mesh of the key has poorly shaped triangles and no Steiner points; the other has Steiner points and all angles between 30° and 120°. The cutaway view at lower right reveals some of the tetrahedral elements inside the mesh.

shapes" typically include elements that are nearly equilateral and equiangular, and typically exclude elements that are long and thin, for example, shaped like a needle or a kite. However, some applications require *anisotropic* elements that are long and thin, albeit with specified orientations and eccentricities, to interpolate fields with anisotropic second derivatives or to model anisotropic physical phenomena such as laminar air flow over an airplane wing.

By our reckoning, the history of mesh generation falls into three periods, conveniently divided by decade. The pioneering work was done by researchers from several branches of engineering, especially mechanics and fluid dynamics, during the 1980s—though as we shall see, the earliest work dates back to at least 1970. This period brought forth most of the techniques used today: the Delaunay, octree, and advancing front methods for mesh generation, and mesh "clean-up" methods for improving an existing mesh. Unfortunately, nearly all the algorithms developed during this period are fragile, and produce unsatisfying meshes when confronted by complex domain geometries and stringent demands on element shape.

Around 1988, these problems attracted the interest of researchers in computational geometry, a branch of theoretical computer science. Whereas most engineers were satisfied with mesh generators that usually work for their chosen domains, computational geometers set a loftier goal: *provably good mesh generation*, the design of algorithms that are mathematically guaranteed to produce a satisfying mesh, even for domain geometries unimagined by the algorithm designer. This work flourished during the 1990s and continues to this day. It is the subject of this book.

During the first decade of the 2000s, mesh generation became bigger than the finite element methods that gave birth to it. Computer animation uses triangulated surface models extensively, and the most novel new ideas for using, processing, and generating meshes often debut at computer graphics conferences. By economic measures, the videogame and motion picture industries probably now exceed the finite element industries as users of meshes.

Meshes today find heavy use in hundreds of other applications, such as aerial land surveying, image processing, geographic information systems, radio propagation analysis, shape matching, population sampling, and multivariate interpolation. Mesh generation has become a truly interdisciplinary topic.

1.1 Meshes and the goals of mesh generation

Meshes are categorized according to their dimensionality and choice of elements. *Triangular meshes*, *tetrahedral meshes*, *quadrilateral meshes*, and *hexahedral meshes* are named according to the shapes of their elements. The two-dimensional elements—triangles and quadrilaterals—serve both in modeling two-dimensional domains and in *surface meshes* embedded in three dimensions, which are prevalent in computer graphics, boundary element methods, and simulations of thin plates and shells.

Tetrahedral elements are the simplest of all polyhedra, having four vertices and four triangular faces. Quadrilateral elements are four-sided polygons; their sides need not be parallel. Hexahedral elements are brick-like polyhedra, each having six quadrilateral faces, but their faces need not be parallel or even planar. This book discusses only *simplicial meshes*—triangular and tetrahedral meshes—which are easier to generate than quadrilateral and hexahedral ones. For some applications, quadrilateral and hexahedral meshes offer more accurate interpolation and approximation. Non-simplicial elements sometimes make life easier for the numerical analyst; simplicial elements nearly always make life easier for the mesh generator. For topological reasons, hexahedral meshes can be extraordinarily difficult to generate for geometrically complicated domains.

Meshes are also categorized as structured or unstructured. A *structured mesh*, such as a regular cubical grid, has the property that its vertices can be numbered so that simple arithmetic suffices to determine which vertices share an element with a selected vertex. This book discusses only *unstructured meshes*, which entail explicitly storing each vertex's neighboring vertices or elements. All the meshes in Figure 1.1 are unstructured. Structured

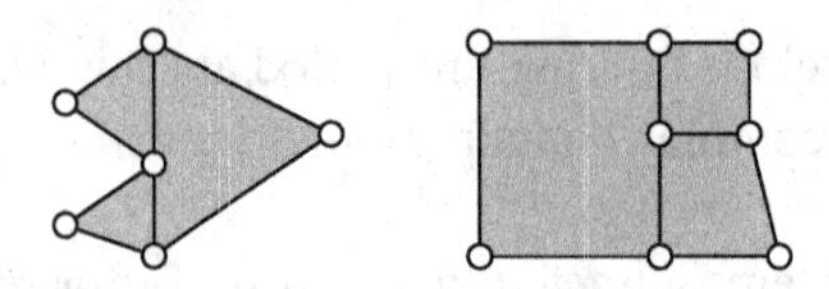

Figure 1.2: Nonconforming elements.

meshes are suitable primarily for domains that have tractable geometries and do not require a strongly graded mesh. Unstructured meshes are much more versatile because of their ability to combine good element shapes with odd domain shapes and element sizes that grade from very small to very large.

For most applications, the elements constituting a mesh must intersect "nicely," meaning that if two elements intersect, their intersection is a vertex or edge or entire face of both. Formally, a mesh must be a *complex*, defined in Section 1.5. The mesh generation problem becomes superficially easier if we permit what finite element practitioners call *nonconforming elements* like those illustrated in Figure 1.2, where an element shares an edge with two other elements each abutting half of that edge. But nonconforming elements rarely alleviate the underlying numerical problems and can be computationally expensive when they do, so they find limited use in unstructured meshes.

The goal of mesh generation is to create elements that *conform* to the shape of the geometric domain and meet constraints on their sizes and shapes. The next two sections discuss domain conformity and element quality.

1.1.1 Domain conformity

Mesh generation algorithms vary in what domains they can mesh and how those domains are specified. The input to a mesh generator might be a simple polygon or polyhedron. Meshing becomes more difficult if the domain can have *internal boundaries* that no element is permitted to cross, such as a boundary between two materials in a heat transfer simulation. Meshing is substantially more difficult for domains that have curved edges and surfaces, called *ridges* and *patches*, which are typically represented by splines, implicit equations, or subdivision surfaces. Each of these kinds of geometry requires a different definition of what it means to *triangulate* a domain. Let us consider these geometries in turn.

A polygon whose boundary is a closed loop of straight edges can be subdivided into triangles whose vertices all coincide with vertices of the polygon; see Section 2.10.1 for a proof of that fact. The set containing those triangles, their edges, and their vertices is called a *triangulation* of the polygon. But as the illustration at top center in Figure 1.1 illustrates, the triangles may be badly shaped. To mesh a polygon with only high-quality triangles, as illustrated at upper right in the figure, a mesh generator usually introduces additional vertices that are not vertices of the polygon. The added vertices are often called *Steiner points*, and the mesh is called a *Steiner triangulation* of the polygon.

Stepping into three dimensions, we discover that polyhedra can be substantially more difficult to triangulate than polygons. It comes as a surprise to learn that many polyhedra do not have triangulations, if a *triangulation* is defined to be a subdivision of a polyhedron into tetrahedra whose vertices are all vertices of the polyhedron. In other words, Steiner points are sometimes mandatory. See Section 4.5 for an example.

Internal boundaries exist to help apply boundary conditions for partial differential equations and to support discontinuities in physical properties, such as differences in heat conductivity in a multi-material simulation. A boundary, whether internal or external, must be represented by a union of edges or faces of the mesh. Elements cannot cross boundaries, and where two materials meet, their meshes must have matching edges and faces. This requirement may seem innocuous, but it makes meshing much harder if the domain has small angles. We define geometric structures called *piecewise linear complexes* to formally treat polygonal and polyhedral domains, like those at upper left and center left in Figure 1.1, in a manner that supports internal boundaries. Piecewise linear complexes and their triangulations are defined in Sections 2.10.1 and 4.5.1.

Curved domains introduce more difficulties. Some applications require elements that curve to match a domain. Others approximate a curved domain with a piecewise linear mesh at the cost of introducing inaccuracies in the shape, the finite element solutions, and the surface normal vectors (which are important for computer graphics). In finite element methods, curved domains are sometimes approximated with elements whose faces are described by parametrized quadratic, cubic, bilinear, or trilinear patches. In this book, the elements are always linear triangles and tetrahedra.

We study algorithms for several types of curved domain: in Chapters 12–14, we study how to mesh smooth surfaces with triangles and how to mesh volumes bounded by smooth surfaces with tetrahedra. Then we mesh more general domains like that at lower left in Figure 1.1, specified by geometric structures called *piecewise smooth complexes*. These complexes are composed of smoothly curved patches and ridges, but patches can meet nonsmoothly at ridges and vertices, and internal boundaries are permitted. Piecewise smooth complexes and their triangulations are defined in Chapter 15.

In this book, we assume that we have mathematically exact representations of domains and ignore the difficulties of numerical robustness and real-world CAD models, but we acknowledge that they are important issues.

1.1.2 Element quality

Most applications of meshes place constraints on both the shapes and sizes of the elements. These constraints come from several sources. First, large angles (near 180°) can cause large interpolation errors. In the finite element method, these errors induce a large *discretization error*—the difference between the computed approximation and the true solution of the PDE. Second, small angles (near 0°) can cause the stiffness matrices associated with the finite element method to be ill-conditioned. Small angles do not harm interpolation accuracy, and many applications can tolerate them. Third, smaller elements offer more accuracy, but cost more computationally. Fourth, small or skinny elements can induce instability in the explicit time integration methods employed by many time-dependent physical simulations. Let us consider these four constraints in turn.

The first constraint forbids large angles, including large plane angles in triangles and large dihedral angles (defined in Section 1.7) in tetrahedra. Most applications of triangulations use them to interpolate a multivariate function whose true value might or might not be

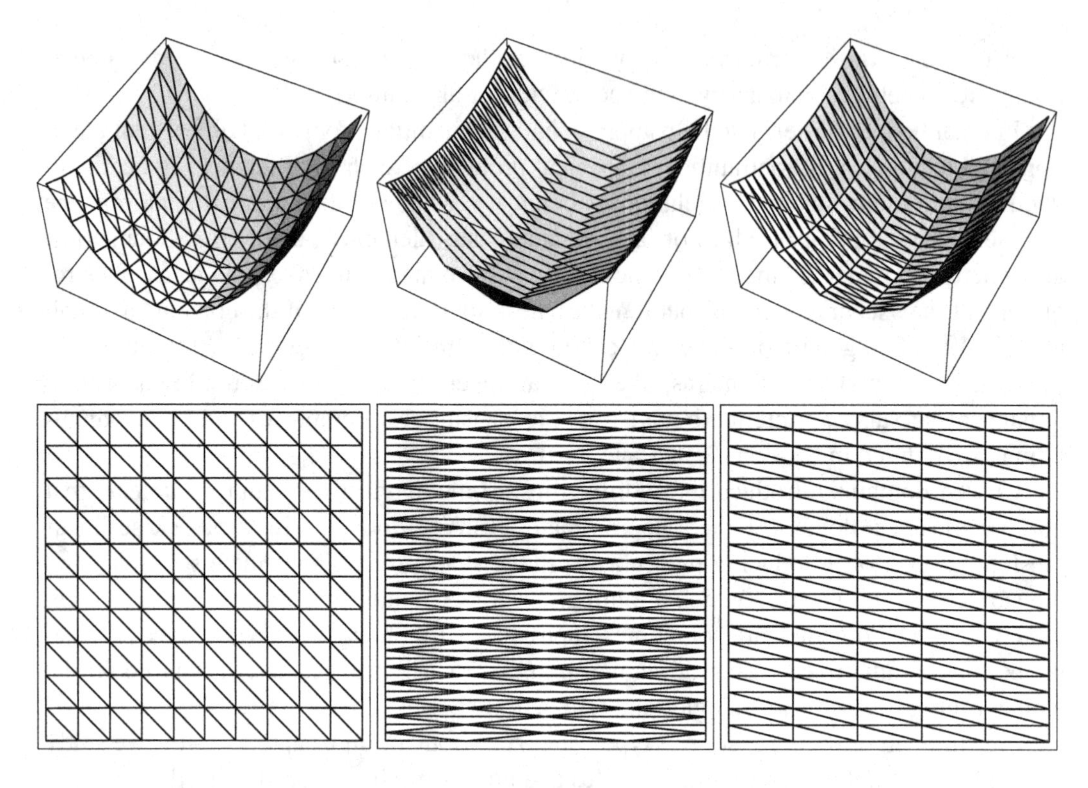

Figure 1.3: An illustration of how large angles, but not small angles, can ruin the interpolated gradients. Each triangulation uses 200 triangles to render a paraboloid.

known. For example, a surveyor may know the altitude of the land at each point in a large sample and use interpolation over a triangulation to approximate the altitude at points where readings were not taken. There are two kinds of *interpolation error* that matter for most applications: the difference between the interpolating function and the true function, and the difference between the gradient of the interpolating function and the gradient of the true function. Element shape is largely irrelevant for the first kind—the way to reduce interpolation error is to use smaller elements.

However, the error in the gradient depends on both the shapes and the sizes of the elements: it can grow arbitrarily large as an element's largest angle approaches 180°. In Figure 1.3, three triangulations, each having 200 triangles, are used to render a paraboloid. The mesh of long thin triangles at right has no angle greater than 90°, and visually performs only slightly worse than the high-quality triangulation at left. The slightly worse performance is because of the longer edge lengths. However, the middle paraboloid looks like a washboard, because the triangles with large angles have very inaccurate gradients.

Figure 1.4 shows why this problem occurs. Let f be a function—perhaps some physical quantity like temperature—linearly interpolated on the illustrated triangle. The values of f at the vertices of the bottom edge are 35 and 65, so the linearly interpolated value of f at the center of the edge is 50. This value is independent of the value associated with the top vertex. As the angle at the upper vertex approaches 180°, the interpolated point with value 50 becomes arbitrarily close to the upper vertex with value 40. Hence, the interpolated gradient ∇f can become arbitrarily large and is clearly specious as an approximation of the

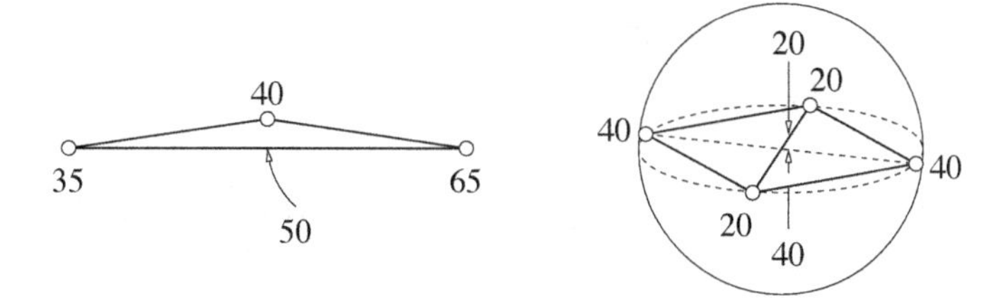

Figure 1.4: As the large angle of the triangle approaches 180°, or the sliver tetrahedron becomes arbitrarily flat, the magnitude of the interpolated gradient becomes arbitrarily large.

true gradient. The same effect is seen between two edges of a sliver tetrahedron that pass near each other, also illustrated in Figure 1.4.

In the finite element method, the discretization error is usually proportional to the error in the gradient, although the relationship between the two depends on the PDE and the order of the basis functions used to discretize it. In surface meshes for computer graphics, large angles cause triangles to have normal vectors that poorly approximate the normal to the true surface, and these can create visual artifacts in rendering. We derive bounds on this approximation in Section 12.7.2.

For tetrahedral elements, usually it is their largest dihedral angles that matter most. Nonconvex quadrilateral and hexahedral elements, with interior angles exceeding 180°, sabotage interpolation and the finite element method.

The second constraint on mesh generators is that many applications forbid small angles, although fewer than those that forbid large angles. If the application is the finite element method, then the eigenvalues of the stiffness matrix associated with the method ideally should be clustered as close together as possible. Matrices with poor eigenvalue spectra affect linear equation solvers by slowing down iterative methods and introducing large roundoff errors into direct methods. The relationship between element shape and matrix conditioning depends on the PDE being solved and the basis functions and test functions used to discretize it, but as a rule of thumb, it is the small angles that are deleterious: the largest eigenvalue of the stiffness matrix approaches infinity as an element's smallest angle approaches zero. Fortunately, most linear equation solvers cope well with a few bad eigenvalues.

The third constraint on mesh generators governs element size. Many mesh generation algorithms take as input not just the domain geometry, but also a space-varying *size field* that specifies the ideal size, and sometimes anisotropy, of an element as a function of its position in the domain. (The size field is often implemented by interpolation over a *background mesh.*) A large number of *fine* (small) elements may be required in one region where they are needed to attain good accuracy—often where the physics is most interesting, as amid turbulence in a fluid flow simulation—whereas other regions might be better served by *coarse* (large) elements, to keep their number small and avoid imposing an overwhelming computational burden on the application. The ideal element in one part of the mesh may vary in volume by a factor of a million or more from the ideal element in another part of the mesh. If elements of uniform size are used throughout the mesh, one must choose a size small enough to guarantee sufficient accuracy in the most demanding portion of the problem domain and thereby incur excessively large computational demands.

Figure 1.5: A mesh of this domain must have a *new* small angle.

A *graded mesh* is one that has large disparities in element size. Ideally, a mesh generator should be able to *grade* from very small to very large elements over a short distance. However, overly aggressive grading introduces skinny elements in the transition region. The size field alone does not determine element size: mesh generators often create elements smaller than specified to maintain good element quality in a graded mesh and to conform to small geometric features of a domain.

Given a coarse mesh—one with relatively few elements—it is typically easy to *refine* it, guided by the size field, to produce another mesh having a larger number of smaller elements. The reverse process is much harder. Hence, mesh generation algorithms often set themselves the goal of being able, in principle, to generate as coarse a mesh as possible.

The fourth constraint forbids unnecessarily small or skinny elements for time-dependent PDEs solved with explicit time integration methods. The stability of explicit time integration is typically governed by the *Courant–Friedrichs–Lewy condition*, which implies that the computational time step must be small enough that a half-wave or other time-dependent signal cannot cross more than one element per time step. Therefore, elements with short edges or short altitudes may force a simulation to take unnecessarily small time steps, at great computational cost, or risk introducing a large dose of spurious energy that causes the simulation to "explode."

These four constraints can be difficult to reconcile. Some meshing problems are impossible. A polygonal domain that has a corner bearing a 1° angle obviously cannot be meshed with triangles whose angles all exceed 30°; but suppose we merely ask that all angles be greater than 30° *except* the 1° angle? This request can always be granted for a polygon with no internal boundaries, but Figure 1.5 depicts a domain composed of two polygons glued together that, surprisingly, provably has no mesh whose *new* angles are all over 30°. Simple polyhedra in three dimensions inherit this hurdle, even without internal boundaries. One of the biggest challenges in mesh generation is three-dimensional domains with small angles and internal boundaries, wherein an arbitrary number of ridges and patches can meet at a single vertex. Chapters 9 and 15 present algorithms for meshing linear and curved domains with these difficulties.

1.2 Delaunay triangulations and Delaunay refinement

This book is about provably good mesh generation algorithms that employ the *Delaunay triangulation*, a geometric structure possessed of mathematical properties uniquely well suited to creating good triangular and tetrahedral meshes. The defining property of a Delaunay triangulation in the plane is that no vertex of the triangulation lies in the interior

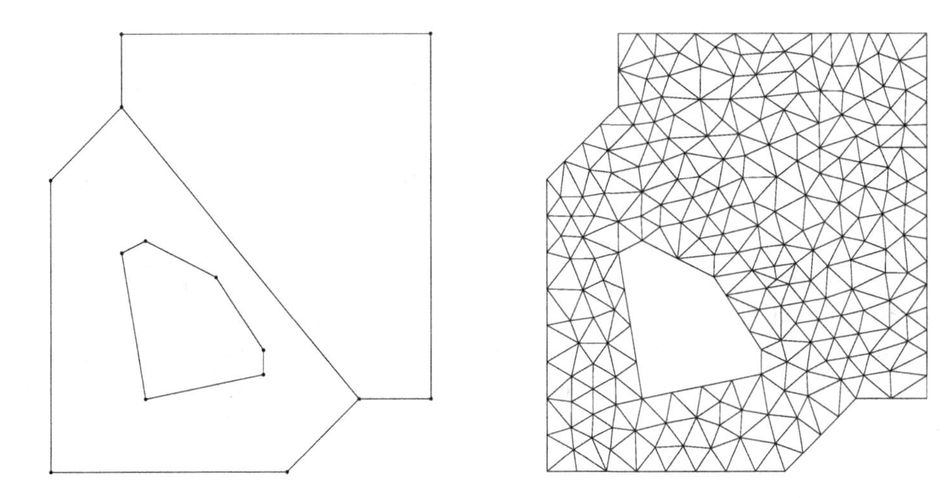

Figure 1.6: A mesh generated by Chew's first Delaunay refinement algorithm.

of any triangle's *circumscribing disk*—the unique circular disk whose boundary touches the triangle's three vertices. In three dimensions, no vertex is enclosed by any tetrahedron's circumscribing sphere. Delaunay triangulations optimize several geometric criteria, including some related to interpolation accuracy.

Delaunay refinement algorithms construct a Delaunay triangulation and refine it by inserting new vertices, chosen to eliminate skinny or oversized elements, while always maintaining the Delaunay property of the mesh. The key to ensuring good element quality is to prevent the creation of unnecessarily short edges. The Delaunay triangulation serves as a guide to finding locations to place new vertices that are far from existing ones, so that short edges and skinny elements are not created needlessly.

As a preview, consider the first provably good Delaunay refinement algorithm, invented by Paul Chew, which takes as input a polygonal domain and generates a triangular mesh whose angles are all between 30° and 120°. (The input polygon may not have an angle less than 30°.) Chew begins by subdividing the polygon's edges so that all the edge lengths are in a range $[h, \sqrt{3}h]$, where h is chosen small enough that such a subdivision exists with no two edge endpoints closer to each other than h. Next, he constructs the *constrained Delaunay triangulation* of the subdivision, defined in Section 2.10.2. Finally, he refines the triangulation by repeatedly choosing a triangle whose circumscribing disk has radius greater than h and inserting a new vertex at the center of the circumscribing disk, until no such triangle survives. The vertex is inserted by an algorithm that maintains the constrained Delaunay property of the mesh and thereby eliminates the skinny triangle. Chew's algorithm is quite useful in practice, but it generates only meshes with uniformly sized triangles, as Figure 1.6 illustrates, and not graded meshes.

The first third of this book lays out the mathematical underpinnings of Delaunay triangulations and the most practical algorithms for constructing them. The second third of this book describes Delaunay refinement algorithms for domains expressed as *piecewise linear complexes*, which generalize polygons and polyhedra to support internal boundaries. The final third of this book describes Delaunay refinement algorithms for curved domains—specifically, smooth surfaces, volumes bounded by smooth surfaces, and piecewise smooth domains that have curved ridges and patches and are represented by *piecewise smooth complexes*.

1.3 A brief history of mesh generation

Three classes of mesh generation algorithms predominate nowadays: advancing front methods, wherein elements crystallize one by one, coalescing from the boundary of a domain to its center; grid, quadtree, and octree algorithms, which overlay a structured background grid and use it as a guide to subdivide a domain; and Delaunay refinement algorithms, the subject of this book. An important fourth class is mesh improvement algorithms, which take an existing mesh and make it better through local optimization. The few fully unstructured mesh generation algorithms that do not fall into one of these four categories are not yet in widespread use.

Automatic unstructured mesh generation for finite element methods began in 1970 with an article by C. O. Frederick, Y. C. Wong, and F. W. Edge entitled "Two-Dimensional Automatic Mesh Generation for Structural Analysis" in the *International Journal for Numerical Methods in Engineering*. This startling paper describes, to the best of our knowledge, the first Delaunay mesh generation algorithm, the first advancing front method, and the first algorithm for Delaunay triangulations in the plane besides slow exhaustive search—all one and the same. The irony of this distinction is that the authors appear to have been unaware that the triangulations they create are Delaunay. Moreover, a careful reading of their paper reveals that their meshes are *constrained* Delaunay triangulations, a sophisticated variant of Delaunay triangulations which we discuss in Section 2.10.2. The paper is not well known, perhaps because it was two decades ahead of its time.

Advancing front methods construct elements one by one, starting from the domain boundary and advancing inward, as illustrated in Figure 1.7—or occasionally outward, as when meshing the air around an airplane. The frontier where elements meet unmeshed domain is called the *front*, which ventures forward until the domain is paved with elements and the front vanishes. Advancing front methods are characterized by exceptionally high quality elements at the domain boundary. The worst elements appear where the front collides with itself, and assuring their quality is difficult, especially in three dimensions; there is no literature on provably good advancing front algorithms. Advancing front methods have been particularly successful in fluid mechanics, because it is easy to place extremely

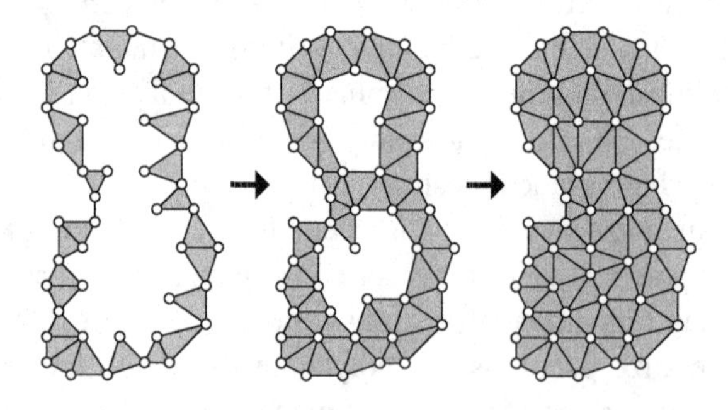

Figure 1.7: Advancing front mesh generation.

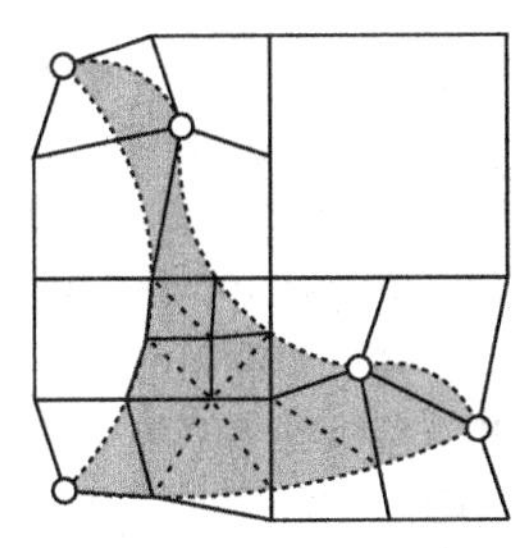

Figure 1.8: A quadtree mesh.

anisotropic elements or specialized elements at the boundary, where they are needed to model phenomena such as laminar air flow.

Most early methods created vertices and then triangulated them in two separate stages. For instance, Frederick, Wong, and Edge use "a magnetic pen to record node point data and a computer program to generate element data." The simple but crucial next insight—arguably, the "true" advancing front technique—was to interleave vertex creation with element creation, so the front can guide the placement of vertices. Alan George took this step in his 1971 doctoral dissertation, but it was forgotten and reinvented several times, and finally became widespread around 1988.

Most Delaunay mesh generators, unlike advancing front methods, create their worst elements near the domain boundary and their best elements in the interior. The early Delaunay mesh generators, like the early advancing front methods, created vertices and triangulated them in two separate stages. The era of modern meshing began in 1987 with the insight, courtesy of William Frey, to use the triangulation as a search structure to decide where to place the vertices. *Delaunay refinement* is the notion of maintaining a Delaunay triangulation while inserting vertices in locations dictated by the triangulation itself. The advantage of Delaunay methods, besides the optimality properties of the Delaunay triangulation, is that they can be designed to have mathematical guarantees: that they will always construct a valid mesh and, at least in two dimensions, that they will never produce skinny elements.

The third class of mesh generators is those that overlay a domain with a background grid whose resolution is small enough that each of its cells overlaps a very simple, easily triangulated portion of the domain, as illustrated in Figure 1.8. A variable-resolution grid, usually a quadtree or octree, yields a graded mesh. Element quality is usually assured by warping the grid so that no short edges appear when the cells are triangulated, or by improving the mesh afterward.

Grid meshers place excellent elements in the domain interior, but the elements near the domain boundary are worse than with other methods. Other disadvantages are the tendency for most mesh edges to be aligned in a few preferred directions, which may influence subsequent finite element solutions, and the difficulty of creating anisotropic elements that are not aligned with the grid. Their advantages are their speed, their ease of parallelism, the fact that some of them have mathematical guarantees, and most notably, their robustness for meshing imprecisely specified geometry and dirty CAD data. Mark Yerry and Mark Shephard published the first quadtree mesher in 1983 and the first octree mesher in 1984.

From nearly the beginning of the field, most mesh generation systems have included a mesh "clean-up" component that improves the quality of a finished mesh. Today, simplicial mesh improvement heuristics offer by far the highest quality of all the methods, and excellent control of anisotropy. Their disadvantages are the requirement for an initial mesh and a lack of mathematical guarantees. (They can guarantee they will not make the mesh worse.)

The ingredients of a mesh improvement method are a set of local transformations, which replace small groups of tetrahedra with other tetrahedra of better quality, and a schedule that searches for opportunities to apply them. *Smoothing* is the act of moving a vertex to improve the quality of the elements adjoining it. Smoothing does not change the connectivity (topology) of the mesh. *Topological transformations* are operations that change the mesh connectivity by removing elements from a mesh and replacing them with a different configuration of elements occupying the same space.

The simplest topological transformation is the *edge flip* in a triangular mesh, which replaces two adjacent triangles with two different triangles. There are analogous transformations for tetrahedra, quadrilaterals, and hexahedra. Simple transformations called *bistellar flips* that act on triangles and tetrahedra are discussed in Section 4.4. In Chapter 10, we describe a provably good algorithm called *sliver exudation* that uses bistellar flips to improve Delaunay meshes.

The story of provably good mesh generation is an interplay of ideas between Delaunay methods and methods based on grids, quadtrees, and octrees. It began in 1988, when Brenda Baker, Eric Grosse, and Conor Rafferty gave an algorithm to triangulate a polygon so that all the new angles in the mesh are between 14° and 90°. They overlay the polygon with a fine square grid, create new vertices at some grid points and at some intersections between grid lines and the polygon boundary, and triangulate them with a complicated case analysis.

The following year, Paul Chew gave a more practical algorithm, which we have described in Section 1.2, that uses Delaunay refinement to guarantee angles between 30° and 120°. In 1992, Dey, Bajaj, and Sugihara generalized Chew's algorithm to generate tetrahedral meshes of convex polyhedral domains. Although their algorithm is guaranteed to eliminate most types of bad tetrahedra, a few bad tetrahedra slip through: a type of tetrahedron called a *sliver* or *kite*.

The canonical sliver is formed by arranging four vertices around the equator of a sphere, equally spaced, then perturbing one of the vertices slightly off the equator, as Figure 1.9 illustrates. A sliver can have dihedral angles arbitrarily close to 0° and 180° yet have no edge that is particularly short. Provably good sliver removal is one of the most difficult theoretical problems in mesh generation, although mesh improvement algorithms beat slivers consistently in practice.

None of the provably good algorithms discussed above produce graded meshes. The 1990 quadtree algorithm of Marshall Bern, David Eppstein, and John Gilbert meshes a polygon so no new angle is less than 18.4°. It has been influential in part because the meshes it produces are not only graded, but *size-optimal*: the number of triangles in a mesh is at most a constant factor times the number in the smallest possible mesh (measured by triangle count) having no angle less than 18.4°. Ironically, the algorithm produces too many triangles to be practical—but only by a constant factor.

In 1992, Scott Mitchell and Stephen Vavasis developed an octree algorithm that offers guarantees on dihedral angles, grading, and size optimality. The bounds are not strong

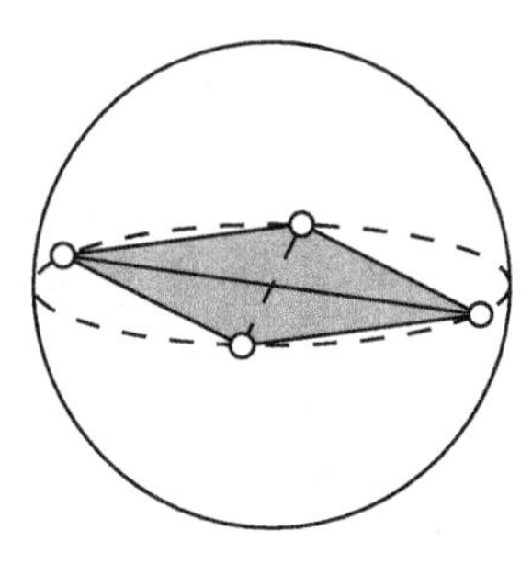

Figure 1.9: The mesh generator's nemesis: a sliver tetrahedron.

enough to be meaningful in practice and are not explicitly stated. Nevertheless, the papers by Bern et al. and Mitchell and Vavasis decidedly broadened the ambitions of provably good meshing.

A groundbreaking 1992 paper by Jim Ruppert on triangular meshing brought guaranteed good grading and size optimality to Delaunay refinement algorithms. Ruppert's algorithm, described in Chapter 6, accepts nonconvex domains with internal boundaries and produces graded meshes of modest size and high quality in practice.

Domains with curved geometries, represented by splines, isosurfaces, or other surface representations, increase the challenge appreciably. Most early algorithms for meshing surfaces work in the parametric space of a spline, but most grid and octree methods work directly in three-dimensional space, as do a few advancing front methods. A 1993 paper by Paul Chew partly generalizes Delaunay triangulations to curved surfaces. He proposes an algorithm that takes a triangulation of a spline patch, flips its edges to make it Delaunay, and refines it. If the initial mesh is fine enough, the triangles in the final mesh are guaranteed to have high quality.

These early works in guaranteed-quality mesh generation launched a rapid escalation of research on the subject, in which we were fortunate to participate.

1.4 A personal history of working in mesh generation

When we came to study mesh generation in the 1990s, we were drawn by the unusually strong way it combines theory and practice, complexity and elegance, and combinatorial and numerical computing. There is a strong tradition of practical meshing algorithms in scientific computing and computer graphics, yet their difficulty and fragility bring up fundamental theoretical questions in approximation theory, surface sampling, topology, algorithm design, numerical computing, and the structure of Delaunay triangulations and their weighted and constrained relatives. Mesh generation demands an understanding of both combinatorial and numerical algorithms, because meshing is geometric and most meshes are used by numerical applications. Lastly, meshes and their applications are as attractive to the eye as their mathematics are to the soul.

Galvanized by the publication of Ruppert's algorithm, Jonathan generalized it to three dimensions in 1997. The tetrahedral Delaunay refinement algorithm described in Chapter 8 accepts nonconvex domains with internal boundaries and offers guaranteed good grading.

However, it is not guaranteed to eliminate slivers, which implies (for technical reasons) that it cannot guarantee size optimality.

It soon became apparent that there are two serious difficulties in developing a truly satisfying meshing algorithm for polyhedral domains. First, domain faces that meet at small angles are particularly difficult to mesh, especially if many edges and faces, including internal boundaries, converge at a point. Second, although Delaunay refinement algorithms naturally eliminate most types of bad tetrahedra, they cannot guarantee that there will be no slivers, and even successful attempts to eliminate slivers in practice tend to overrefine the mesh. Researchers have made progress on both problems, but they are still areas of active research.

It is sometimes impossible to place high-quality tetrahedra at the apex of a small domain angle, so a mesh generation algorithm must know when and where to relax its guarantees on tetrahedron quality. In Chapter 9, we present a new algorithm that uses a variant of the Delaunay triangulation called a *weighted* Delaunay triangulation to help enforce domain conformity near small angles. The algorithm includes contributions from all three of us and several other collaborators, as we have collectively worked on this problem for over a decade.

In 1999, Siu-Wing and Tamal participated in the development of a provably good technique called *sliver exudation* for removing the worst slivers from a Delaunay mesh. Like our method for treating small domain angles, sliver exudation uses a weighted Delaunay triangulation; it removes slivers by shifting the weights of the vertices. We describe this technique in Chapter 10, and how to combine it with Delaunay refinement in Chapter 11. Since the original paper, Jonathan has joined the collaboration and together we have tightened the analysis considerably.

Surface meshing has been a particularly absorbing and rewarding research topic for us. It has compelled researchers to bring topology and approximation theory into mesh generation to help prove that certain meshes are topologically and geometrically accurate representations of curved domains. In 1997, Herbert Edelsbrunner and Nimish Shah took a large step forward by introducing the *restricted Delaunay triangulation*, a subcomplex of the three-dimensional Delaunay triangulation that serves as a surface mesh under the right conditions. Specifically, they prove a result known as the Topological Ball Theorem, which states that if the intersection of each face of a Voronoi diagram with a surface is a topological ball of the right dimension, then the restricted Delaunay triangulation is topologically equivalent to the surface. We define restricted Delaunay triangulations in Section 13.1 and state the Topological Ball Theorem in Section 13.2.

Provably good surface meshing draws on ideas in sampling theory originally developed for the problem of reconstructing the shape of a three-dimensional object from a finite set of points sampled from its surface by a laser scanner or stereo photography. In a seminal work from 1999, Nina Amenta and Marshall Bern use sampling theory and the Topological Ball Theorem to show that if a smooth surface is sampled sufficiently densely, the Delaunay tetrahedralization of the sample points includes a subset of triangles that accurately reconstruct the surface, by both topological and geometric criteria.

The recognition of these remarkable connections prompted us and other researchers to develop surface meshing algorithms with topological and geometric guarantees. Tamal collaborated with Ho-Lun Cheng, Herbert Edelsbrunner, and John Sullivan in 2001 to develop a Delaunay refinement algorithm that chooses sample points and computes topologically correct triangular meshes for a class of smooth surfaces called *skin surfaces*. This work includes further developments in sampling theory that suggest how to choose vertices to ensure that the preconditions of the Topological Ball Theorem hold for more general classes of surfaces. Independently in 2003, Jean-Daniel Boissonnat and Steve Oudot developed a similar sampling theory and a simple Delaunay refinement algorithm for a more general class of smooth surfaces. We devote Chapters 12 and 13 to developing an updated sampling theory for smooth surfaces and restricted Delaunay triangulations. In Chapter 14, we study mesh generation algorithms that depend on this theory, including several algorithms for generating a triangular mesh of a smooth surface, and an algorithm for generating a tetrahedral mesh of a volume bounded by a smooth surface.

Meshing is yet more difficult for curved domains that are only piecewise smooth, with smooth surface patches meeting along smoothly curved ridges. With Edgar Ramos, Siu-Wing and Tamal introduced the first provably good algorithm for such domains in 2007. It uses a weighted Delaunay triangulation to enforce domain conformity at the corners and creases (ridges) of the domain, and motivates the development of additional sampling theory to ensure the topological correctness of the mesh. In the years following, this algorithm was made more practical with additional contributions from Josh Levine. Our final chapter describes a considerably updated and improved version of this algorithm.

As part of our research, we have developed several mesh generation packages that are publicly available on the web. Most of the images of meshes in this book were generated by these programs. Jonathan Shewchuk's program TRIANGLE[1] robustly constructs constrained Delaunay triangulations and high-quality triangular meshes in the plane, using Ruppert's algorithm to generate the latter. In 2003, TRIANGLE received the James Hardy Wilkinson Prize in Numerical Software. Bryan Klingner and Jonathan Shewchuk also offer a tetrahedral mesh improvement program STELLAR[2] that employs algorithms not discussed in this book (as they do not use Delaunay triangulations).

In collaboration with Edgar Ramos and Tathagata Ray, Siu-Wing and Tamal developed an algorithm for generating tetrahedral meshes of polyhedral domains with small angles and another algorithm for remeshing polygonal surfaces. Tathagata Ray implemented these two algorithms and released the programs QUALMESH[3] and SURFREMESH[4]. SURFREMESH is a precursor of the more practical algorithm DELSURF we describe in Chapter 14. Together with Josh Levine, Tamal designed an algorithm for generating triangular and tetrahedral meshes of piecewise smooth complexes. Josh Levine implemented the algorithm and released the program DELPSC[5], which is a precursor of the algorithm we describe in Chapter 15. We have taken the liberty of including illustrations generated by these programs throughout the book as prototypes, even though we have subsequently improved many of the algorithms.

---

[1] http://www.cs.cmu.edu/~quake/triangle.html

[2] http://www.cs.berkeley.edu/~jrs/stellar

[3] http://www.cse.ohio-state.edu/~tamaldey/qualmesh.html

[4] http://www.cse.ohio-state.edu/~tamaldey/surfremesh.html

[5] http://www.cse.ohio-state.edu/~tamaldey/delpsc.html

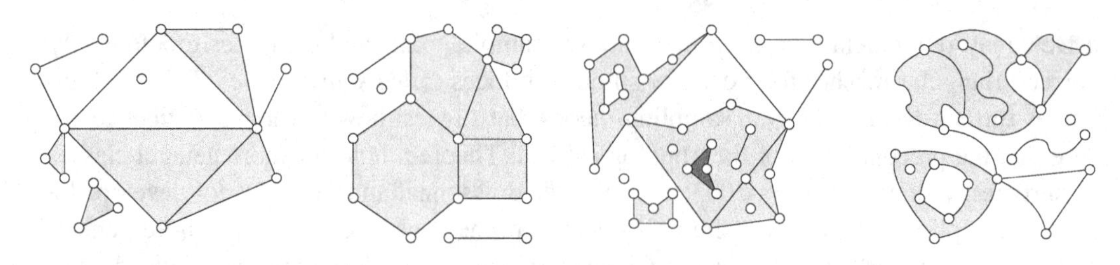

Figure 1.10: From left to right, a simplicial complex, a polyhedral complex, a piecewise linear complex, and a piecewise smooth complex. The shaded areas are triangles, convex polygons, linear 2-cells, and smooth 2-cells, respectively. In the piecewise linear complex, observe that several linear cells have holes, one of which is filled by another linear cell (darkly shaded).

1.5 Simplices, complexes, and polyhedra

Tetrahedra, triangles, edges, and vertices are instances of *simplices*. In this book, we represent meshes and the domains we wish to mesh as *complexes*. There are several different types of complexes, illustrated in Figure 1.10, which all share two common properties. First, a complex is a set that contains not only volumes such as tetrahedra, but also the facets, edges, and vertices of those volumes. Second, the cells in a complex must intersect each other according to specified rules, which depend on the type of complex.

The simplest type of complex is a *simplicial complex*, which contains only simplices. All the mesh generation algorithms in this book produce simplicial complexes. More general are *polyhedral complexes*, composed of convex polyhedra; these "polyhedra" can be of any dimension from zero on up. The most important polyhedral complexes we study in this book are the famous *Voronoi diagram*, defined in Section 7.1, and the *Delaunay subdivision*, defined in Section 2.2.

We use two other kinds of complexes to specify domains to be triangulated. *Piecewise linear complexes*, defined in Sections 2.10.1 and 4.5.1, differ from polyhedral complexes by permitting nonconvex polyhedra and by relaxing the rules of intersection of those polyhedra. *Piecewise smooth complexes*, defined in Section 15.1, generalize straight edges and flat facets to curved ridges and patches.

To a mathematician, a "triangle" is a set of points, which includes all the points inside the triangle as well as the points on the three edges. Likewise, a polyhedron is a set of points covering its entire volume. A complex is a set of sets of points. We define these and other geometric structures in terms of affine hulls and convex hulls. Simplices, convex polyhedra, and their faces are convex sets of points. A point set C is *convex* if for every pair of points $p, q \in C$, the line segment pq is included in C.

Definition 1.1 (affine hull; flat). Let $X = \{x_1, x_2, \ldots, x_k\}$ be a set of points in $\mathbb{R}^d$. An *affine combination* of the points in X is a point p that can be written $p = \sum_{i=1}^{k} w_i x_i$ for a set of scalar *weights* w_i such that $\sum_{i=1}^{k} w_i = 1$. A point p is *affinely independent* of X if it is not an affine combination of points in X. The points in X are *affinely independent* if no point in X is an affine combination of the others. In $\mathbb{R}^d$, no more than $d + 1$ points can be affinely independent. The *affine hull* of X, denoted $\operatorname{aff} X$, is the set of all affine combinations of

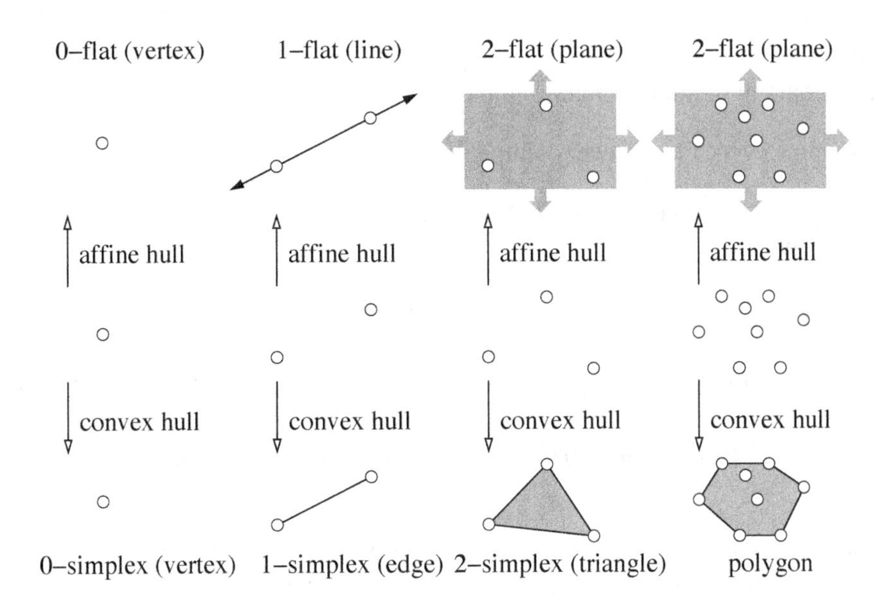

Figure 1.11: Examples of affine hulls and convex hulls in the plane.

points in X, as illustrated in Figure 1.11. A *k-flat*, also known as an *affine subspace*, is the affine hull of $k + 1$ affinely independent points; so a 0-flat is a vertex, a 1-flat is a line, a 2-flat is a plane, etc. A $(d - 1)$-flat in $\mathbb{R}^d$ is called a *hyperplane*. A k-flat is said to have *dimension k*.

Definition 1.2 (convex hull). A *convex combination* of the points in X is a point that can be written as an affine combination with all the weights nonnegative; i.e. $w_i \geq 0$ for all i. The *convex hull* of X, denoted conv X, is the set of all convex combinations of points in X, as illustrated in Figure 1.11. An alternative definition is that conv X is the most exclusive convex point set such that $X \subseteq \text{conv}\, X$.

Simplices and convex polyhedra are convex hulls of finite point sets, with k-simplices being the simplest possible k-dimensional polyhedra. One way that mathematical language deviates from lay usage is that a "face" of a polyhedron can be of any dimension; mathematicians use "facet" to denote what a layman calls a "face."

Definition 1.3 (simplex). A *k-simplex* τ is the convex hull of a set X of $k + 1$ affinely independent points. In particular, a 0-simplex is a *vertex*, a 1-simplex is an *edge*, a 2-simplex is a *triangle*, and a 3-simplex is a *tetrahedron*. A k-simplex is said to have *dimension k*. A *face* of τ is a simplex that is the convex hull of a nonempty subset of X. Faces of τ come in all dimensions from zero[6] (τ's vertices) to k; τ is a face of τ. A *proper face* of τ is a simplex that is the convex hull of a proper subset of X; i.e. any face except τ. In particular, the $(k - 1)$-faces of τ are called *facets* of τ; τ has $k + 1$ facets. For instance, the facets of a tetrahedron are its four triangular faces.

[6]Some writers use the convention that the empty set is a simplex of dimension -1 and a face of every simplex, albeit not a proper face. We make no use of this convention.

Definition 1.4 (simplicial complex). A *simplicial complex* $\mathcal{T}$, also known as a *triangulation*, is a set containing finitely[7] many simplices that satisfies the following two restrictions.

- $\mathcal{T}$ contains every face of every simplex in $\mathcal{T}$.
- For any two simplices $\sigma, \tau \in \mathcal{T}$, their intersection $\sigma \cap \tau$ is either empty or a face of both σ and τ.

Convex polyhedra are as easy to define as simplices, but their faces are trickier. Whereas the convex hull of a subset of a simplex's vertices is a face of the simplex, the convex hull of an arbitrary subset of a cube's vertices is usually not a face of the cube. The faces of a polyhedron are defined below in terms of *supporting hyperplanes*; observe that the definition of a face of a polyhedron below is consistent with the definition of a face of a simplex above.

Definition 1.5 (convex polyhedron). A *convex polyhedron* is the convex hull of a finite point set. A convex polyhedron whose affine hull is a k-flat is called a *k-polyhedron* and is said to have *dimension k*. A 0-polyhedron is a vertex, a 1-polyhedron is an edge, and a 2-polyhedron is a *convex polygon*. The *proper faces* of a convex polyhedron C are the polyhedra that can be generated by taking the intersection of C with a hyperplane that intersects C's boundary but not C's interior; such a hyperplane is called a *supporting hyperplane* of C. For example, the proper faces of a cube are six squares, twelve edges, and eight vertices. The *faces* of C are the proper faces of C and C itself. The *facets* of a k-polyhedron are its $(k-1)$-faces.

A polyhedral complex imposes exactly the same restrictions as a simplicial complex.

Definition 1.6 (polyhedral complex). A *polyhedral complex* $\mathcal{P}$ is a set containing finitely many convex polyhedra that satisfies the following two restrictions.

- $\mathcal{P}$ contains every face of every polyhedron in $\mathcal{P}$.
- For any two polyhedra $C, D \in \mathcal{P}$, their intersection $C \cap D$ is either empty or a face of both C and D.

To support Voronoi diagrams, we will later extend Definition 1.5 to permit polyhedra that are *unbounded*—that is, infinitely large. Specifically, Section 7.1 redefines convex polyhedra as intersections of halfspaces instead of convex hulls of points.

Piecewise linear complexes are sets of polyhedra that are not necessarily convex. We call these polyhedra *linear cells*.

Definition 1.7 (linear cell). A *linear k-cell* is the pointwise union of a finite number of convex k-polyhedra, all included in some common k-flat. A linear 0-cell is a vertex, a linear 2-cell is sometimes called a *polygon*, and a linear 3-cell is sometimes called a *polyhedron*.

---

[7]Topologists usually define complexes so they have countable cardinality. We restrict complexes to finite cardinality to avoid some interesting quirks, like the possibility that a polygon with a 1° angle can be meshed with a countably infinite set of triangles having no angle less than 20°.

Thus, we can build nonconvex polyhedra by uniting convex ones. For $k \geq 1$, a linear k-cell can have multiple connected components. These do no harm; removing a linear cell from a complex and replacing it with its connected components, or vice versa, makes no material difference. To simplify the exposition, we will forbid disconnected linear 1-cells in our complexes; i.e. the only linear 1-cells we use are edges. For $k \geq 2$, a linear cell can be only tenuously connected; e.g. a union of two squares that intersect at a single point is a linear 2-cell, even though it is not a simple polygon.

Another difference between linear cells and convex polyhedra is that we define the faces of a linear cell in a fundamentally different way that supports configurations like those in Figures 1.2 and 1.10. A linear cell's faces are not an intrinsic property of the linear cell alone, but depend on the complex that contains it. We defer the details to Section 2.10.1, where we define piecewise linear complexes.

Piecewise smooth complexes are sets of cells we call *smooth cells*, which are similar to linear cells except that they are not linear, but are smooth manifolds. See Chapter 15 for details.

Two cells in a complex are said to *adjoin* each other if they intersect each other, which implies that they have a face in common or one is a face of the other. Two cells that do not adjoin each other are *disjoint*.

A complex (or a mesh) is a representation of a domain. The former is a set of sets of points, and the latter is a set of points. The following operator collapses the former to the latter.

Definition 1.8 (underlying space). The *underlying space* of a complex $\mathcal{P}$, denoted $|\mathcal{P}|$, is the pointwise union of its cells; that is, $|\mathcal{P}| = \bigcup_{C \in \mathcal{P}} C$.

Ideally, a complex provided as input to a mesh generation algorithm and the mesh produced as output should cover exactly the same points. This ideal is not always possible—for example, if we are generating a linear tetrahedral mesh of a curved domain. When it is achieved, we call it *exact conformity*.

Definition 1.9 (exact conformity). A complex $\mathcal{T}$ *exactly conforms* to a complex $\mathcal{P}$ if $|\mathcal{T}| = |\mathcal{P}|$ and every cell in $\mathcal{P}$ is a union of cells in $\mathcal{T}$. We also say that $\mathcal{T}$ is a *subdivision* of $\mathcal{P}$.

1.6 Metric space topology

This section introduces basic notions from point set topology that underlie triangulations and other complexes. These notions are prerequisites for more sophisticated topological ideas—manifolds, homeomorphism, and isotopy—introduced in Chapter 12 to study algorithms for meshing domains with curved boundaries. A complex of linear elements cannot exactly conform to a curved domain, which raises the question of what it means for a triangulation to be a mesh of such a domain. To a layman, the word *topology* evokes visions of "rubber-sheet topology": the idea that if you bend and stretch a sheet of rubber, it changes shape but always preserves the underlying structure of how it is connected to itself. Homeomorphisms offer a rigorous way to state that a mesh preserves the topology of a domain, and isotopy offers a rigorous way to state that the domain can be deformed into the shape of the linear mesh without ever colliding with itself.

Topology begins with a set $\mathbb{T}$ of points—perhaps the points constituting the d-dimensional Euclidean space $\mathbb{R}^d$ or perhaps the points on the surface of a volume such as a coffee mug. We suppose that there is a *metric* $d(p, q)$ that specifies the scalar *distance* between every pair of points $p, q \in \mathbb{T}$. In the Euclidean space $\mathbb{R}^d$ we choose the Euclidean distance. On the surface of the coffee mug, we could choose the Euclidean distance too; alternatively, we could choose the *geodesic distance*, namely, the length of the shortest path from p to q on the mug's surface.

Let us briefly review the Euclidean metric. We write points in $\mathbb{R}^d$ as $p = (p_1, p_2, \ldots, p_d)$, where each p_i is a real-valued *coordinate*. The *Euclidean inner product* of two points $p, q \in \mathbb{R}^d$ is $\langle p, q\rangle = \sum_{i=1}^d p_i q_i$. The *Euclidean norm* of a point $p \in \mathbb{R}^d$ is $\|p\| = \langle p, p\rangle^{1/2} = \left(\sum_{i=1}^d p_i^2\right)^{1/2}$, and the *Euclidean distance* between two points $p, q \in \mathbb{R}^d$ is $d(p, q) = \|p - q\| = \left(\sum_{i=1}^d (p_i - q_i)^2\right)^{1/2}$. We also use the notation $d(\cdot, \cdot)$ to express minimum distances between point sets $P, Q \subseteq \mathbb{T}$,

$$\begin{aligned} d(p, Q) &= \inf\{d(p, q) : q \in Q\} \text{ and} \\ d(P, Q) &= \inf\{d(p, q) : p \in P, q \in Q\}. \end{aligned}$$

The heart of topology is the question of what it means for a set of points—say, a squiggle drawn on a piece of paper—to be *connected.* After all, two distinct points cannot be adjacent to each other; they can only be connected to another by an uncountably infinite bunch of intermediate points. Topologists solve that mystery with the idea of *limit points.*

Definition 1.10 (limit point). Let $Q \subseteq \mathbb{T}$ be a point set. A point $p \in \mathbb{T}$ is a *limit point* of Q, also known as an *accumulation point* of Q, if for every real number $\epsilon > 0$, however tiny, Q contains a point $q \neq p$ such that $d(p, q) < \epsilon$.

In other words, there is an infinite sequence of points in Q that get successively closer and closer to p—without actually being p—and get arbitrarily close. Stated succinctly, $d(p, Q \setminus \{p\}) = 0$. Observe that it doesn't matter whether $p \in Q$ or not.

Definition 1.11 (connected). Let $Q \subseteq \mathbb{T}$ be a point set. Imagine coloring every point in Q either red or blue. Q is *disconnected* if there exists a coloring having at least one red point and at least one blue point, wherein no red point is a limit point of the blue points, and no blue point is a limit point of the red points. A disconnected point set appears at left in Figure 1.12. If no such coloring exists, Q is *connected*, like the point set at right in Figure 1.12.

In this book, we frequently distinguish between closed and open point sets. Informally, a triangle in the plane is *closed* if it contains all the points on its edges, and *open* if it excludes all the points on its edges, as illustrated in Figure 1.13. The idea can be formally extended to any point set.

Definition 1.12 (closure; closed; open). The *closure* of a point set $Q \subseteq \mathbb{T}$, denoted $\mathrm{Cl}\, Q$, is the set containing every point in Q and every limit point of Q. A point set Q is *closed* if $Q = \mathrm{Cl}\, Q$, i.e. Q contains all its limit points. The *complement* of a point set Q is $\mathbb{T} \setminus Q$. A point set Q is *open* if its complement is closed, i.e. $\mathbb{T} \setminus Q = \mathrm{Cl}\, (\mathbb{T} \setminus Q)$.

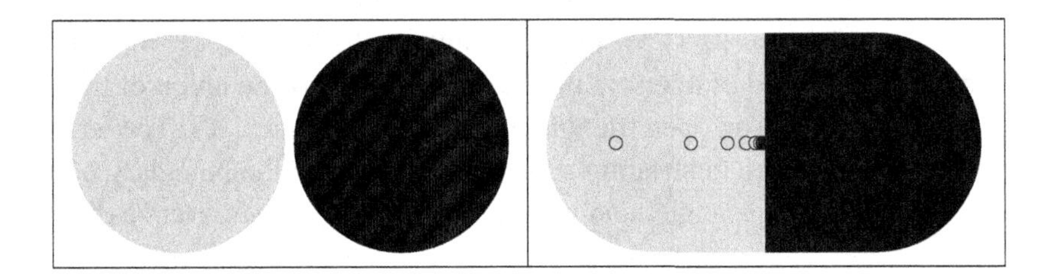

Figure 1.12: The disconnected point set at left can be partitioned into two connected subsets, which are shaded differently here. The point set at right is connected. The dark point at its center is a limit point of the lightly shaded points.

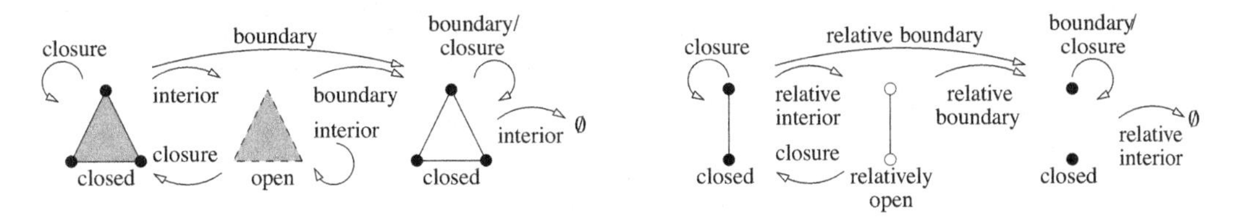

Figure 1.13: Closed, open, and relatively open point sets in the plane. Dashed edges and open circles indicate points missing from the point set.

For example, let $(0, 1)$ denote an *open interval* on the real number line—the set containing every $r \in \mathbb{R}$ such that $r > 0$ and $r < 1$—and let $[0, 1]$ denote a *closed interval* $(0, 1) \cup \{0\} \cup \{1\}$. The numbers zero and one are both limit points of the open interval, so $\text{Cl}\,(0, 1) = [0, 1] = \text{Cl}\,[0, 1]$. Therefore, $[0, 1]$ is closed and $(0, 1)$ is not. The numbers zero and one are also limit points of the complement of the closed interval, $\mathbb{R} \setminus [0, 1]$, so $(0, 1)$ is open, but $[0, 1]$ is not.

The terminology is misleading because "closed" and "open" are not opposites. In every nonempty metric space $\mathbb{T}$, there are at least two point sets that are both closed and open: $\emptyset$ and $\mathbb{T}$. The interval $(0, 1]$ on the real number line is neither open nor closed.

The definition of *open set* hides a subtlety that often misleads newcomers to point set topology: a triangle τ that is missing the points on its edges, and therefore is open in the two-dimensional metric space $\text{aff}\,\tau$, is not open in the metric space $\mathbb{R}^3$. Every point in τ is a limit point of $\mathbb{R}^3 \setminus \tau$, because we can find sequences of points that approach τ from the side. In recognition of this quirk, a simplex $\sigma \subset \mathbb{R}^d$ is said to be *relatively open* if it is open relative to its affine hull. It is commonplace to abuse terminology by writing "open simplex" for a simplex that is only relatively open, and we sometimes follow this convention in this book. Particularly useful is the concept of an "open edge," an edge that is missing its endpoints, illustrated in Figure 1.13.

Informally, the boundary of a point set Q is the set of points where Q meets its complement $\mathbb{T} \setminus Q$. The interior of Q contains all the other points of Q. Limit points provide formal definitions.

Definition 1.13 (boundary; interior)**.** The *boundary* of a point set Q in a metric space $\mathbb{T}$, denoted $\text{Bd}\,Q$, is the intersection of the closures of Q and its complement; i.e. $\text{Bd}\,Q = \text{Cl}\,Q \cap \text{Cl}\,(\mathbb{T} \setminus Q)$. The *interior* of Q, denoted $\text{Int}\,Q$, is $Q \setminus \text{Bd}\,Q = Q \setminus \text{Cl}\,(\mathbb{T} \setminus Q)$.

For example, $\mathrm{Bd}\,[0,1] = \{0,1\} = \mathrm{Bd}\,(0,1)$ and $\mathrm{Int}\,[0,1] = (0,1) = \mathrm{Int}\,(0,1)$. The boundary of a triangle (closed or open) in the Euclidean plane is the union of the triangle's three edges, and its interior is an open triangle, illustrated in Figure 1.13. The terms *boundary* and *interior* have the same misleading subtlety as open sets: the boundary of a triangle embedded in $\mathbb{R}^3$ is the whole triangle, and its interior is the empty set. Hence the following terms.

Definition 1.14 (relative boundary; relative interior). The *relative boundary* of a convex polyhedron $C \subset \mathbb{R}^d$ is its boundary with respect to the metric space of its affine hull—that is, $\mathrm{Cl}\, C \cap \mathrm{Cl}\,((\mathrm{aff}\, C) \setminus C)$. The *relative interior* of C is C minus its relative boundary.

Again, we often abuse terminology by writing *boundary* for relative boundary and *interior* for relative interior. The same subtlety arises with curved ridges and surface patches, but these have fundamentally different definitions of *boundary* and *interior* which we give in Section 12.3.

Definition 1.15 (bounded; compact). The *diameter* of a point set Q is $\sup_{p,q \in Q} d(p,q)$. The set Q is *bounded* if its diameter is finite, or *unbounded* if its diameter is infinite. A point set Q in a metric space is *compact* if it is closed and bounded.

As we have defined them, simplices and polyhedra are bounded, but in Section 7.1 we will see how to define unbounded polyhedra, which arise in Voronoi diagrams. Besides simplices and polyhedra, the point sets we use most in this book are balls.

Definition 1.16 (Euclidean ball). In $\mathbb{R}^d$, the *Euclidean d-ball* with center c and radius r, denoted $B(c,r)$, is the point set $B(c,r) = \{p \in \mathbb{R}^d : d(p,c) \leq r\}$. A 1-ball is an edge, and a 2-ball is called a *disk*. A *unit ball* is a ball with radius 1. The boundary of the d-ball is called the *Euclidean $(d-1)$-sphere* and denoted $S(c,r) = \{p \in \mathbb{R}^d : d(p,c) = r\}$. For example, a circle is a 1-sphere, and a layman's "sphere" in $\mathbb{R}^3$ is a 2-sphere. If we remove the boundary from a ball, we have the *open Euclidean d-ball* $B_o(c,r) = \{p \in \mathbb{R}^d : d(p,c) < r\}$.

The foregoing text introduces point set topology in terms of metric spaces. Surprisingly, it is possible to define all the same concepts without the use of a metric, point coordinates, or any scalar values at all. Section 12.1 discusses *topological spaces*, a mathematical abstraction for representing the topology of a point set while excluding all information that is not topologically essential. In this book, all our topological spaces have metrics.

1.7 How to measure an element

Here, we describe ways to measure the size, angles, and quality of a simplicial element, and we introduce some geometric structures associated with simplices—most importantly, their circumballs and circumcenters.

Definition 1.17 (circumball). Let τ be a simplex embedded in $\mathbb{R}^d$. A *circumball*, or *circumscribing ball*, of τ is a d-ball whose boundary passes through every vertex of τ, illustrated in Figure 1.14. Its boundary, a $(d-1)$-sphere, is called a *circumsphere*, or *circumscribing sphere*, of τ. A *closed circumball* includes its boundary—the circumsphere—and an

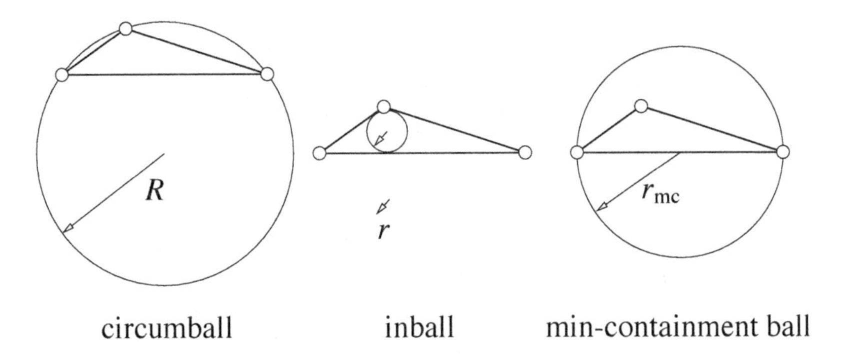

Figure 1.14: Three balls associated with a triangle.

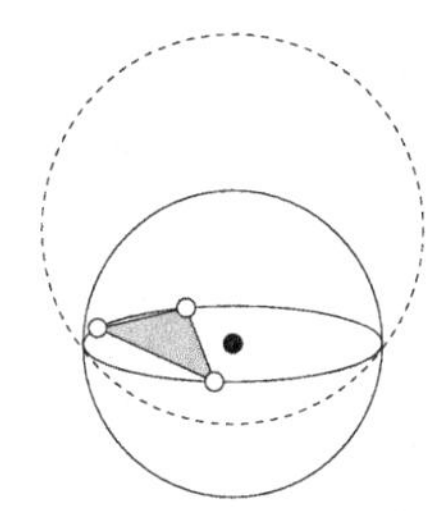

Figure 1.15: A triangle, two circumballs of the triangle of which the smaller (solid) is the triangle's diametric ball, the triangle's circumdisk (the equatorial cross-section of the diametric ball), and the triangle's circumcenter.

open circumball excludes it. If τ is a k-simplex, the *k-circumball* of τ is the unique k-ball whose boundary passes through every vertex of τ, and its relative boundary is the $(k-1)$-*circumsphere* of τ. We sometimes call a 2-circumball a *circumdisk* and a 1-circumsphere a *circumcircle.*

If τ is a d-simplex in $\mathbb{R}^d$, it has one unique circumsphere and circumball; but if τ has dimension less than d, it has an infinite set of circumspheres and circumballs. Consider a triangle τ in $\mathbb{R}^3$, for example. There is only one circumdisk of τ, whose boundary passes through τ's three vertices, but τ has infinitely many circumballs, and the intersection of any of those circumballs with τ's affine hull is τ's circumdisk. The smallest of these circumballs is special, because its center lies on τ's affine hull, it has the same radius as τ's circumdisk, and τ's circumdisk is its equatorial cross-section. We call τ's smallest circumball, illustrated in Figure 1.15, its *diametric ball.*

Definition 1.18 (diametric ball; circumcenter)**.** The *diametric ball* of a simplex τ is the circumball of τ with the smallest radius. The *circumcenter* of τ is the point at the center of τ's diametric ball, which always lies on aff τ. The *circumradius* of τ is the radius of τ's diametric ball.

The significance of circumcenters in Delaunay refinement algorithms is that the best place to insert a new vertex into a mesh is often at the circumcenter of a poorly shaped element, domain boundary triangle, or domain boundary edge. In a Delaunay mesh, these circumcenters are locally far from other mesh vertices, so inserting them does not create overly short edges.

Other balls associated with simplicial elements are the inball and the min-containment ball, both illustrated in Figure 1.14.

Definition 1.19 (inball). The *inball*, or *inscribed ball*, of a k-simplex τ is the largest k-ball $B \subset \tau$. Observe that B is tangent to every facet of τ. The *incenter* of τ is the point at the center of B, and the *inradius* of τ is the radius of B.

Definition 1.20 (min-containment ball). The *min-containment ball*, or *minimum enclosing ball*, of a k-simplex τ is the smallest k-ball $B \supset \tau$.

The min-containment ball of τ is always a diametric ball of a face of τ; that face could be of any dimension from an edge up to τ itself.

Finite element practitioners often represent the size of an element by the length of its longest edge, but one could argue that the radius of its min-containment ball is a slightly better measure, because there are sharp error bounds for piecewise linear interpolation over simplicial elements that are directly proportional to the squares of the radii of their min-containment balls. Details appear in Section 4.3.

A *quality measure* is a map from elements to scalar values that estimates the suitability of an element's shape independently of its size. The most obvious quality measures of a triangle are its smallest and largest angles, and a tetrahedron can be judged by its dihedral angles. We denote the *plane angle* between two vectors $\mathbf{u}$ and $\mathbf{v}$ as

$$\angle(\mathbf{u}, \mathbf{v}) = \arccos \frac{\mathbf{u} \cdot \mathbf{v}}{\|\mathbf{u}\| \, \|\mathbf{v}\|}.$$

We compute an angle $\angle xyz$ of a triangle as $\angle(x - y, z - y)$.

For a pair of intersecting lines or line segments ℓ_1 and ℓ_2, we generally measure the acute angle between them, denoted $\angle_a(\ell_1, \ell_2)$. When we replace ℓ_1 or ℓ_2 with a vector $\mathbf{v}$, the affine hull of $\mathbf{v}$ is implied; $\angle_a(\mathbf{u}, \mathbf{v})$ denotes the acute angle between the affine hulls of $\mathbf{u}$ and $\mathbf{v}$. Thus, $\angle_a$ disregards the vector orientation whereas $\angle$ does not. These angles satisfy a *triangle inequality*

$$\angle_a(\ell_1, \ell_2) \leq \angle_a(\ell_1, \ell_3) + \angle_a(\ell_3, \ell_2).$$

A *dihedral angle* is a measure of the angle separating two planes or polygons in $\mathbb{R}^3$—for example, the facets of a tetrahedron or 3-polyhedron. Suppose that two flat facets meet at an edge yz, where y and z are points in $\mathbb{R}^3$. Let w be a point lying on one of the facets, and let x be a point lying on the other. It is helpful to imagine the tetrahedron $wxyz$. The dihedral angle separating the two facets is the same angle separating wyz and xyz, namely, $\angle(\mathbf{u}, \mathbf{v})$ where $\mathbf{u} = (y - w) \times (z - w)$ and $\mathbf{v} = (y - x) \times (z - x)$ are vectors normal to wyz and xyz.

Elements can go bad in different ways, and it is useful to distinguish types of skinny elements. There are two kinds of skinny triangles, illustrated in Figure 1.16: needles, which have one edge much shorter than the others, and caps, which have an angle near 180° and a large circumdisk. Figure 1.17 offers a taxonomy of types of skinny tetrahedra. The tetrahedra in the top row are skinny in one dimension and fat in two. Those in the bottom row are skinny in two dimensions and fat in one. Spears, spindles, spades, caps, and slivers

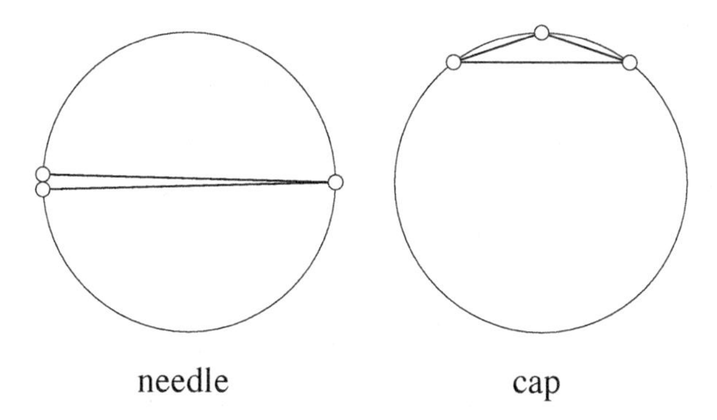

Figure 1.16: Skinny triangles have circumdisks larger than their shortest edges.

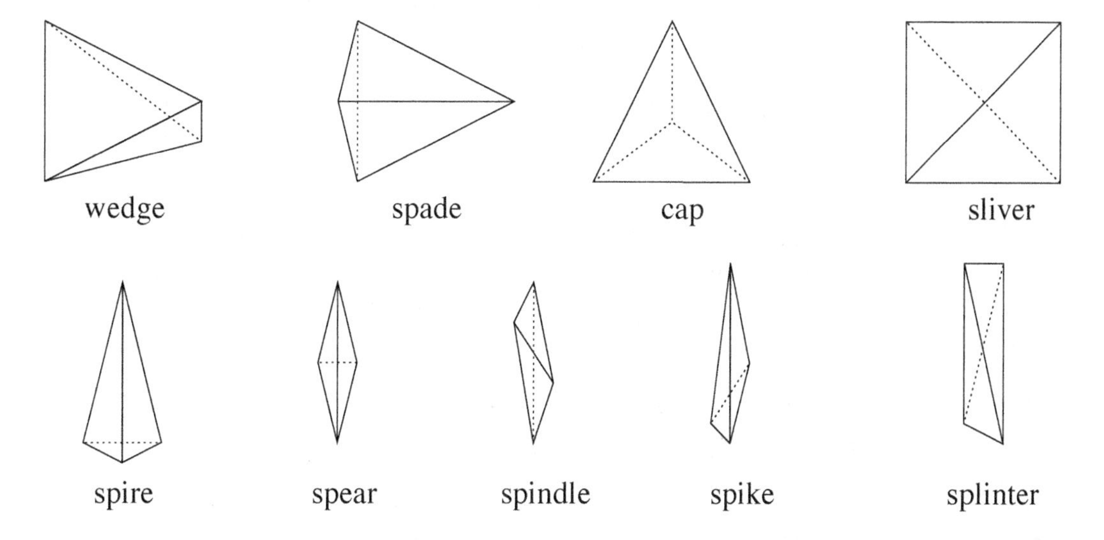

Figure 1.17: A taxonomy of skinny tetrahedra.

have a dihedral angle near 180°; the others may or may not. Spikes, splinters, and all the tetrahedra in the top row have a dihedral angle near 0°; the others may or may not. The cap, which has a vertex quite close to the center of the opposite triangle, is notable for a large solid angle, near 360°. Spikes also can have a solid angle arbitrarily close to 360°, and all the skinny tetrahedra can have a solid angle arbitrarily close to zero.

There are several surprises. The first is that spires, despite being skinny, can have all their dihedral angles between 60° and 90°, even if two edges are separated by a plane angle near 0°. Spires with good dihedral angles are harmless in many applications, and are indispensable at the tip of a needle-shaped domain, but some applications eschew them anyway. The second surprise is that a spear or spindle tetrahedron can have a dihedral angle near 180° without having a small dihedral angle. By contrast, a triangle with an angle near 180° must have an angle near 0°.

For many purposes—mesh improvement, for instance—it is desirable to have a single quality measure that punishes both angles near 0° and angles near 180°, and perhaps spires as well. Most quality measures are designed to reach one extreme value for an equilateral

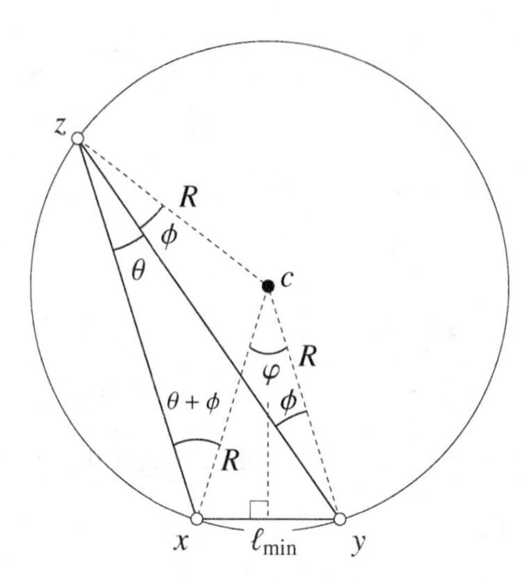

Figure 1.18: Relationships between the circumradius R, shortest edge $\ell_{\min}$, and smallest angle θ.

triangle or tetrahedron, and a value at the opposite extreme for a *degenerate* element—a triangle whose vertices are collinear, or a tetrahedron whose vertices are coplanar. In this book, the most important quality measure is the *radius-edge ratio*, because Delaunay refinement algorithms naturally tend to improve it.

Definition 1.21 (radius-edge ratio)**.** The *radius-edge ratio* of a simplex τ is $R/\ell_{\min}$, where R is τ's circumradius and $\ell_{\min}$ is the length of its shortest edge.

We would like the radius-edge ratio to be as small as possible; it ranges from ∞ for most degenerate simplices down to $1/\sqrt{3} \doteq 0.577$ for an equilateral triangle or $\sqrt{6}/4 \doteq 0.612$ for an equilateral tetrahedron. But is it a good estimate of element quality?

In two dimensions, the answer is yes. A triangle's radius-edge ratio is related to its smallest angle $\theta_{\min}$ by the formula

$$\frac{R}{\ell_{\min}} = \frac{1}{2 \sin \theta_{\min}}.$$

Figure 1.18 illustrates how this identity is derived for a triangle xyz with circumcenter c. Observe that the triangles ycz and xcz are isosceles, so their apex angles are $\angle ycz = 180°-2\phi$ and $\angle xcz = 180° - 2\phi - 2\theta$. Therefore, $\varphi = 2\theta$ and $\ell_{\min} = 2R \sin \theta$. This reasoning holds even if ϕ is negative.

The smaller a triangle's radius-edge ratio, the larger its smallest angle. The angles of a triangle sum to 180°, so the triangle's largest angle is at most $180° - 2\theta_{\min}$; hence an upper bound on the radius-edge ratio places bounds on both the smallest and largest angles.

In three dimensions, however, the radius-edge ratio is a flawed measure. It screens out all the tetrahedra in Figure 1.17 except slivers. A degenerate sliver can have a radius-edge ratio as small as $1/\sqrt{2} \doteq 0.707$, which is not far from the 0.612 of an equilateral tetrahedron. Delaunay refinement algorithms are guaranteed to remove all tetrahedra with large radius-edge ratios, but they do not promise to remove all slivers.

There are other quality measures that screen out all the skinny tetrahedra in Figure 1.17, including slivers and spires, but Delaunay refinement does not promise to bound these measures. A popular measure is r/R, where r is τ's inradius and R is its circumradius. This measure is sometimes called the *aspect ratio* or the *radius ratio*. It obtains a maximum value of $1/2$ for an equilateral triangle or $1/3$ for an equilateral tetrahedron, and a minimum value of zero for a degenerate element. This implies that it approaches zero as any dihedral angle separating τ's faces approaches 0° or 180°, any plane angle separating τ's edges approaches 0° or 180°, or any solid angle at τ's vertices approaches 0° or 360°.

For a triangle τ, the aspect ratio is related to the smallest angle $\theta_{\min}$ by the inequalities

$$2\sin^2\frac{\theta_{\min}}{2} \leq \frac{r}{R} \leq 2\tan\frac{\theta_{\min}}{2},$$

which implies that the aspect ratio approaches zero as $\theta_{\min}$ approaches zero, and vice versa.

Two unfortunate properties of the circumradius are that it is relatively expensive to compute for a tetrahedron, and it can be numerically unstable. A tiny perturbation of the position of one vertex of a skinny tetrahedron can induce an arbitrarily large change in its circumradius. Both the radius-edge ratio and the aspect ratio inherit these problems. In these respects, a better quality measure for tetrahedra is the *volume-length measure* V/ℓ_{rms}^3, where V is the volume of a tetrahedron and ℓ_{rms} is the root-mean-squared length of its six edges. It obtains a maximum value of $1/(6\sqrt{2})$ for an equilateral tetrahedron and a minimum value of zero for a degenerate tetrahedron. The volume-length measure is numerically stable and faster to compute than a tetrahedron's circumradius. It has proven itself as a filter against all poorly shaped tetrahedra and as an objective function for mesh improvement algorithms, especially optimization-based smoothing.

1.8 Notes and exercises

This chapter's opening quote comes from Thompson [215]. An excellent source for many aspects of mesh generation not covered by this book is the *Handbook of Grid Generation* [216], which includes many chapters on the generation of structured meshes, chapters that describe advancing front methods in unusual detail by Peraire, Peiró, and Morgan [169] and Marcum [142], and a fine survey of quadrilateral and hexahedral meshing by Schneiders [186]. Further surveys of the mesh generation literature are supplied by Bern and Eppstein [16] and Thompson and Weatherill [217]. Boissonnat, Cohen-Steiner, Mourrain, Rote, and Vegter [27] survey algorithms for surface meshing.

For evidence that the discretization error and the error in the gradient under piecewise linear interpolation grow with a triangle's largest angle, see Synge [212], Babuška and Aziz [12], and Jamet [117]. For similar evidence for a tetrahedron's largest dihedral angle, see Křížek [125]. The association between the largest eigenvalue of a stiffness matrix and the smallest angle of an element is noted by Fried [101] and Bank and Scott [14]. All these connections are summarized and elaborated by Shewchuk [202]. The Courant–Friedrichs–Lewy condition for stable explicit time integration is, not surprisingly, by Courant, Friedrichs, and Lewy [66]. Adaptive mesh refinement is surveyed by Oden and Demkowicz [161]. There is a large literature on how to numerically evaluate the quality of an element; see Field [95] for a survey.

Advancing front methods that create vertices and triangulate them in two separate stages include those by Frederick, Wong, and Edge [97]; Cavendish [38]; and Lo [139]. Early advancing front methods that interleave vertex creation and element creation include triangular mesh generators by George [102], Sadek [182], and Peraire, Vahdati, Morgan, and Zienkiewicz [170]; tetrahedral meshers by Löhner and Parikh [140] and Peraire, Peiró, Formaggia, Morgan, and Zienkiewicz [168]; a quadrilateral mesher by Blacker and Stephenson [21]; and a hexahedral mesher by Blacker and Meyers [20].

Delaunay mesh generators that create vertices and triangulate them in two separate; stages include those by Frederick et al. [97]; Cavendish, Field, and Frey [39]; and Jameson, Baker, and Weatherill [116]. The first Delaunay refinement algorithm we know of that interleaves the two operations is by Frey [100].

Yerry and Shephard [226, 227] published the first quadtree and octree meshers. Readers not familiar with quadtrees and octrees may consult Samet's book [183].

The simplest and most famous way to smooth an interior vertex is to move it to the centroid of the vertices that adjoin it. This method, which dates back at least to Kamel and Eisenstein [121] in 1970, is called *Laplacian smoothing* because of its interpretation as a Laplacian finite difference operator. It usually works well for triangular meshes, but it is unreliable for tetrahedra, quadrilaterals, and hexahedra. More sophisticated optimization-based smoothers began to appear in the 1990s [164, 37, 163]. Slower but better smoothing is provided by the nonsmooth optimization algorithm of Freitag, Jones, and Plassmann [98], which can optimize the worst element in a group—for instance, maximizing the minimum dihedral angle among the tetrahedra that share a specified vertex. For some quality measures, optimal mesh smoothing can be done with generalized linear programming [4].

Mesh improvement is usually driven by a schedule that searches the mesh for elements that can be improved by local transformations, ideally as quickly as possible. Canann, Muthukrishnan, and Phillips [36] provide a fast triangular mesh improvement schedule. Sophisticated schedules for tetrahedral mesh improvement are provided by Joe [120], Freitag and Ollivier-Gooch [99], and Klingner and Shewchuk [124]. For a list of flips for quadrilateral and hexahedral meshes, see Bern, Eppstein, and Erickson [17]. Kinney [122] describes mesh improvement methods for quadrilateral meshes. There does not seem to have been much work on applying hexahedral flips.

The first provably good mesh generation algorithm, by Baker, Grosse, and Rafferty [13], employs a square grid. The first provably good Delaunay refinement algorithm in the plane is by Chew [59], and the most successful is by Ruppert [178, 180]. The first provably good three-dimensional Delaunay refinement algorithm is by Dey, Bajaj, and Sugihara [74]. For a proof that the domain in Figure 1.5 has no mesh whose new angles all exceed 30°, see Shewchuk [196].

The first mesh generator offering provably good grading and size optimality is the quadtree algorithm of Bern, Eppstein, and Gilbert [18]. Neugebauer and Diekmann [158] improve the algorithm by replacing square quadrants with rhomboids. They produce triangles with angles between 30° and 90°, many of them equilateral. The first tetrahedral mesh generator offering size optimality is the octree algorithm of Mitchell and Vavasis [151]. Remarkably, Mitchell and Vavasis [152] extended their mathematical guarantees to meshes of polyhedra of any dimensionality by using d-dimensional 2^d-trees.

The first paper to suggest a generalization of the Delaunay property to meshes of curved surfaces in three dimensions and the first algorithm offering a guarantee on the aspect ratios of the triangles in a surface mesh are by Chew [61]. See the bibliographical notes in Section 14.6 and the aforementioned survey by Boissonnat et al. [27] for a discussion of subsequent surface meshing algorithms. Guaranteed-quality triangular mesh generators for two-dimensional domains with curved boundaries include those by Boivin and Ollivier-Gooch [32] and Pav and Walkington [166]. Labelle and Shewchuk [127] provide a provably good triangular mesh generator that produces anisotropic meshes in the plane, and Cheng, Dey, Ramos, and Wenger [54] generalize it to generate anisotropic meshes of curved surfaces in three-dimensional space.

Bibliographic information for the developments discussed in Section 1.4 is provided in the notes of the chapters listed there, so we omit details here. Publications noted in that section include papers by Shewchuk [197, 198]; Cheng, Dey, Edelsbrunner, Facello, and Teng [49]; Cheng and Dey [48]; Edelsbrunner and Shah [92]; Amenta and Bern [3]; Cheng, Dey, Edelsbrunner, and Sullivan [47]; Boissonnat and Oudot [29]; Cheng, Dey, and Ramos [51]; Cheng, Dey, and Levine [50]; and Dey and Levine [77]. Companion papers are available for each of the programs TRIANGLE [196], STELLAR [124], QUALMESH [52], SURFREMESH [53], and DELPSC [77].

Books by Hocking and Young [113], Munkres [155], and Weeks [223] are standard texts on point set topology, giving detailed definitions of topological spaces and maps. Books by Hatcher [110] and Stillwell [209] are good sources for algebraic and combinatorial topology; they describe simplicial complexes and their use in triangulations of topological spaces. Some useful definitions in computational topology are collected in the survey paper by Dey, Edelsbrunner, and Guha [75]. Books by Hadwiger [108] and Ziegler [228] are good sources for the mathematics of polyhedra and polytopes. A recent book by De Loera, Rambau, and Santos [68] surveys the mathematical properties of triangulations. Hadwiger popularized Definition 1.7 for nonconvex polyhedra, which we call linear cells. The notion of a piecewise linear complex was introduced by Miller, Talmor, Teng, Walkington, and Wang [149]. Piecewise smooth complexes were introduced by Cheng, Dey, and Ramos [51]. The classification of tetrahedra with tiny angles in Figure 1.17 is adapted from Cheng, Dey, Edelsbrunner, Facello, and Teng [49].

Miller, Talmor, Teng, and Walkington [148] pointed out that the radius-edge ratio is the most natural and elegant measure for analyzing Delaunay refinement algorithms. The use of the incenter-circumcenter ratio as a quality measure was suggested by Cavendish, Field, and Frey [39]. The volume-length measure was suggested by Parthasarathy, Graichen, and Hathaway [163]. See Klingner and Shewchuk [124] for evidence of its utility and the instability of the incenter-circumcenter ratio in mesh improvement algorithms.

Exercises

1. Let X be a point set, not necessarily finite, in $\mathbb{R}^d$. Prove that the following two definitions of the convex hull of X are equivalent.

    - The set of all points that are convex combinations of the points in X.
    - The intersection of all convex sets that include X.

2. Suppose we change the second condition in Definition 1.4, which defines *simplicial complex*, to

    - For any two simplices $\sigma, \tau \in \mathcal{T}$, their intersection $\sigma \cap \tau$ is either empty or a simplex in $\mathcal{T}$.

   Give an illustration of a set that is a simplicial complex under this modified definition but not under the true definition.

3. In every metric space $\mathbb{T}$, the point sets $\emptyset$ and $\mathbb{T}$ are both closed and open.

   (a) Give an example of a metric space that has more than two sets that are both closed and open, and list all of those sets.

   (b) Explain the relationship between the idea of connectedness and the number of sets that are both closed and open.

4. Prove that for every subset X of a metric space, $\text{Cl}\,\text{Cl}\,X = \text{Cl}\,X$. In other words, augmenting a set with its limit points does not give it more limit points.

5. Show that among all triangles whose longest edge has length one, the circumradius approaches infinity if and only if the largest angle approaches 180°, whereas the inradius approaches zero if and only if the smallest angle approaches 0°.

6. One quality measure for a simplex is its minimum altitude divided by the length of its longest edge, which unfortunately is also called the *aspect ratio*. For triangles, prove that this ratio approaches zero if and only if the smallest angle approaches zero. For tetrahedra, prove that this ratio approaches zero if a dihedral angle approaches 0° or 180°, but that the converse is not true.

7. Another measure, which has been seriously proposed in the literature as a quality measure for a tetrahedron by researchers who will remain anonymous, is the length of its longest edge divided by its circumradius. Explain why this is a bad quality measure. In a few words, what triangle shapes maximize it? What is the worst tetrahedron shape that maximizes it?

8. The classification of tetrahedra shown in Figure 1.17 is qualitative. To make it precise, set two thresholds ρ and ℓ, and call a radius-edge ratio large if it exceeds ρ and an edge length small if it is less than ℓ. Describe how to use ρ and ℓ to decide into which class in Figure 1.17 a tetrahedron falls.

Chapter 2

Two-dimensional Delaunay triangulations

The Delaunay triangulation is a geometric structure that engineers have used for meshes since mesh generation was in its infancy. In two dimensions, it has a striking advantage: among all possible triangulations of a fixed set of points, the Delaunay triangulation maximizes the minimum angle. It also optimizes several other geometric criteria related to interpolation accuracy. If it is our goal to create a triangulation without small angles, it seems almost silly to consider a triangulation that is not Delaunay. Delaunay triangulations have been studied thoroughly, and excellent algorithms are available for constructing and updating them.

A constrained triangulation is a triangulation that enforces the presence of specified edges—for example, the boundary of a nonconvex object. A constrained Delaunay triangulation relaxes the Delaunay property just enough to recover those edges, while enjoying optimality properties similar to those of a Delaunay triangulation. Constrained Delaunay triangulations are nearly as popular as their unconstrained ancestors.

This chapter surveys two-dimensional Delaunay triangulations, constrained Delaunay triangulations, weighted Delaunay triangulations, and their geometric properties.

2.1 Triangulations of a planar point set

The word *triangulation* usually refers to a simplicial complex, but it has multiple meanings when we discuss a *triangulation of* some geometric entity that is being triangulated. There are triangulations of point sets, polygons, polyhedra, and many other structures. Consider points in the plane (or in any Euclidean space).

Definition 2.1 (triangulation of a point set)**.** Let S be a finite set of points in the plane. A *triangulation of* S is a simplicial complex $\mathcal{T}$ such that S is the set of vertices in $\mathcal{T}$, and the union of all the simplices in $\mathcal{T}$ is the convex hull of S—that is, $|\mathcal{T}| = \text{conv}\, S$.

Does every point set have a triangulation? Yes. Consider the *lexicographic triangulation* illustrated in Figure 2.1. To construct one, sort the points *lexicographically* (that is, by x-coordinate, ordering points with the same x-coordinate according to their y-coordinates),

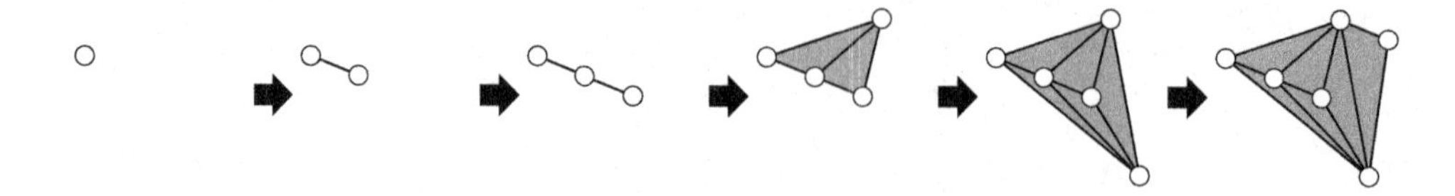

Figure 2.1: Incremental construction of a lexicographic triangulation.

yielding a sorted sequence $v_1, v_2, \ldots, v_n$ of points. Define the lexicographic triangulation $\mathcal{T}_i$ of the first i points by induction as follows. The first triangulation is $\mathcal{T}_1 = \{v_1\}$. Each subsequent triangulation is $\mathcal{T}_i = \mathcal{T}_{i-1} \cup \{v_i\} \cup \{\text{conv}(\{v_i\} \cup \sigma) : \sigma \in \mathcal{T}_{i-1}$ and the relative interior of $\text{conv}(\{v_i\} \cup \sigma)$ intersects no simplex in $\mathcal{T}_{i-1}\}$.

Even if the points in S are all collinear, there is a triangulation of S: $\mathcal{T}_n$ contains n vertices, $n - 1$ collinear edges connecting them, and no triangles.

A triangulation of n points in the plane has at most $2n - 5$ triangles and $3n - 6$ edges as a consequence of Euler's formula. With no change, Definition 2.1 defines triangulations of point sets in higher-dimensional Euclidean spaces as well.

2.2 The Delaunay triangulation

The *Delaunay triangulation* of a point set S, introduced by Boris Nikolaevich Delaunay in 1934, is characterized by the *empty circumdisk property*: no point in S lies in the interior of any triangle's circumscribing disk; recall Definition 1.17.

Definition 2.2 (Delaunay). In the context of a finite point set S, a triangle is *Delaunay* if its vertices are in S and its open circumdisk is *empty*—i.e. contains no point in S. Note that any number of points in S can lie on a Delaunay triangle's circumcircle. An edge is *Delaunay* if its vertices are in S and it has at least one empty open circumdisk. A *Delaunay triangulation* of S, denoted Del S, is a triangulation of S in which every triangle is Delaunay, as illustrated in Figure 2.2.

One might wonder whether every point set has a Delaunay triangulation, and how many Delaunay triangulations a point set can have. The answer to the first question is "yes." Section 2.3 gives some intuition for why this is true, and Section 2.5 gives a proof.

The Delaunay triangulation of S is unique if and only if no four points in S lie on a common empty circle, a fact proved in Section 2.7. Otherwise, there are Delaunay triangles

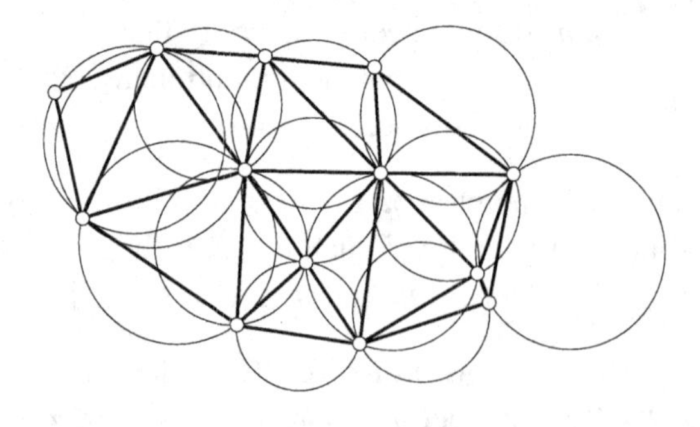

Figure 2.2: Every triangle in a Delaunay triangulation has an empty open circumdisk.

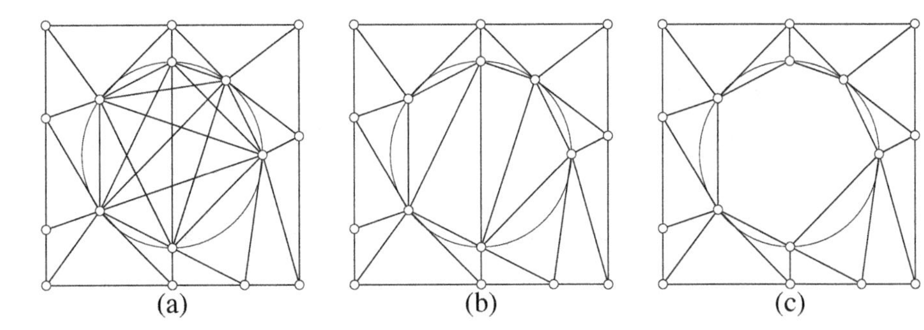

Figure 2.3: Three ways to define a Delaunay structure in the presence of cocircular vertices. (a) Include all the Delaunay simplices. (b) Choose a subset of Delaunay simplices that constitutes a triangulation. (c) Exclude all crossing Delaunay edges, and fuse overlapping Delaunay triangles into Delaunay polygons.

and edges whose interiors intersect, as illustrated in Figure 2.3(a). Most applications omit some of these triangles and edges so that the survivors form a simplicial complex, as in Figure 2.3(b). Depending on which Delaunay simplices one keeps and which one discards, one obtains different Delaunay triangulations.

It is sometimes useful to unite the intersecting triangles into a single polygon, depicted in Figure 2.3(c). The *Delaunay subdivision* obtained this way is a polyhedral complex, rather than a simplicial complex. It has the advantage of being the geometric dual of the famous *Voronoi diagram*, discussed in Section 7.1.

Clearly, a simplex's being Delaunay does not guarantee that it is in *every* Delaunay triangulation of a point set. But a slightly stronger property does provide that guarantee.

Definition 2.3 (strongly Delaunay)**.** In the context of a finite point set S, a triangle τ is *strongly Delaunay* if its vertices are in S and its *closed* circumdisk contains no point in S except the vertices of τ. An edge e is *strongly Delaunay* if its vertices are in S and it has at least one closed circumdisk that contains no point in S except the vertices of e. Every point in S is a strongly Delaunay vertex.

Every Delaunay triangulation of S contains every strongly Delaunay simplex, a fact proved in Section 2.7. The Delaunay subdivision contains the strongly Delaunay edges and triangles, and no others.

Consider two examples of strongly Delaunay edges. First, every edge on the boundary of a triangulation of S is strongly Delaunay. Figure 2.4 shows why. Second, the edge connecting a point $v \in S$ to its nearest neighbor $w \in S$ is strongly Delaunay, because the smallest closed disk containing v and w does not contain any other point in S. Therefore, every Delaunay triangulation connects every vertex to its nearest neighbor.

2.3 The parabolic lifting map

Given a finite point set S, the *parabolic lifting map* transforms the Delaunay subdivision of S into faces of a convex polyhedron in three dimensions, as illustrated in Figure 2.5.

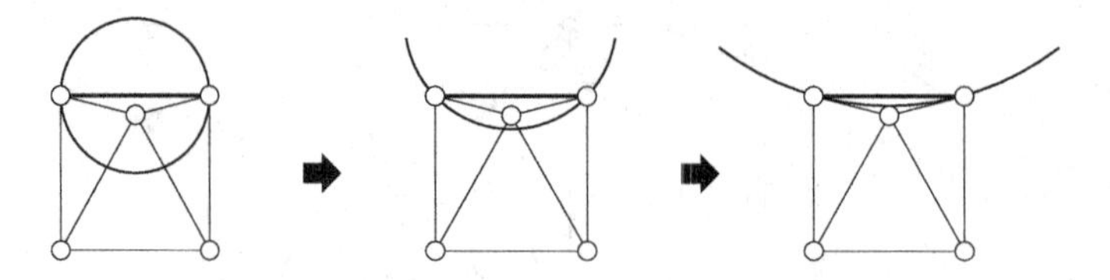

Figure 2.4: Every edge on the boundary of a convex triangulation is strongly Delaunay, because it is always possible to find an empty disk that contains its endpoints and no other vertex.

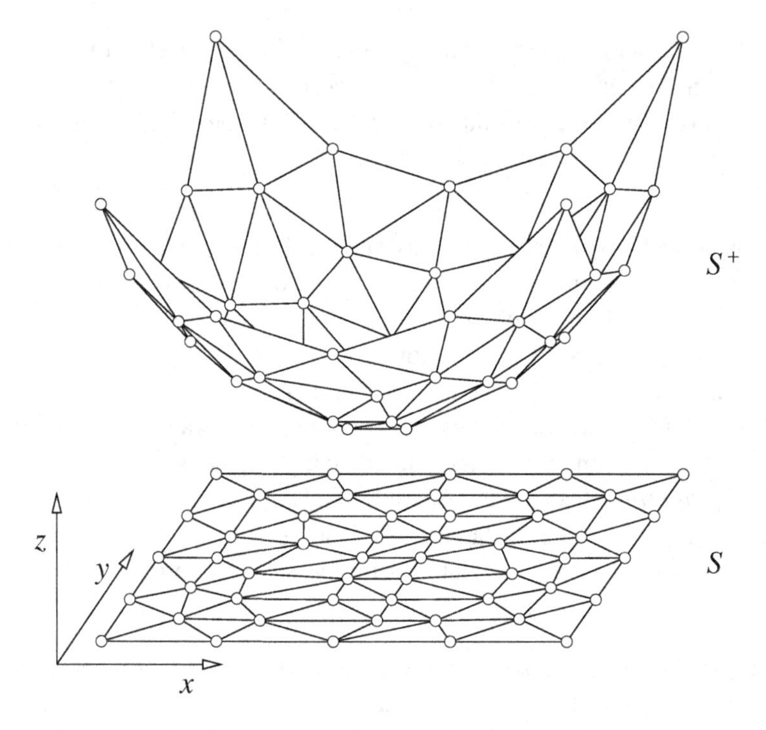

Figure 2.5: The parabolic lifting map.

This relationship between Delaunay triangulations and convex hulls has two consequences. First, it makes many properties of the Delaunay triangulation intuitive. For example, from the fact that every finite point set has a polyhedral convex hull, it follows that every finite point set has a Delaunay triangulation. Second, it brings to mesh generation the power of a huge literature on polytope theory and algorithms. For example, every convex hull algorithm is a Delaunay triangulation algorithm!

The parabolic lifting map sends each point $p = (x, y) \in \mathbb{R}^2$ to a point $p^+ = (x, y, x^2 + y^2) \in \mathbb{R}^3$. Call p^+ the *lifted companion* of p.

Consider the convex hull conv S^+ of the lifted points $S^+ = \{v^+ : v \in S\}$. Figure 2.5 illustrates its downward-facing faces. Formally, a face f of conv S^+ is *downward-facing* if no point in conv S^+ is directly below any point in f, with respect to the z-axis. Call the collection of downward-facing faces the *underside* of conv S^+. Projecting the underside of conv S^+ to the x-y plane (by discarding every point's z-coordinate) yields the Delaunay subdivision of S. If S has more than one Delaunay triangulation, this Delaunay subdivision has non-triangular polygons, like the hexagon in Figure 2.3(c). Triangulating these polygonal faces yields a Delaunay triangulation.

For a simplex σ in the plane, its *lifted companion* σ^+ is the simplex embedded in $\mathbb{R}^3$ whose vertices are the lifted companions of the vertices of σ. Note that σ^+ is flat and does not curve to hug the paraboloid. The following lemmas show that every Delaunay simplex's lifted companion is included in a downward-facing face of conv S^+.

Lemma 2.1 (Lifting Lemma). *Let C be a circle in the plane. Let $C^+ = \{p^+ : p \in C\}$ be the ellipse obtained by lifting C to the paraboloid. Then the points of C^+ lie on a plane h, which is not parallel to the z-axis. Furthermore, every point p inside C lifts to a point p^+ below h, and every point p outside C lifts to a point p^+ above h. Therefore, testing whether a point p is inside, on, or outside C is equivalent to testing whether the lifted point p^+ is below, on, or above h.*

Proof. Let o and r be the center and radius of C, respectively. Let p be a point in the plane. The z-coordinate of p^+ is $\|p\|^2$. By expanding $d(o,p)^2$, we have the identity $\|p\|^2 = 2\langle o,p\rangle - \|o\|^2 + d(o,p)^2$. With o and r fixed and $p \in \mathbb{R}^2$ varying, the equation $z = 2\langle o,p\rangle - \|o\|^2 + r^2$ defines a plane h in $\mathbb{R}^3$, not parallel to the z-axis. For every point $p \in C$, $d(o,p) = r$, so $C^+ \subset h$. For every point $p \notin C$, if $d(o,p) < r$, then the lifted point p^+ lies below h, and if $d(o,p) > r$, then p^+ lies above h. □

Proposition 2.2. *Let σ be a simplex whose vertices are in S, and let σ^+ be its lifted companion. Then σ is Delaunay if and only if σ^+ is included in some downward-facing face of* conv S^+. *The simplex σ is strongly Delaunay if and only if σ^+ is a downward-facing face of* conv S^+.

Proof. If σ is Delaunay, σ has a circumcircle C that encloses no point in S. Let h be the unique plane in $\mathbb{R}^3$ that includes C^+. By the Lifting Lemma (Lemma 2.1), no point in S^+ lies below h. Because the vertices of σ^+ are in C^+, $h \supset \sigma^+$. Therefore, σ^+ is included in a downward-facing face of the convex hull of S^+. If σ is strongly Delaunay, every point in S^+ lies above h except the vertices of σ^+. Therefore, σ^+ is a downward-facing face of the convex hull of S^+. The converse implications follow by reversing the argument. □

The parabolic lifting map works equally well for Delaunay triangulations in three or more dimensions; the Lifting Lemma (Lemma 2.1) and Proposition 2.2 generalize to higher dimensions without any new ideas. Proposition 2.2 implies that any algorithm for constructing the convex hull of a point set in $\mathbb{R}^{d+1}$ can construct the Delaunay triangulation of a point set in $\mathbb{R}^d$.

2.4 The Delaunay Lemma

Perhaps the most important result concerning Delaunay triangulations is the *Delaunay Lemma*, proved by Boris Delaunay himself. It provides an alternative characterization of the Delaunay triangulation: a triangulation whose edges are *locally Delaunay*.

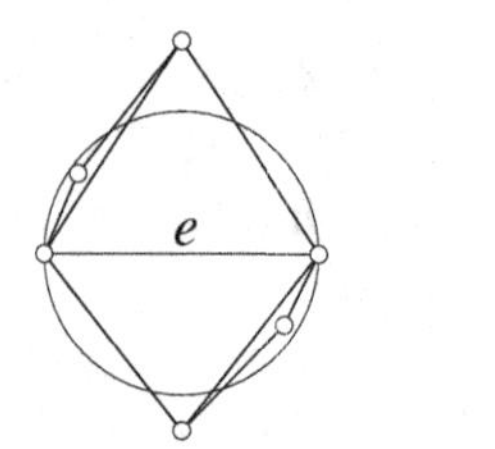

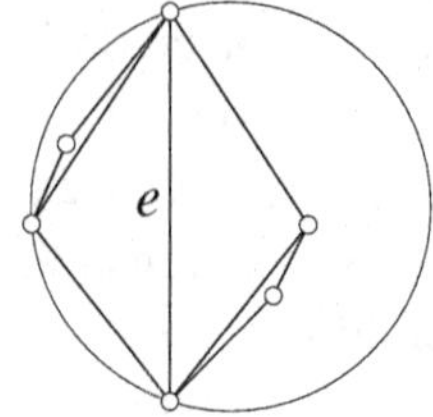

Figure 2.6: At left, e is locally Delaunay. At right, e is not.

Definition 2.4 (locally Delaunay). Let e be an edge in a triangulation $\mathcal{T}$ in the plane. If e is an edge of fewer than two triangles in $\mathcal{T}$, then e is said to be *locally Delaunay*. If e is an edge of exactly two triangles τ_1 and τ_2 in $\mathcal{T}$, then e is said to be *locally Delaunay* if it has an open circumdisk containing no vertex of τ_1 nor τ_2. Equivalently, the open circumdisk of τ_1 contains no vertex of τ_2. Equivalently, the open circumdisk of τ_2 contains no vertex of τ_1.

Figure 2.6 shows two different triangulations of six vertices. In the triangulation at left, the edge e is locally Delaunay, because the depicted circumdisk of e does not contain either vertex opposite e. Nevertheless, e is not Delaunay, thanks to other vertices in e's circumdisk. In the triangulation at right, e is not locally Delaunay; every open circumdisk of e contains at least one of the two vertices opposite e.

The Delaunay Lemma has several uses. First, it provides a linear-time algorithm to determine whether a triangulation of a point set is Delaunay: simply test whether every edge is locally Delaunay. Second, it implies a simple algorithm for producing a Delaunay triangulation called the *flip algorithm* (Section 2.5). The flip algorithm helps to prove that Delaunay triangulations have useful optimality properties. Third, the Delaunay Lemma helps to prove the correctness of other algorithms for constructing Delaunay triangulations.

As with many properties of Delaunay triangulations, the lifting map provides intuition for the Delaunay Lemma. On the lifting map, the Delaunay Lemma is essentially the observation that a simple polyhedron is convex if and only if its has no *reflex edge*. A reflex edge is an edge where the polyhedron is locally nonconvex; that is, two adjoining triangles meet along that edge at a dihedral angle greater than 180°, measured through the interior of the polyhedron. If a triangulation has an edge that is not locally Delaunay, that edge's lifted companion is a reflex edge of the lifted triangulation (by the Lifting Lemma, Lemma 2.1).

Lemma 2.3 (Delaunay Lemma). *Let $\mathcal{T}$ be a triangulation of a point set S. The following three statements are equivalent.*

(i) *Every triangle in $\mathcal{T}$ is Delaunay (i.e. $\mathcal{T}$ is Delaunay).*

(ii) *Every edge in $\mathcal{T}$ is Delaunay.*

(iii) *Every edge in $\mathcal{T}$ is locally Delaunay.*

Proof. If the points in S are all collinear, S has only one triangulation, which trivially satisfies all three properties.

Otherwise, let e be an edge in $\mathcal{T}$; e is an edge of at least one triangle $\tau \in \mathcal{T}$. If τ is Delaunay, τ's open circumdisk is empty, and because τ's circumdisk is also a circumdisk

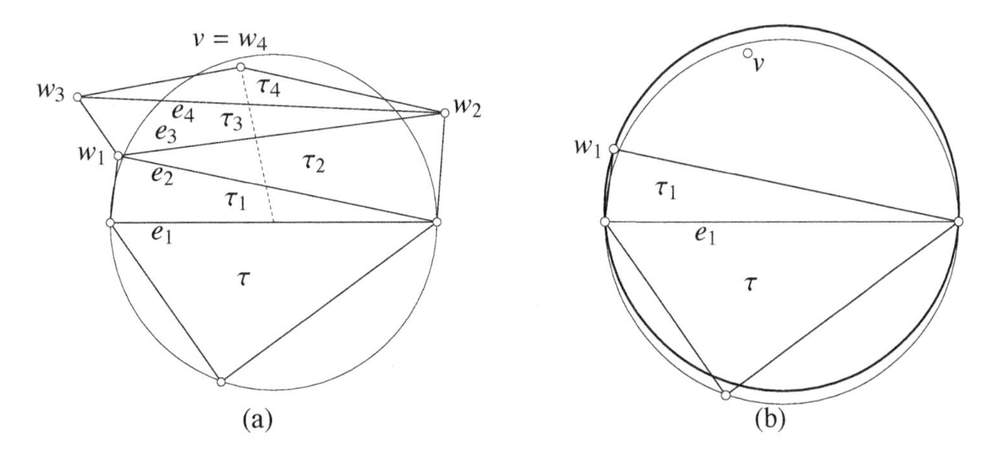

Figure 2.7: (a) Because τ's open circumdisk contains v, some edge between v and τ is not locally Delaunay. (b) Because v lies above e_1 and in τ's open circumdisk, and because w_1 lies outside τ's open circumdisk, v must lie in τ_1's open circumdisk.

of e, e is Delaunay. Therefore, Property (i) implies Property (ii). If an edge is Delaunay, it is clearly locally Delaunay too, so Property (ii) implies Property (iii). The proof is complete if Property (iii) implies Property (i). Of course, this is the hard part.

Suppose that every edge in $\mathcal{T}$ is locally Delaunay. Suppose for the sake of contradiction that Property (i) does not hold. Then some triangle $\tau \in \mathcal{T}$ is not Delaunay, and some vertex $v \in S$ is inside τ's open circumdisk. Let e_1 be the edge of τ that separates v from the interior of τ, as illustrated in Figure 2.7(a). Without loss of generality, assume that e_1 is oriented horizontally, with τ below e_1.

Draw a line segment ℓ from the midpoint of e_1 to v—see the dashed line in Figure 2.7(a). If the line segment intersects some vertex other than v, replace v with the lowest such vertex and shorten ℓ accordingly. Let $e_1, e_2, e_3, \ldots, e_m$ be the sequence of triangulation edges (from bottom to top) whose relative interiors intersect ℓ. Because $\mathcal{T}$ is a triangulation of S, every point on the line segment lies either in a single triangle or on an edge. Let w_i be the vertex above e_i that forms a triangle τ_i in conjunction with e_i. Observe that $w_m = v$.

By assumption, e_1 is locally Delaunay, so w_1 lies outside the open circumdisk of τ. As Figure 2.7(b) shows, it follows that the open circumdisk of τ_1 includes the portion of τ's open circumdisk above e_1 and, hence, contains v. Repeating this argument inductively, we find that the open circumdisks of $\tau_2, \ldots, \tau_m$ contain v. But $w_m = v$ is a vertex of τ_m, which contradicts the claim that v is in the open circumdisk of τ_m. □

2.5 The flip algorithm

The *flip algorithm* has at least three uses: it is a simple algorithm for computing a Delaunay triangulation, it is the core of a constructive proof that every finite set of points in the plane has a Delaunay triangulation, and it is the core of a proof that the Delaunay triangulation optimizes several geometric criteria when compared with all other triangulations of the same point set.

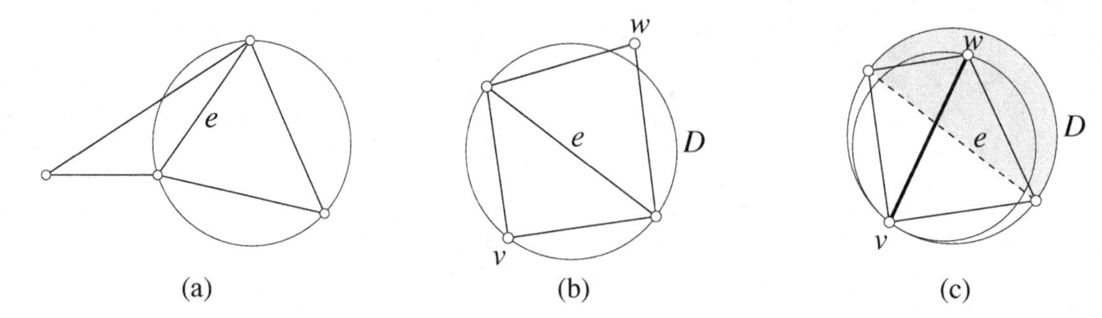

Figure 2.8: (a) In this nonconvex quadrilateral, e cannot be flipped, and e is locally Delaunay. (b) The edge e is locally Delaunay. (c) The edge e is not locally Delaunay. The edge created by a flip of e is locally Delaunay.

Let S be a point set to be triangulated. The flip algorithm begins with any triangulation $\mathcal{T}$ of S; for instance, the lexicographic triangulation described in Section 2.1. The Delaunay Lemma tells us that $\mathcal{T}$ is Delaunay if and only if every edge in $\mathcal{T}$ is locally Delaunay. The flip algorithm repeatedly chooses any edge that is not locally Delaunay and *flips* it.

The union of two triangles that share an edge is a quadrilateral, and the shared edge is a diagonal of the quadrilateral. To flip an edge is to replace it with the quadrilateral's other diagonal, as illustrated in Figure 2.6. An edge flip is legal only if the two diagonals cross each other—equivalently, if the quadrilateral is strictly convex. Fortunately, unflippable edges are always locally Delaunay, as Figure 2.8(a) shows.

Proposition 2.4. *Let e be an edge in a triangulation of S. Either e is locally Delaunay, or e is flippable and the edge created by flipping e is locally Delaunay.*

Proof. Let v and w be the vertices opposite e. Consider the quadrilateral defined by e, v, and w, illustrated in Figure 2.8. Let D be the open disk whose boundary passes through v and the vertices of e.

If w is outside D, as in Figure 2.8(b), then the empty circumdisk D demonstrates that e is locally Delaunay.

Otherwise, w is in the section of D bounded by e and opposite v. This section is shaded in Figure 2.8(c). The quadrilateral is thus strictly convex, so e is flippable. Furthermore, the open disk that is tangent to D at v and has w on its boundary does not contain the vertices of e, because D includes it, as Figure 2.8(c) demonstrates. Therefore, the edge vw is locally Delaunay. □

Proposition 2.4 shows that the flip algorithm can flip any edge that is not locally Delaunay, thereby creating an edge that is. Unfortunately, the outer four edges of the quadrilateral might discover that they are no longer locally Delaunay, even if they were locally Delaunay before the flip. If the flip algorithm repeatedly flips edges that are not locally Delaunay, will it go on forever? The following proposition says that it won't.

Proposition 2.5. *Given a triangulation of n points, the flip algorithm terminates after $O(n^2)$ edge flips, yielding a Delaunay triangulation.*

Proof. Let $\mathcal{T}$ be the initial triangulation provided as input to the flip algorithm. Let $\mathcal{T}^+ = \{\sigma^+ : \sigma \in \mathcal{T}\}$ be the initial triangulation lifted to the parabolic lifting map; $\mathcal{T}^+$ is a simplicial complex embedded in $\mathbb{R}^3$. If $\mathcal{T}$ is Delaunay, then $\mathcal{T}^+$ triangulates the underside of conv S^+; otherwise, by the Lifting Lemma (Lemma 2.1), the edges of $\mathcal{T}$ that are not locally Delaunay lift to reflex edges of $\mathcal{T}^+$.

By Proposition 2.4, an edge flip replaces an edge that is not locally Delaunay with one that is. In the lifted triangulation $\mathcal{T}^+$, a flip replaces a reflex edge with a convex edge. Let Q be the set containing the four vertices of the two triangles that share the flipped edge. Then conv Q^+ is a tetrahedron whose upper faces are the pre-flip simplices and whose lower faces are the post-flip simplices. Imagine the edge flip as the act of gluing the tetrahedron conv Q^+ to the underside of $\mathcal{T}^+$.

Each edge flip monotonically lowers the lifted triangulation, so once flipped, an edge can never reappear. The flip algorithm can perform no more than $n(n-1)/2$ flips—the number of edges that can be defined on n vertices—so it must terminate. But the flip algorithm terminates only when every edge is locally Delaunay. By the Delaunay Lemma, the final triangulation is Delaunay. □

The fact that the flip algorithm terminates helps to prove that point sets have Delaunay triangulations.

Proposition 2.6. *Every finite set of points in the plane has a Delaunay triangulation.*

Proof. Section 2.1 demonstrates that every finite point set has at least one triangulation. By Proposition 2.5, the application of the flip algorithm to that triangulation produces a Delaunay triangulation. □

An efficient implementation of the flip algorithm requires one extra ingredient. How quickly can one find an edge that is not locally Delaunay? To repeatedly test every edge in the triangulation would be slow. Instead, the flip algorithm maintains a list of edges that might not be locally Delaunay. The list initially contains every edge in the triangulation. Thereafter, the flip algorithm iterates the following procedure until the list is empty, whereupon the algorithm halts.

- Remove an edge from the list.
- Check whether the edge is still in the triangulation, and if so, whether it is locally Delaunay.
- If the edge is present but not locally Delaunay, flip it, and add the four edges of the flipped quadrilateral to the list.

The list may contain multiple copies of the same edge, but they do no harm.

Implemented this way, the flip algorithm runs in $O(n + k)$ time, where n is the number of vertices (or triangles) of the triangulation and k is the number of flips performed. In the worst case, $k = \Theta(n^2)$, giving $O(n^2)$ running time. But there are circumstances where the flip algorithm is fast in practice. For instance, if the vertices of a Delaunay mesh are perturbed by small displacements during a physical simulation, it might take only a small

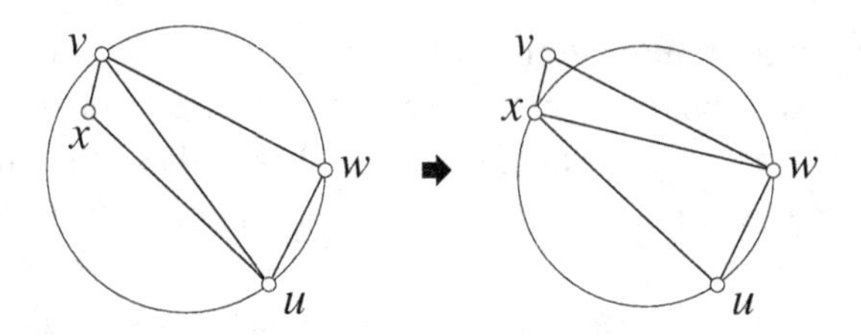

Figure 2.9: A Delaunay flip increases the angle opposite edge uw and, if $\angle wxu$ is acute, reduces the size of the circumdisk of the triangle adjoining that edge.

number of flips to restore the Delaunay property. In this circumstance, the flip algorithm probably outperforms any algorithm that reconstructs the triangulation from scratch.

2.6 The optimality of the Delaunay triangulation

Delaunay triangulations are valuable in part because they optimize several geometric criteria: the smallest angle, the largest circumdisk, and the largest min-containment disk. Recall from Definition 1.20 that the *min-containment disk* of a triangle is the smallest closed disk that includes it. For a triangle with no obtuse angle, the circumdisk and the min-containment disk are the same, but for an obtuse triangle, the min-containment disk is smaller.

Proposition 2.7. *Flipping an edge that is not locally Delaunay increases the minimum angle and reduces the largest circumdisk among the triangles changed by the flip.*

Proof. Let uv be the flipped edge, and let wvu and xuv be the triangles deleted by the flip, so wxu and xwv are the triangles created by the flip.

The angle opposite the edge uw is $\angle wvu$ before the flip, and $\angle wxu$ after the flip. As Figure 2.9 illustrates, because the open circumdisk of wvu contains x, the latter angle is greater than the former angle by the Inscribed Angle Theorem, a standard fact about circle geometry that was known to Euclid. Likewise, the flip increases the angles opposite wv, vx, and xu.

Each of the other two angles of the new triangles, $\angle xuw$ and $\angle wvx$, is a sum of two pre-flip angles that merge when uv is deleted. It follows that all six angles of the two post-flip triangles exceed the smallest of the four angles in which uv participates before the flip.

Suppose without loss of generality that the circumdisk of wxu is at least as large as the circumdisk of xwv, and that $\angle wxu \leq \angle uwx$, implying that $\angle wxu$ is acute. Because the open circumdisk of wvu contains x, it is larger than the circumdisk of wxu, as illustrated in Figure 2.9. It follows that the largest pre-flip circumdisk is larger than the largest post-flip circumdisk. □

In Section 4.3, we show that a Delaunay flip never increases the largest min-containment disk among the triangles changed by the flip. These local results imply a global optimality result.

Theorem 2.8. *Among all the triangulations of a point set, there is a Delaunay triangulation that maximizes the minimum angle in the triangulation, a Delaunay triangulation that* *minimizes the largest circumdisk, and a Delaunay triangulation that minimizes the largest min-containment disk.*

PROOF. Each of these properties is locally improved when an edge that is not locally Delaunay is flipped—or at least not worsened, in the case of min-containment disks. There is at least one optimal triangulation $\mathcal{T}$. If $\mathcal{T}$ has an edge that is not locally Delaunay, flipping that edge produces another optimal triangulation. When the flip algorithm runs with $\mathcal{T}$ as its input, every triangulation it iterates through is optimal by induction, and by Proposition 2.5, that includes a Delaunay triangulation. □

Theorem 2.8 is not the strongest statement we can make, but it is easy to prove. With more work, one can show that *every* Delaunay triangulation of a point set optimizes these criteria. See Exercise 3 for details.

Unfortunately, the only optimality property of Theorem 2.8 that generalizes to Delaunay triangulations in dimensions higher than two is the property of minimizing the largest min-containment ball. However, the list of optimality properties in Theorem 2.8 is not complete. In the plane, the Delaunay triangulation maximizes the mean inradius of its triangles and minimizes a property called the *roughness* of a piecewise linearly interpolated function, which is the integral over the triangulation of the square of the gradient. Section 4.3 discusses criteria related to interpolation error for which Delaunay triangulations of any dimension are optimal.

Another advantage of the Delaunay triangulation arises in numerical discretizations of the Laplacian operator ∇^2. Solutions to Dirichlet boundary value problems associated with Laplace's equation $\nabla^2\varphi = 0$ satisfy a *maximum principle*: the maximum value of φ always occurs on the domain boundary. Ideally, an approximate solution found by a numerical method should satisfy a discrete maximum principle, both for physical realism and because it helps to prove strong convergence properties for the numerical method and to bound its error. A piecewise linear finite element discretization of Laplace's equation over a Delaunay triangulation in the plane satisfies a discrete maximum principle. Moreover, the stiffness matrix is what is called a *Stieltjes matrix* or an *M-matrix*, which implies that it can be particularly stable in numerical methods such as incomplete Cholesky factorization. These properties extend to three-dimensional Delaunay triangulations for some finite volume methods but, unfortunately, not for the finite element method.

2.7 The uniqueness of the Delaunay triangulation

The strength of a strongly Delaunay simplex is that it appears in *every* Delaunay triangulation of a point set. If a point set has multiple Delaunay triangulations, they differ only in their choices of simplices that are merely Delaunay. Hence, if a point set is *generic*—if it has no four cocircular points—it has only one Delaunay triangulation.

Let us prove these facts. Loosely speaking, the following proposition says that strongly Delaunay simplices intersect nicely.

Proposition 2.9. *Let σ be a strongly Delaunay simplex, and let τ be a Delaunay simplex. Then $\sigma \cap \tau$ is either empty or a shared face of both σ and τ.*

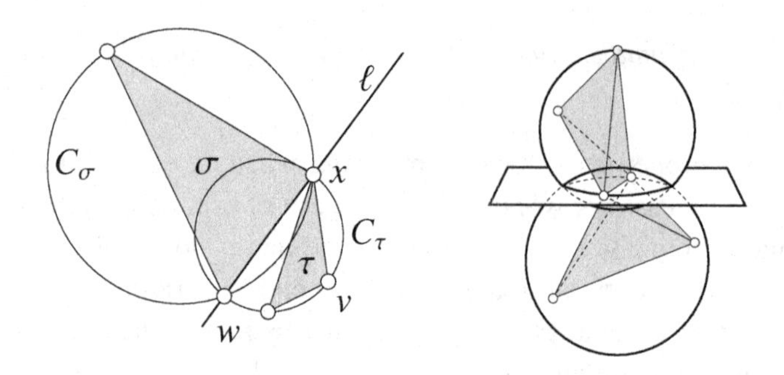

Figure 2.10: A strongly Delaunay simplex σ intersects a Delaunay simplex τ at a shared face of both. The illustration at right foreshadows the fact that this result holds in higher dimensions too.

Proof. If τ is a face of σ, the proposition follows immediately. Otherwise, τ has a vertex v that σ does not have. Because τ is Delaunay, it has an empty circumcircle C_τ. Because σ is strongly Delaunay, it has an empty circumcircle C_σ that does not pass through v, illustrated in Figure 2.10. But v lies on C_τ, so $C_\sigma \neq C_\tau$.

The intersection of circumcircles $C_\sigma \cap C_\tau$ contains zero, one, or two points. In the first two cases, the proposition follows easily, so suppose it is two points w and x, and let ℓ be the unique line through w and x. On one side of ℓ, an arc of C_σ encloses an arc of C_τ, and because C_σ is empty, no vertex of τ lies on this side of ℓ. Symmetrically, no vertex of σ lies on the other side of ℓ. Therefore, $\sigma \cap \tau \subset \ell$. It follows that $\sigma \cap \ell$ is either $\emptyset$, $\{w\}$, $\{x\}$, or the edge wx. The same is true of $\tau \cap \ell$ and, therefore, of $\sigma \cap \tau$. □

Proposition 2.9 leads us to see that if a point set has several Delaunay triangulations, they differ only by the simplices that are not strongly Delaunay.

Proposition 2.10. *Every Delaunay triangulation of a point set contains every strongly Delaunay simplex.*

Proof. Let $\mathcal{T}$ be any Delaunay triangulation of a point set S. Let σ be any strongly Delaunay simplex. Let p be a point in the relative interior of σ.

Some Delaunay simplex τ in $\mathcal{T}$ contains the point p. By Proposition 2.9, $\sigma \cap \tau$ is a shared face of σ and τ. But $\sigma \cap \tau$ contains p, which is in the relative interior of σ, so $\sigma \cap \tau = \sigma$. Therefore, σ is a face of τ, so $\sigma \in \mathcal{T}$. □

An immediate consequence of this proposition is that "most" point sets—at least, most point sets with randomly perturbed real coordinates—have just one Delaunay triangulation.

Theorem 2.11. *Let S be a point set. Suppose no four points in S lie on a common empty circle. Then S has one unique Delaunay triangulation.*

Proof. By Proposition 2.6, S has at least one Delaunay triangulation. Because no four points lie on a common empty circle, every Delaunay simplex is strongly Delaunay. By Proposition 2.10, every Delaunay triangulation of S contains every Delaunay simplex. By definition, no Delaunay triangulation contains a triangle that is not Delaunay. Hence, the

Delaunay triangulation is uniquely defined as the set of all Delaunay triangles and their faces. □

Theorem 2.11 does not preclude the possibility that all the vertices might be collinear. In that case, the vertices have a unique triangulation that has edges but no triangles and is vacuously Delaunay.

2.8 The weighted Delaunay triangulation

The parabolic lifting map connects Delaunay triangulations with convex hulls. It also suggests a generalization of Delaunay triangulations in which lifted vertices are not required to lie on the paraboloid. This observation is exploited by several mesh generation algorithms.

The simplest version of this idea begins with a planar point set S and assigns each point an arbitrary *height* to which it is lifted in $\mathbb{R}^3$. Imagine taking the convex hull of the points lifted to $\mathbb{R}^3$ and projecting its underside down to the plane, yielding a convex subdivision called the *weighted Delaunay subdivision*. If some of the faces of this subdivision are not triangular, they may be triangulated arbitrarily, and S has more than one *weighted Delaunay triangulation*.

In recognition of the special properties of the parabolic lifting map, it is customary to endow each point $v \in S$ with a scalar *weight* ω_v that represents how far its height deviates from the paraboloid. Specifically, v's height is $\|v\|^2 - \omega_v$, and its lifted companion is

$$v^+ = (v_x, v_y, v_x^2 + v_y^2 - \omega_v).$$

A positive weight thus implies that the point's lifted companion is below the paraboloid; a negative weight implies above. The reason weight is defined in opposition to height is so that increasing a vertex's weight will tend to increase its influence on the underside of the convex hull conv S^+.

Given a weight assignment $\omega : S \to \mathbb{R}$, we denote the *weighted point set* $S[\omega]$. If the weight of a vertex v is so small that its lifted companion v^+ is not on the underside of conv $S[\omega]^+$, as illustrated in Figure 2.11, it does not appear in the weighted Delaunay subdivision at all, and v is said to be *submerged* or *redundant*. Submerged vertices create some confusion of terminology, because a weighted Delaunay triangulation of S is not necessarily a triangulation of S—it might omit some of the vertices in S.

If every vertex in S has a weight of zero, every weighted Delaunay triangulation of S is a Delaunay triangulation of S. No vertex is submerged because every point on the paraboloid is on the underside of the convex hull of the paraboloid.

The weighted analog of a Delaunay simplex is called a *weighted Delaunay* simplex, and the analog of a circumdisk is a *witness* plane.

Definition 2.5 (weighted Delaunay triangulation; witness). Let $S[\omega]$ be a weighted point set in $\mathbb{R}^3$. A simplex σ whose vertices are in S is *weighted Delaunay* if σ^+ is included in a downward-facing face of conv $S[\omega]^+$. In other words, there exists a non-vertical plane $h_\sigma \subset \mathbb{R}^3$ such that $h_\sigma \supset \sigma^+$ and no vertex in $S[\omega]^+$ lies below h_σ. The plane h_σ is called a *witness* to the weighted Delaunay property of σ. A *weighted Delaunay triangulation* of

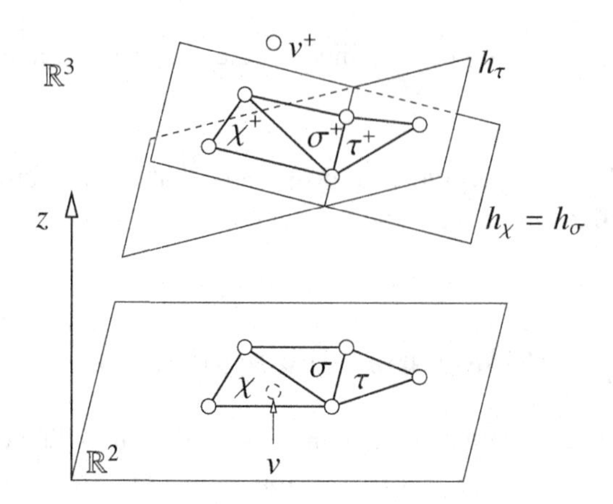

Figure 2.11: The triangles χ, σ, and τ are all weighted Delaunay, but only τ is strongly weighted Delaunay. Triangles χ and σ have the same witness plane $h_\chi = h_\sigma$, and τ has a different witness h_τ. The vertex v is submerged.

$S[\omega]$, denoted Del $S[\omega]$, is a triangulation of a subset of S such that $|\text{Del}\, S[\omega]| = \text{conv}\, S$ and every simplex in Del $S[\omega]$ is weighted Delaunay with respect to $S[\omega]$.

Figure 2.11 illustrates three weighted Delaunay triangles and their witnesses. All their edges and vertices are weighted Delaunay as well, but the submerged vertex v is not weighted Delaunay. A triangle has a unique witness, but an edge or vertex can have an infinite number of witnesses.

The weighted analog of a strongly Delaunay simplex is a *strongly weighted Delaunay* simplex.

Definition 2.6 (strongly weighted Delaunay). A simplex σ is *strongly weighted Delaunay* if σ^+ is a downward-facing face of conv $S[\omega]^+$ and no vertex in $S[\omega]^+$ lies on σ^+ except the vertices of σ^+. In other words, there exists a non-vertical plane $h_\sigma \subset \mathbb{R}^3$ such that $h_\sigma \supset \sigma^+$ and in $S[\omega]^+$ lies above h_σ, except the vertices of σ^+. The plane h_σ is a *witness* to the strongly weighted Delaunay property of σ.

Of the three triangles in Figure 2.11, only τ is strongly weighted Delaunay. All the edges are strongly weighted Delaunay except the edge shared by χ and σ. All the vertices are strongly weighted Delaunay except v.

Proposition 2.2 shows that if all the weights are zero, "weighted Delaunay" is equivalent to "Delaunay" and "strongly weighted Delaunay" is equivalent to "strongly Delaunay." If a simplex σ is weighted Delaunay, it appears in at least one weighted Delaunay triangulation of S. If σ is strongly weighted Delaunay, it appears in *every* weighted Delaunay triangulation of S (by a generalization of Proposition 2.10).

Definition 2.7 (generic). A weighted point set $S[\omega]$ in $\mathbb{R}^2$ is *generic* if no four points in $S[\omega]^+$ lie on a common non-vertical plane in $\mathbb{R}^3$.

If a weighted point set is generic, then every weighted Delaunay simplex is strongly weighted Delaunay, and the point set has exactly one weighted Delaunay triangulation. For points with weight zero, this definition is equivalent to the statement that no four points are cocircular.

2.9 Symbolic weight perturbations

Some algorithms for constructing Delaunay triangulations, like the gift-wrapping algorithm described in Section 3.11, have difficulties triangulating point sets that have multiple Delaunay triangulations. These problems can be particularly acute in three or more dimensions. One way to make the points generic is to perturb them in space so that their Delaunay triangulation is unique. A better way is to assign the points infinitesimal weights such that they have one unique weighted Delaunay triangulation. Because the weights are infinitesimal, that triangulation is also an ordinary Delaunay triangulation.

To put this idea on firm mathematical ground, replace the infinitesimals with tiny, finite weights that are *symbolic*—their magnitudes are not explicitly specified. Given a point set $S = \{v_1, v_2, \ldots, v_n\}$, assign vertex v_i a weight of ϵ^i, where $\epsilon > 0$ is presumed to be so small that making it smaller would not change the weighted Delaunay triangulation. There is no need to compute an ϵ that satisfies this presumption; it is enough to know that such an ϵ exists.

An intuitive way to understand these weights is to imagine the result of a procedure that perturbs the vertex weights one at a time. Initially every vertex has a weight of zero, and the Delaunay subdivision may have some polygons that are not triangular. Perturb the weight of each vertex $v_1, v_2, \ldots$ in turn to subdivide the non-triangular polygons adjoining the perturbed vertex. The perturbation of vertex v_i's weight is chosen so that v_i^+ is not coplanar with any three other lifted vertices, and it is chosen sufficiently small that if v_i^+ was above the affine hull of three other lifted vertices, it remains above that affine hull after the perturbation. Therefore, every polygon adjoining v_i in the Delaunay subdivision after the perturbation must be a triangle, and no face not adjoining v_i is changed by the perturbation. Both these goals are achieved by choosing each weight perturbation to be infinitesimally smaller than all the previous weight perturbations—for instance, a weight of ϵ^i in the limit as $\epsilon > 0$ approaches zero.

Proposition 2.12. *Let S be a set of points in the plane. Let $\omega : S \to \mathbb{R}$ be the weight assignment described above. The following statements hold.*

- *If a simplex is strongly Delaunay with respect to S, it is strongly weighted Delaunay with respect to $S[\omega]$.*
- *If a simplex is strongly weighted Delaunay with respect to $S[\omega]$, it is Delaunay with respect to S.*
- *There is exactly one weighted Delaunay triangulation of $S[\omega]$, which is a Delaunay triangulation of S.*

Proof. Let σ be a simplex that is strongly Delaunay with respect to S. Then σ is strongly weighted Delaunay with respect to $S[0]$, and some plane $h \supset \sigma^+$ is a witness to that fact. If σ is a triangle, vertical perturbations of the vertices of σ^+ induce a unique perturbation of h. If σ is an edge or a vertex, choose one or two arbitrary points in h that are affinely independent of σ^+ and fix them so that a perturbation of σ^+ uniquely perturbs h.

Every vertex in S^+ lies above h except the vertices of σ^+, which lie on h. If ϵ is sufficiently small, the vertically perturbed points $S[\omega]^+$ preserve this property: every vertex in $S[\omega]^+$ lies above the perturbed witness plane for the perturbed σ^+, except the vertices of the perturbed σ^+. Therefore, σ is strongly weighted Delaunay with respect to $S[\omega]$ too, confirming the first statement of the proposition.

If a simplex σ is strongly weighted Delaunay with respect to $S[\omega]$, then every vertex in $S[\omega]^+$ lies above some witness plane $h \supset \sigma^+$ except the vertices of σ^+. If ϵ is sufficiently small, the vertically perturbed points S^+ nearly preserve this property: every vertex in S^+ lies above or on the perturbed witness plane for the perturbed σ^+. (If this were not so, some vertex would have moved from below the affine hull of three other vertices to above their affine hull; this can be prevented by making ϵ smaller.) This confirms the second statement of the proposition.

If ϵ is sufficiently small, no four vertices of $S[\omega]^+$ lie on a common non-vertical plane, so every face of the weighted Delaunay subdivision of $S[\omega]$ is a triangle, and the weighted Delaunay triangulation of $S[\omega]$ is unique. Every simplex of this triangulation is Delaunay with respect to S, so it is a Delaunay triangulation of S. □

The converse of the second statement of Proposition 2.12 is not true: a simplex that is Delaunay with respect to S is not necessarily weighted Delaunay with respect to $S[\omega]$. This is not surprising; the purpose of the weight perturbations is to break coplanarities in S^+ and eliminate some of the Delaunay simplices so that the Delaunay triangulation is unique.

An important advantage of symbolic perturbations is that it is easy to simulate them in software—see Exercise 2 in Chapter 3—and they do not introduce the numerical problems associated with finite, numerical perturbations. Software for constructing Delaunay triangulations can ignore the symbolic perturbations until it encounters four cocircular vertices—if constructing a weighted Delaunay triangulation, four vertices whose lifted companions are coplanar. In that circumstance only, the software must simulate the circumstance where the four lifted vertices are perturbed so they are not coplanar.

2.10 Constrained Delaunay triangulations in the plane

As planar Delaunay triangulations maximize the minimum angle, do they solve the problem of triangular mesh generation? No, for two reasons illustrated in Figure 2.12. First, skinny triangles might appear anyway. Second, the Delaunay triangulation of a domain's vertices might not respect the domain's boundary. Both these problems can be solved by introducing additional vertices, as illustrated.

An alternative solution to the second problem is to use a *constrained Delaunay triangulation* (CDT). A CDT is defined with respect to a set of points and *segments* that demarcate the domain boundary. Every segment is required to become an edge of the CDT. The triangles of a CDT are *not* required to be Delaunay; instead, they must be *constrained Delaunay*, a property that partly relaxes the empty circumdisk property.

One virtue of a CDT is that it can respect arbitrary segments without requiring the insertion of any additional vertices besides the vertices of the segments. Another is that the CDT inherits the Delaunay triangulation's optimality: among all triangulations of a point set *that include all the segments*, the CDT maximizes the minimum angle, minimizes the largest circumdisk, and minimizes the largest min-containment disk.

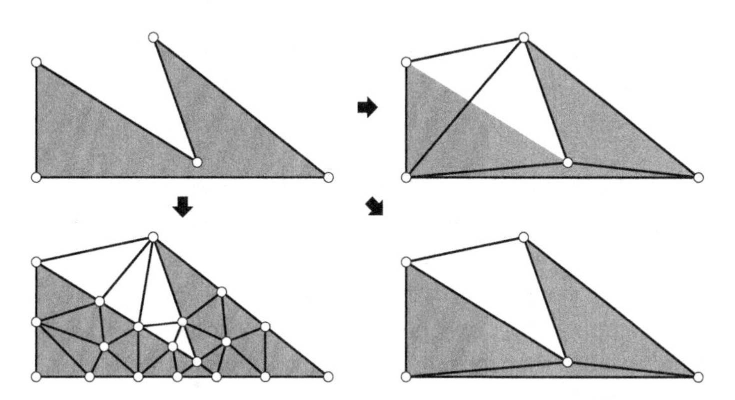

Figure 2.12: The Delaunay triangulation (upper right) may omit domain edges and contain skinny triangles. A Steiner Delaunay triangulation (lower left) can fix these faults by introducing new vertices. A constrained Delaunay triangulation (lower right) fixes the first fault without introducing new vertices.

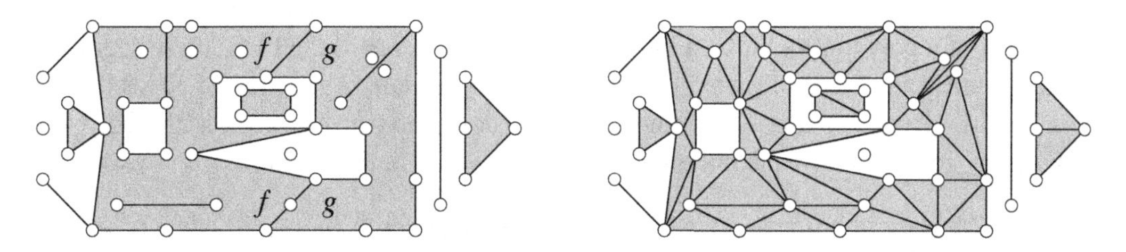

Figure 2.13: A two-dimensional piecewise linear complex and its constrained Delaunay triangulation. Each polygon may have holes, slits, and vertices in its interior.

2.10.1 Piecewise linear complexes and their triangulations

The domain over which a CDT is defined (and the input to a CDT construction algorithm) is not just a set of points; it is a complex composed of points, edges, and polygons, illustrated in Figure 2.13. The purpose of the edges is to dictate that triangulations of the complex must contain those edges. The purpose of the polygons is to specify the region to be triangulated. The polygons are linear 2-cells (recall Definition 1.7), which are not necessarily convex and may have holes.

Definition 2.8 (piecewise linear complex). In the plane, a *piecewise linear complex* (PLC) $\mathcal{P}$ is a finite set of linear cells—vertices, edges, and polygons—that satisfies the following properties.

- The vertices and edges in $\mathcal{P}$ form a simplicial complex. That is, $\mathcal{P}$ contains both vertices of every edge in $\mathcal{P}$, and the relative interior of an edge in $\mathcal{P}$ intersects no vertex in $\mathcal{P}$ nor any other edge in $\mathcal{P}$.
- For each polygon f in $\mathcal{P}$, the boundary of f is a union of edges in $\mathcal{P}$.
- If two polygons in $\mathcal{P}$ intersect, their intersection is a union of edges and vertices in $\mathcal{P}$.

The edges in a PLC $\mathcal{P}$ are called *segments* to distinguish them from other edges in a triangulation of $\mathcal{P}$. The *underlying space* of a PLC $\mathcal{P}$, denoted $|\mathcal{P}|$, is the union of its contents; that is, $|\mathcal{P}| = \bigcup_{f \in \mathcal{P}} f$. Usually, the underlying space is the domain to be triangulated.[1]

Figure 2.13 shows a PLC and a triangulation of it. Observe that the intersection of the linear 2-cells f and g has multiple connected components, including two line segments and one isolated point, which are not collinear. The faces of the complex that represent this intersection are three edges and six vertices.

Every simplicial complex and every polyhedral complex is a PLC. But PLCs are more general, and not just because they permit nonconvex polygons. As Figure 2.13 illustrates, segments and isolated vertices can float in a polygon's interior; they constrain how the polygon can be triangulated. One purpose of these floating constraints is to permit the application of boundary conditions at appropriate locations in a mesh of a PLC.

Whereas the faces of a simplex are defined in a way that depends solely on the simplex, and the faces of a convex polyhedron are too, the faces of a polygon are defined in a fundamentally different way that depends on both the polygon and the PLC it is a part of. An edge of a polygon might be a union of several segments in the PLC; these segments and their vertices are faces of the polygon. A PLC may contain segments and edges that lie in the relative interior of a polygon; these are also considered to be faces of the polygon, because they constrain how the polygon can be subdivided into triangles.

Definition 2.9 (face of a linear cell). The *faces* of a linear cell f (polygon, edge, or vertex) in a PLC $\mathcal{P}$ are the linear cells in $\mathcal{P}$ that are subsets of f, including f itself. The *proper faces* of f are all the faces of f except f.

A triangulation of $\mathcal{P}$ must cover every polygon and include every segment.

Definition 2.10 (triangulation of a planar PLC). Let $\mathcal{P}$ be a PLC in the plane. A *triangulation of* $\mathcal{P}$ is a simplicial complex $\mathcal{T}$ such that $\mathcal{P}$ and $\mathcal{T}$ have the same vertices, $\mathcal{T}$ contains every edge in $\mathcal{P}$ (and perhaps additional edges), and $|\mathcal{T}| = |\mathcal{P}|$.

It is not difficult to see that a simplex can appear in a triangulation of $\mathcal{P}$ only if it *respects* $\mathcal{P}$. (See Exercise 4.)

Definition 2.11 (respect). A simplex σ *respects* a PLC $\mathcal{P}$ if $\sigma \subseteq |\mathcal{P}|$ and for every $f \in \mathcal{P}$ that intersects σ, $f \cap \sigma$ is a union of faces of σ.

Proposition 2.13. *Every simple polygon has a triangulation. Every PLC in the plane has a triangulation too.*

Proof. Let P be a simple polygon. If P is a triangle, it clearly has a triangulation. Otherwise, consider the following procedure for triangulating P. Let $\angle uvw$ be a corner of P having an interior angle less than 180°. Two such corners are found by letting v be the lexicographically least or greatest vertex of P.

[1]If one takes the vertices and edges of a planar PLC and discards the polygons, one has a simplicial complex in the plane with no triangles. This complex is called a *planar straight line graph* (PSLG). Most publications about CDTs take a PSLG as the input and assume that the CDT should cover the PSLG's convex hull. PLCs are more expressive, as they can restrict the triangulation to a nonconvex region of the plane.

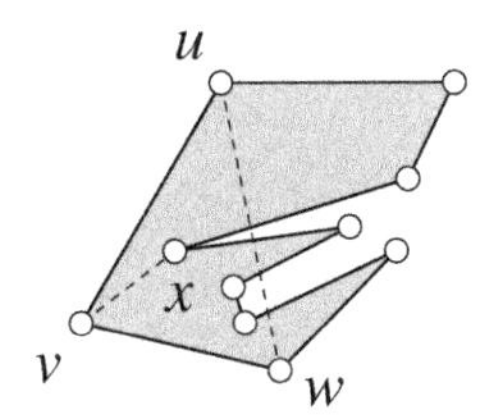

Figure 2.14: The edge vx cuts this simple polygon into two simple polygons.

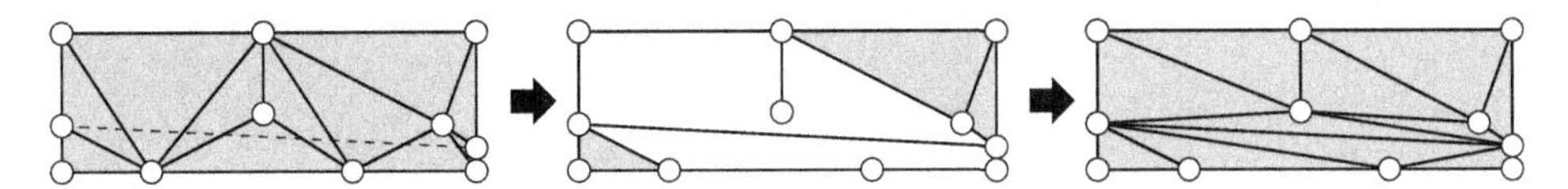

Figure 2.15: Inserting a segment into a triangulation.

If the open edge uw lies strictly in P's interior, then cutting uvw from P yields a polygon having one edge fewer; triangulate it recursively. Otherwise, uvw contains at least one vertex of P besides u, v, and w, as illustrated in Figure 2.14. Among those vertices, let x be the vertex farthest from the line aff uw. The open edge vx must lie strictly in P's interior, because if it intersected an edge of P, that edge would have a vertex further from aff uw. Cutting P at vx produces two simple polygons, each with fewer edges than P; triangulate them recursively. In either case, the procedure produces a triangulation of P.

Let $\mathcal{P}$ be a planar PLC. Consider the following procedure for triangulating $\mathcal{P}$. Begin with an arbitrary triangulation of the vertices in $\mathcal{P}$, such as the lexicographic triangulation described in Section 2.1. Examine each segment in $\mathcal{P}$ to see if it is already an edge of the triangulation. Insert each missing segment into the triangulation by deleting all the edges and triangles that intersect its relative interior, creating the new segment, and retriangulating the two polygonal cavities thus created (one on each side of the segment), as illustrated in Figure 2.15. The cavities might not be simple polygons, because they might have edges dangling in their interiors, as shown. But it is straightforward to verify that the procedure discussed above for triangulating a simple polygon works equally well for a cavity with dangling edges.

The act of inserting a segment never deletes another segment, because two segments in $\mathcal{P}$ cannot cross. Therefore, after every segment is inserted, the triangulation contains all of them. Finally, delete any simplices not included in $|\mathcal{P}|$. □

Definition 2.10 does not permit $\mathcal{T}$ to have vertices absent from $\mathcal{P}$, but mesh generation usually entails adding new vertices to guarantee that the triangles have high quality. This motivates the notion of a Steiner triangulation.

Definition 2.12 (Steiner triangulation of a PLC)**.** Let $\mathcal{P}$ be a PLC. A *Steiner triangulation of* $\mathcal{P}$, also known as a *conforming triangulation of* $\mathcal{P}$ or a *mesh of* $\mathcal{P}$, is a simplicial complex $\mathcal{T}$ such that $\mathcal{T}$ contains every vertex in $\mathcal{P}$ and possibly more, every edge in $\mathcal{P}$ is a union of edges in $\mathcal{T}$, and $|\mathcal{T}| = |\mathcal{P}|$. The new vertices in $\mathcal{T}$, absent from $\mathcal{P}$, are called *Steiner points*. A *Steiner Delaunay triangulation* of $\mathcal{P}$, also known as a *conforming Delaunay triangulation* of $\mathcal{P}$, is a Steiner triangulation of $\mathcal{P}$ in which every simplex is Delaunay.

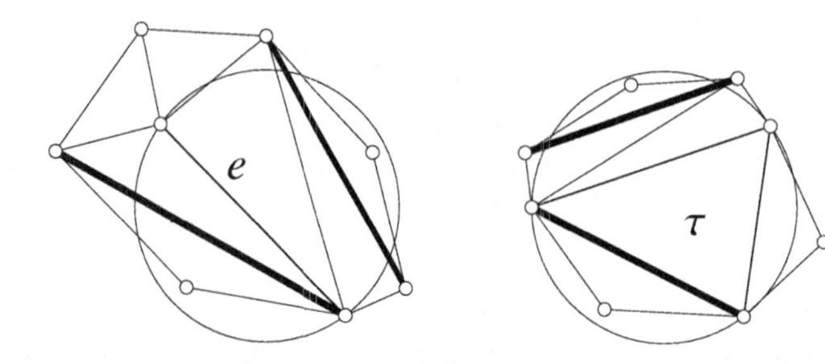

Figure 2.16: The edge e and triangle τ are constrained Delaunay. Bold lines represent segments.

2.10.2 The constrained Delaunay triangulation

Constrained Delaunay triangulations (CDTs) offer a way to force a triangulation to respect the edges in a PLC without introducing new vertices, while maintaining some of the advantages of Delaunay triangulations. However, it is necessary to relax the requirement that all triangles be Delaunay. The terminology can be confusing: whereas every Steiner Delaunay triangulation is a Delaunay triangulation (of some point set), constrained Delaunay triangulations generally are not.

Recall the Delaunay Lemma: a triangulation of a point set is Delaunay if and only if every edge is locally Delaunay. Likewise, there is a Constrained Delaunay Lemma (Section 2.10.3) that offers the simplest definition of a CDT: a triangulation of a PLC is constrained Delaunay if and only if every edge is locally Delaunay *or* a segment. Thus, a CDT differs from a Delaunay triangulation in three ways: it is not necessarily convex, it is required to contain the edges in a PLC, and those edges are exempted from being locally Delaunay.

The defining characteristic of a CDT is that every triangle is constrained Delaunay, as defined below.

Definition 2.13 (visibility). Two points x and y are *visible* to each other if the line segment xy respects $\mathcal{P}$; recall Definition 2.11. We also say that x and y can *see* each other. A linear cell in $\mathcal{P}$ that intersects the relative interior of xy but does not include xy is said to *occlude* the visibility between x and y.

Definition 2.14 (constrained Delaunay). In the context of a PLC $\mathcal{P}$, a simplex σ is *constrained Delaunay* if $\mathcal{P}$ contains the vertices of σ, σ respects $\mathcal{P}$, and there is an open circumdisk of σ that contains no vertex in $\mathcal{P}$ that is visible from a point in the relative interior of σ.

Figure 2.16 illustrates examples of a constrained Delaunay edge e and a constrained Delaunay triangle τ. Bold lines indicate PLC segments. Although e has no empty circumdisk, the depicted open circumdisk of e contains no vertex that is visible from the relative interior of e. There are two vertices in the disk, but both are hidden behind segments. Hence, e is constrained Delaunay. Similarly, the open circumdisk of τ contains two vertices, but both are hidden from the interior of τ by segments, so τ is constrained Delaunay.

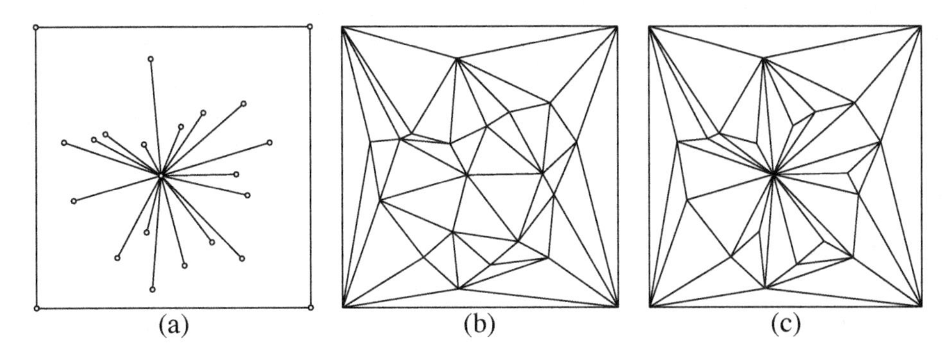

Figure 2.17: (a) A piecewise linear complex. (b) The Delaunay triangulation of its vertices. (c) Its constrained Delaunay triangulation.

Definition 2.15 (constrained Delaunay triangulation). A *constrained Delaunay triangulation* (CDT) of a PLC $\mathcal{P}$ is a triangulation of $\mathcal{P}$ in which every triangle is constrained Delaunay.

Figure 2.17 illustrates a PLC, a Delaunay triangulation of its vertices, and a constrained Delaunay triangulation of the PLC. In the CDT, every triangle is constrained Delaunay, every edge that is not a PLC segment is constrained Delaunay, and every vertex is trivially constrained Delaunay.

CDTs and Steiner Delaunay triangulations are two different ways to force a triangulation to conform to the boundary of a geometric domain. CDTs partly sacrifice the Delaunay property for the benefit of requiring no new vertices. For mesh generation, new vertices are usually needed anyway to obtain good triangles, so many Delaunay meshing algorithms use Steiner Delaunay triangulations. But some algorithms use a hybrid of CDTs and Steiner Delaunay triangulations because it helps to reduce the number of new vertices. A *Steiner CDT* or *conforming CDT* of $\mathcal{P}$ is a Steiner triangulation of $\mathcal{P}$ in which every triangle is constrained Delaunay.

2.10.3 Properties of the constrained Delaunay triangulation

For every property of Delaunay triangulations discussed in this chapter, there is an analogous property of constrained Delaunay triangulations. This section summarizes them. Proofs are omitted, but each of them is a straightforward extension of the corresponding proof for Delaunay triangulations.

The Delaunay Lemma generalizes to CDTs, and provides a useful alternative definition: a triangulation of a PLC $\mathcal{P}$ is a CDT if and only if every one of its edges is locally Delaunay *or* a segment in $\mathcal{P}$.

Lemma 2.14 (Constrained Delaunay Lemma). *Let $\mathcal{T}$ be a triangulation of a PLC $\mathcal{P}$. The following three statements are equivalent.*

- *Every triangle in $\mathcal{T}$ is constrained Delaunay (i.e. $\mathcal{T}$ is constrained Delaunay).*
- *Every edge in $\mathcal{T}$ not in $\mathcal{P}$ is constrained Delaunay.*
- *Every edge in $\mathcal{T}$ not in $\mathcal{P}$ is locally Delaunay.*

One way to construct a constrained Delaunay triangulation of a PLC $\mathcal{P}$ is to begin with any triangulation of $\mathcal{P}$. Apply the flip algorithm, modified so that it never flips a segment: repeatedly choose any edge of the triangulation that is not in $\mathcal{P}$ and not locally Delaunay, and flip it. When no such edge survives, the Constrained Delaunay Lemma tells us that the triangulation is constrained Delaunay.

Proposition 2.15. *Given a triangulation of a PLC having n vertices, the modified flip algorithm (which never flips a PLC segment) terminates after $O(n^2)$ edge flips, yielding a constrained Delaunay triangulation.*

Proposition 2.16. *Every PLC has a constrained Delaunay triangulation.*

The CDT has the same optimality properties as the Delaunay triangulation, except that the optimality is with respect to a smaller set of triangulations—those that include the PLC's edges.

Theorem 2.17. *Among all the triangulations of a PLC, every constrained Delaunay triangulation maximizes the minimum angle in the triangulation, minimizes the largest circumdisk, and minimizes the largest min-containment disk.*

A sufficient but not necessary condition for the CDT to be unique is that no four vertices are cocircular.

Theorem 2.18. *If a PLC is generic—no four of its vertices lie on a common circle—then the PLC has one unique constrained Delaunay triangulation, which contains every constrained Delaunay simplex.*

2.11 Notes and exercises

Delaunay triangulations and the Delaunay Lemma were introduced by Boris Delaunay's seminal 1934 paper [69]. The relationship between Delaunay triangulations and convex hulls was discovered by Brown [34], who proposed a different lifting map that projects the points onto a sphere. The parabolic lifting map of Seidel [190, 90] is numerically better behaved than the spherical lifting map.

The flip algorithm, the incremental insertion algorithm, and the Delaunay triangulation's property of maximizing the minimum angle were all introduced in a classic paper by Charles Lawson [131]. D'Azevedo and Simpson [67] show that two-dimensional Delaunay triangulations minimize the largest circumdisk and the largest min-containment disk. Lambert [129] shows that the Delaunay triangulation maximizes the mean inradius (equivalently, the sum of inradii) of its triangles. Rippa [175] shows that it minimizes the roughness (defined in Section 2.6) of a piecewise linearly interpolated function, and Powar [172] gives a simpler proof. Ciarlet and Raviart [63] show that for Dirichlet boundary value problems on Laplace's equation, finite element discretizations with piecewise linear elements over a triangulation in the plane with no obtuse angles have solutions that satisfy a discrete maximum principle. The result extends easily to all Delaunay triangulations in the plane, even ones with obtuse angles, but it is not clear who first made this observation. Miller, Talmor,

Teng, and Walkington [148] extend the result to a finite volume method that uses a three-dimensional Delaunay triangulation and its Voronoi dual, with Voronoi cells as control volumes.

The symbolic weight perturbation method of Section 2.9 originates with Edelsbrunner and Mücke [89, Section 5.4].

Constrained Delaunay triangulations in the plane were mathematically formalized by Lee and Lin [133] in 1986, though algorithms that unwittingly construct CDTs appeared much earlier [97, 159]. Lee and Lin extend to CDTs Lawson's proof that Delaunay triangulations maximize the minimum angle.

Exercises

1. Draw the Delaunay triangulation of the following point set.

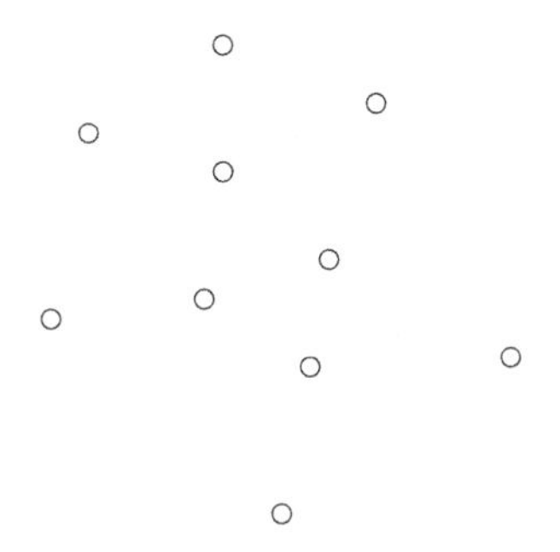

2. Let P and Q be two disjoint point sets in the plane. (Think of them as a red point set and a black point set.) Let $p \in P$ and $q \in Q$ be two points from these sets that minimize the Euclidean distance $d(p, q)$. Prove that pq is an edge of $\mathrm{Del}\,(P \cup Q)$. This observation leads easily to an $O(n \log n)$-time algorithm for finding p and q, the red-black closest pair.

3. Let S be a point set in the plane. S may have subsets of four or more cocircular points, so S may have many Delaunay triangulations.

   (a) Prove that it is possible to transform a triangulation of a convex polygon to any other triangulation of the same polygon by a sequence of edge flips.

   (b) Prove that it is possible to flip from any Delaunay triangulation of S to any other Delaunay triangulation of S, such that every intermediate triangulation is also Delaunay.

   (c) Prove that *every* Delaunay triangulation of S maximizes its minimum angle—there is no triangulation of S whose smallest angle is greater.

4. Show that a simplex can appear in a triangulation of a PLC $\mathcal{P}$ (Definition 2.10) only if it respects $\mathcal{P}$ (Definition 2.11).

5. Recall that every Delaunay triangulation of a point set contains every strongly Delaunay edge, but there is no such guarantee for Delaunay edges that are not strongly

Delaunay. Show constructively that for any PLC $\mathcal{P}$, *every* constrained Delaunay edge is in at least one CDT of $\mathcal{P}$. Hint: See Exercise 3(b).

6. Prove Lemma 2.14, the Constrained Delaunay Lemma.

7. Recall that a triangle τ is *constrained Delaunay* with respect to a PLC $\mathcal{P}$ if its vertices are in $\mathcal{P}$, it respects $\mathcal{P}$, and the open circumdisk of τ contains no vertex in $\mathcal{P}$ that is visible from a point in τ's interior.

   Let τ be a triangle that satisfies the first two of those three conditions. Let q be a point in the interior of τ. Prove that if no vertex of $\mathcal{P}$ in τ's open circumdisk is visible from q, then no vertex of $\mathcal{P}$ in τ's open circumdisk is visible from *any* point in the interior of τ, so τ is constrained Delaunay.

Chapter 3

Algorithms for constructing Delaunay triangulations

There are three classic types of algorithm for constructing Delaunay triangulations besides the flip algorithm.

Gift-wrapping algorithms construct Delaunay triangles one at a time, using the previously computed triangles as a seed on which new triangles crystallize. They are closely related to *advancing front methods* for mesh generation. Gift-wrapping generalizes easily to CDTs and to higher dimensions, and it is easy to implement, but it is difficult to make fast. Section 3.11 describes a basic gift-wrapping algorithm that triangulates n points in the plane in $O(n^2)$ worst-case time, or a PLC in the plane with n vertices and m segments in $O(n^2 m)$ time. There is a plane-sweeping gift-wrapping algorithm that runs in $O(n \log n)$ time, but we omit it because it does not extend to three dimensions.

Divide-and-conquer algorithms partition a set of points into two halves separated by a line, recursively compute the Delaunay triangulation of each subset, and merge the two triangulations into one. A divide-and-conquer algorithm was the first to run in $O(n \log n)$ time, which is optimal in the worst case. This class of algorithms includes the fastest planar Delaunay triangulator in practice today, but the divide-and-conquer strategy is not fast in three dimensions.

Incremental insertion algorithms insert vertices into a Delaunay triangulation one at a time, always restoring the Delaunay property to the triangulation before inserting another vertex. Some incremental insertion algorithms have an expected running time that is optimal for worst-case points sets: $O(n \log n)$ in the plane and $O(n^2)$ in three dimensions. The fastest three-dimensional Delaunay triangulators in practice are in this class.

The difference between a Delaunay triangulation algorithm and a modern Delaunay mesh generator is that the former is given all the vertices at the outset, whereas the latter uses the triangulation to decide where to place additional vertices, making incremental insertion obligatory. Therefore, Sections 3.3–3.7 study incremental insertion in depth. See also Section 5.3, which introduces a more sophisticated vertex ordering method called a *biased randomized insertion order* that offers the fastest speed in practice, especially for large points sets.

All three types of algorithms extend to constrained Delaunay triangulations. There are a divide-and-conquer algorithm and a plane-sweeping gift-wrapping algorithm that both run in $O(n \log n)$ time, but because they are complicated, these algorithms are rarely implemented. See the bibliographical notes (Section 3.12) for citations.

The most commonly used CDT construction method in practice is incremental insertion: first construct a Delaunay triangulation of the PLC's vertices, then insert the PLC's segments one by one. The algorithms commonly used to perform segment insertion in practice are slow, but a specialized incremental algorithm described in Section 3.10 runs in expected $O(n \log n + n \log^2 m)$ time. Realistic PLCs have few segments long enough to cross many edges, and it is typical to observe $O(n \log n)$ running time in practice.

3.1 The orientation and incircle predicates

Most geometric algorithms perform a mix of combinatorial and numerical computations. The numerical computations are usually packaged as *geometric primitives* of two types: *geometric constructors* that create new entities, such as the point where two specified lines intersect, and *geometric predicates* that determine relationships among entities, such as whether or not two lines intersect at all. Many Delaunay triangulation algorithms require just two predicates, called the *orientation* and *incircle* tests.

Let a, b, and c be three points in the plane. Consider a function $\textsc{Orient2D}(a, b, c)$ that returns a positive value if the points a, b, and c are arranged in counterclockwise order (see Figure 3.1, left), a negative value if the points are in clockwise order, and zero if the points are collinear. Another interpretation, important for many geometric algorithms, is that $\textsc{Orient2D}$ returns a positive value if a lies to the left of the line $\operatorname{aff} bc$ relative to an observer standing at b and facing c. The orientation test can be implemented as a matrix determinant that computes the signed area of the parallelogram determined by the vectors $a - c$ and $b - c$.

$$\textsc{Orient2D}(a, b, c) = \begin{vmatrix} a_x & a_y & 1 \\ b_x & b_y & 1 \\ c_x & c_y & 1 \end{vmatrix} \tag{3.1}$$

$$= \begin{vmatrix} a_x - c_x & a_y - c_y \\ b_x - c_x & b_y - c_y \end{vmatrix}. \tag{3.2}$$

To extend these expressions to higher dimensions, add more rows and columns representing additional points and coordinate axes. Given four points a, b, c, and d in $\mathbb{R}^3$, define $\textsc{Orient3D}(a, b, c, d)$ to be the signed volume of the parallelepiped determined by the vectors $a - d$, $b - d$, and $c - d$. It is positive if the points occur in the orientation illustrated in Figure 3.1, negative if they occur in the mirror-image orientation, and zero if the four points are coplanar. You can apply a *right-hand rule*: orient your right hand with fingers curled to

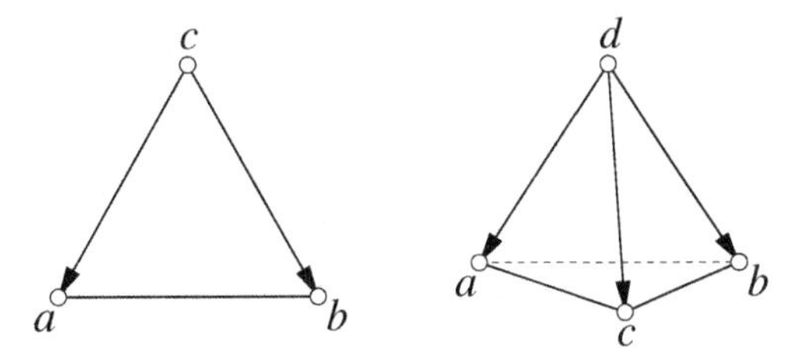

Figure 3.1: A triangle and a tetrahedron, both having positive orientation.

follow the circular sequence *bcd*. If your thumb points toward *a*, ORIENT3D is positive.

$$\text{ORIENT3D}(a,b,c,d) = \begin{vmatrix} a_x & a_y & a_z & 1 \\ b_x & b_y & b_z & 1 \\ c_x & c_y & c_z & 1 \\ d_x & d_y & d_z & 1 \end{vmatrix} = \begin{vmatrix} a_x - d_x & a_y - d_y & a_z - d_z \\ b_x - d_x & b_y - d_y & b_z - d_z \\ c_x - d_x & c_y - d_y & c_z - d_z \end{vmatrix}.$$

Given four points a, b, c, and d in $\mathbb{R}^2$, the *incircle test* INCIRCLE(a,b,c,d) returns a positive value if d lies inside the unique (and possibly degenerate) circle through a, b, and c, assuming that the latter three points occur in counterclockwise order around the circle. INCIRCLE returns zero if and only if all four points lie on a common circle or line. INCIRCLE is derived from ORIENT3D and the Lifting Lemma (Lemma 2.1), which shows that testing whether a point is inside a circle is equivalent to an orientation test on the points lifted by the parabolic lifting map.

$$\text{INCIRCLE}(a,b,c,d) = \begin{vmatrix} a_x & a_y & a_x^2 + a_y^2 & 1 \\ b_x & b_y & b_x^2 + b_y^2 & 1 \\ c_x & c_y & c_x^2 + c_y^2 & 1 \\ d_x & d_y & d_x^2 + d_y^2 & 1 \end{vmatrix} \tag{3.3}$$

$$= \begin{vmatrix} a_x - d_x & a_y - d_y & (a_x - d_x)^2 + (a_y - d_y)^2 \\ b_x - d_x & b_y - d_y & (b_x - d_x)^2 + (b_y - d_y)^2 \\ c_x - d_x & c_y - d_y & (c_x - d_x)^2 + (c_y - d_y)^2 \end{vmatrix}. \tag{3.4}$$

These expressions also extend easily to higher dimensions. Let a, b, c, d, and e be five points in $\mathbb{R}^3$, with the first four ordered so that ORIENT3D(a,b,c,d) is positive. The function INSPHERE(a,b,c,d,e) is positive if e lies inside the sphere passing through a, b, c, and d; negative if e lies outside the sphere; and zero if all five points are cospherical or coplanar.

$$\textsc{InSphere}(a,b,c,d,e) = \begin{vmatrix} a_x & a_y & a_z & a_x^2+a_y^2+a_z^2 & 1 \\ b_x & b_y & b_z & b_x^2+b_y^2+b_z^2 & 1 \\ c_x & c_y & c_z & c_x^2+c_y^2+c_z^2 & 1 \\ d_x & d_y & d_z & d_x^2+d_y^2+d_z^2 & 1 \\ e_x & e_y & e_z & e_x^2+e_y^2+e_z^2 & 1 \end{vmatrix}$$
$$= \begin{vmatrix} a_x-e_x & a_y-e_y & a_z-e_z & (a_x-e_x)^2+(a_y-e_y)^2+(a_z-e_z)^2 \\ b_x-e_x & b_y-e_y & b_z-e_z & (b_x-e_x)^2+(b_y-e_y)^2+(b_z-e_z)^2 \\ c_x-e_x & c_y-e_y & c_z-e_z & (c_x-e_x)^2+(c_y-e_y)^2+(c_z-e_z)^2 \\ d_x-e_x & d_y-e_y & d_z-e_z & (d_x-e_x)^2+(d_y-e_y)^2+(d_z-e_z)^2 \end{vmatrix}. \quad (3.5)$$

Orient2D, Orient3D, InCircle, and InSphere have the symmetry property that interchanging any two of their parameters reverses their sign. If the points a, b, c occur in clockwise order, InCircle behaves as if the circle's outside were its inside. Likewise, if Orient3D(a, b, c, d) is negative, the sign returned by InSphere is reversed.

Expressions (3.1) and (3.2) can be shown to be equivalent by simple algebraic transformations, as can Expressions (3.3) and (3.4) with a little more effort. Expressions (3.2) and (3.4) should be strongly preferred over Expressions (3.1) and (3.3) for fixed precision floating-point computation, because they lose far less accuracy to roundoff error. Ideally, some form of exact arithmetic should be used to perform these tests, or the triangulation algorithms cannot be guaranteed to work correctly.

3.2 A dictionary data structure for triangulations

Two data structures are commonly used to represent triangulations: edge-based data structures, of which the best known is the doubly connected edge list, and triangle-based data structures. What these two data structures have in common is that records represent edges or triangles, and the records store pointers that point at neighboring edges or triangles. Many implementations of triangulation algorithms read and change these pointers directly; experience shows that these implementations are difficult to code and debug.

Here, we advocate an interface that does not expose pointers to the algorithms that use the triangulation data structure. Through this interface, triangulation algorithms access the triangulation in a natural way, by adding or deleting triangles specified as triples of vertices. It is wholly the responsibility of the triangulation storage library to determine triangle adjacencies and to correctly maintain any pointers it uses internally. This policy improves programmer productivity and simplifies debugging.

The interface appears in Figure 3.2. Two procedures, AddTriangle and DeleteTriangle, create and delete triangles by specifying the vertices of a triangle, ordered so that all the triangles stored in the data structure have positive orientation. The data structure enforces the invariant that only two triangles may adjoin an edge and only one may be on each side of the edge. Therefore, if the data structure contains a positively oriented triangle uvw and an application calls AddTriangle(u, v, x), the triangle uvx is rejected and the data structure does not change. However, the triangulation storage library never computes the geometric orientation of a triangle, and it does not even need to know the vertex coordinates; rather,

Procedure	Purpose
AddTriangle(u, v, w)	Add a positively oriented triangle uvw
DeleteTriangle(u, v, w)	Delete a positively oriented triangle uvw
Adjacent(u, v)	Return a vertex w such that uvw is a positively oriented triangle
Adjacent2Vertex(u)	Return vertices v, w such that uvw is a positively oriented triangle

Figure 3.2: An interface for a triangulation data structure.

it treats orientation as a purely combinatorial property of a triangle, and it trusts the caller to pass in vertices with the correct ordering.

At least two query operations are supported. The procedure Adjacent(u, v) returns a vertex w if the triangulation includes a positively oriented triangle uvw, or the empty set otherwise. Adjacent(u, v) and Adjacent(v, u) identify different triangles, on opposite sides of the edge uv. The procedure Adjacent2Vertex(u) identifies an arbitrary triangle having vertex u, or returns the empty set if no such triangle exists.

A fast way to implement Adjacent efficiently is to store each triangle uvw three times in a hash table, keyed on the directed edges uv, vw, and wu. A hash table can query edges and store triangles in expected constant time. The bibliographical notes at the end of this chapter cite a more space-efficient implementation.

Unfortunately, it takes substantial additional memory to guarantee that the Adjacent2Vertex query will run fast. Many algorithms for Delaunay triangulation and meshing can be implemented without it, so we recommend using Adjacent2Vertex as little as possible, and settling for a slow but memory-efficient implementation, perhaps even searching the entire hash table. A good compromise implementation is to maintain an array that stores, for each vertex u, a vertex v such that the most recently added triangle adjoining u also had v for a vertex. When Adjacent2Vertex(u) is invoked, it looks up the edges uv and vu in the hash table to find an adjoining triangle in expected constant time. The flaw in this implementation is that the triangles having edge uv may have been subsequently deleted, in which case a triangle adjoining u must be found some other way (e.g. searching the entire hash table). However, observe that this catastrophe will not occur if every triangle deletion is followed by triangle creations that cover all the same vertices—which is true of most of the algorithms discussed in this book.

3.3 Inserting a vertex into a Delaunay triangulation

Consider inserting a new vertex u into a Delaunay triangulation. If a triangle's open circumdisk contains u, that triangle is no longer Delaunay, so it must be deleted. This suggests the *Bowyer–Watson algorithm.*

BowyerWatson(u)

1. Find one triangle whose open circumdisk contains u.

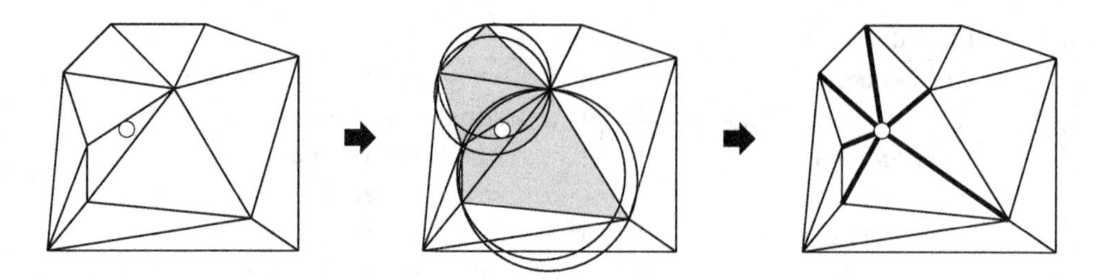

Figure 3.3: The Bowyer–Watson algorithm in the plane. At left, a Delaunay triangulation and a new vertex to insert. At center, every triangle whose open circumdisk contains the new vertex is shaded. These triangles are no longer Delaunay. At right, the shaded triangles disappear, replaced by new triangles that connect the new vertex to the edges of the cavity.

2. Find all the others (in time linear in their number) by a depth-first search in the triangulation.
3. Delete them all, evacuating a polyhedral cavity, which is shaded in Figure 3.3.
4. For each edge of this cavity, create a new triangle joining it with u, as illustrated.

The first step of BowyerWatson is called *point location*. Most Delaunay mesh generation algorithms generate new vertices in the circumdisks of badly shaped or oversized triangles, in which case point location is free. However, point location is not free for the domain vertices provided as input to the mesh generator. Locating these points in the triangulation is sometimes the most costly and complicated part of the incremental insertion algorithm. Incremental insertion is really a class of Delaunay triangulation algorithms, differentiated by their point location methods. Section 3.6 describes a point location method that helps a randomized incremental insertion algorithm to construct the Delaunay triangulation of n vertices in expected $O(n \log n)$ time. Section 5.5 describes a point location method that seems to be even faster in practice, albeit only if the vertices are carefully ordered as described in Section 5.3.

The following three propositions demonstrate the correctness of the Bowyer–Watson algorithm if a correct point location algorithm is available. The first proposition shows that the deleted triangles—those that are no longer Delaunay—constitute a star-shaped polygon. This fact guarantees that a depth-first search (Step 2 of BowyerWatson) will find all the triangles that are no longer Delaunay, and that Steps 3 and 4 of BowyerWatson yield a simplicial complex.

Proposition 3.1. *The union of the triangles whose open circumdisks contain u is a connected star-shaped polygon, meaning that for every point p in the polygon, the polygon includes the line segment pu.*

Proof. Prior to the insertion of u, the triangulation is Delaunay, so all of its edges are locally Delaunay. Let τ be a triangle whose open circumdisk contains u. Let p be any point

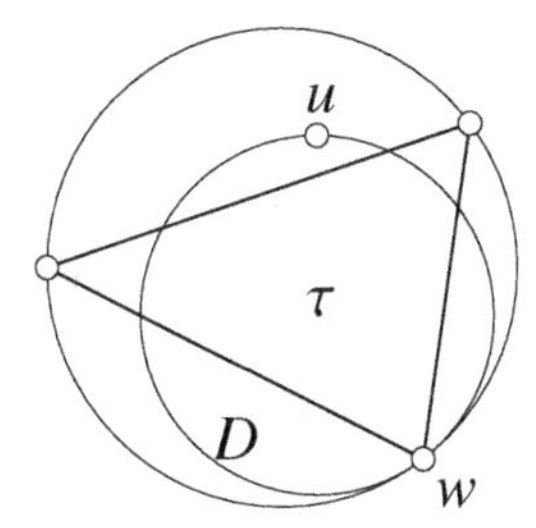

Figure 3.4: Because τ was Delaunay before u was inserted, uw is strongly Delaunay.

in τ. By the same inductive reasoning employed in the proof of the Delaunay Lemma (Theorem 2.3), every triangle that intersects the open line segment pu also has u in its open circumdisk. The result follows. □

The key to proving that the updated triangulation is Delaunay is to show that all its edges are Delaunay and apply the Delaunay Lemma. The following proposition shows that every newly created edge is strongly Delaunay and, therefore, appears in every Delaunay triangulation of the vertices.

Proposition 3.2. *Let u be a newly inserted vertex. Let τ be a triangle that is deleted because its open circumdisk contains u. Let w be a vertex of τ. Then the edge uw is strongly Delaunay.*

Proof. See Figure 3.4. The open circumdisk of τ contains no vertex but u. Let D be the circumdisk whose boundary is tangent to τ's circumdisk at w and passes through u. The disk D demonstrates that uw is strongly Delaunay. □

Proposition 3.3. *A triangulation produced by applying the Bowyer–Watson algorithm to a Delaunay triangulation is Delaunay.*

Proof. It follows from Proposition 3.1 that the update produces a triangulation of the point set augmented with the new point. All the surviving old triangles are Delaunay; otherwise they would have been deleted. It follows that their edges are Delaunay too. By Proposition 3.2, all of the newly created edges are Delaunay as well. By the Delaunay Lemma, the new triangulation is Delaunay. □

3.4 Inserting a vertex outside a Delaunay triangulation

The Bowyer–Watson algorithm works only if the newly inserted vertex lies in the triangulation. However, there is an elegant way to represent a triangulation so that the algorithm, with almost no changes, can insert a vertex outside the triangulation equally well. Imagine that every edge on the boundary of the triangulation adjoins a *ghost triangle*, as illustrated in Figure 3.5. The third vertex of every ghost triangle is the *ghost vertex*, a vertex "at infinity" shared by every ghost triangle. Every ghost triangle has two *ghost edges* that adjoin the ghost vertex. A triangle that is not a ghost is called a *solid triangle*.

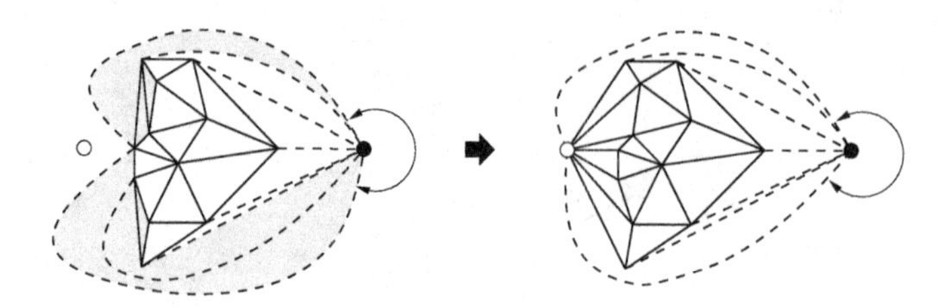

Figure 3.5: Inserting a vertex outside the triangulation. The black dot is the ghost vertex. The white dot is a new vertex to be inserted. The circular arrow indicates two ghost edges that are really the same edge. Three ghost triangles and three solid triangles (shaded) are deleted and replaced with two new ghost triangles and six new solid triangles.

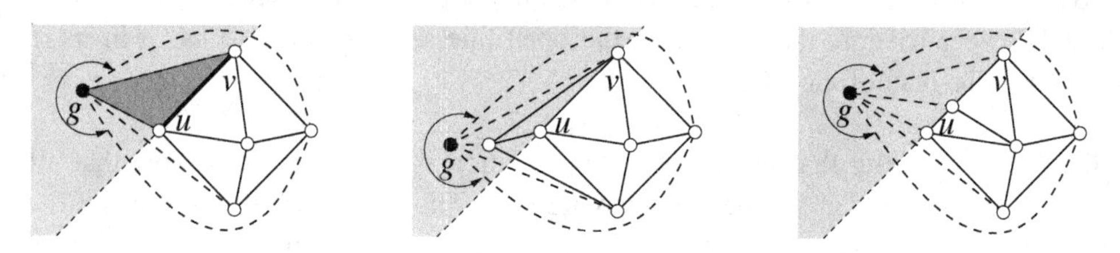

Figure 3.6: The ghost triangle uvg is deleted if a new vertex is inserted in the shaded open halfplane (as at center) or on uv (as at right). The union of the open halfplane and uv is the *outer halfplane* of uvg.

The ghost triangles are explicitly stored in the triangulation data structure. They are not merely cosmetic; they make it possible for the Bowyer–Watson algorithm to efficiently traverse the triangulation boundary, and thus they are essential to obtaining an incremental insertion algorithm with optimal running time.

Consider an edge uv on the boundary of a triangulation, directed clockwise around the boundary. Define a positively oriented ghost triangle uvg, where g is the ghost vertex. Like any other triangle, uvg has a circumdisk—albeit a degenerate one—and must be deleted if a new vertex is inserted in it. The definition of *circumdisk* is a bit tricky, though. The circumcircle degenerates to the line aff uv, which divides the plane into two halfplanes.

There are two cases in which the ghost triangle uvg must be deleted (i.e. uv is no longer a boundary edge of the triangulation), both illustrated in Figure 3.6: if a vertex is inserted in the open halfplane on the other side of aff uv from the triangulation, or if a vertex is inserted on the open edge uv. Call the union of these two regions the *outer halfplane* of uv. It is neither an open nor a closed halfplane, but something in between. It is the set of points in the open circumdisk of uvg in the limit as g moves away from the triangulation.

For weighted Delaunay triangulations, the rules are slightly more complicated. A ghost triangle is deleted if a vertex is inserted in its outer halfplane, except perhaps if the vertex lies on the solid edge of the ghost triangle, in which case the ghost triangle is deleted if the new vertex is not submerged.

Ghost triangles have an intuitive interpretation in terms of the lifting map. Imagine that in $\mathbb{R}^3$, the solid triangles are lifted to the paraboloid, and the ghost triangles and ghost edges are vertical—parallel to the z-axis. By magic, the ghost vertex is interpreted as being directly above every other vertex at an infinite height. The faces of the convex hull of this three-dimensional point set, including the magic ghost vertex, are in one-to-one correspondence with the faces and ghost faces of the Delaunay triangulation.

The following pseudocode shows in more detail a recursive implementation of the Bowyer–Watson algorithm, omitting the point location step. It interleaves the second, third, and fourth steps of BowyerWatson, thereby achieving simplicity and speed although obscuring the algorithm's workings. The parameters to InsertVertex are a vertex u to insert and a positively oriented triangle vwx whose open circumdisk contains u. In this pseudocode, all the triangles are positively oriented and all the edges are oriented (i.e. the vertex order matters).

InsertVertex(u, vwx)

1. Call DeleteTriangle(v, w, x).
2. Call DigCavity(u, vw), DigCavity(u, wx), and DigCavity(u, xv) to identify the other deleted triangles and insert new triangles.

DigCavity(u, vw)

1. Let x = Adjacent(w, v); wvx is the triangle on the other side of edge vw from u.
2. If $x = \emptyset$, then return, because the triangle has already been deleted.
3. If InCircle(u, v, w, x) > 0, then uvw and wvx are not Delaunay, so call DeleteTriangle(w, v, x). Call DigCavity(u, vx) and DigCavity(u, xw) to recursively identify more deleted triangles and insert new triangles, and return.
4. Otherwise, vw is an edge of the polygonal cavity, so call AddTriangle(u, v, w).

Take care to note that Step 2 of DigCavity is checking for the circumstance where the edge wv adjoins the growing cavity, and the test $x = \emptyset$ is *not* true when x is the ghost vertex. The use of a ghost vertex is needed for Step 2 to be correct. However, Step 2 is necessary only to compute a weighted Delaunay triangulation and can be omitted for an ordinary Delaunay triangulation, though we will not justify that claim.

Step 3 of DigCavity requires a modification to the InCircle test discussed in Section 3.1: if wvx is a ghost triangle, then replace the formula (3.4) with a test of whether u lies in the outer halfplane of wvx. To adapt the code for a weighted Delaunay triangulation, replace (3.4) with Orient3D(u^+, v^+, w^+, x^+), which tests whether x^+ is below the witness plane of the triangle uvw. The parameter vwx of InsertVertex must be a triangle whose witness plane is above u^+. Sometimes a vertex u is submerged, in which case no such triangle exists and the calling procedure must not to try to insert u.

A popular alternative to ghost triangles is to enclose the input vertices in a giant triangular bounding box, illustrated in Figure 3.7. After all the vertices have been inserted, every triangle having a bounding box vertex is deleted. The difficulty with this approach is that the bounding box vertices may leave concave divots in the triangulation if they are too close, and it is difficult to determine how far away they need to be. One solution to this problem is to compute a weighted Delaunay triangulation, assigning the three bounding box

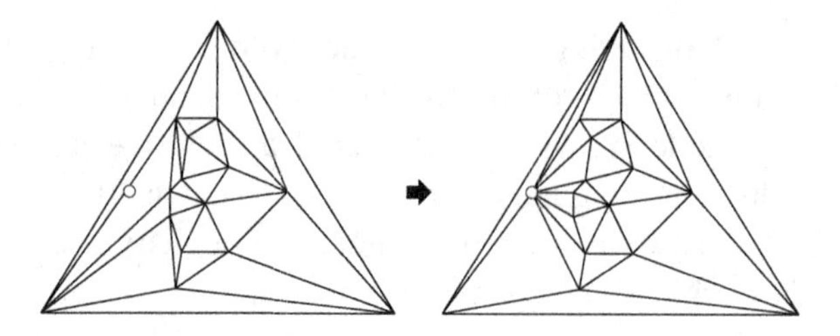

Figure 3.7: Enclosing the vertices in a large triangular bounding box. In practice, the box would be much, much larger.

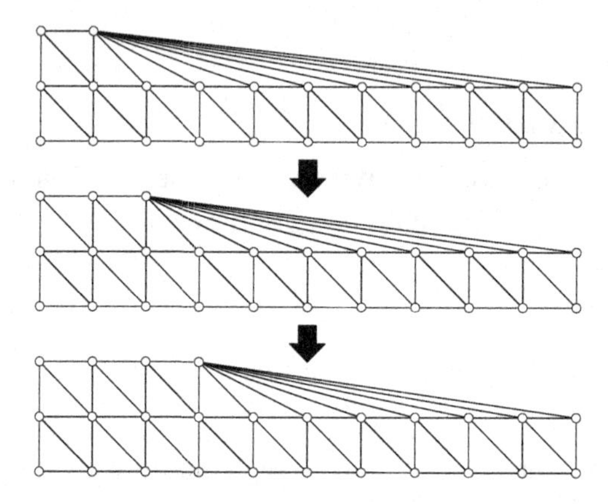

Figure 3.8: Each vertex insertion can delete $\Theta(n)$ triangles and create $\Theta(n)$ others.

vertices weights of negative infinity. These three infinite weights must be incomparable—say, ∞, 2^{∞}, and $2^{2^{\infty}}$—so that ORIENT3D tests involving two of the bounding box vertices operate consistently. This approach seems to run more slowly (perhaps by 10%) than the ghost triangle implementation. Another solution is to use the segment insertion algorithm described in Section 3.10 to fill the divots.

3.5 The running time of vertex insertion

How expensive is vertex insertion, leaving out the cost of point location? This section considers two cases: the worst case and the expected case when vertices are inserted in random order. The latter case is a part of an incremental insertion algorithm that computes the Delaunay triangulation of n vertices in expected $O(n \log n)$ time, and it also introduces an elegant algorithm analysis technique called *backward analysis*.

Figure 3.8 illustrates the worst case. A single vertex insertion can delete $\Theta(n)$ triangles and create $\Theta(n)$ others, taking $\Theta(n)$ time. Moreover, this dismal performance can be repeated for $\Theta(n)$ successive vertex insertions. Therefore, the incremental insertion algorithm for constructing a Delaunay triangulation takes $\Theta(n^2)$ time if the vertices and their insertion order are chosen badly. The grid arrangement and vertex ordering in the figure are common in practice.

Fortunately, there are better ways to order the vertex insertion operations. The *randomized incremental insertion algorithm* inserts the vertices in random order, with each permutation of the vertices being equally likely. Surprisingly, the *expected* number of triangles created by each successive vertex insertion operation is less than six, as Proposition 3.4 below shows. The catch is that all the vertices must be known in advance, so that a random permutation can be computed. The randomized algorithm is excellent for creating an initial triangulation of the vertices of a domain, but its analysis does not apply to the vertices that are subsequently generated during mesh generation, because their order cannot be randomized. Nevertheless, the proposition provides intuition for why constant-time vertex insertion is so commonly observed in mesh generation.

Proposition 3.4. *Let S be a set of n vertices in the plane. Let $(v_1, v_2, \ldots, v_n)$ be a permutation of S chosen uniformly at random from the set of all such permutations. For $i \in [0, n]$, let $\mathcal{T}_i$ be the Delaunay triangulation constructed by inserting the first i vertices in order. When v_i is inserted into $\mathcal{T}_{i-1}$, the expected number of new triangles (including ghost triangles) created is less than six. An expected total of $O(n)$ triangles are created and deleted during the n vertex insertions that construct $\mathcal{T}_n$.*

This proposition is most easily proved with *backward analysis*, a remarkable analysis technique that Raimund Seidel summarizes thusly: "*Analyze an algorithm as if it was running backwards in time, from output to input.*" Imagine that instead of inserting a randomly chosen vertex into $\mathcal{T}_{i-1}$, the algorithm is deleting a randomly chosen vertex from $\mathcal{T}_i$. Because a random permutation written backward is still a random permutation, each vertex in $\mathcal{T}_i$ is deleted with equal probability.

Proof. For every vertex v of $\mathcal{T}_i$, the number of triangles adjoining v, including ghost triangles, is equal to the degree of v, counting one ghost edge if v is on the boundary of the triangulation. When v_i is inserted into $\mathcal{T}_{i-1}$ to construct $\mathcal{T}_i$, every new triangle created has v_i for a vertex. Therefore, the expected number of new triangles created is equal to the expected degree of v_i.

There is one technical difficulty: if four vertices of $\mathcal{T}_i$ lie on a common empty circle, then $\mathcal{T}_i$ depends on the order in which the vertices are inserted. Thus, let $\mathcal{C}_i$ be the Delaunay subdivision of $\{v_1, v_2, \ldots, v_i\}$, wherein triangles in $\mathcal{T}_i$ sharing a common circumdisk are merged into a polygon. Recall from Section 2.2 that $\mathcal{C}_i$ contains the strongly Delaunay edges of $\mathcal{T}_i$ and no others, and is therefore unique. By Proposition 3.2, every edge adjoining v_i in $\mathcal{T}_i$ is strongly Delaunay, so the degree of v_i in $\mathcal{T}_i$ is equal to the degree of v_i in $\mathcal{C}_i$.

Because the permutation is chosen uniformly at random, each vertex of $\mathcal{C}_i$ is equally likely to be v_i. The expected degree of a randomly chosen vertex in $\mathcal{C}_i$ (or any planar graph) is less than six, by the following reasoning.

Let $i+1$, e, and f denote the number of vertices, edges, and triangles of $\mathcal{T}_i$, respectively, with the ghost vertex, ghost edges, and ghost triangles included. By Euler's formula, $i + 1 - e + f = 2$. Each triangle has three edges, and each edge is shared by two triangles, so $2e = 3f$. Eliminating f from Euler's formula gives $e = 3i - 3$. Each edge has two vertices, so the total number of edge-vertex incidences is $6i - 6$, and the average degree of a non-ghost vertex in $\mathcal{T}_i$ is less than $6 - 6/i$. Hence, the average degree of a non-ghost vertex in $\mathcal{C}_i$ is less than $6 - 6/i$.

Each vertex insertion creates, in expectation, fewer than six new triangles, so the expected total number of triangles created during the n vertex insertions is $O(n)$. A triangle cannot be deleted unless it is created first, so the expected total number of triangles deleted is also $O(n)$. □

Proposition 3.4 bounds not only the number of structural changes, but also the running time of the depth-first search in the Bowyer–Watson algorithm. This search visits all the triangles that are deleted and all the triangles that share an edge with a deleted triangle. The depth-first search takes time linear in the number of visited triangles and, therefore, linear in the number of deleted triangles.

It follows that the expected running time of the randomized incremental insertion algorithm, *excluding* point location, is $O(n)$. We shall see that point location is the dominant cost of the algorithm.

A general fact about randomized algorithms is that there is a chance that they will run much, much more slowly than their expected running time, but the probability of that is exceedingly small. If the incremental insertion algorithm gets unlucky and endures a slow vertex insertion like those depicted in Figure 3.8, other, faster vertex insertions will probably make up for it. The probability that many such slow vertex insertions will occur in one run is tiny, but it can happen.

3.6 Optimal point location by a conflict graph

This section describes a point location method of Ken Clarkson and Peter Shor that enables the randomized incremental insertion algorithm to construct a Delaunay triangulation of a set S of n points in expected $O(n \log n)$ time. All the points must be known from the beginning so they can be inserted in a random order. The point location method is appropriate for triangulating the vertices of a domain, but not for subsequently adding new vertices that are generated on the fly.

Let $(v_1, v_2, \ldots, v_n)$ be a permutation of S chosen uniformly at random from the set of all such permutations. For $i \in [0, n]$, let $\mathcal{T}_i$ be the Delaunay triangulation of the first i vertices.

A *conflict* is a vertex-triangle pair (w, τ) consisting of an uninserted vertex w (i.e. $w \in S$ but $w \notin \mathcal{T}_i$) and a triangle $\tau \in \mathcal{T}_i$ whose open circumdisk contains w. Ghost triangles can participate in conflicts; (w, τ) is a conflict if τ is a ghost triangle whose outer halfplane contains w. A conflict implies that τ will not survive after w is inserted. Each vertex insertion creates some conflicts—those involving the newly created triangles—and eliminates some others—those involving the deleted triangles.

A *conflict graph* is a bipartite graph that connects uninserted vertices to conflicting triangles. In some geometric algorithms, conflict graphs record every conflict, but for speed and simplicity, the present algorithm records just one conflict for each uninserted vertex. The data structure representing the conflict graph supports fast queries in both directions: for each vertex it stores one conflicting triangle, which can be queried in constant time, and for each triangle $\tau \in \mathcal{T}_i$ it stores a *conflict list* of all the uninserted vertices that choose τ as their conflicting triangle.

When a vertex is inserted, the conflict graph performs point location by immediately identifying one triangle that will be deleted. The other conflicting triangles are found by depth-first search as usual. The real work of the point location method is to update the

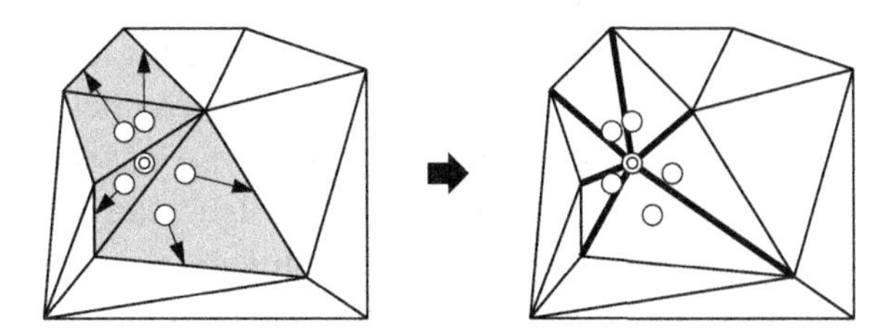

Figure 3.9: Redistributing uninserted vertices to new conflict triangles by shooting rays away from the newly inserted vertex.

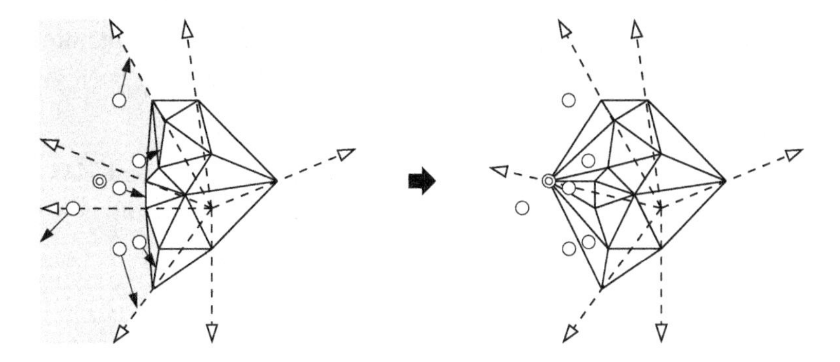

Figure 3.10: Partition the plane outside the triangulation into unbounded, three-sided ghost "triangles." Redistribute uninserted vertices to new conflict triangles, which may be ghost triangles or solid triangles.

conflict graph after those triangles are deleted and new ones are created. Every uninserted vertex that has chosen a deleted triangle as its conflict must be *redistributed* to a new conflict triangle. Because each triangle records a list of vertices that choose it, the conflict graph quickly identifies the vertices that must be redistributed. However, the obvious redistribution method, testing each vertex for a conflict with every newly created triangle, is not fast enough. The redistribution algorithm we present here requires that for each uninserted vertex w, the conflict graph records a conflict (w, τ) such that $w \in \tau$. If two triangles contain w, a solid triangle is preferred over a ghost.

Consider the operation that inserts v_i into $\mathcal{T}_{i-1}$ to construct $\mathcal{T}_i$. Let $w \in S$ be a vertex not present in $\mathcal{T}_i$. Let $\tau \in \mathcal{T}_{i-1}$ be the triangle recorded in the conflict graph as w's conflict, so $w \in \tau$. If τ survives in $\mathcal{T}_i$, the conflict (w, τ) remains in the graph. If τ is deleted, find the edge of the cavity struck by the ray $\overrightarrow{v_i w}$, as illustrated in Figure 3.9, and let σ be the new triangle adjoining that cavity edge; clearly $w \in \sigma$. Replace the conflict (w, τ) with (w, σ) in the conflict graph. Finding the edge entails tracing the ray through a sequence of deleted triangles.

To extend this point location method to uninserted vertices that lie outside the triangulation, treat the ghost edges as rays that point directly away from a central point in the interior of the triangulation, as illustrated in Figure 3.10. The ghost edges partition the portion of the plane outside the triangulation into unbounded, convex ghost "triangles," each having one solid edge and two diverging rays in its boundary. Each uninserted vertex outside the triangulation chooses as its conflict the unbounded ghost triangle that contains it.

The procedure InsertVertexAtConflict below inserts a vertex into a Delaunay triangulation. It calls a procedure RedistributeList that updates the conflict graph by redistributing a triangle's vertices. Unlike the procedure InsertVertex in Section 3.4, InsertVertexAtConflict cannot delete triangles during the depth-first search step, because RedistributeList needs the old triangles to efficiently redistribute uninserted vertices to their new conflict triangles. Instead, the algorithm maintains a list D of triangles to be deleted and a list C of triangles to be created after the vertices are redistributed.

InsertVertexAtConflict(u)

1. Let vwx be the triangle stored in the conflict graph as containing the vertex u, with vwx positively oriented. Mark vwx as having been visited.
2. Initialize the sets $D = \{vwx\}$ and $C = \emptyset$.
3. Call MarkCavity(u, vw, D, C), MarkCavity(u, wx, D, C), and MarkCavity(u, xv, D, C) to identify the other deleted triangles and the new triangles.
4. For each triangle $vwx \in D$, call RedistributeList(u, vwx).
5. For each triangle $vwx \in D$, call DeleteTriangle(v, w, x).
6. For each triangle $vwx \in C$, call AddTriangle(v, w, x).

MarkCavity(u, vw, D, C)

1. Let x = Adjacent(w, v); wvx is the triangle on the other side of edge vw from u.
2. If wvx is marked as having been visited, then return.
3. If InCircle(u, v, w, x) $>$ 0, then uvw and wvx are not Delaunay. Mark wvx as having been visited and append wvx to D. Call MarkCavity(u, vx, D, C) and MarkCavity(u, xw, D, C) to identify more deleted and new triangles, and return.
4. Otherwise, vw is an edge of the polygonal cavity, so insert the triangle uvw into C and return.

RedistributeList(u, vwx)

1. For each vertex y in the conflict list for vwx
   (a) Starting from the triangle vwx, walk from triangle to triangle from the point y along the ray $\overrightarrow{uy}$, stopping at the first triangle abc not marked as visited. The edge ab through which $\overrightarrow{uy}$ enters abc is an edge of the polygonal cavity. (See discussion below.)
   (b) Add the conflict (y, uba) to the conflict graph: designate uba as y's conflict triangle and append y to the conflict list for uba.
2. Deallocate the conflict list for vwx.

A hash table can map triangles to their conflict lists and visitation marks in expected constant time. Each triangle's entry in the hash table should be created when the triangle is added to the set C in Step 4 of MarkCavity, and can optionally be removed when the triangle is deleted in Step 5 of InsertVertexAtConflict.

The pseudocode excludes some technicalities related to ghost triangles. In Step 3 of MarkCavity, wvx may be a ghost triangle, in which case the InCircle test is replaced with a test of whether u lies in the outer halfplane of wvx. In Step 1(a) of RedistributeList, if vwx is a ghost triangle, the walk along the ray $\overrightarrow{uy}$ may travel through ghost triangles and across ghost edges represented as rays; the walking implementation must treat them correctly. Moreover, the ray might reach infinity without ever striking an unmarked triangle (see the leftmost vertex in Figure 3.10), in which case the vertex y lies in one of the two newly created ghost triangles.

With a few changes, the pseudocode can construct a weighted Delaunay triangulation. Replace the InCircle test with Orient3D(u^+, v^+, w^+, x^+). An uninserted vertex can be submerged, in which case it has no conflicts, so change Step 1 of InsertVertexAtConflict to return immediately if u has no conflict triangle. Likewise, Step 1(b) of RedistributeList should run only if y actually conflicts with the triangle uba that contains it. If uba is a solid triangle, y conflicts with uba if Orient3D(u^+, b^+, a^+, y^+) > 0; otherwise, y is submerged and has no conflicts. If uba is a ghost triangle, y conflicts with uba if y does not lie on the solid edge of uba; otherwise, test y for a conflict with the solid triangle that shares the edge.

Recall from Section 3.5 that all operations done by the algorithm *except point location* run in expected $O(n)$ time. The dominant cost of the algorithm is the cost of redistributing vertices. A crucial observation is that every triangle that Step 1(a) of RedistributeList walks through conflicts with y (see Exercise 4). Therefore, the cost of redistributing a vertex is at worst proportional to the number of deleted triangles it conflicts with—in other words, the number of conflicts it participates in that are eliminated. (All conflicts are counted here, not just the conflicts that are recorded in the conflict graph.) Hence, the total running time of the algorithm is at worst proportional to the total number of conflicts that are eliminated during a run of the algorithm from start to finish. The algorithm has no conflicts at the beginning (no triangles) nor at the end (no uninserted vertices), so this number is equal to the total number of conflicts that are *created* during a run of the algorithm from start to finish. Let us see how to bound this latter number.

The following proposition is short but, in its time, it reflected a breakthrough in the analysis of randomized incremental construction of geometric structures. To estimate how many conflicts an average vertex participates in, simply observe that *if* that vertex were inserted next, the expected number of triangles deleted would be constant.

Proposition 3.5. *At any time during the algorithm, the expected number of conflicts that a randomly chosen uninserted vertex participates in is less than four.*

Proof. By Proposition 3.4, when a triangulation $\mathcal{T}_{i-1}$ is updated by the insertion of a randomly chosen uninserted vertex, the expected number of new triangles created (including ghost triangles) is less than six. The number of new triangles is always two greater than the number of deleted triangles. Therefore, the expected number of triangles in $\mathcal{T}_{i-1}$ (including ghost triangles) that the vertex conflicts with is less than four. □

Proposition 3.6. *When the algorithm inserts v_i into $\mathcal{T}_{i-1}$ to construct $\mathcal{T}_i$, the expected number of conflicts created is less than $12(n-i)/i$.*

Proof. The proof uses backward analysis, which we introduced in Section 3.5 to prove Proposition 3.4. Imagine the algorithm running backward in time: it deletes a vertex v_i chosen uniformly at random from $\mathcal{T}_i$, yielding $\mathcal{T}_{i-1}$. Each triangle in $\mathcal{T}_i$ has three vertices, so the probability that any given solid triangle is deleted is $3/i$. The probability is $2/i$ for ghost triangles, as the ghost vertex is never deleted. Therefore, the probability that any given conflict disappears is at most $3/i$.

By Proposition 3.5 and the linearity of expectation, the expected number of conflicts over all the triangles in $\mathcal{T}_i$ is less than $4(n-i)$, because there are $n-i$ uninserted vertices. Hence, the expected number of conflicts that disappear when v_i is deleted is less than $4(n-i)\cdot(3/i) = 12(n-i)/i$. □

Proposition 3.6 is more subtle than it looks, because the random events are not all independent. When v_i is deleted, the probability that one particular triangle will be deleted is not independent of the probability that another will be deleted. Likewise, the number of conflicts in which one particular vertex participates is not independent of the number in which another vertex participates. Nevertheless, we can bound the expected total number of conflicts, because linearity of expectation does not require independence among the vertices. We can bound the expected number of conflicts that disappear, because the probability that a triangle disappears is independent of the number of conflicts in which it participates.

Theorem 3.7. *The randomized incremental insertion algorithm for constructing a Delaunay triangulation of n points in the plane runs in expected $O(n \log n)$ time.*

Proof. The cost of all the point location operations is proportional to the total number of conflicts created during the algorithm's run. By Proposition 3.6, that number is less than $\sum_{i=1}^{n} 12\,(n-i)\,/i = 12n\left(\sum_{i=1}^{n} 1/i\right) - 12n = O(n \log n)$. □

3.7 The incremental insertion algorithm

All the ingredients for an efficient incremental insertion algorithm have been set out in the previous sections. There is still one missing piece, though: how to get the algorithm started. A potential hazard is that, as they are presented here, neither the Bowyer–Watson method nor the conflict graph work correctly if all the vertices inserted so far are collinear, because uninserted vertices that are collinear with the inserted vertices have no conflicts.

In practice, the simplest way to bootstrap the algorithm is to choose three arbitrary vertices that are not collinear, construct the triangle having those three vertices, construct three ghost triangles, compute a conflict graph, then repeatedly call InsertVertexAtConflict to insert the remaining vertices one by one in random order. However, the requirement that the first three vertices not be collinear violates the assumption that all permutations of the vertices are equally likely. It is still possible to prove that the randomized algorithm runs in expected $O(n \log n)$ time; adapting the proof is left as Exercise 5.

We conclude by pointing out that for a fast implementation, we recommend an incremental insertion algorithm that combines a biased randomized insertion order, described in Section 5.3, with walking point location, described in Section 5.5, over point location by conflict graph. The former is easier to implement than a conflict graph, and it appears to be faster in practice, albeit not in theory. However, a conflict graph offers the security of a guarantee on its expected running time, and its analysis offers insight as to why our recommended algorithm is also fast in practice.

3.8 Deleting a vertex from a Delaunay triangulation

Vertex deletion is an operation that reverses vertex insertion—it updates a Delaunay triangulation, so it has one less vertex and is still Delaunay. It is more difficult than vertex insertion. The deletion of a vertex also deletes all the triangles adjoining the vertex, evacuating a polygonal cavity. The problem is to efficiently retriangulate the cavity with Delaunay triangles. The most common retriangulation method in practice is a gift-wrapping algorithm (see Section 3.11) that takes $O(k^2)$ time, where k is the degree of the vertex to be deleted. This section discusses a faster algorithm by Paul Chew that runs in expected $O(k)$ time. It is a randomized incremental insertion algorithm that uses a simpler point location method than a conflict graph.

The simplest interpretation of the algorithm is as a method to compute the Delaunay triangulation of a convex polygon. We will see that it can also retriangulate the cavity evacuated when a vertex is deleted from a Delaunay triangulation, even though the cavity might not be convex. For now, suppose it is convex. Let S be a sequence listing the k vertices of a convex polygon in counterclockwise order.

The algorithm begins by generating a random permutation of S that dictates the order in which the vertices will be inserted. It constructs a triangle from the first three vertices of the permutation, then uses the Bowyer–Watson algorithm to insert the remaining vertices one by one. By Proposition 3.4, the algorithm runs in expected $O(k)$ time, excluding the cost of point location.

Just before a vertex u is inserted, it lies outside the growing triangulation, but only one triangulation edge vw separates u from the triangulation's interior. Point location is the task of identifying the edge vw. To insert u, retriangulate the cavity formed by taking the union of uvw and all the triangles whose open circumdisks contain u, as illustrated in the lower half of Figure 3.11.

The point location, like backward analysis, is done by imagining the incremental insertion algorithm running backward in time. Specifically, imagine taking the input polygon and removing vertices one by one, reversing the random permutation of S, yielding a shrinking sequence of convex polygons, as illustrated in the upper half of Figure 3.11. Removing a vertex u has the effect of joining its neighbors v and w with an edge vw, which is the edge sought for point location.

The procedure ConvexDT below implements the cavity retriangulation algorithm. It takes as its parameter the counterclockwise sequence S of polygon vertices. ConvexDT performs all point location in advance, before constructing any triangles. It begins by creating a circularly-, doubly-linked list of the polygon vertices. Then it removes vertices from

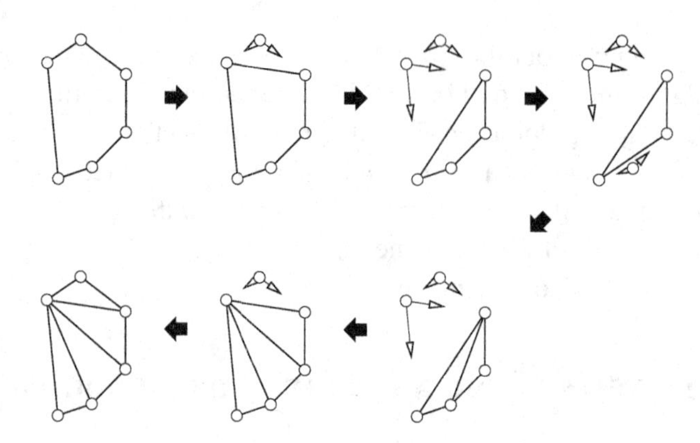

Figure 3.11: Chew's algorithm for computing the Delaunay triangulation of a convex polygon removes vertices from the polygon in random order to precompute the information needed for point location, then inserts the vertices in the opposite order.

the linked list in a random order until only three vertices remain. This preprocessing takes $O(k)$ time. The algorithm constructs a triangle from the three surviving vertices, then inserts the other vertices in the reverse of the order in which they were removed.

ConvexDT(S)

1. Write $S = (v_0, v_1, \ldots, v_{k-1})$, the polygon vertices in counterclockwise order. Let $\pi[0, 1, \ldots, k-1]$ be a permutation of $0, 1, \ldots, k-1$ chosen uniformly at random.
2. Construct a circularly-, doubly-linked list L of the polygon vertices in S by setting $\text{next}[i] \leftarrow (i+1) \bmod k$ and $\text{prev}[i] \leftarrow (i-1) \bmod k$ for each $i \in [0, k-1]$.
3. For $i \leftarrow k-1$ down to 3
   (a) Remove $v_{\pi[i]}$ from L by setting $\text{next}[\text{prev}[\pi[i]]] \leftarrow \text{next}[\pi[i]]$ and setting $\text{prev}[\text{next}[\pi[i]]] \leftarrow \text{prev}[\pi[i]]$. Do not change $\text{next}[\pi[i]]$ or $\text{prev}[\pi[i]]$, which will be used later for point location.
4. Call AddTriangle($v_{\pi[0]}, v_{\text{next}[\pi[0]]}, v_{\text{prev}[\pi[0]]}$)
5. For $i \leftarrow 3$ to $k-1$
   (a) Call ConvexInsertVertex($v_{\pi[i]}, v_{\text{next}[\pi[i]]} v_{\text{prev}[\pi[i]]}$) to insert $v_{\pi[i]}$.

ConvexInsertVertex(u, vw)

1. Let $x = $ Adjacent(w, v); wvx is the triangle on the other side of edge vw from u.
2. If $x \in S$ and InCircle(u, v, w, x) > 0, then uvw and wvx are not Delaunay, so call DeleteTriangle(w, v, x). Call ConvexInsertVertex(u, vx) and ConvexInsertVertex(u, xw) to identify more deleted triangles and insert new triangles. Return.
3. Otherwise, the edge vw remains Delaunay, so call AddTriangle(u, v, w).

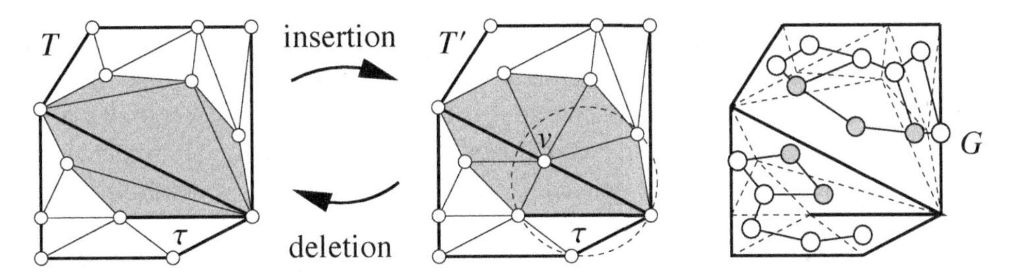

Figure 3.12: Inserting or deleting a vertex v in a CDT. Bold edges are segments. The shaded polygon is the union of the deleted/created triangles. Simplices not intersecting the interior of the shaded polygon are constrained Delaunay before and after. When v is inserted, depth-first search on the graph G identifies the deleted triangles. Observe that although v lies in τ's open circumdisk, v's insertion does not delete τ because G does not connect τ to any deleted triangle.

The pseudocode does not use ghost triangles, as they confer no benefit here. Step 2 of ConvexInsertVertex tests whether $x \in S$ to prevent the algorithm from breaking out of the polygon. If the polygon is triangulated in isolation, this test is equivalent to testing that $x \neq \emptyset$. If the polygon is a cavity in a larger triangulation, the test ensures that x is a vertex of the cavity. Reasonably implemented, this test should take $O(1)$ time.

To adapt the code to construct a weighted Delaunay triangulation, replace the InCircle test in Step 2 of ConvexInsertVertex with Orient3D(u^+, v^+, w^+, x^+).

It is easy to see that the algorithm still works correctly if the convexity is not strict—that is, three or more consecutive vertices in S are collinear. It is more difficult to see that the algorithm still works correctly if the polygon is a nonconvex cavity evacuated by the deletion of a vertex v from a Delaunay triangulation. This insight comes by imagining the Delaunay triangulation lifted by the parabolic lifting map. Then Delaunay vertex deletion is the problem of deleting a vertex v^+ from the convex hull of the lifted vertices, and retriangulating the hole thus created. Although the hole might not be convex when projected to the xy-plane, the hole is convex from the point of view of v^+, and the algorithm is best understood from that point of view. The test InCircle(u, v, w, x) in Step 2 of ConvexInsertVertex is equivalent to Orient3D(u^+, v^+, w^+, x^+), so Step 2 is testing for triangular facets that cannot be part of the convex hull of S^+ because u^+ is underneath them.

3.9 Inserting or deleting a vertex in a CDT

To "insert a vertex into a CDT" is to take as input a CDT of some PLC $\mathcal{P}$ and a new vertex v to insert, and produce a CDT of $\mathcal{P} \cup \{v\}$. An implementation might also support the insertion of a vertex v on a segment $s \in \mathcal{P}$, in which case the algorithm subdivides s into two subsegments s_1 and s_2 having vertex v, and produces the CDT of $\mathcal{P} \cup \{v, s_1, s_2\} \setminus \{s\}$.

With a small change, the Bowyer–Watson algorithm can insert a vertex v into a CDT, as Figure 3.12 illustrates. The change, of course, is that the algorithm deletes the triangles that are no longer *constrained* Delaunay. Fortunately, it is possible to enumerate those triangles without performing expensive visibility tests. To accomplish that, the first step—point location—finds the triangle that contains v. There may be two such triangles, if v lies on a triangulation edge. The depth-first search that identifies triangles that are no longer constrained Delaunay (Step 2 of BowyerWatson) is modified so that it never walks across a segment. As Figure 3.12 shows, this restriction suffices to ensure that only triangles whose interiors are visible from v will be deleted. If v lies on a segment in $\mathcal{P}$, depth-first searches must be run from both of the two adjoining triangles. Steps 3 and 4 of BowyerWatson do not change. In a straightforward extension of the proofs in Section 3.3, one can show that the depth-first search finds all the triangles that are no longer constrained Delaunay, the cavity is always star-shaped, and the algorithm works correctly.

Some mesh generation algorithms require an operation that deletes a vertex from a CDT. Usually, a vertex cannot be deleted if it is a vertex of a segment, but an implementation might support one exception, an operation that is the reverse of inserting a vertex on a segment: if a vertex v is a vertex of exactly two segments s_1 and s_2 and they are collinear, then the deletion of v fuses s_1 and s_2 into a single segment s and produces the CDT of $\mathcal{P} \cup \{s\} \setminus \{v, s_1, s_2\}$.

The algorithm for deleting a vertex from a Delaunay triangulation described in Section 3.8 works without change to delete a vertex v from a CDT $\mathcal{P}$ if v is not a vertex of any segment in $\mathcal{P}$. It is only slightly more difficult to treat the second case, where two segments are fused into one. The deletion of v yields two cavities, one on each side of the fused segment s, each of which can be retriangulated independently by the algorithm of Section 3.8.

3.10 Inserting a segment into a CDT

To "insert a segment into a CDT" is to take as input a CDT of a PLC $\mathcal{P}$ and a new segment s to insert, and produce a CDT of $\mathcal{P} \cup \{s\}$. It is only meaningful if $\mathcal{P} \cup \{s\}$ is a valid PLC—that is, $\mathcal{P}$ already contains the vertices of s (otherwise, they must be inserted first, as described in Section 3.9), and the relative interior of s intersects no segment or vertex in $\mathcal{P}$. This section presents a segment insertion algorithm similar to Chew's algorithm for computing a Delaunay triangulation of a convex polygon. Its expected running time is linear in the number of edges the segment crosses.

If m segments are inserted in random order, it is known that the expected number of deleted edges is $O(n \log^2 m)$. By constructing a Delaunay triangulation of a PLC's n vertices and inserting its m segments in random order, one can construct its CDT in expected $O(n \log n + n \log^2 m)$ time. Although this running time is not optimal, the algorithm is easier to implement than the known algorithms that run in $O(n \log n)$ time, and it can take advantage of the fastest available software for Delaunay triangulation.

Let $\mathcal{T}$ be a CDT of $\mathcal{P}$. If $s \in \mathcal{T}$, then $\mathcal{T}$ is also the CDT of $\mathcal{P} \cup \{s\}$. Otherwise, the algorithm begins by deleting from $\mathcal{T}$ the edges and triangles that intersect the relative interior of s. All the other simplices in $\mathcal{T}$ remain constrained Delaunay after s is inserted. Next, the algorithm adds s to the triangulation and retriangulates the two polygonal cavities on each side of s with constrained Delaunay triangles; recall Figure 2.15.

Let P be one of the two polygonal cavities; its edges include s. We retriangulate P with a randomized incremental insertion algorithm called CavityCDT, which is similar to the

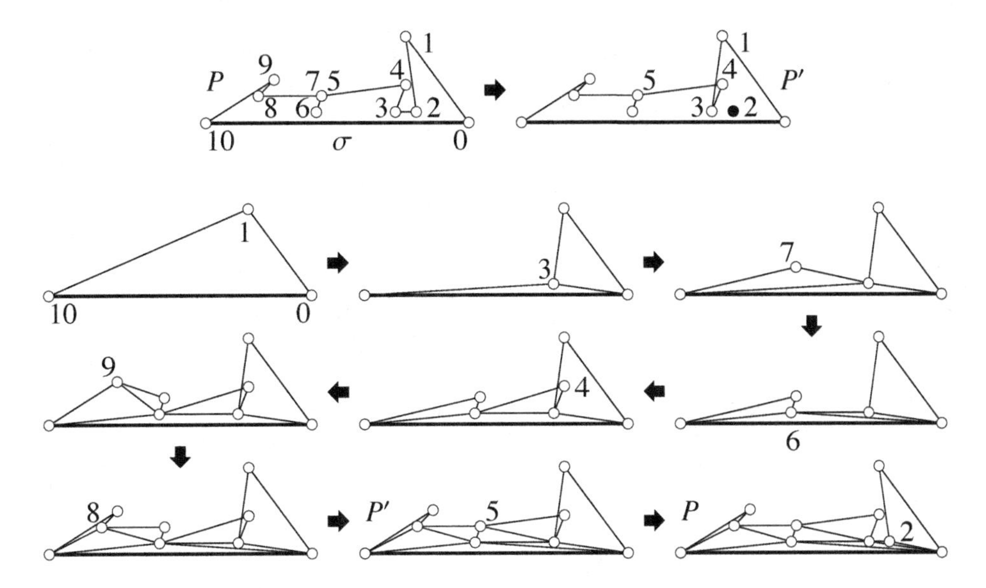

Figure 3.13: Computing the constrained Delaunay triangulation of a cavity obtained by inserting a segment s. The cavity has a repeated vertex, numbered 5 and 7, because of the dangling edge adjoining it. The deletion of vertex 2 creates a self-intersection, but the algorithm works correctly anyway.

algorithm ConvexDT in Section 3.8. Be forewarned that CavityCDT *cannot* compute the CDT of an arbitrary polygon; it depends upon the special nature of the cavities evacuated by segment insertion for its correctness.

CavityCDT differs from ConvexDT in several ways to account for the fact that P is not convex. First, the vertices of the segment s are inserted first. Second, P might have edges dangling in its interior, as illustrated in Figure 3.13. In this case, imagine an ant walking a counterclockwise circuit of P's interior without crossing any edges; it will visit one or more vertices of P twice. We split each such vertex into two copies and pretend they are two separate vertices, like vertices 5 and 7 in the figure. (In rare circumstances, there may be more than two copies of a vertex.) Third, the algorithm maintains the invariant that after each vertex insertion, the computed triangulation is the CDT of the polygon whose boundary is the subsequence of vertices inserted so far. Because the algorithm maintains a CDT and not merely a Delaunay triangulation, a newly inserted vertex sometimes causes a triangle to be deleted not because the new vertex lies in the triangle's circumdisk, but because the two polygon edges adjoining the new vertex cut through the triangle; for example, the insertion of vertex 8 in Figure 3.13 deletes a triangle whose circumdisk does not contain it. (In the pseudocode below, Step 2 of CavityInsertVertex accounts for this possibility with an orientation test.)

Fourth, an intermediate subpolygon might be *self-intersecting*. Observe in Figure 3.13 that deleting vertex 2 from the cavity P creates a subpolygon P' in which the edge connecting vertices 1 and 3 crosses the edge connecting vertices 4 and 5, and the polygon's interior angle at vertex 3 exceeds 360°. P' is not actually a polygon by Definition 1.7, although it is a polygon in the conventional sense of a looped chain of edges; and its triangulation in Figure 3.13 (bottom center) is not a simplicial complex, because it has overlapping triangles, but it is almost a CDT in the sense of having all the combinatorial properties of a triangulation of a polygon and having every edge locally Delaunay.

The incremental insertion algorithm works even when these self-intersecting subpolygons arise, subject to one caveat: it will not correctly insert a vertex at which the polygon's internal angle is 360° or greater. For example, it cannot create P in Figure 3.13 by inserting vertex 6 last, nor create P' by inserting vertex 3 last. These vertex insertions are anticipated and averted during the algorithm's point location step, when random vertices are removed from P one by one until the subpolygon is reduced to a triangle. Hence, the random permutation by which the vertices are inserted is not chosen uniformly from all permutations of the vertices.

For the sake of speed, CavityCDT does not compute internal angles. Instead, let ℓ be the affine hull of the segment s. It is a property of the cavities created by segment insertion that a subpolygon vertex can have an internal angle of 360° or greater only if that vertex is closer to ℓ than both its neighbors on the subpolygon chain. CavityCDT declines to remove from P any vertex with the latter property.

The pseudocode for CavityCDT, below, is similar to that of ConvexDT, and it calls a recursive procedure CavityInsertVertex analogous to ConvexInsertVertex. It takes as its parameter a sequence S listing the vertices of the polygonal cavity in counterclockwise order, in which vertices may be repeated as discussed above to indicate dangling edges. Be forewarned that CavityCDT, unlike ConvexDT, cannot use the same triangulation data structure as the triangulation in which the segment s is being inserted, because CavityCDT sometimes temporarily creates triangles that conflict with those outside the cavity. CavityCDT requires the use of a separate, empty triangulation data structure, and the final triangles must subsequently be added to the main triangulation.

CavityCDT(S)

1. Write $S = (v_0, v_1, \ldots, v_{k-1})$, the vertices in counterclockwise order around the cavity, where $v_{k-1}v_0$ is the newly inserted segment. Let $\pi[1, 2, \ldots, k-2]$ be a permutation of $1, 2, \ldots, k-2$ chosen uniformly at random.
2. Construct a circularly-, doubly-linked list L of the polygon vertices in S by setting next[i] $\leftarrow (i+1) \bmod k$ and prev[i] $\leftarrow (i-1) \bmod k$ for each $i \in [0, k-1]$.
3. For $i \leftarrow k-2$ down to 2
   (a) While $v_{\pi[i]}$ is closer to $\ell = \text{aff } v_0 v_{k-1}$ than both its neighbors in L are, i.e. $d(v_{\pi[i]}, \ell) < d(v_{\text{prev}[\pi[i]]}, \ell)$ and $d(v_{\pi[i]}, \ell) < d(v_{\text{next}[\pi[i]]}, \ell)$,
      (i) Draw an integer j uniformly at random from $[1, i-1]$.
      (ii) Change the permutation π by swapping $\pi[i]$ with $\pi[j]$.
   (b) Remove $v_{\pi[i]}$ from L by setting next[prev[$\pi[i]$]] $\leftarrow$ next[$\pi[i]$] and setting prev[next[$\pi[i]$]] $\leftarrow$ prev[$\pi[i]$]. Do not change next[$\pi[i]$] or prev[$\pi[i]$], which will be used later for point location.
4. Call AddTriangle($v_0, v_{\pi[1]}, v_{k-1}$).
5. For $i \leftarrow 2$ to $k-2$

(a) Call CavityInsertVertex($v_{\pi[i]}, v_{\text{next}[\pi[i]]} v_{\text{prev}[\pi[i]]}$) to insert $v_{\pi[i]}$.

CavityInsertVertex(u, vw)

1. Let x = Adjacent(w, v); wvx is the triangle on the other side of edge vw from u.
2. If $x = \emptyset$ or (Orient2D(u, v, w) > 0 and InCircle(u, v, w, x) ≤ 0), then uvw is constrained Delaunay, so call AddTriangle(u, v, w) and return.
3. Otherwise, wvx is not constrained Delaunay, so call DeleteTriangle(w, v, x). Call CavityInsertVertex(u, vx) and CavityInsertVertex(u, xw) to identify more deleted triangles and insert new triangles.

Steps 2–3 of CavityCDT perform point location, and Steps 4–5 construct the triangulation. To implement Step 3(a), observe that a cavity vertex v_i is closer to ℓ than another cavity vertex v_j if and only if Orient2D(v_0, v_i, v_{k-1}) $<$ Orient2D(v_0, v_j, v_{k-1}), where v_0 and v_{k-1} are the endpoints of s. The values Orient2D(v_0, v_i, v_{k-1}) for each i can be precomputed once at the start.

We omit the lengthy proof of correctness of CavityCDT, but analyze its running time.

Proposition 3.8. *Given a k-vertex cavity,* CavityCDT *runs in expected $O(k)$ time.*

Proof. Every triangulation of an k-vertex polygon has $k-3$ interior edges. At least $(k-1)/2$ vertices are eligible to be the first vertex deleted during point location and the last vertex inserted. One of those vertices is chosen uniformly at random; in expectation it adjoins at most $2(k-3)/((k-1)/2) < 4$ interior edges. The cost of CavityInsertVertex is proportional to the number of edges adjoining the vertex it inserts, so the expected cost of inserting the last vertex is constant. The same reasoning holds for the other vertices. Summing this cost over all the vertices yields an expected linear running time. □

3.11 The gift-wrapping algorithm

This section describes a basic gift-wrapping algorithm for constructing Delaunay triangulations and CDTs. Unadorned gift-wrapping is slow, but it is easy to implement and fast enough to retriangulate small cavities in a triangulation. It is particularly useful for computing the CDT of a small nonconvex polygon, and it remains the most practical algorithm for deleting a vertex from a three-dimensional Delaunay triangulation, unless the vertex has very high degree.

Be forewarned that if the input point set or PLC has four cocircular vertices, gift-wrapping can make decisions that are mutually inconsistent and fail to construct a valid triangulation. Figure 3.14 depicts a simple, unconstrained example where Delaunay gift-wrapping fails. Gift-wrapping can be modified to handle these inputs by symbolically perturbing the vertex weights as described in Section 2.9 so the input is generic, or by identifying groups of cocircular vertices that can see each other and triangulating them all at once.

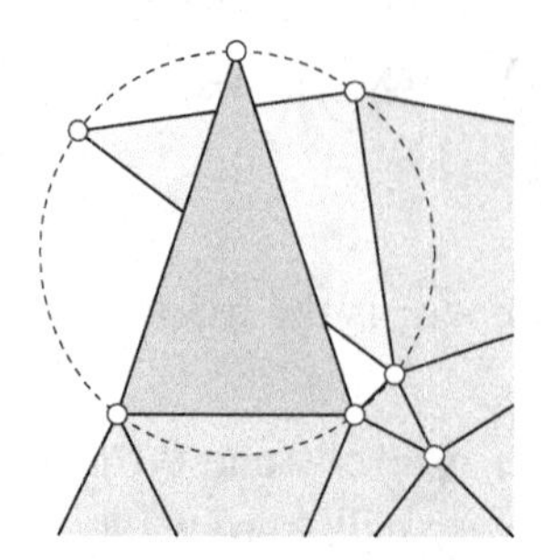

Figure 3.14: A gift-wrapping failure because of cocircular vertices.

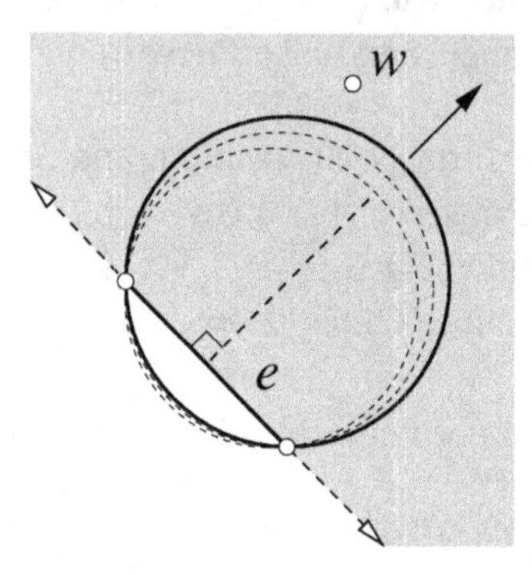

Figure 3.15: An empty circumdisk of e, expanding in search of a vertex w.

Gift-wrapping algorithms rely on a simple procedure called a *gift-wrapping step*, which constructs a triangle adjoining a specified edge. Let $e = uv$ be an oriented edge. The *front* of e is the open halfplane to the left of e, and a positively oriented triangle uvw is said to be *in front of* e. The *back* of e is the open halfplane to the right of e, and a positively oriented triangle vux is said to be *behind* e.

During the execution of a gift-wrapping algorithm, an oriented edge constructed by the algorithm is said to be *unfinished* if the algorithm has not yet identified the triangle in front of the edge. A gift-wrapping step *finishes* the edge by constructing that triangle, or by determining that there is no such triangle because the edge is on the boundary of the domain.

An edge e has an infinite number of circumdisks, any one of which can be continuously deformed into any other such that every intermediate disk is also a circumdisk of e. Imagine beginning with a circumdisk that contains no vertex in front of e, then deforming it so it expands in front of e and shrinks behind e, always remaining a circumdisk of e, as illustrated in Figure 3.15. As the circumdisk deforms, its center always lies on e's bisector.

Eventually, the expanding portion of the circumdisk might touch a vertex w that is visible from the relative interior of $e = uv$, in which case the gift-wrapping step constructs $\tau = uvw$, thereby finishing e. Proposition 3.9 below shows that τ is constrained Delaunay if e is constrained Delaunay or a segment. Alternatively, the expanding portion of the circumdisk might never touch a vertex, in which case e is on the boundary of the convex hull of the vertices.

Although the expanding circumdisk gives the right intuition for which vertex is chosen, the algorithm that implements a gift-wrapping step works the opposite way, by shrinking the front of the circumdisk: it scans through the vertices in front of e and remembers which vertex, so far, minimizes the portion of the circumdisk in front of e.

The gift-wrapping step FINISH, below, takes as parameters an oriented edge uv, the set S of vertices in a PLC $\mathcal{P}$, and the set K of segments in $\mathcal{P}$. FINISH returns the Delaunay triangle or constrained Delaunay triangle in front of uv, or $\emptyset$ if no such triangle exists. One gift-wrapping step takes $O(n)$ time for a Delaunay triangulation, or $O(nm)$ time for a CDT, where $n = |S|$ and $m = |K|$. Condition (c) in Step 2 accounts for the factor of m.

FINISH(uv, S, K)

1. Let $\tau = \emptyset$. Let p be any point in the relative interior of uv.
2. For each vertex $w \in S$, if all of the following conditions hold:
   (a) w is in front of uv,
   (b) $\tau = \emptyset$ or the open circumdisk of τ contains w, and
   (c) no segment $s \in K$ occludes the visibility between w and p,

   then set $\tau \leftarrow$ the positively oriented triangle uvw.
3. Return τ.

Proposition 3.9. *Suppose that $\mathcal{P}$ is generic. If the edge uv is constrained Delaunay or a segment in $\mathcal{P}$, and the front of uv does not adjoin the exterior of $|\mathcal{P}|$, then the algorithm* FINISH *returns a constrained Delaunay triangle.*

PROOF. Because $\mathcal{P}$ is generic, $\mathcal{P}$ has one unique CDT $\mathcal{T}$ by Theorem 2.18, and every constrained Delaunay edge is *strongly* constrained Delaunay, so $uv \in \mathcal{T}$. Because the front of uv does not adjoin the exterior of $|\mathcal{P}|$, some triangle $uvx \in \mathcal{T}$ is in front of uv. Because uvx is constrained Delaunay, x is visible from the point p chosen in Step 1 of FINISH. No vertex is simultaneously in front of uv, in the open circumdisk of uvx, and visible from p, because such a vertex would also be visible from the interior of uvx, contradicting the fact that uvw is constrained Delaunay.

It follows that FINISH returns uvx. Once Step 2 has selected $\tau = uvx$, it can never replace uvx with another triangle, because we have established that no vertex meets the conditions. If Step 2 initially selects another triangle $\tau = uvw$, it will eventually replace uvw with uvx because w is outside the open circumdisk of uvx; this and the fact that $\mathcal{P}$ is generic imply that x is inside the open circumdisk of uvw. □

The gift-wrapping algorithm begins with the PLC segments, upon which the constrained Delaunay triangles crystallize one by one. The core of the algorithm is a loop that selects an unfinished edge and finishes it by invoking the procedure FINISH. Often, a new triangle finishes more than one unfinished edge. To detect this circumstance, the algorithm maintains the unfinished edges in a dictionary (e.g. a hash table) so they can be quickly looked up by their vertex indices. The data structure in Section 3.2 is easily modified to serve this purpose while also storing the triangulation. The gift-wrapping algorithm GIFTWRAPCDT, below, takes as parameters the set S of vertices in a PLC $\mathcal{P}$ and the set K of segments in $\mathcal{P}$.

GiftWrapCDT(S, K)

1. Initialize an empty dictionary to store oriented edges. For each segment $s \in K$ that adjoins a PLC polygon on one side only, insert s into the dictionary, oriented so its front adjoins a polygon.
2. Remove an oriented edge e from the dictionary.
3. Let τ = Finish(e, S, K).
4. If $\tau \neq \emptyset$
   (a) Call AddTriangle(τ).
   (b) For each oriented edge f of τ other than e
      (i) If f is in the dictionary, then remove f from it.
      (ii) Otherwise, reverse the orientation of f (so f faces away from τ) and insert f into the dictionary.
5. If the dictionary is not empty, go to Step 2.

The algorithm can construct Delaunay triangulations too, but the GiftWrapCDT pseudocode assumes that $\mathcal{P}$ contains at least one segment that can serve as a seed upon which to build the triangulation. When there are no segments, seed the algorithm by constructing one strongly Delaunay edge—an arbitrary vertex and its nearest neighbor will do—and entering it (twice, with both orientations) in the dictionary.

The algorithm takes $O(n^2)$ time for a Delaunay triangulation, or $O(n^2m)$ time for a CDT, where $n = |S|$ is the number of vertices and $m = |K|$ is the number of segments. These are unimpressive speeds, especially in comparison with the randomized incremental insertion and divide-and-conquer algorithms. See the bibliographical notes for several ways to speed up gift-wrapping that make it more practical.

3.12 Notes and exercises

The first published Delaunay triangulation algorithm we know of appears in a 1967 article by J. Desmond Bernal and John Finney [19]. Bernal was a father of structural biology who discovered the structure of graphite and was awarded a Stalin Peace Prize. Finney, Bernal's last Ph.D. student, implemented a program that produces a three-dimensional Voronoi diagram and used it to characterize the structures of liquids, amorphous metal alloys, protein molecules, and random packings. Finney's is the brute force algorithm: test every possible tetrahedron (every combination of four vertices) to see if its circumball is empty, taking $O(n^5)$ time—or more generally, $O(n^{d+2})$ time for d-dimensional Delaunay triangulations.

Gift-wrapping—also called *graph traversal*, *pivoting*, and *incremental search*—is an obvious algorithm that is rediscovered frequently [41, 97, 144, 211, 213]. See Section 1.3 for a discussion of the earliest publication, in 1970 by Frederick, Wong, and Edge [97]. The bottleneck of gift-wrapping is the cost of identifying new triangles, so the fastest gift-wrapping algorithms are differentiated by sweep orderings for constructing the triangles [192, 96] or sophisticated vertex search strategies [84, 210]. *Sweepline algorithms* construct the triangles in a disciplined order, making it possible to determine which vertex finishes each edge without an exhaustive search. Fortune [96] developed such an algorithm for Delaunay triangulations. Seidel [193] extended it to CDTs. Both algorithms run in $O(n \log n)$ time.

Another way to avoid an exhaustive search is to subdivide the plane into square buckets, record the vertices in their respective buckets, and finish each edge by searching through the buckets in an appropriate order. Dwyer [84] shows that if the vertices are distributed uniformly at random in a disk, this technique finishes each face in $O(1)$ expected time, so an entire Delaunay triangulation can be constructed in $O(n)$ expected time. Moreover, the algorithm extends to higher dimensions, still with expected linear running time! Unfortunately, this method does not extend easily to CDTs, and not all real-world point sets are so nicely distributed. It is easy to construct a point set for which most of the points fall into one bucket.

The first Delaunay triangulation algorithm to run in $O(n \log n)$ time was the 1975 divide-and-conquer algorithm of Shamos and Hoey [195], subsequently simplified by Lee and Schachter [134] and Guibas and Stolfi [106] and sped up by Dwyer [83]. Chew [58] extended the algorithm to CDTs, maintaining the $O(n \log n)$ running time.

In 1977, Lawson [131] invented the first algorithm for inserting a vertex into a Delaunay triangulation and restoring the Delaunay property. His algorithm uses edge flips. See Section 5.9 for bibliographic information on the Bowyer–Watson algorithm, which is slightly faster than Lawson's because it does not create triangles that will immediately be deleted. In one of the greatest papers of computational geometry, Clarkson and Shor [64] introduce conflict graphs and show (among many other things) that the randomized incremental insertion algorithm with a conflict graph runs in expected $O(n \log n)$ time in the plane, and in expected $O(n^{\lceil d/2 \rceil})$ time for Delaunay triangulations in $d \geq 3$ dimensions. The simpler analysis in Section 3.6 comes from a charming paper by Seidel [194] that popularized backward analysis; he credits this analysis to Kenneth Clarkson. The quote in Section 3.5 is taken from this paper. In its original presentation, the Clarkson–Shor algorithm is an incremental convex hull algorithm, so it works for weighted Delaunay triangulations too.

In the decision-tree model of computation, no Delaunay triangulation algorithm can run in less than $\Theta(n \log n)$ time, because there is a linear-time reduction of sorting to triangulation. But faster algorithms are possible in other models of computation, just as radix sort can be said to run in linear time. Chan and Pătraşcu [40] were the first to beat the $O(n \log n)$ bound. The current state of the art is an expected $O(n)$-time incremental algorithm by Buchin and Mulzer [35], which relies on the linear-time computation of a nearest-neighbor graph to assist with point location.

Chew [60] invented the expected linear-time algorithm for deleting a vertex from a Delaunay triangulation given in Section 3.8. Chew's analysis is the source of the argument in the proof of Proposition 3.4 and the first known use of backward analysis in computational geometry. Klein and Lingas [123] point out the extension of Chew's vertex deletion algorithm to weighted Delaunay triangulations and three-dimensional convex hulls.

The segment insertion algorithm of Section 3.10 is by Shewchuk and Brown [205]. The expected $O(n \log^2 m)$ bound on the number of edges deleted by m segment insertions in random order is proved by Agarwal, Arge, and Yi [1].

The paper by Guibas and Stolfi [106], which includes detailed pseudocode, is an indispensable aid for anyone wishing to implement the divide-and-conquer or incremental insertion algorithms. Shewchuk [196] notes that implementations of algorithms for Delaunay triangulation and mesh generation run about twice as fast with triangle-based data structures than with Guibas and Stolfi's edge-based data structures. Blelloch et al. [23, 22] point out the usefulness of a dictionary data structure (Section 3.2) for programmer productivity and easier debugging. For a substantially more space-efficient representation, see Blandford, Blelloch, Clemens, and Kadow [22].

Exercises

1. Write the pseudocode for an algorithm that takes as input a point set in the plane and produces its lexicographic triangulation, defined in Section 2.1. Your pseudocode should include calls to the predicate ORIENT2D.

2. Let us consider how to implement the symbolic weight perturbations described in Section 2.9. Let $S = \{v_1, v_2, \ldots, v_n\}$ be a set of vertices in the plane. S may have subsets of four or more cocircular vertices, so S may have many Delaunay triangulations. Define the lifting map $v = (x, y) \mapsto v^+ = (x, y, x^2 + y^2 - \omega_v)$, where ω_v is the weight of vertex v. Vertex v_i has weight ϵ^i for an extremely small $\epsilon > 0$.

   We modify an algorithm for computing Delaunay triangulations by replacing each call to INCIRCLE(u, v, w, x) with a call to ORIENT3D(u^+, v^+, w^+, x^+). However, we want to use symbolic weights instead of assigning ϵ an explicit numerical weight, because numerical roundoff makes computation with a tiny ϵ impractical. So we begin by computing ORIENT3D on the *un*perturbed points (i.e. with the weights set to zero, therefore equivalent to INCIRCLE). Proposition 2.12 tells us we only need to take the weights into account if that result is zero, in which case we must perform one or more additional tests to disambiguate the sign of ORIENT3D on the perturbed points. Describe those additional tests.

   Hint: Observe that ORIENT3D is a *linear* function of the weights, so it is easy to determine how changing a weight will change the function, even if we treat ORIENT3D as a black box.

3. An interesting property of the incremental insertion algorithm is that it is robust against point sets that have many cocircular points, unlike the gift-wrapping algorithm. Show that even if the incremental insertion algorithm does not explicitly simulate the symbolic weight perturbations described in Section 2.9, it nevertheless produces the Delaunay triangulation consistent with those perturbations, where each point is assigned a positive weight infinitesimally smaller than all the previously inserted points.

4. Section 3.6 claims that when the incremental insertion algorithm with a conflict graph inserts a new vertex u and redistributes a vertex y from a deleted triangle to a newly created triangle by walking from y along the ray $\overrightarrow{uy}$, all the deleted triangles that intersect $\overrightarrow{uy} \setminus uy$ conflict with y as well as u. Prove it.

5. Section 3.7 claims that the expected running time of the randomized incremental insertion algorithm is still $O(n \log n)$ even if the first three inserted vertices are chosen arbitrarily rather than randomly. Explain why this is true. In particular, explain how Propositions 3.4, 3.5, and 3.6 must be modified to account for the possibility that the first three vertices might be bad choices.

6. (a) Design a simplified gift-wrapping algorithm for retriangulating a cavity evacuated when a segment is inserted into a CDT. The input to the algorithm is identical to the input to CavityCDT: the sequence of vertices in counterclockwise order around the cavity. Vertices may be repeated to indicate dangling edges. The algorithm should run in $O(k^2)$ time, where k is the length of the sequence of vertices.

   Your algorithm should begin gift-wrapping from the new segment and use recursion to continue the process. Exploit the structure of the cavity to simplify the gift-wrapping algorithm as much as possible—in particular, there should be no need for a dictionary of unfinished facets.

   (b) Because of the special structure of the cavity—every vertex originally participated in a triangle whose interior intersects the new segment—the visibility tests (Step 2(c) of Finish) are unnecessary in this algorithm. Prove that with the visibility tests omitted, the procedure still produces constrained Delaunay triangles. (You may assume that every edge of the cavity is constrained Delaunay or a segment in the larger triangulation.)

7. Consider an alternative proof of Proposition 3.9 that works even if $\mathcal{P}$ is not generic. Let τ be a triangle whose vertices are vertices in a PLC $\mathcal{P}$. Let e be an edge of τ that respects $\mathcal{P}$, and let p be a point in the relative interior of e. Suppose that $\mathcal{P}$ contains no vertex that is in front of e, in τ's open circumdisk, and visible from p.

   (a) Prove that $\mathcal{P}$ contains no vertex that is in front of e, in τ's open circumdisk, and visible from *any* point in the interior of τ. (Note that this problem is similar to Exercise 7 in Section 2.11.)

   (b) Prove that τ respects $\mathcal{P}$.

   (c) Prove that if e is constrained Delaunay or a segment in $\mathcal{P}$, then τ is constrained Delaunay.

Chapter 4

Three-dimensional Delaunay triangulations

Three-dimensional triangulations are sometimes called tetrahedralizations. Delaunay tetrahedralizations are not quite as effective as planar Delaunay triangulations at producing elements of good quality, but they are nearly as popular in the mesh generation literature as their two-dimensional cousins. Many properties of Delaunay triangulations in the plane generalize to higher dimensions, but many of the optimality properties do not. Notably, Delaunay tetrahedralizations do not maximize the minimum angle (whether plane angle or dihedral angle). Figure 4.1 depicts a three-dimensional counterexample. The hexahedron at the top is the convex hull of its five vertices. The Delaunay triangulation of those vertices, to the left, includes a thin tetrahedron known as a *sliver* or *kite*, whose vertices are nearly coplanar and whose dihedral angles can be arbitrarily close to 0° and 180°. A triangulation of the same vertices that is not Delaunay, at lower right, has better quality.

This chapter surveys Delaunay triangulations and constrained Delaunay triangulations in three—and occasionally higher—dimensions. Constrained Delaunay triangulations generalize uneasily to three dimensions, because there are polyhedra that do not have any tetrahedralization at all.

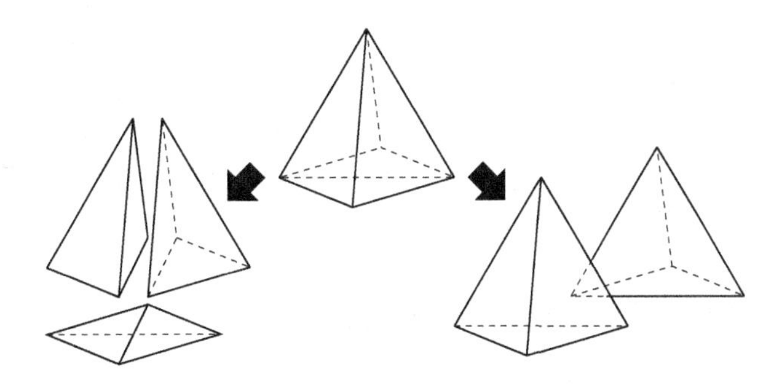

Figure 4.1: This hexahedron has two tetrahedralizations. The Delaunay tetrahedralization at left includes an arbitrarily thin sliver tetrahedron. The non-Delaunay tetrahedralization at right consists of two nicely shaped tetrahedra.

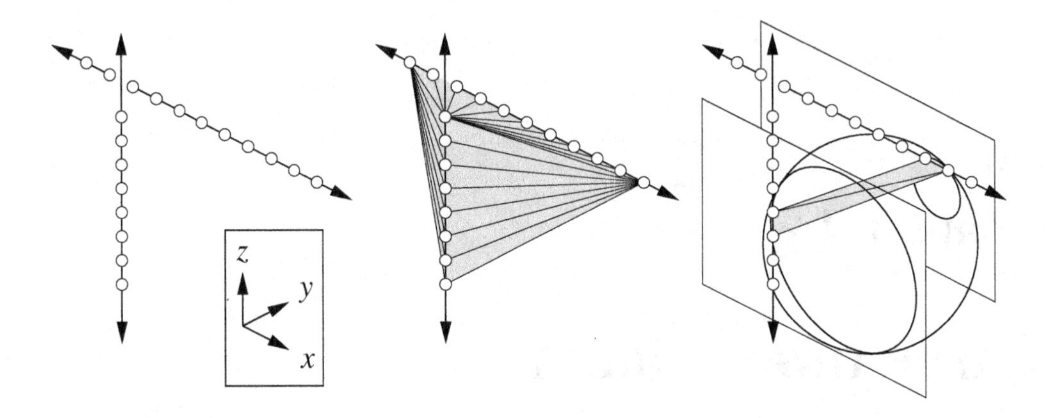

Figure 4.2: At center, the Delaunay tetrahedralization of the points at left. At right, the circumball of one Delaunay tetrahedron with two cross-sections showing it is empty.

4.1 Triangulations of a point set in $\mathbb{R}^d$

Definition 2.1 in Section 2.1 defines a triangulation of a set of points to be a simplicial complex whose vertices are the points and whose union is the convex hull of the points. With no change, the definition holds in any finite dimension d. Figures 4.1–4.4 illustrate triangulations of point sets in three dimensions. Every finite point set in $\mathbb{R}^d$ has a triangulation; for example, the lexicographic triangulation of Section 2.1 also generalizes to higher dimensions with no change.

Let S be a set of n points in $\mathbb{R}^d$. Recall from Section 2.1 that if all the points in S are collinear, they have one triangulation having n vertices and $n-1$ collinear edges connecting them. This is true regardless of d; the triangulation is one-dimensional, although it is embedded in $\mathbb{R}^d$. More generally, if the affine hull of S is k-dimensional, then every triangulation of S is a k-dimensional triangulation embedded in $\mathbb{R}^d$: the simplicial complex has at least one k-simplex but no $(k+1)$-simplex.

The *complexity* of a triangulation is its total number of simplices of all dimensions. Whereas a planar triangulation of n points has $O(n)$ triangles and edges, a surprising property of higher-dimensional triangulations is that they can have superlinear complexity. Figure 4.2 shows a triangulation of n points that has $\Theta(n^2)$ edges and tetrahedra. Every vertex lies on one of two non-intersecting lines, and there is one tetrahedron for each pairing of an edge on one line and an edge on the other. This is the *only* triangulation of these points, and it is Delaunay. In general, a triangulation of n vertices in $\mathbb{R}^3$ has at most $(n^2-3n-2)/2$ tetrahedra, at most n^2-3n triangles, and at most $(n^2-n)/2$ edges. An n-vertex triangulation in $\mathbb{R}^d$ can have a maximum of $\Theta(n^{\lceil d/2 \rceil})$ d-simplices.

4.2 The Delaunay triangulation in $\mathbb{R}^d$

Delaunay triangulations generalize easily to higher dimensions. Let S be a finite set of points in $\mathbb{R}^d$, for $d \geq 1$. Let σ be a k-simplex (for any $k \leq d$) whose vertices are in S. The simplex σ is *Delaunay* if there exists an open circumball of σ that contains no point in S. Clearly, every face of a Delaunay simplex is Delaunay too. The simplex σ is *strongly*

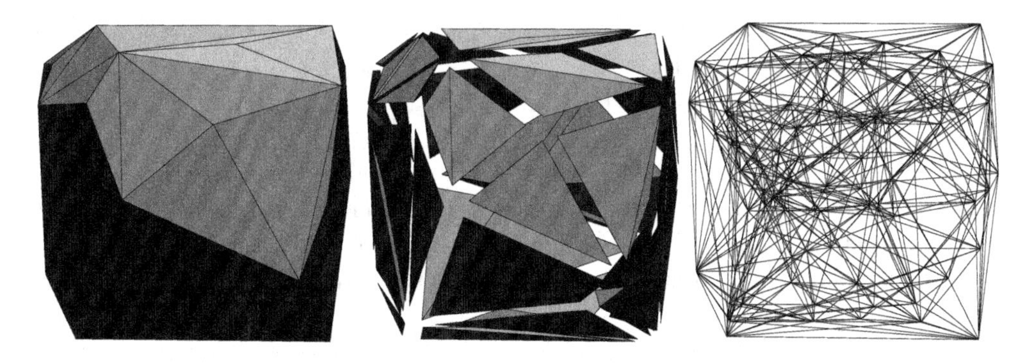

Figure 4.3: Three renderings of a Delaunay tetrahedralization.

Delaunay if there exists a closed circumball of σ that contains no point in S except the vertices of σ. Every point in S is trivially a strongly Delaunay vertex.

Definition 4.1 (Delaunay triangulation). Let S be a finite point set in $\mathbb{R}^d$, and let k be the dimension of its affine hull. A *Delaunay triangulation* Del S of S is a triangulation of S in which every k-simplex is Delaunay—and therefore, every simplex is Delaunay.

Figure 4.2 depicts a Delaunay tetrahedralization and the empty circumball of one of its tetrahedra. Figure 4.3 depicts a more typical Delaunay tetrahedralization, with complexity linear in the number of vertices.

The parabolic lifting map generalizes to higher dimensions too. It maps each point $p = (p_1, p_2, \ldots, p_d) \in \mathbb{R}^d$ to its *lifted companion*, the point $p^+ = (p_1, p_2, \ldots, p_d, p_1^2 + p_2^2 + \cdots + p_d^2)$ in $\mathbb{R}^{d+1}$. Consider the $(d + 1)$-dimensional convex hull of the lifted points, $S^+ = \{v^+ : v \in S\}$. Projecting the downward-facing faces of conv S^+ to $\mathbb{R}^d$ yields a polyhedral complex called the *Delaunay subdivision* of S. If S is *generic*, its Delaunay subdivision is simplicial and S has exactly one Delaunay triangulation.

Definition 4.2 (generic). Let S be a point set in $\mathbb{R}^d$. Let k be the dimension of the affine hull of S. The set S is *generic* if no $k + 2$ points in S lie on the boundary of a single d-ball.

If S if not generic, its Delaunay subdivision may have non-simplicial faces; recall Figure 2.3. In that case, S has multiple Delaunay triangulations, which differ according to how the non-simplicial faces are triangulated.

Whereas each non-simplicial face in a two-dimensional Delaunay subdivision can be triangulated independently, in higher dimensions the triangulations are not always independent. Figure 4.4 illustrates a set of twelve points in $\mathbb{R}^3$ whose Delaunay subdivision includes two cubic cells that share a square 2-face. The square face can be divided into two triangles in two different ways, and each cube can be divided into five or six tetrahedra in several ways, but they are not independent: the triangulation of the square face constrains how both cubes are triangulated.

A *least-vertex triangulation* provides one way to safely subdivide a polyhedral complex into a simplicial complex. To construct it, triangulate the 2-faces through the d-faces in order of increasing dimension. To triangulate a non-simplicial k-face f, subdivide it into k-simplices of the form conv $(v \cup g)$, where v is the lexicographically minimum vertex of

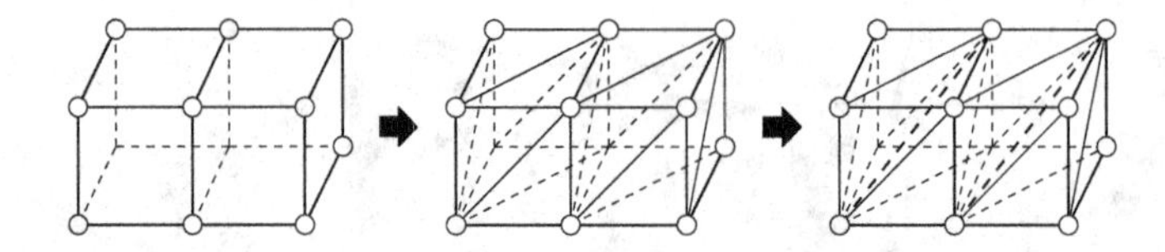

Figure 4.4: A Delaunay subdivision comprising two cubic cells and their faces. The least-vertex Delaunay triangulation subdivides each 2-face into triangles adjoining the face's lexicographically minimum vertex, and likewise subdivides each 3-face into tetrahedra.

f, and g varies over the $(k-1)$-simplices on f's subdivided boundary that do not contain v. The choice of the lexicographically minimum vertex of each face ensures that the face triangulations are compatible with each other. The least-vertex triangulation is consistent with the weight perturbations described in Section 2.9.

Many properties of planar Delaunay triangulations discussed in Chapter 2 generalize to higher dimensions. A few of them are summarized below. Proofs are omitted, but each of them is a straightforward extension of the corresponding proof for two dimensions.

Recall that a *facet* of a polyhedral complex is a $(d-1)$-face, and a facet of a triangulation is a $(d-1)$-simplex. The forthcoming Delaunay Lemma provides an alternative definition of a Delaunay triangulation: a triangulation of a point set in which every facet is locally Delaunay. A facet f in a triangulation $\mathcal{T}$ is said to be *locally Delaunay* if it is a face of fewer than two d-simplices in $\mathcal{T}$, or it is a face of exactly two d-simplices τ_1 and τ_2 and it has an open circumball that contains no vertex of τ_1 nor τ_2. Equivalently, the open circumball of τ_1 contains no vertex of τ_2. Equivalently, the open circumball of τ_2 contains no vertex of τ_1.

Lemma 4.1 (Delaunay Lemma). *Let $\mathcal{T}$ be a triangulation of a finite, d-dimensional set S of points in $\mathbb{R}^d$. The following three statements are equivalent.*

- *Every d-simplex in $\mathcal{T}$ is Delaunay (i.e. $\mathcal{T}$ is Delaunay).*
- *Every facet in $\mathcal{T}$ is Delaunay.*
- *Every facet in $\mathcal{T}$ is locally Delaunay.* □

As in the plane, a generic point set has exactly one Delaunay triangulation, composed of every strongly Delaunay simplex. The following three propositions have essentially the same proofs as in Section 2.7.

Proposition 4.2. *Let σ be a strongly Delaunay simplex, and let τ be a Delaunay simplex. Then $\sigma \cap \tau$ is either empty or a shared face of both σ and τ.*

Proposition 4.3. *Every Delaunay triangulation of a point set contains every strongly Delaunay simplex.*

Theorem 4.4. *A generic point set has exactly one Delaunay triangulation.*

4.3 The optimality of the Delaunay triangulation in $\mathbb{R}^d$

Some optimality properties of Delaunay triangulations hold in any dimension. Consider the use of triangulations for piecewise linear interpolation of a quadratic multivariate function. If the function is isotropic—of the form $\alpha\|p\|^2 + \langle a, p\rangle + \beta$ for $p \in \mathbb{R}^d$—then the Delaunay triangulation minimizes the interpolation error measured in the L_q-norm for every $q \geq 1$, compared with all other triangulations of the same points. (If the function is not isotropic, but is parabolic rather than hyperbolic, then the optimal triangulation is a weighted Delaunay triangulation in which the function determines the vertex heights.)

Delaunay triangulations also minimize the radius of the largest min-containment ball of their simplices (recall Definition 1.20). This result implies a third optimality result, also related to multivariate piecewise linear interpolation. Suppose one must choose a triangulation to interpolate an unknown function, and one wishes to minimize the largest pointwise error in the domain. After one chooses the triangulation, an adversary will choose the worst possible smooth function for the triangulation to interpolate, subject to a fixed upper bound on the absolute curvature (i.e. second directional derivative) of the function anywhere in the domain. The Delaunay triangulation is the optimal choice.

To better understand these three optimality properties, consider multivariate piecewise linear interpolation on a triangulation $\mathcal{T}$ of a point set S. Let $\mathcal{T}^+ = \{\sigma^+ : \sigma \in \mathcal{T}\}$ be the triangulation lifted by the parabolic lifting map; $\mathcal{T}^+$ is a simplicial complex embedded in $\mathbb{R}^{d+1}$. Think of $\mathcal{T}^+$ as inducing a continuous piecewise linear function $\mathcal{T}^+(p)$ that maps each point $p \in \operatorname{conv} S$ to a real value.

How well does $\mathcal{T}^+$ approximate the paraboloid? Let $e(p) = \mathcal{T}^+(p) - \|p\|^2$ be the error in the interpolated function $\mathcal{T}^+$ as an approximation of the paraboloid $\|p\|^2$. At each vertex $v \in S$, $e(v) = 0$. Because $\|p\|^2$ is convex, the error satisfies $e(p) \geq 0$ for all $p \in \operatorname{conv} S$.

Proposition 4.5. *At every point $p \in \operatorname{conv} S$, every Delaunay triangulation $\mathcal{T}$ of S minimizes $\mathcal{T}^+(p)$, and therefore minimizes the interpolation error $e(p)$, among all triangulations of S. Hence, every Delaunay triangulation of S minimizes $\|e\|_{L_q}$ for every Lebesgue norm L_q, and every other norm monotonic in e.*

Proof. If $\mathcal{T}$ is Delaunay, then $\mathcal{T}^+$ is the set of faces of the underside of the convex hull $\operatorname{conv} S^+$ of the lifted vertices (or a subdivision of those faces if some of them are not simplicial). No simplicial complex in $\mathbb{R}^{d+1}$ whose vertices are all in S^+ can pass through any point below $\operatorname{conv} S^+$. □

Proposition 4.6. *Let o and r be the circumcenter and circumradius of a d-simplex σ. Let o_{mc} and r_{mc} be the center and radius of the min-containment ball of σ. Let q be the point in σ nearest o. Then $o_{\mathrm{mc}} = q$ and $r_{\mathrm{mc}}^2 = r^2 - d(o,q)^2$.*

Proof. Let τ be the face of σ whose relative interior contains q. The face τ is not a vertex, because the vertices of σ are σ's furthest points from o. Because q is the point in τ nearest o and because q is in the relative interior of τ, the line segment oq is orthogonal to τ. (This is true even if $\tau = \sigma$, in which case $o - q$ is the zero vector.) This fact, plus the fact that o is equidistant from all the vertices of τ, implies that q is equidistant from all the vertices of τ (as Figure 4.5 demonstrates). Let r be the distance between q and any vertex of τ. As $q \in \tau$,

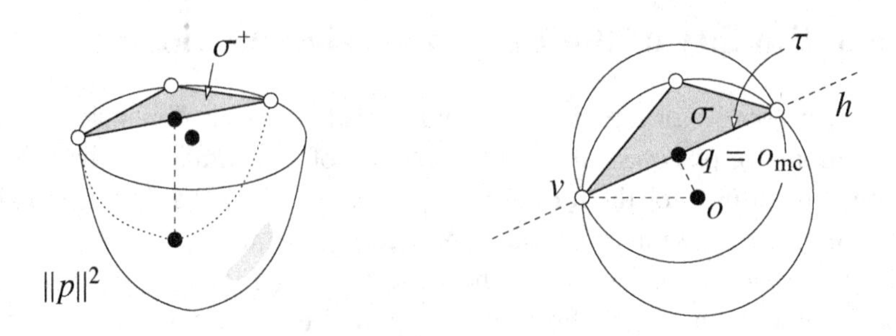

Figure 4.5: Left: within σ, the error $e(p)$ is maximized at the point nearest the circumcenter of σ. Right: top view of σ, its circumdisk, and its min-containment disk.

there is no containment ball of τ (or σ) with radius less than r because q cannot move in any direction without moving away from some vertex of τ. Therefore, q and r are the center and radius of the min-containment ball of τ.

By the following reasoning, σ has the same min-containment ball as τ. If $q = o$, this conclusion is immediate. Otherwise, let h be the hyperplane through q orthogonal to oq. Observe that $\tau \subset h$. No point in σ is on the same side of h as o: if there were such a point w, there would be a point in σ (between w and q) closer to o than q, contradicting the fact that q is closest. The hyperplane h cuts the closed circumball of σ into two pieces, and the piece that includes σ is included in the min-containment ball of τ. Therefore, q and r are the center and radius of the min-containment ball of σ.

Let v be any vertex of τ. Pythagoras' Theorem on the triangle oqv (see Figure 4.5) yields $r_{\mathrm{mc}}^2 = r^2 - d(o,q)^2$. □

Proposition 4.7. *Every Delaunay triangulation of S minimizes the largest min-containment ball, compared with all other triangulations of S.*

Proof. Over any single d-simplex σ, there is an explicit expression for $e(p)$. Recall from the proof of the Lifting Lemma (Lemma 2.1) that the hyperplane h_σ that includes σ^+ is defined by the function $h_\sigma(p) = 2\langle o, p\rangle - \|o\|^2 + r^2$, where o and r are the circumcenter and circumradius of σ and $p \in \mathbb{R}^d$ varies freely. Hence, for all $p \in \sigma$,

$$\begin{aligned} e(p) &= \mathcal{T}^+(p) - \|p\|^2 \\ &= h_\sigma(p) - \|p\|^2 \\ &= 2\langle o, p\rangle - \|o\|^2 + r^2 - \|p\|^2 \\ &= r^2 - d(o,p)^2. \end{aligned}$$

Figure 4.5 (left) illustrates the functions $h_\sigma(p)$ and $\|p\|^2$ over a triangle σ. The error $e(p)$ is the vertical distance between the two functions. At which point p in σ is $e(p)$ largest? At the point nearest the circumcenter o, because $d(o,p)^2$ is smallest there. (The error is maximized at $p = o$ if o is in σ; Figure 4.5 gives an example where it is not.) Let o_{mc} and r_{mc} be the center and radius of the min-containment ball of σ, respectively. By Proposition 4.6, the point in σ nearest o is o_{mc}, and $e(o_{\mathrm{mc}}) = r^2 - d(o, o_{\mathrm{mc}})^2 = r_{\mathrm{mc}}^2$.

It follows that the square of the min-containment radius of σ is $r_{\mathrm{mc}}^2 = \max_{p \in \sigma} e(p)$, and thus $\max_{p \in \mathrm{conv}\, S} e(p)$ is the squared radius of the largest min-containment ball of the entire triangulation $\mathcal{T}$. By Proposition 4.5, the Delaunay triangulation minimizes this quantity among all triangulations of S. □

The optimality of the Delaunay triangulation for controlling the largest min-containment radius dovetails nicely with an error bound for piecewise linear interpolation derived by Waldron. Let $\mathcal{F}(c)$ be the space of scalar functions defined over conv S that are C^1-continuous and whose absolute curvature nowhere exceeds c. In other words, for every $f \in \mathcal{F}(c)$, every point $p \in \mathrm{conv}\, S$, and every unit direction vector $\vec{u}$, the magnitude of the second directional derivative $f''_{\vec{u}}(p)$ is at most c. This is a common starting point for analyses of piecewise linear interpolation error.

Let f be a function in $\mathcal{F}(c)$. Let $\sigma \subseteq \mathrm{conv}\, S$ be a simplex (of any dimensionality) with min-containment radius r_{mc}. Let h_σ be a linear function that interpolates f at the vertices of σ. Waldron shows that for all $p \in \sigma$, the absolute error $|e(p)| = |h_\sigma(p) - f(p)|$ is at most $cr_{\mathrm{mc}}^2/2$. Furthermore, this bound is sharp: for every simplex σ with min-containment radius r_{mc}, there is a function $f \in \mathcal{F}(c)$ and a point $p \in \sigma$ such that $|e(p)| = cr_{\mathrm{mc}}^2/2$. That function is $f(p) = c\|p\|^2/2$, as illustrated in Figure 4.5, and that point is $p = o_{\mathrm{mc}}$.

Proposition 4.8. *Every Delaunay triangulation* $\mathcal{T}$ *of* S *minimizes* $\max_{f \in \mathcal{F}(c)} \max_{p \in \mathrm{conv}\, S} |\mathcal{T}^+(p) - f(p)|$, *the worst-case pointwise interpolation error, among all triangulations of* S.

Proof. Per Waldron, for any triangulation $\mathcal{T}$, $\max_{f \in \mathcal{F}(c)} \max_{p \in \mathrm{conv}\, S} |\mathcal{T}^+(p) - f(p)| = cr_{\max}^2/2$, where $r_{\max}$ is the largest min-containment radius among all the simplices in $\mathcal{T}$. The result follows immediately from Proposition 4.7. □

One of the reasons for the longstanding popularity of Delaunay triangulations is that, as Propositions 4.5 and 4.8 show, the Delaunay triangulation is an optimal piecewise linear interpolating surface. Of course, $e(p)$ is not the only criterion for the merit of a triangulation used for interpolation. Many applications require that the interpolant approximate the gradient, i.e. $\nabla\mathcal{T}^+(p)$ must approximate $\nabla f(p)$ well. For the goal of approximating $\nabla f(p)$ in three or more dimensions, the Delaunay triangulation is sometimes far from optimal even for simple functions such as the paraboloid $f(p) = \|p\|^2$. This is why eliminating slivers is a crucial problem in Delaunay mesh generation.

4.4 Bistellar flips and the flip algorithm

The flip algorithm described in Section 2.5 extends to three or more dimensions, but unfortunately, it does not always produce a Delaunay triangulation. The natural generalizations of edge flips are *bistellar flips*, operations that replace one set of simplices with another set filling the same volume. Figure 4.6 illustrates the bistellar flips in one, two, and three dimensions. The three-dimensional analogs of edge flips are called *2-3 flips* and *3-2 flips*. The names specify the numbers of tetrahedra deleted and created, respectively.

The upper half of the figure depicts basic bistellar flips, which retriangulate the convex hull of $d+2$ vertices in $\mathbb{R}^d$ by replacing a collection of k d-simplices with $d+2-k$ different

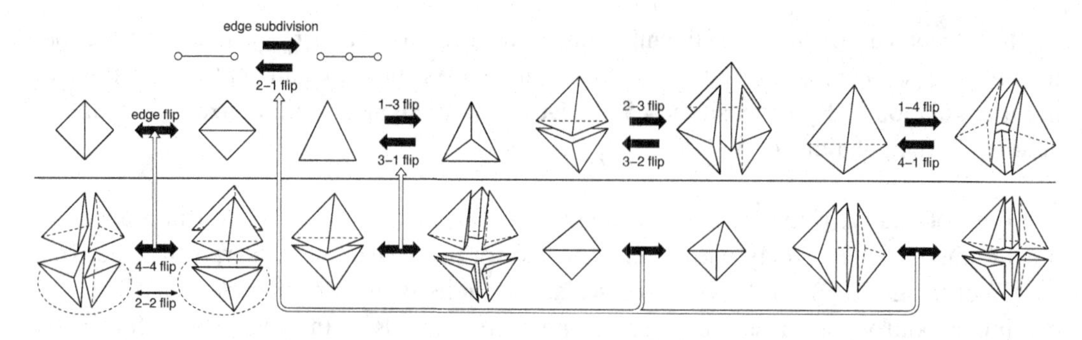

Figure 4.6: Basic bistellar flips in one, two, and three dimensions appear above the line. Extended bistellar flips appear below the line. White arrows connect extended flips to the lower-dimensional flips on which they are based. The 2-2 flip at bottom left typically involves two coplanar triangular faces on the boundary of a domain, whereas the 4-4 flip occurs when the corresponding faces are in a domain interior. Edge subdivisions and their inverses can also occur on a domain boundary, as at bottom right, or in a domain interior.

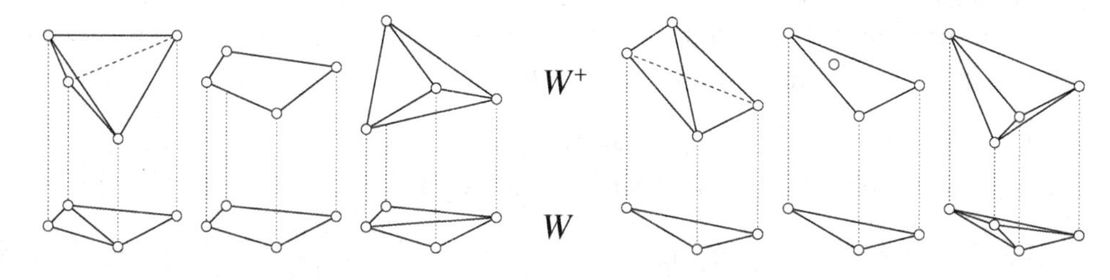

Figure 4.7: If the convex hull of four points in $\mathbb{R}^3$ is a tetrahedron, the facets on its underside determine one planar triangulation, and the facets on its upper side determine another.

d-simplices. In three dimensions, there are four basic flips: 1-4, 2-3, 3-2, and 4-1. The 1-4 flip inserts a vertex, and the 4-1 flip deletes one.

The lower half of the figure depicts extended bistellar flips, in which a lower-dimensional basic flip transforms higher-dimensional simplices, possibly many of them. For example, consider bisecting an edge of a tetrahedralization, as illustrated at the lower right of the figure. In essence, this operation is a one-dimensional 1-2 flip, replacing one edge with two. But every tetrahedron that includes the subdivided edge is subdivided into two tetrahedra, and the number of tetrahedra that share the edge can be arbitrarily large. Therefore, some extended flips can delete and create arbitrarily many d-simplices.

An intuitive way to understand the basic bistellar flips in $\mathbb{R}^d$ is through the faces of a simplex in $\mathbb{R}^{d+1}$, as illustrated in Figure 4.7. Let W be a set of $d + 2$ points in $\mathbb{R}^d$, and let W^+ be the same points lifted by the parabolic lifting map in $\mathbb{R}^{d+1}$. Call the $(d + 1)$th coordinate axis (along which the points are lifted) the *vertical axis*. Assume the points in W are not cospherical. Then the points in W^+ are not cohyperplanar, and conv W^+ is a $(d + 1)$-simplex—call it the *W-simplex*. Each facet of the W-simplex can be placed in one of three classes: vertical (parallel to the vertical axis), lower (facing the negative end of the vertical axis), or upper (those that would get wet if rain were falling).

The W-simplex suggests two different triangulations of the region conv W: the Delaunay triangulation by projecting the lower facets of conv W^+ to $\mathbb{R}^d$, and a non-Delaunay triangulation by projecting the upper facets. (If the points in W^+ are cohyperplanar, W has two Delaunay triangulations.) The act of replacing one such triangulation with the other is a bistellar flip. If the W-simplex has no vertical facet, the flip is basic. These are the only two ways to triangulate conv W such that all the vertices are in W. One of the two triangulations might omit a vertex of W—witness the 3-1 flip, which deletes a vertex.

An extended bistellar flip in $\mathbb{R}^d$ is built on a j-dimensional basic flip as follows. Consider a basic flip that replaces a set T_D of k j-simplices with a set T_C of $j+2-k$ j-simplices. Let the *join* $\tau * \zeta$ of two disjoint simplices τ and ζ be the simplex conv $(\tau \cup \zeta)$ having all the vertices of τ and ζ, and let the *join* of two sets T and Z of simplices be the set of simplices $\{\tau * \zeta : \tau \in T \text{ and } \zeta \in Z\}$. Let Z be a set of $(d - j - 1)$-simplices such that every member of $T_D * Z$ is a nondegenerate d-simplex and $T_D * Z$ contains every d-simplex in $\mathcal{T}$ with a face in T_D. If such a Z exists, the act of replacing the simplices $T_D * Z$ with the simplices $T_C * Z$ is an extended flip. For example, in the edge bisection illustrated at the lower right of Figure 4.6, T_D contains just the bisected edge, T_C contains the two edges the bisection yields, and Z contains one edge for each bisected tetrahedron.

The flip algorithm requires a solution to the following problem. A triangulation $\mathcal{T}$ has a facet f that is not locally Delaunay. What is the appropriate flip to eliminate f? Let W be the set containing the d vertices of f and the two additional vertices of the two d-simplices having f for a face. The fact that f is not locally Delaunay implies that f^+ lies on the upper surface of the W-simplex conv W^+. The upper facets of the W-simplex indicate which d-simplices the flip algorithm should delete from $\mathcal{T}$ (including the two adjoining f), and the lower facets indicate which d-simplices should replace them. The procedure Flip in Figure 4.8 identifies these tetrahedra for $d = 3$. If the W-simplex has vertical facets, Flip performs an extended flip.

The difficulty is that the simplices to be deleted might not all be in $\mathcal{T}$. Figure 4.9 illustrates circumstances where the flip algorithm wishes to perform a flip, but cannot. At left, the shaded triangle is not locally Delaunay. The right flip to remove it is a 3-2 flip, but the flip is possible only if the third tetrahedron is present. If four or more tetrahedra share the bold edge, the flip is blocked, at least until another flip creates the missing tetrahedron. The flip algorithm can get stuck in a configuration where *every* locally non-Delaunay triangle's removal is blocked, and the algorithm cannot make further progress toward the Delaunay triangulation. This is not a rare occurrence in practice.

If the flip algorithm is asked to compute a *weighted* Delaunay triangulation, it must sometimes perform a flip that deletes a vertex, such as the 4-1 flip illustrated at right in Figure 4.9. Such a flip is possible only if all the simplices the flip is meant to delete are present. Even in the plane, there are circumstances where every desired 3-1 flip is blocked and the flip algorithm is stuck.

Extended flips are even more delicate; they require not only that $T_D * Z \subseteq \mathcal{T}$, but also that $T_D * Z$ contains *every* tetrahedron in $\mathcal{T}$ that has a face in T_D.

Whether it succeeds or gets stuck, the running time of the flip algorithm is $O(n^{1+\lfloor d/2 \rfloor})$, by the same reasoning explained in Section 2.5: a flip can be modeled as the act of gluing a $(d + 1)$-simplex to the underside of the lifted triangulation, and a triangulation in $\mathbb{R}^{d+1}$ has at most $O(n^{1+\lfloor d/2 \rfloor})$ simplices.

One of the most important open problems in combinatorial geometry asks whether the *flip graph* is connected. For a specified point set, the flip graph has one node for every

```
FLIP(𝒯, uvw)
1.  x ← ADJACENT(u, v, w)
2.  y ← ADJACENT(w, v, u)
3.  T_D ← {wvuy, uvwx}       { delete tetrahedra wvuy and uvwx }
4.  T_C ← ∅
5.  V ← ∅
6.  For (a, b, c) ← (u, v, w), (v, w, u), and (w, u, v)
7.        α ← ORIENT3D(a, b, x, y)
8.        If α > 0
9.              T_D ← T_D ∪ {abxy}       { delete tetrahedron abxy }
10.       else if α < 0
11.             T_C ← T_C ∪ {abyx}       { create tetrahedron abyx }
12.       else V ← V ∪ {c}       { abxy is degenerate and c is not part of the basic flip }
    { perform a flip that replaces T_D with T_C }
13. If V = ∅      { basic flip: 2-3 flip or 3-2 flip or 4-1 flip }
          { note: if any simplex in T_D is absent from 𝒯, the flip is blocked }
14.       For each τ ∈ T_D
15.             Call DELETETETRAHEDRON(τ)
16.       For each τ ∈ T_C
17.             Call ADDTETRAHEDRON(τ)
18. else      { extended flip: 2-2 flip or 4-4 flip or 3-1 flip or 6-2 flip or edge merge }
19.       j ← 3 − |V|
20.       Remove the vertices in V from every simplex in T_D and T_C, yielding j-simplices
21.       σ ← a j-simplex in T_D
22.       For each (d − j − 1)-simplex ζ such that σ ∗ ζ is a tetrahedron in 𝒯
23.             For each τ ∈ T_D
                      { note: if τ ∗ ζ is absent from 𝒯, the flip is blocked }
24.                   Call DELETETETRAHEDRON(τ ∗ ζ)
25.             For each τ ∈ T_C
26.                   Call ADDTETRAHEDRON(τ ∗ ζ)
```

Figure 4.8: Algorithm for performing a tetrahedral bistellar flip. The parameters to FLIP are a triangulation $\mathcal{T}$ and a facet $uvw \in \mathcal{T}$ to flip. FLIP assumes uvw can be flipped; comments identify places where this assumption could fail. DELETETETRAHEDRON and ADDTETRAHEDRON are described in Section 5.1.

triangulation of those points. Two nodes are connected by an edge if one triangulation can be transformed into the other by a single bistellar flip, excluding those flips that create or delete a vertex. For every planar point set, its flip graph is connected—in other words, any triangulation of the points can be transformed into any other triangulation of the same points by a sequence of edge flips. One way to see this is to recall that Proposition 2.5 states that every triangulation can be flipped to the Delaunay triangulation. However, there exist point sets in five or more dimensions whose flip graphs are not connected; they have triangulations that cannot be transformed to Delaunay by a sequence of bistellar flips. The

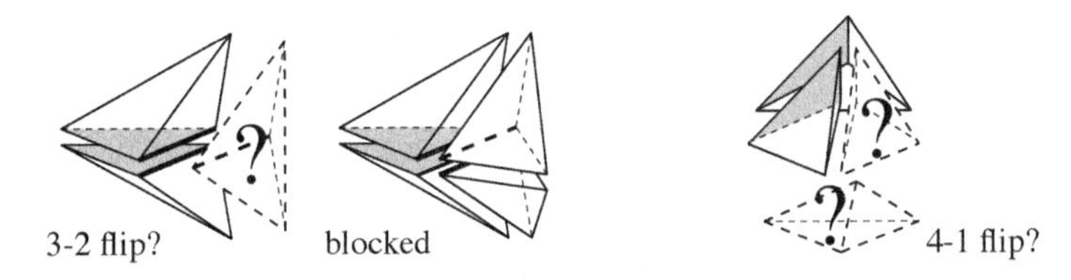

Figure 4.9: The shaded facet at left is not locally Delaunay. It can be removed by a 3-2 flip, but only if the third tetrahedron is present; the flip is blocked if more than three tetrahedra share the central edge (bold). The shaded facet at right can be removed by a 4-1 flip if the other two tetrahedra are present.

question remains open in three and four dimensions. But even if all flip graphs for three-dimensional point sets are connected, flipping facets that are locally non-Delaunay does not suffice to find the Delaunay triangulation.

Despite the failure of the flip algorithm for three-dimensional Delaunay triangulations and weighted two-dimensional Delaunay triangulations, some Delaunay triangulation algorithms rely on bistellar flips, including several incremental vertex insertion algorithms and an algorithm for inserting a polygon into a CDT, the latter described in Section 5.8. In particular, if a new vertex is introduced into a Delaunay triangulation by a simple 1-4 flip (or by subdividing a facet or edge of the triangulation), and the flip algorithm is run before the triangulation is changed in any other way, the flip algorithm is guaranteed to restore the Delaunay property without getting stuck.

4.5 Three-dimensional constrained Delaunay triangulations

Constrained Delaunay triangulations generalize to three or more dimensions, but whereas every piecewise linear complex in the plane has a CDT, not every three-dimensional PLC has one. Worse yet, there exist simple polyhedra that do not have triangulations at all—that is, they cannot be subdivided into tetrahedra without creating new vertices (i.e. tetrahedron vertices that are not vertices of the polyhedron).

E. Schönhardt furnishes an example depicted in Figure 4.10. The easiest way to envision this polyhedron is to begin with a triangular prism. Imagine grasping the prism so that its bottom triangular face cannot move, while twisting the top triangular face so it rotates slightly about its center while remaining horizontal. This rotation breaks each of the three square faces into two triangular faces along a diagonal *reflex edge*—an edge at which the polyhedron is locally nonconvex. After this transformation, the upper left corner and lower right corner of each (former) square face are separated by a reflex edge and are no longer visible to each other within the polyhedron. Any four vertices of the polyhedron include two separated by a reflex edge; thus, any tetrahedron whose vertices are vertices of the polyhedron does not lie entirely within the polyhedron. Therefore, Schönhardt's polyhedron cannot be triangulated without additional vertices. It can be subdivided into tetrahedra with the addition of one vertex at its center.

Adding to the difficulty, it is NP-hard to determine whether a polyhedron has a triangulation, or whether it can be subdivided into tetrahedra with only k additional vertices for an arbitrary constant k.

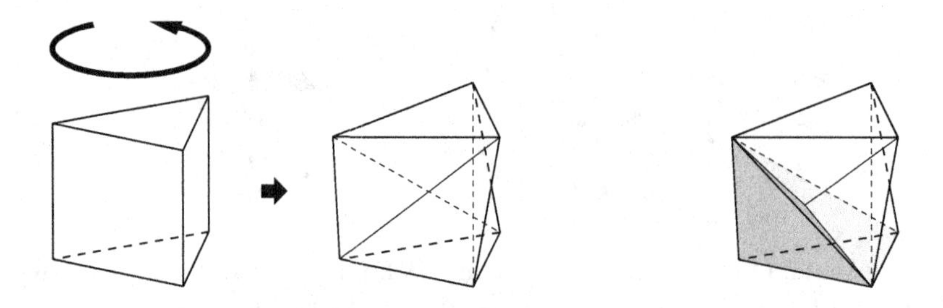

Figure 4.10: Schönhardt's untetrahedralizable polyhedron (center) is formed by rotating one end of a triangular prism (left), thereby creating three diagonal reflex edges. The convex hull of any four polyhedron vertices (right) sticks out.

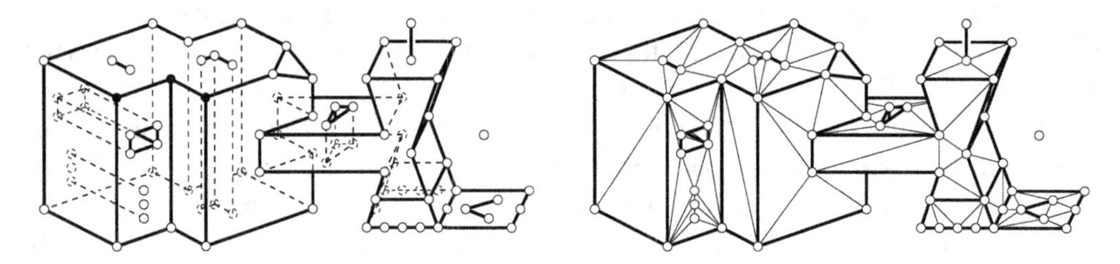

Figure 4.11: A three-dimensional piecewise linear complex and its constrained Delaunay triangulation. Each polygon and polyhedron may have holes, slits, and vertices in its relative interior. Each polyhedron may also have polygons in its interior.

The following sections discuss triangulations and CDTs of polyhedra and PLCs in three dimensions. It is possible to refine any polyhedron or PLC by adding new vertices on its edges so that it has a constrained Delaunay triangulation. This fact makes CDTs useful in three dimensions.

4.5.1 Piecewise linear complexes and their triangulations in $\mathbb{R}^d$

The domain over which a general-dimensional CDT is defined is a general-dimensional piecewise linear complex, which is a set of linear cells—vertices, edges, polygons, and polyhedra—as illustrated in Figure 4.11. The linear cells constrain how the complex can be triangulated: each linear cell in the complex must be a union of simplices in the triangulation. The union of the linear cells specifies the region to be triangulated.

Definition 4.3 (piecewise linear complex). A *piecewise linear complex* (PLC) $\mathcal{P}$ is a finite set of linear cells that satisfies the following properties.

- The vertices and edges in $\mathcal{P}$ form a simplicial complex.
- For each linear cell $f \in \mathcal{P}$, the boundary of f is a union of linear cells in $\mathcal{P}$.
- If two distinct linear cells $f, g \in \mathcal{P}$ intersect, their intersection is a union of linear cells in $\mathcal{P}$, all having lower dimension than at least one of f or g.

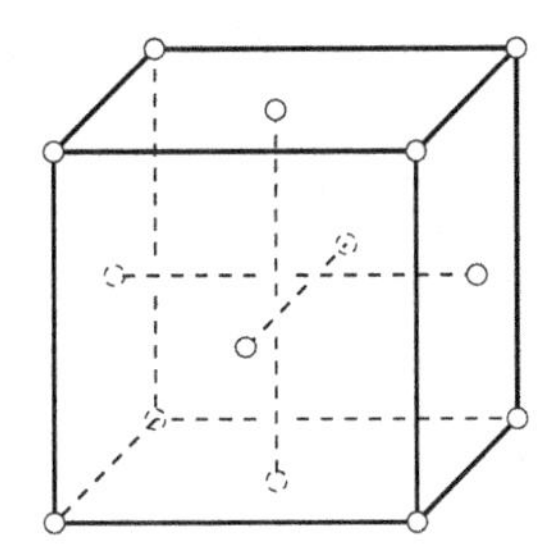

Figure 4.12: A convex PLC with no triangulation.

As in the plane, the edges in $\mathcal{P}$ are called *segments*. Its *underlying space* is $|\mathcal{P}| = \bigcup_{f \in \mathcal{P}} f$, which is usually the domain to be triangulated. The *faces* of a linear cell $f \in \mathcal{P}$ are the linear cells in $\mathcal{P}$ that are subsets of f, including f itself.

A triangulation of a PLC must cover every polyhedron, respect every polygon, and include every segment and vertex.

Definition 4.4 (triangulation of a PLC). Let $\mathcal{P}$ be a PLC. A *triangulation of* $\mathcal{P}$ is a simplicial complex $\mathcal{T}$ such that $\mathcal{P}$ and $\mathcal{T}$ have the same vertices, every linear cell in $\mathcal{P}$ is a union of simplices in $\mathcal{T}$, and $|\mathcal{T}| = |\mathcal{P}|$.

Because this definition does not allow $\mathcal{T}$ to have new vertices absent from $\mathcal{P}$, every edge in $\mathcal{P}$ must appear in $\mathcal{T}$. However, the polygons in $\mathcal{P}$ may be subdivided into triangles in $\mathcal{T}$.

Schönhardt's polyhedron shows that not every PLC has a triangulation. Every convex polyhedron has a triangulation; what about convex polyhedra with internal segments? Figure 4.12 illustrates a PLC with no triangulation, consisting of a cube inside which three orthogonal segments pass by each other but do not intersect. If any one of the segments is omitted, the PLC has a triangulation. This example shows that, unlike with planar triangulations, it is not always possible to insert a new edge into a tetrahedralization.

Because some polyhedra and PLCs do not have triangulations, Steiner triangulations are even more important in three dimensions than in the plane.

Definition 4.5 (Steiner triangulation of a PLC). Let $\mathcal{P}$ be a PLC. A *Steiner triangulation of* $\mathcal{P}$, also known as a *conforming triangulation of* $\mathcal{P}$ or a *mesh of* $\mathcal{P}$, is a simplicial complex $\mathcal{T}$ such that $\mathcal{T}$ contains every vertex in $\mathcal{P}$ and possibly more, every linear cell in $\mathcal{P}$ is a union of simplices in $\mathcal{T}$, and $|\mathcal{T}| = |\mathcal{P}|$. The new vertices in $\mathcal{T}$, not present in $\mathcal{P}$, are called *Steiner points*. A *Steiner Delaunay triangulation* of $\mathcal{P}$, also known as a *conforming Delaunay triangulation* of $\mathcal{P}$, is a Steiner triangulation of $\mathcal{P}$ in which every simplex is Delaunay.

Every n-vertex polyhedron has a Steiner triangulation with at most $O(n^2)$ vertices, found by constructing a *vertical decomposition* of the polyhedron. The same is true for PLCs of complexity n. Unfortunately, there are polyhedra for which it is not possible to do better; Figure 4.13 depicts Chazelle's polyhedron, which has n vertices and $O(n)$ edges, but cannot be divided into fewer than $\Theta(n^2)$ convex bodies. The worst-case complexity of subdividing a polyhedron is related to its number of reflex edges: there is an algorithm that divides any polyhedron with r reflex edges into $O(n + r^2)$ tetrahedra, and some polyhedra with r reflex edges cannot be divided into fewer than $\Omega(n + r^2)$ convex bodies.

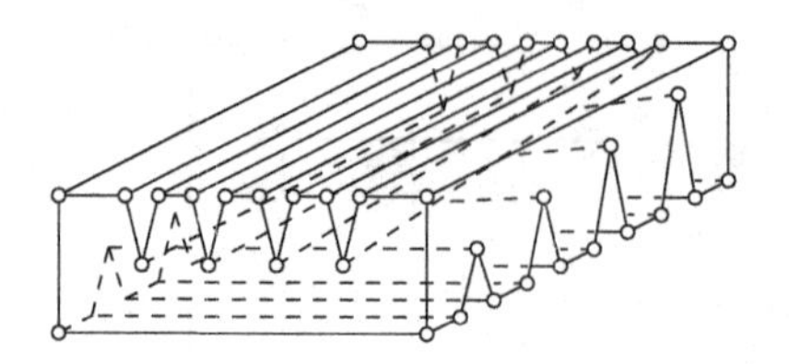

Figure 4.13: Chazelle's polyhedron.

It appears likely, though it is proved only in two dimensions, that there exist PLCs whose smallest Steiner Delaunay triangulations are asymptotically larger than their smallest Steiner triangulations. There are algorithms that can find a Steiner Delaunay tetrahedralization of any three-dimensional polyhedron, but they might introduce a superpolynomial number of new vertices. No known algorithm for finding Steiner Delaunay tetrahedralizations is guaranteed to introduce only a polynomial number of new vertices, and no algorithm of any complexity has been offered for four- or higher-dimensional Steiner Delaunay triangulations. Moreover, the existing algorithms all seem to introduce an unnecessarily large number of vertices near small domain angles. These problems can be partly remediated by Steiner CDTs.

4.5.2 The constrained Delaunay triangulation in $\mathbb{R}^3$

Three-dimensional constrained Delaunay triangulations aspire to retain most of the advantages of Delaunay triangulations while respecting constraints. But Figures 4.10, 4.12, and 4.13 demonstrate that some PLCs, even some polyhedra, have no triangulation at all. Moreover, some polyhedra that do have triangulations do not have CDTs. Nevertheless, CDTs are useful because, if we are willing to add new vertices, every three-dimensional PLC has a Steiner CDT, and a Steiner CDT might require many fewer vertices than a Steiner Delaunay triangulation.

As in the plane, there are several equivalent definitions of *constrained Delaunay triangulation* in three dimensions. The simplest is that a CDT is a triangulation of a PLC in which every facet not included in a PLC polygon is locally Delaunay. A CDT differs from a Delaunay triangulation in three ways: it is not necessarily convex, it is required to respect a PLC, and the facets of the CDT that are included in PLC polygons are exempt from being locally Delaunay.

Recall from Definition 2.11 that a simplex σ *respects* a PLC $\mathcal{P}$ if $\sigma \subseteq |\mathcal{P}|$ and for every $f \in \mathcal{P}$ that intersects σ, $f \cap \sigma$ is a union of faces of σ. By Definition 2.13, two points x and y are *visible* to each other if xy respects $\mathcal{P}$. A linear cell in $\mathcal{P}$ that intersects the relative interior of xy but does not include xy *occludes* the visibility between x and y. The primary definition of CDT specifies that every tetrahedron is constrained Delaunay, defined as follows.

Definition 4.6 (constrained Delaunay). In the context of a PLC $\mathcal{P}$, a simplex σ is *constrained Delaunay* if $\mathcal{P}$ contains the vertices of σ, σ respects $\mathcal{P}$, and there is an open circumball of σ that contains no vertex in $\mathcal{P}$ that is visible from any point in the relative interior of σ.

Figure 4.14 depicts a constrained Delaunay tetrahedron τ. Every face of τ whose relative interior intersects the polygon f is included in f, so τ respects $\mathcal{P}$. The open circumball

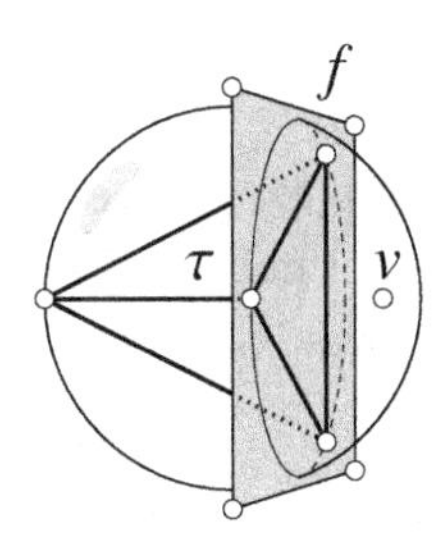

Figure 4.14: A constrained Delaunay tetrahedron τ.

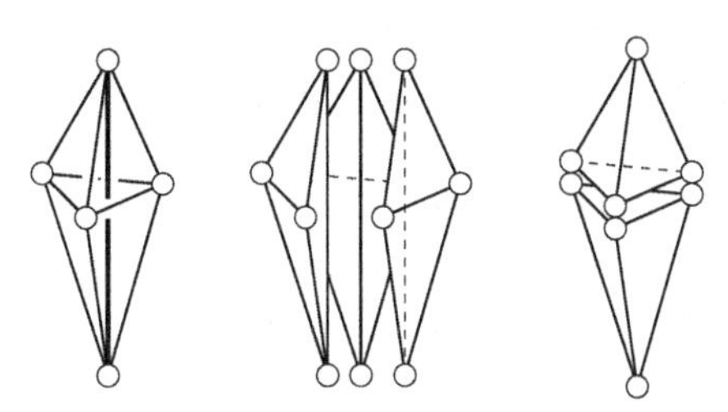

Figure 4.15: Left: a PLC with no CDT. Center: the sole tetrahedralization of this PLC. Its three tetrahedra are not constrained Delaunay. Right: the two Delaunay tetrahedra do not respect the central segment.

of τ contains one vertex v, but v is not visible from any point in the interior of τ, because f occludes its visibility.

Definition 4.7 (constrained Delaunay triangulation)**.** Let $\mathcal{P}$ be a three-dimensional PLC. A *constrained Delaunay triangulation* (CDT) of $\mathcal{P}$ is a triangulation of $\mathcal{P}$ in which every tetrahedron is constrained Delaunay, and every dangling triangle (i.e. not a face of any tetrahedron) is also constrained Delaunay.

Figure 4.11 illustrates a PLC and its CDT. Observe that the PLC has a polygon that is not a face of any polyhedron; this face is triangulated with constrained Delaunay triangles.

Figure 4.15 illustrates a PLC that has no CDT because of a segment that runs vertically through the domain interior. There is only one tetrahedralization of this PLC—composed of three tetrahedra encircling the central segment—and its tetrahedra are not constrained Delaunay, because each of them has a visible vertex in its open circumball. Whereas polygons usually block enough visibility to ensure their presence in a CDT, segments usually do not. But segments can dictate that a CDT does not exist at all. If the central segment in Figure 4.15 is removed, the PLC has a CDT made up of two tetrahedra.

A *Steiner CDT* or *conforming CDT* of $\mathcal{P}$ is a Steiner triangulation of $\mathcal{P}$ in which every tetrahedron is constrained Delaunay, and every dangling triangle (i.e. not a face of any tetrahedron) is also constrained Delaunay. A PLC with no CDT has a Steiner CDT, but one or more Steiner points must be added on its segments. For example, the PLC in Figure 4.15 has a Steiner CDT with one Steiner point on its central segment.

4.5.3 The CDT Theorem

Although not all piecewise linear complexes have constrained Delaunay triangulations, there is an easy-to-test, sufficient (but not necessary) condition that guarantees that a CDT exists. A three-dimensional PLC $\mathcal{P}$ is *edge-protected* if every edge in $\mathcal{P}$ is strongly Delaunay.

Theorem 4.9 (CDT Theorem). *Every edge-protected PLC has a CDT.* □

It is not sufficient for every edge in $\mathcal{P}$ to be Delaunay. If all six vertices of Schönhardt's polyhedron lie on a common sphere, then all of its edges (and all its faces) are Delaunay, but it still has no tetrahedralization. It is not possible to place the vertices of Schönhardt's polyhedron so that all three of its reflex edges are strongly Delaunay, though any two may be.

What if a PLC that one wishes to triangulate is not edge-protected? One can make it edge-protected by adding vertices on its segments—a task that any Delaunay mesh generation algorithm must do anyway. The augmented PLC has a CDT, which is a Steiner CDT of the original PLC.

Figure 4.16 illustrates the difference between using a Delaunay triangulation and using a CDT for mesh generation. With a Delaunay triangulation, the mesh generator must insert new vertices that guarantee that every segment is a union of Delaunay (preferably strongly Delaunay) edges, and every polygon is a union of Delaunay (preferably strongly Delaunay) triangles. With a CDT, new vertices must be inserted that guarantee that every segment is a union of strongly Delaunay edges; but then the augmented PLC is edge-protected, and the CDT Theorem guarantees that the polygons can be recovered without inserting any additional vertices. The advantage of a CDT is that many fewer vertices might be required.

Testing whether a PLC $\mathcal{P}$ is edge-protected is straightforward. Form the Delaunay triangulation of the vertices in $\mathcal{P}$. If a segment $\sigma \in \mathcal{P}$ is missing from the triangulation, then σ is not strongly Delaunay, and $\mathcal{P}$ is not edge-protected. If σ is present, it is Delaunay. If the symbolic perturbations described in Section 2.9 are used to make the vertices in $\mathcal{P}$ generic, then every Delaunay edge is strongly Delaunay; so if every segment in $\mathcal{P}$ is present, $\mathcal{P}$ is

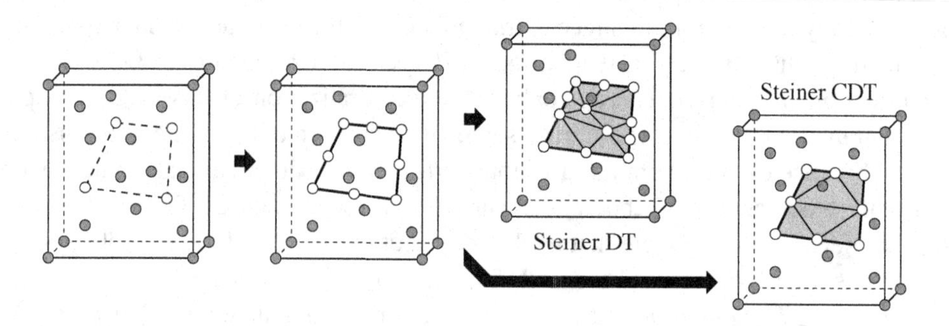

Figure 4.16: Comparison of Steiner Delaunay triangulations and Steiner CDTs. For clarity, vertices inside each box are shown, but tetrahedra are not. For both types of triangulations, missing segments are recovered by inserting new vertices until each segment is a union of strongly Delaunay edges. In a Steiner Delaunay triangulation, additional vertices are inserted until each polygon is a union of strongly Delaunay triangles. In a Steiner CDT, no additional vertices need be inserted; the polygons are recovered by computing a CDT.

edge-protected. (If symbolic perturbations are not used, then testing whether a Delaunay segment σ is strongly Delaunay is equivalent to determining whether the Voronoi polygon dual to σ is nondegenerate.)

4.5.4 Properties of the constrained Delaunay triangulation in $\mathbb{R}^3$

This section summarizes the properties of three-dimensional CDTs.

The Delaunay Lemma for three-dimensional CDTs provides an alternative definition of CDT: a triangulation of a PLC $\mathcal{P}$ is a CDT if and only if every one of its facets is locally Delaunay *or* is included in a polygon in $\mathcal{P}$.

Lemma 4.10 (Constrained Delaunay Lemma). *Let $\mathcal{P}$ be a PLC in which every linear cell is a face of some polyhedron in $\mathcal{P}$, so there are no dangling polygons. Let $\mathcal{T}$ be a triangulation of $\mathcal{P}$. The following three statements are equivalent.*

(i) *Every tetrahedron in $\mathcal{T}$ is constrained Delaunay (i.e. $\mathcal{T}$ is constrained Delaunay).*

(ii) *Every facet in $\mathcal{T}$ not included in a polygon in $\mathcal{P}$ is constrained Delaunay.*

(iii) *Every facet in $\mathcal{T}$ not included in a polygon in $\mathcal{P}$ is locally Delaunay.*

A constrained Delaunay triangulation $\mathcal{T}$ of $\mathcal{P}$ induces a two-dimensional triangulation of each polygon $f \in \mathcal{P}$, namely, $\mathcal{T}|_f = \{\sigma \in \mathcal{T} : \sigma \subseteq f\}$. Statement (ii) above implies that the triangles in $\mathcal{T}|_f$ need not be constrained Delaunay with respect to $\mathcal{P}$—but they *are* constrained Delaunay with respect to the polygon f, in the following sense.

Proposition 4.11. *Let $\mathcal{T}$ be a CDT of a three-dimensional PLC $\mathcal{P}$. Let $f \in \mathcal{P}$ be a polygon. Let $\mathcal{T}|_f$ be the set of simplices in $\mathcal{T}$ that are included in f. Let $\mathcal{P}|_f$ be the set of faces of f (including f itself); $\mathcal{P}|_f$ is a two-dimensional PLC embedded in three-dimensional space. Then $\mathcal{T}|_f$ is a CDT of $\mathcal{P}|_f$.*

A PLC is *generic* if its vertices are generic. A generic PLC has a unique CDT, if it has one at all.

Proposition 4.12. *A generic piecewise linear complex has at most one constrained Delaunay triangulation.*

A consequence of Propositions 4.11 and 4.12 is that, if a PLC is generic, a CDT construction algorithm can begin by computing the two-dimensional CDTs of the polygons, then use them to help compute the three-dimensional CDT of the PLC, secure in the knowledge that the polygon triangulations will match the volume triangulation.

CDTs inherit the optimality properties of Delaunay triangulations described in Section 4.3, albeit with respect to a smaller set of triangulations, namely, the triangulations of a PLC. However, if a PLC has no CDT, finding the optimal triangulation is an open problem.

Proposition 4.13. *Let $\mathcal{P}$ be a PLC. If $\mathcal{P}$ has a CDT, then every CDT of $\mathcal{P}$ minimizes the largest min-containment ball, compared with all other triangulations of $\mathcal{P}$. Every CDT of $\mathcal{P}$ also optimizes the criteria discussed in Propositions 4.5 and 4.8.*

4.6 Notes and exercises

The upper bound of $\Theta(n^{\lceil d/2 \rceil})$ simplices in an n-vertex triangulation follows from McMullen's celebrated Upper Bound Theorem [145] of 1970. Seidel [191] gives a one-paragraph proof of the asymptotic bound.

Rajan [173] shows that the Delaunay triangulation minimizes the largest min-containment ball in any dimensionality, thereby generalizing the two-dimensional result of D'Azevedo and Simpson [67] and yielding Proposition 4.7. For an algebraic proof of Proposition 4.6 based on quadratic program duality, see Lemma 3 of Rajan [173]. Rippa [176] shows that the Delaunay triangulation in the plane minimizes the piecewise linear interpolation error for bivariate functions of the form $Ax^2 + Ay^2 + Bx + Cy + D$, measured in the L_q-norm for every $q \geq 1$, and Melissaratos [146] generalizes Rippa's result to higher dimensions, yielding Proposition 4.5. Shewchuk [204] extends all these optimality results to CDTs. The error bound for piecewise linear interpolation given in Section 4.3 is by Waldron [221].

Lawson [132] proves the claim from Section 4.4 that there are only two triangulations of the configuration of vertices involved in a basic bistellar flip. An earlier paper by Lawson [130] shows that for every planar point set, the flip graph is connected. Santos [184, 185] gives examples of point sets in five or more dimensions whose flip graphs are not connected. Joe [118] gives an example of a tetrahedralization for which the flip algorithm is stuck and can make no progress toward the Delaunay tetrahedralization. Edelsbrunner and Shah [91] give an example of a triangulation in the plane and a set of weights for which the flip algorithm is stuck and can make no progress toward the weighted Delaunay triangulation. The fact that the flip algorithm does not get stuck after a single vertex is introduced into a Delaunay triangulation by subdivision is proved by Joe [119] and by Edelsbrunner and Shah [91] for weighted Delaunay triangulations.

Schönhardt's polyhedron was discovered by Schönhardt [187], and Chazelle's polyhedron by Chazelle [44]. The NP-hardness of determining whether a polyhedron has a triangulation, cited in Section 4.5, is proved by Ruppert and Seidel [181]. Chazelle [44] proposes the vertical decomposition of a polyhedron, and Chazelle and Palios [45] give an algorithm that subdivides any n-vertex polyhedron with r reflex edges into $O(n + r^2)$ tetrahedra. This bound is optimal for the worst polyhedra.

The notion of a PLC was proposed by Miller, Talmor, Teng, Walkington, and Wang [149].[1]

Algorithms for computing a Steiner Delaunay triangulation of a PLC include those by Murphy, Mount, and Gable [156]; Cohen-Steiner, Colin de Verdière, and Yvinec [65]; Cheng and Poon [56]; Cheng, Dey, Ramos, and Ray [52]; and Rand and Walkington [174]. None has a polynomial bound on the number of new vertices.

CDTs were generalized to three or more dimensions by Shewchuk [204], whose paper

[1] Miller et al. call it a *piecewise linear system*, but their construction is so obviously a complex that a change in name seems obligatory. Our definition is different from that of Miller et al., but nearly equivalent, with one true difference: Miller et al. do not impose the restriction that the vertices and edges form a simplicial complex; they permit vertices to lie in the relative interior of an edge. Disallowing such vertices simplifies our presentation while entailing no essential loss of generality, because edges with vertices in their relative interiors can be subdivided into edges that obey the restriction.

includes proofs of the CDT Theorem and the properties of three-dimensional CDTs given in Section 4.5.4.

Exercises

1. Definition 4.3 of *piecewise linear complex* implies that if the interior of a segment intersects the interior of a polygon, the segment is entirely included in the polygon. Prove it.

2. Show that the edges and triangular faces of a strongly Delaunay tetrahedron are strongly Delaunay.

3. Prove Proposition 4.2. Consult Figure 2.10 for inspiration.

4. Prove Proposition 4.11.

5. Exercise 7 in Chapter 2 asks for a proof of a fact about constrained Delaunay triangles in the plane. Give a counterexample that demonstrates that the analogous fact is *not* true of constrained Delaunay tetrahedra in three dimensions.

6. Design an algorithm that adds vertices to a three-dimensional PLC so that the augmented PLC has a CDT.

Chapter 5

Algorithms for constructing Delaunay triangulations in $\mathbb{R}^3$

The most popular algorithms for constructing Delaunay triangulations in $\mathbb{R}^3$ are incremental insertion and gift-wrapping algorithms, both of which generalize to three or more dimensions with little difficulty. This chapter reprises those algorithms, with attention to the aspects that are different in three dimensions. In particular, the analysis of the running time of point location with a conflict graph is more complicated in three dimensions than in the plane. We use this gap as an opportunity to introduce a more sophisticated vertex ordering and its analysis. Instead of fully randomizing the order in which vertices are inserted, we recommend using a *biased randomized insertion order* that employs just enough randomness to ensure that the expected running time is the worst-case optimal $O(n^2)$—or better yet, $O(n \log n)$ time for the classes of point sets most commonly triangulated in practice—while maintaining enough spatial locality that implementations of the algorithm use the memory hierarchy more efficiently. This vertex ordering, combined with a simpler point location method, yields the fastest three-dimensional Delaunay triangulators in practice.

CDTs have received much less study in three dimensions than in two. There are two classes of algorithm available: gift-wrapping and incremental polygon insertion. Gift-wrapping is easier to implement; it is not much different in three dimensions than in two. It runs in $O(nh)$ time for Delaunay triangulations and $O(nmh)$ time for CDTs, where n is the number of vertices, m is the total complexity of the PLC's polygons, and h is the number of tetrahedra produced.

Perhaps the fastest three-dimensional CDT construction algorithm in practice is similar to the one we advocate in two dimensions. First, construct a Delaunay triangulation of the PLC's vertices, then insert its polygons one by one with a flip algorithm described in Section 5.8. This algorithm constructs a CDT in $O(n^2 \log n)$ time, though there are reasons to believe it will run in $O(n \log n)$ time on most PLCs in practice. Be forewarned, however, that this algorithm only works on edge-protected PLCs. This is rarely a fatal restriction, because a mesh generation algorithm that uses CDTs should probably insert vertices on the PLC's edges to make it edge-protected and ensure that it has a CDT.

Procedure	Purpose
AddTetrahedron(u, v, w, x)	Add a positively oriented tetrahedron $uvwx$
DeleteTetrahedron(u, v, w, x)	Delete a positively oriented tetrahedron $uvwx$
Adjacent(u, v, w)	Return a vertex x such that $uvwx$ is a positively oriented tetrahedron
Adjacent2Vertex(u)	Return vertices v, w, x such that $uvwx$ is a positively oriented tetrahedron

Figure 5.1: An interface for a three-dimensional triangulation data structure.

5.1 A dictionary data structure for tetrahedralizations

Figure 5.1 summarizes an interface for storing a tetrahedral complex, analogous to the interface for planar triangulations in Section 3.2. Two procedures, AddTetrahedron and DeleteTetrahedron, specify a tetrahedron to be added or deleted by listing its vertices with a positive orientation, as described in Section 3.1. The procedure Adjacent recovers the tetrahedron adjoining a specified oriented triangular face, or returns $\emptyset$ if there is no such tetrahedron. The vertices of a tetrahedron may include the ghost vertex. The data structure enforces the invariant that only two tetrahedra may adjoin a triangular face, and only one on each side of the face.

The simplest fast implementation echoes the implementation described in Section 3.2. Store each tetrahedron $uvwx$ four times in a hash table, keyed on the oriented faces uvw, uxv, uwx, and vxw. Then the first three procedures run in expected $O(1)$ time. To support Adjacent2Vertex queries, an array stores, for each vertex u, a triangle uvw such that the most recently added tetrahedron adjoining u has uvw for a face. As Section 3.2 discusses, these queries take expected $O(1)$ time in most circumstances, but not when the most recently added tetrahedron adjoining u has subsequently been deleted.

The interface and data structure extend easily to permit the storage of triangles or edges that are not part of any tetrahedron, but it does not support fast adjacency queries on edges.

5.2 Delaunay vertex insertion in $\mathbb{R}^3$

The Bowyer–Watson algorithm extends in a straightforward way to three (or more) dimensions. Recall that the algorithm inserts a vertex u into a Delaunay triangulation in four steps. First, find one tetrahedron whose open circumball contains u (point location). Second, a depth-first search in the triangulation finds all the other tetrahedra whose open circumballs contain u, in time proportional to their number. Third, delete these tetrahedra, as illustrated in Figure 5.2. The union of the deleted tetrahedra is a star-shaped polyhedral cavity. Fourth, for each triangular face of the cavity, create a new tetrahedron joining it with u, as illustrated.

To support inserting vertices that lie outside the triangulation, each triangular face on the boundary of the triangulation adjoins a *ghost tetrahedron* analogous to the ghost triangles of Section 3.4, having three solid vertices and a ghost vertex g. A tetrahedron that is not a ghost is called *solid*. Let vwx be a boundary triangle, oriented so its back adjoins

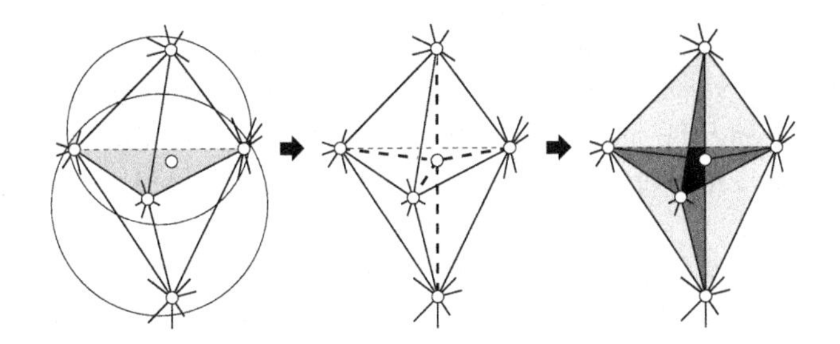

Figure 5.2: The Bowyer–Watson algorithm in three dimensions. A new vertex falls in the open circumballs of the two tetrahedra illustrated at left. These tetrahedra may be surrounded by other tetrahedra, which for clarity are not shown. The two tetrahedra and the face they share (shaded) are deleted. At center, the five new Delaunay edges. At right, the nine new Delaunay triangles—one for each edge of the cavity. Six new tetrahedra are created—one for each facet of the cavity.

a positively oriented solid tetrahedron $xwvy$. The incremental insertion algorithm stores a positively oriented ghost tetrahedron $vwxg$ in the triangulation data structure.

When a new vertex is inserted, there are two cases in which $vwxg$ must be deleted, i.e. vwx is no longer a boundary triangle: if a vertex is inserted in the open halfspace in front of vwx, or if a newly inserted vertex lies in the open circumdisk of vwx (i.e. it is coplanar with vwx and in its open diametric ball). Call the union of these two regions the *outer halfspace* of vwx. It is the set of points in the open circumball of $vwxg$ in the limit as g moves away from the triangulation.

The following pseudocode details the Bowyer–Watson algorithm in three dimensions, omitting point location. The parameters to InsertVertex3D are a vertex u to insert and a positively oriented tetrahedron $vwxy$ whose open circumball contains u. In this pseudocode, all the triangles and tetrahedra are oriented, so the vertex order matters.

InsertVertex3D($u, vwxy$)

1. Call DeleteTetrahedron(v, w, x, y).
2. Call DigCavity3D(u, xwv), DigCavity3D(u, yvw), DigCavity3D(u, vyx), and DigCavity3D(u, wxy) to identify the other deleted tetrahedra and insert new tetrahedra.

DigCavity3D(u, wxy)

1. Let z = Adjacent(w, x, y); $wxyz$ is the tetrahedron on the other side of facet wxy from u.
2. If $z = \emptyset$, then return, because the tetrahedron has already been deleted.
3. If InSphere(u, w, x, y, z) $>$ 0, then $uwxy$ and $wxyz$ are not Delaunay, so call DeleteTetrahedron(w, x, y, z). Call DigCavity3D(u, wxz), DigCavity3D(u, xyz), and DigCavity3D(u, ywz) to recursively identify more deleted tetrahedra and insert new tetrahedra, and return.
4. Otherwise, wxy is a face of the polyhedral cavity, so call AddTetrahedron (u, w, x, y).

The correctness of InsertVertex3D depends on the use of a ghost vertex. In particular, Step 2 of DigCavity3D should not confuse the ghost vertex with $\emptyset$; the former marks the triangulation exterior and the latter marks the cavity. Unlike with the planar algorithm InsertVertex in Section 3.4, Step 2 is necessary for both unweighted and weighted Delaunay triangulations.

Step 3 requires a modification to the InSphere test discussed in Section 3.1: if $wxyz$ is a ghost tetrahedron, then replace the formula (3.5) with a test of whether u lies in the outer halfspace of $wxyz$. To adapt the code for a weighted Delaunay triangulation, replace (3.5) with Orient4D(u^+, w^+, x^+, y^+, z^+), which tests whether u^+ is below the witness plane of $wxyz$. The parameter $vwxy$ of InsertVertex3D must be a tetrahedron whose witness plane is above u^+.

How expensive is vertex insertion, leaving out the cost of point location? The insertion of a single vertex into an n-vertex Delaunay triangulation can delete $\Theta(n^2)$ tetrahedra if the triangulation is the one depicted in Figure 4.2. However, a single vertex insertion can only create $\Theta(n)$ tetrahedra: observe that the boundary of the cavity is a planar graph, so the cavity has fewer than $2n$ boundary triangles.

It follows that during a sequence of n vertex insertion operations, at most $\Theta(n^2)$ tetrahedra are created. A tetrahedron can only be deleted if it is first created, so at most $\Theta(n^2)$ tetrahedra are deleted, albeit possibly most of them in a single vertex insertion. For the worst points sets, randomizing the vertex insertion order does not improve these numbers.

A special case that occurs frequently in practice—by all accounts it seems to be the norm—is the circumstance where the Delaunay triangulation has complexity linear, rather than quadratic, in the number of vertices, and moreover the intermediate triangulations produced during incremental insertion have expected linear complexity. For point sets with this property, a random insertion order guarantees that each vertex insertion will create and delete an expected constant number of tetrahedra, just as it does in the plane, and we shall see that the randomized incremental insertion algorithm with a conflict graph runs in expected $O(n \log n)$ time. This running time is often observed in practice, even in higher dimensions. Be forewarned, however, that there are point sets for which the final triangulation has linear complexity but the intermediate triangulations have expected quadratic complexity, thereby slowing down the algorithm dramatically.

Even for worst-case point sets, randomization helps to support fast point location. Recall that, excluding the point location step, the Bowyer–Watson algorithm runs in time proportional to the number of tetrahedra it deletes and creates, so the running time of the three-dimensional incremental insertion algorithm, *excluding* point location, is $O(n^2)$. With a conflict graph and a random insertion order, point location is no more expensive than this, so the randomized incremental insertion algorithm achieves a worst-case optimal expected running time of $O(n^2)$.

5.3 Biased randomized insertion orders

The advantage of inserting vertices in random order is that it guarantees that the expected running time of point location is optimal, and that pathologically slow circumstances like those illustrated in Figure 3.8 are unlikely to happen. But there is a serious disadvantage: randomized vertex insertions tend to interact poorly with the memory hierarchy in modern computers, especially virtual memory. Ideally, data structures representing tetrahedra and vertices that are close together geometrically should be close together in memory—a property called *spatial locality*—for better cache and virtual memory performance.

Fortunately, the permutation of vertices does not need to be uniformly random for the running time to be asymptotically optimal. A *biased randomized insertion order* (BRIO) is a permutation of the vertices that has strong spatial locality but retains enough randomness to obtain an expected running time of $O(n^2)$. Experiments show that a BRIO greatly improves the efficiency of the memory hierarchy—especially virtual memory.

Experiments also show that incremental insertion achieves superior running times in practice when it uses a BRIO but replaces the conflict graph with a point location method that simply walks from the previously inserted vertex toward the next inserted vertex; see Section 5.5. Although walking point location does not offer a strong theoretical guarantee on running time as a conflict graph does, this incremental insertion algorithm is perhaps the most attractive in practice, as it combines excellent observed speed with a simple implementation.

Let n be the number of vertices to triangulate. A BRIO orders the vertices in a sequence of *rounds* numbered zero through $\lceil \log_2 n \rceil$. Each vertex is assigned to the final round, round $\lceil \log_2 n \rceil$, with probability $1/2$. The remaining vertices are assigned to the second-last round with probability $1/2$, and so on. Each vertex is assign to round zero with probability $(1/2)^{\lceil \log_2 n \rceil} \leq 1/n$. The incremental insertion algorithm begins by inserting the vertices in round zero, then round one, and so on to round $\lceil \log_2 n \rceil$.

Within any single round, the vertices can be arranged in any order without threatening the worst-case expected running time of the algorithm, as Section 5.4 proves. Hence, we order the vertices within each round to create as much spatial locality as possible. One way to do this is to insert the vertices in the order they are encountered on a space-filling curve such as a Hilbert curve or a z-order curve. Another way is to store the vertices in an octree or k-d tree, refined so each leaf node contains only a few vertices; then order the vertices by a traversal of the tree. (Octree traversal is one way to sort vertices along a Hilbert or z-order curve.)

The tendency of vertices that are geometrically close together to be close together in the ordering does not necessarily guarantee that the data structures associated with them will be close together in memory. Nevertheless, experiments show that several popular Delaunay triangulation programs run faster with a BRIO than with a vertex permutation chosen uniformly at random, especially when the programs run out of main memory and have to resort to virtual memory.

Whether one uses the traditional randomized incremental insertion algorithm or a BRIO, one faces the problem of bootstrapping the algorithm, as discussed in Section 3.7. The most practical approach is to choose four affinely independent vertices, construct their Delaunay triangulation (a single tetrahedron), create four adjoining ghost tetrahedra, construct a conflict graph, and insert the remaining vertices in a random order (a uniformly chosen permutation or a BRIO). Even if the four bootstrap vertices are not chosen randomly, it is possible to prove that the expected asymptotic running time of the algorithm is not compromised.

5.4 Optimal a conflict graph in $\mathbb{R}^3$

Section 3.6 describes how to use a conflict graph to perform point location. Conflict graphs generalize to higher dimensions, yielding a randomized incremental insertion algorithm that constructs three-dimensional Delaunay triangulations in expected $O(n^2)$ time, which is optimal in the worst case. Moreover, the algorithm runs in expected $O(n \log n)$ time in the special case where the Delaunay triangulation of a random subset of the input points has expected linear complexity.

Conflict graphs and the vertex redistribution algorithm extend to three or more dimensions straightforwardly, with no new ideas needed. A conflict is a vertex-tetrahedron pair consisting of an uninserted vertex and a tetrahedron whose open circumball contains it. For each uninserted vertex, the conflict graph records a tetrahedron that contains the vertex. If an uninserted vertex lies outside the growing triangulation, its conflict is an unbounded, convex ghost "tetrahedron," each having one solid triangular facet on the triangulation boundary and three unbounded ghost facets. The ghost edges and ghost facets diverge from a point in the interior of the triangulation (recall Figure 3.10). For each tetrahedron, the conflict graph records a list of the uninserted vertices that choose that tetrahedron as their conflict. When a vertex is inserted into the triangulation, the conflict graph is updated exactly as described in Section 3.6.

Let us analyze the running time. No backward analysis is known for a BRIO, so the analysis given here differs substantially from that of Section 3.6. This analysis requires us to assume that the point set is generic, although the algorithm is just as fast for point sets that are not.

Let S be a generic set of n points in $\mathbb{R}^3$, not all coplanar. Let σ be a tetrahedron whose vertices are in S. We call σ a *j-tetrahedron* if its open circumball contains j vertices in S.

Because S is generic, Del S is composed of all the 0-tetrahedra of S. However, the incremental insertion algorithm transiently constructs many other tetrahedra that do not survive to the end; these are 1-tetrahedra, 2-tetrahedra, and so forth. We wish to determine, for each j, how many j-tetrahedra exist and what the probability is that any one of them is constructed. From this we can determine the expected number of j-tetrahedra that appear.

The *triggers* of a j-tetrahedron are its four vertices, and the *stoppers* of a j-tetrahedron are the j vertices its open circumball contains. A tetrahedron is constructed if and only if all of its triggers precede all of its stoppers in the insertion order. The probability of that decreases rapidly as j increases.

Let f_j be the total number of j-tetrahedra, which depends on S, and let $F_j = \sum_{i=0}^{j} f_i$ be the total number of i-tetrahedra for all $i \leq j$. Del S contains at most $O(n^2)$ tetrahedra, so $F_0 = f_0 = O(n^2)$. Unfortunately, it is a difficult open problem to find a tight bound on the number f_j; we must settle for a tight bound on the number F_j. The following proposition is an interesting use of the probabilistic method. It exploits the fact that if we compute the Delaunay triangulation of a random subset of S, we know the probability that any given j-tetrahedron will appear in the triangulation.

Proposition 5.1. *For a generic set S of n points, $F_j = O(j^2 n^2)$. If the Delaunay triangulation of a random r-point subset of S has expected $O(r)$ complexity for every $r < n$, then $F_j = O(j^3 n)$.*

Proof. Let R be a random subset of S, where each point in S is chosen with probability $1/j$. (This is a magical choice, best understood by noting that for any j-tetrahedron, the expected number of its stoppers in R is one.) Let $r = |R|$. Observe that r is a random variable with a binomial distribution. Therefore, the expected complexity of Del R is $O(E[r^2]) = O(E[r]^2) = O(n^2/j^2)$.

Let σ be a k-tetrahedron for an arbitrary k. Because S is generic, Del R contains σ if and only if R contains all four of σ's triggers but none of its k stoppers. The probability of that is

$$p_k = \left(\frac{1}{j}\right)^4 \left(1 - \frac{1}{j}\right)^k.$$

If $k \leq j$, then

$$p_k \geq \left(\frac{1}{j}\right)^4 \left(1 - \frac{1}{j}\right)^j \geq \left(\frac{1}{j}\right)^4 \frac{1}{e},$$

where e is the base of the natural logarithm. This inequality follows from the identity $\lim_{j\to\infty}(1 + x/j)^j = e^x$.

The expected number of tetrahedra in Del R is

$$E[\text{size}(\text{Del } R)] = \sum_{k=0}^{n} p_k f_k \geq \sum_{k=0}^{j} p_k f_k \geq \sum_{k=0}^{j} \left(\frac{1}{j}\right)^4 \frac{f_k}{e} = \frac{F_j}{e j^4}.$$

Recall that this quantity is $O(n^2/j^2)$. Therefore,

$$F_j \leq e j^4 E[\text{size}(\text{Del } R)] = O(j^2 n^2).$$

If the expected complexity of the Delaunay triangulation of a random subset of S is linear, the expected complexity of Del R is $O(E[r]) = O(n/j)$, so $F_j = O(j^3 n)$. □

Proposition 5.1 places an upper bound on the number of j-tetrahedra with j small. Next, we wish to know the probability that any particular j-tetrahedron will be constructed during a run of the randomized incremental insertion algorithm.

Proposition 5.2. *If the permutation of vertices is chosen uniformly at random, then the probability that a specified j-tetrahedron is constructed during the randomized incremental insertion algorithm is less than $4!/j^4$.*

Proof. A j-tetrahedron appears only if its four triggers (vertices) appear in the permutation before its j stoppers. This is the probability that if four vertices are chosen randomly from the $j + 4$ triggers and stoppers, they will all be triggers; namely,

$$\frac{1}{\binom{j+4}{4}} = \frac{j!4!}{(j+4)!} < \frac{4!}{j^4}.$$

□

Proposition 5.2 helps to bound the expected running time of the standard randomized incremental insertion algorithm, but we need a stronger proposition to bound the running time with a biased randomized insertion order.

Proposition 5.3. *If the permutation of vertices is a BRIO, as described in Section 5.3, then the probability q_j that a specified j-tetrahedron will be constructed during the incremental insertion algorithm is less than*

$$\frac{16 \cdot 4! + 1}{j^4}.$$

PROOF. Let σ be a j-tetrahedron. Let i be the round in which its first stopper is inserted. The incremental insertion algorithm can create σ only if its last trigger is inserted in round i or earlier.

Consider a trigger that is inserted in round i or earlier for any $i \neq 0$. The probability of that trigger being inserted before round i is $1/2$. Therefore, the probability that all four triggers are inserted in round $i \neq 0$ or earlier is exactly 2^4 times greater than the probability that all four triggers are inserted before round i. Therefore, the probability q_j that the algorithm creates σ is bounded as follows.

$$\begin{aligned}
q_j &\leq \text{Prob[last trigger round} \leq \text{first stopper round]} \\
&\leq 2^4\,\text{Prob[last trigger round} < \text{first stopper round]} + \text{Prob[last trigger round} = 0] \\
&\leq 16\,\text{Prob[all triggers appear before all stoppers in} \\
&\qquad\qquad \text{a uniform random permutation]} + (1/2)^{4\lceil \log_2 n \rceil} \\
&< 16\frac{4!}{j^4} + \frac{1}{n^4} \\
&\leq \frac{16 \cdot 4! + 1}{j^4}.
\end{aligned}$$

The fourth line of this inequality follows from Proposition 5.2. □

Theorem 5.4. *The expected running time of the randomized incremental insertion algorithm on a generic point set S in three dimensions, whether with a uniformly random permutation or a BRIO, is $O(n^2)$. If the Delaunay triangulation of a random r-point subset of S has expected $O(r)$ complexity for every $r \leq n$, then the expected running time is $O(n \log n)$.*

PROOF. Section 3.6 shows that the total running time of the algorithm is proportional to the number of tetrahedra created plus the number of conflicts created. The expected number of tetrahedra created is $\sum_{j=0}^{n} q_j f_j$, with q_j defined as in Proposition 5.3. Because a j-tetrahedron participates in j conflicts, the expected number of conflicts created is $\sum_{j=0}^{n} j q_j f_j$. Hence, the expected running time is

$$O\left(\sum_{j=0}^{n} q_j f_j + \sum_{j=0}^{n} j q_j f_j\right) \leq O\left(n^2 + \sum_{j=1}^{n} j q_j f_j\right).$$

The $O(n^2)$ term accounts for the first term of the first summation, for which $j = 0$ and $q_j = 1$, because the algorithm constructs every 0-tetrahedron. The second summation is bounded as follows.

$$
\begin{aligned}
\sum_{j=1}^{n} jq_j f_j &< \sum_{j=1}^{n} \frac{16 \cdot 4! + 1}{j^3}(F_j - F_{j-1}) \\
&= (16 \cdot 4! + 1)\left(\sum_{j=1}^{n} \frac{F_j}{j^3} - \sum_{k=1}^{n} \frac{F_{k-1}}{k^3}\right) \\
&= (16 \cdot 4! + 1)\left(\sum_{j=1}^{n-1} \frac{F_j}{j^3} + \frac{F_n}{n^3} - \sum_{k=2}^{n} \frac{F_{k-1}}{k^3} - F_0\right) \\
&= (16 \cdot 4! + 1)\left(\sum_{j=1}^{n-1} \left(\frac{F_j}{j^3} - \frac{F_j}{(j+1)^3}\right) + \frac{F_n}{n^3} - F_0\right) \\
&< (16 \cdot 4! + 1)\left(\sum_{j=1}^{n-1} \frac{F_j}{j^3(j+1)^3}\left((j+1)^3 - j^3\right) + \frac{F_n}{n^3}\right) \\
&= O\left(\sum_{j=1}^{n-1} \frac{j^2 n^2}{j^6} j^2 + n\right) \\
&= O\left(n^2 \sum_{j=1}^{n-1} \frac{1}{j^2} + n\right) \\
&= O(n^2).
\end{aligned}
$$

The last line follows from Euler's identity $\sum_{j=1}^{\infty} 1/j^2 = \pi^2/6$. If the expected complexity of the Delaunay triangulation of a random subset of S is linear, $F_j = O(j^3 n)$, so the expected running time is $O(n \sum_{j=1}^{n-1}(1/j)) = O(n \log n)$. □

5.5 Point location by walking

In conjunction with a BRIO, a simple point location method called *walking* appears to outperform conflict graphs in practice, although there is no guarantee of a fast running time. A walking point location algorithm simply traces a straight line through the triangulation, visiting tetrahedra that intersect the line as illustrated in Figure 5.3, until it arrives at a tetrahedron that contains the new vertex. In conjunction with a vertex permutation chosen uniformly at random (rather than with a BRIO), walking point location visits many tetrahedra and is very slow. But walking is fast in practice if it follows two guidelines: the vertices should be inserted in an order that has much spatial locality, such as a BRIO, and each walk should begin at the most recently created solid tetrahedron. Then the typical walk visits a small constant number of tetrahedra.

To avoid a long walk between rounds of a BRIO, the vertex order (e.g. the tree traversal or the direction of the space-filling curve) should be reversed on even-numbered rounds, so each round begins near where the previous round ends.

Researchers have observed that the three-dimensional incremental insertion algorithm with a BRIO and walking point location appears to run in linear time, not counting the

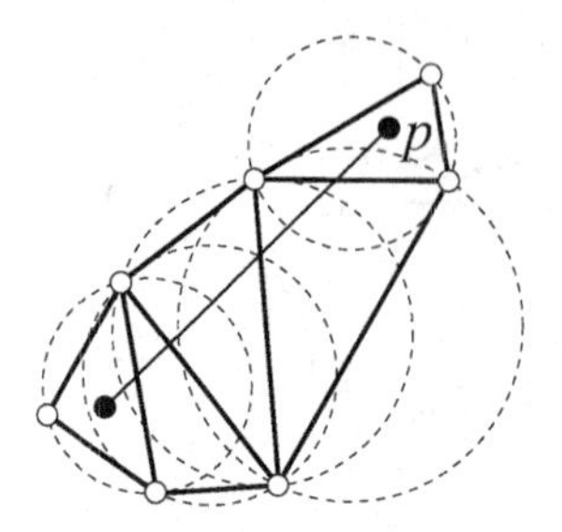

Figure 5.3: Walking to the triangle that contains p.

initial $O(n \log n)$-time computation of a BRIO. This observation holds whether they use a BRIO or a spatial ordering generated by traversing an octree with no randomness at all. Randomness is often unnecessary in practice—frequently, simply sorting the vertices along a space-filling curve will yield excellent speed—but because points sets such as the one illustrated in Figure 3.8 are common in practice, we recommend choosing a BRIO to prevent the possibility of a pathologically slow running time.

5.6 The gift-wrapping algorithm in $\mathbb{R}^3$

The simplest algorithm for retriangulating the cavity evacuated when a vertex is deleted from a three-dimensional Delaunay triangulation or CDT, or when a polygon is inserted or deleted in a CDT, is gift-wrapping. (See the bibliographic notes for more sophisticated vertex deletion algorithms, also based on gift-wrapping, that are asymptotically faster in theory.) The gift-wrapping algorithm described in Section 3.11 requires few new ideas to work in three (or more) dimensions. The algorithm constructs tetrahedra one at a time and maintains a dictionary of unfinished triangular facets. The pseudocode for FINISH and GIFTWRAPCDT can be adapted, with triangles replaced by tetrahedra, oriented edges replaced by oriented facets, and circumdisks replaced by circumballs.

The biggest change is that triangles, not segments, seed the algorithm. But the polygons in a PLC are not always triangles. Recall from Proposition 4.11 that a CDT of a PLC $\mathcal{P}$ induces a two-dimensional CDT of each polygon in $\mathcal{P}$. To seed the three-dimensional gift-wrapping algorithm, one can compute the two-dimensional CDT of a polygon (or every polygon), then enter each CDT triangle (twice, with both orientations) in the dictionary.

To gift-wrap a Delaunay triangulation, seed the algorithm with one strongly Delaunay triangle. One way to find one is to choose an arbitrary input point and its nearest neighbor. For the third vertex of the triangle, choose the input point that minimizes the radius of the circle through the three vertices. If the set of input points is generic, the triangle having these three vertices is strongly Delaunay.

If the input (point set or PLC) is not generic, gift-wrapping is in even greater danger in three dimensions than in the plane. Whereas the planar gift-wrapping algorithm can handle subsets of four or more cocircular points by identifying them and giving them special treatment, no such approach works reliably in three dimensions. Imagine a point set that includes six points lying on a common empty sphere. Suppose that gift-wrapping inadvertently tetrahedralizes the space around these points so they are the vertices of a hollow cavity shaped like Schönhardt's polyhedron (from Section 4.5). The algorithm will be unable to fill the cavity. By far the most practical solution is to symbolically perturb the points so that they are generic, as discussed in Section 2.9. The same perturbation should also be used to compute the two-dimensional CDTs of the PLC's polygons.

Another difficulty is that the input PLC might not have a CDT, in which case gift-wrapping will fail in one of two ways. One possibility is that the algorithm will fail to finish an unfinished facet, even though there is a vertex in front of that facet, because no vertex in front of that facet is visible from the facet's interior. This failure is easy to detect. The second possibility is that the algorithm will finish a facet by constructing a tetrahedron that is not constrained Delaunay, either because the tetrahedron's open circumball contains a visible vertex, or because the tetrahedron intersects the preexisting simplices wrongly (not in a complex). An attempt to gift-wrap Schönhardt's polyhedron brings about the last fate. The algorithm becomes substantially slower if it tries to detect these failures. Perhaps a better solution is to run the algorithm only on PLCs that are edge-protected or otherwise known to have CDTs.

A strange property of the CDT is that it is NP-hard to determine whether a three-dimensional PLC has a CDT, if the PLC is not generic. However, a polynomial-time algorithm is available for generic PLCs: run the gift-wrapping algorithm, and check whether it succeeded.

Gift-wrapping takes $O(nh)$ time for a Delaunay triangulation, or $O(nmh)$ time for a CDT, where n is the number of input points, m is the total complexity of the input polygons, and h is the number of tetrahedra in the CDT; h is usually linear in n, but could be quadratic in the worst case.

5.7 Inserting a vertex into a CDT in $\mathbb{R}^3$

Section 3.9 describes how to adapt the Bowyer–Watson vertex insertion algorithm to CDTs in the plane. The same adaptions work for three-dimensional CDTs, but there is a catch: even if a PLC $\mathcal{P}$ has a CDT, an augmented PLC $\mathcal{P} \cup \{v\}$ might not have one. This circumstance can be diagnosed after the depth-first search step of the Bowyer–Watson algorithm in one of two ways: by the fact that the cavity is not star-shaped, thus one of the newly created tetrahedra has nonpositive orientation, or by the fact that a segment or polygon runs through the interior of the cavity. An implementation can check explicitly for these circumstances and signal that the vertex v cannot be inserted.

5.8 Inserting a polygon into a CDT

To "insert a polygon into a CDT" is to take as input the CDT $\mathcal{T}$ of some PLC $\mathcal{P}$ and a new polygon f to insert, and produce the CDT of $\mathcal{P}^f = \mathcal{P} \cup \{f\}$. It is only meaningful if $\mathcal{P}^f$ is a valid PLC—which implies that f's boundary is a union of segments in $\mathcal{P}$, among other things. It is only possible if $\mathcal{P}^f$ has a CDT. If $\mathcal{P}$ is edge-protected, then $\mathcal{P}^f$ is edge-protected (polygons play no role in the definition of *edge-protected*), and both have CDTs. But if $\mathcal{P}$ is not edge-protected, it is possible that $\mathcal{P}$ has a CDT and $\mathcal{P}^f$ does not; see Exercise 5.

The obvious algorithm echoes the segment insertion algorithm in Section 3.10: delete the tetrahedra whose interiors intersect f. All the simplices not deleted are still constrained Delaunay. Then retriangulate the polyhedral cavities on either side of f with constrained Delaunay simplices, perhaps by gift-wrapping. (Recall Figure 2.15.) Note that if $\mathcal{P}^f$ has no CDT, the retriangulation step will fail.

This section describes an alternative algorithm that uses bistellar flips to achieve the same result. One can construct a three-dimensional CDT of an n-vertex, ridge-protected PLC $\mathcal{P}$ in $O(n^2 \log n)$ time by first constructing a Delaunay triangulation of the vertices in $\mathcal{P}$ in expected $O(n^2)$ time with the randomized incremental insertion algorithm, then inserting the polygons one by one with the flip-based algorithm. For most PLCs that arise in practice, this CDT construction algorithm is likely to run in $O(n \log n)$ time and much faster than gift-wrapping.

The algorithm exploits the fact that when the vertices of the triangulation are lifted by the parabolic lifting map, every locally Delaunay facet lifts to a triangle where the lifted triangulation is locally convex. Say that a facet g, shared by two tetrahedra σ and τ, is *locally weighted Delaunay* if the lifted tetrahedra σ^+ and τ^+ adjoin each other at a dihedral angle, measured from above, of less than 180°. In other words, the interior of τ^+ lies above the affine hull of σ^+, and vice versa.

The algorithm's main idea is to move some of the lifted vertices vertically, continuously, and linearly so they rise above the paraboloid, and use bistellar flips to dynamically maintain local convexity as they rise. Recall from Figure 4.7 that a tetrahedral bistellar flip is a transition between the upper and lower faces of a 4-simplex. If the vertices of that simplex are moving vertically at different (but constant) speeds, they may pass through an instantaneous state in which the five vertices of the 4-simplex are cohyperplanar, whereupon the lower and upper faces are exchanged. In the facet insertion algorithm, this circumstance occurs when two lifted tetrahedra σ^+ and τ^+ that share a triangular facet g^+ become cohyperplanar, whereupon the algorithm uses the procedure FLIP from Section 4.4 to perform a bistellar flip that deletes g.

The algorithm for inserting a polygon f into a triangulation $\mathcal{T}$ begins by identifying a region R: the union of the tetrahedra in $\mathcal{T}$ whose interiors intersect f and thus must be deleted. The polygon insertion algorithm only performs flips in the region R. Let h be the plane aff f. Call the vertices in $\mathcal{P}$ on one (arbitrary) side of h *left vertices*, and the vertices on the other side *right vertices*. Vertices on h are neither. The flip algorithm linearly increases the heights of the vertices according to their distance from h, and uses flips to maintain locally weighted Delaunay facets in R as the heights change. Figure 5.4 is a sequence of snapshots of the algorithm at work.

Assign each vertex $v \in \mathcal{P}$ a time-varying *height* of $v_z(\kappa) = \|v\|^2 + \kappa d(v, h)$, where κ is the time and $d(v, h)$ is the Euclidean distance of v from h. (This choice of $d(\cdot, \cdot)$ is pedagogically useful but numerically poor; a better choice for implementation is to let $d(v, h)$ be the distance of v from h along one coordinate axis, preferably the axis most nearly perpendicular to h. This distance is directly proportional to the Euclidean distance, but can be computed without radicals.)

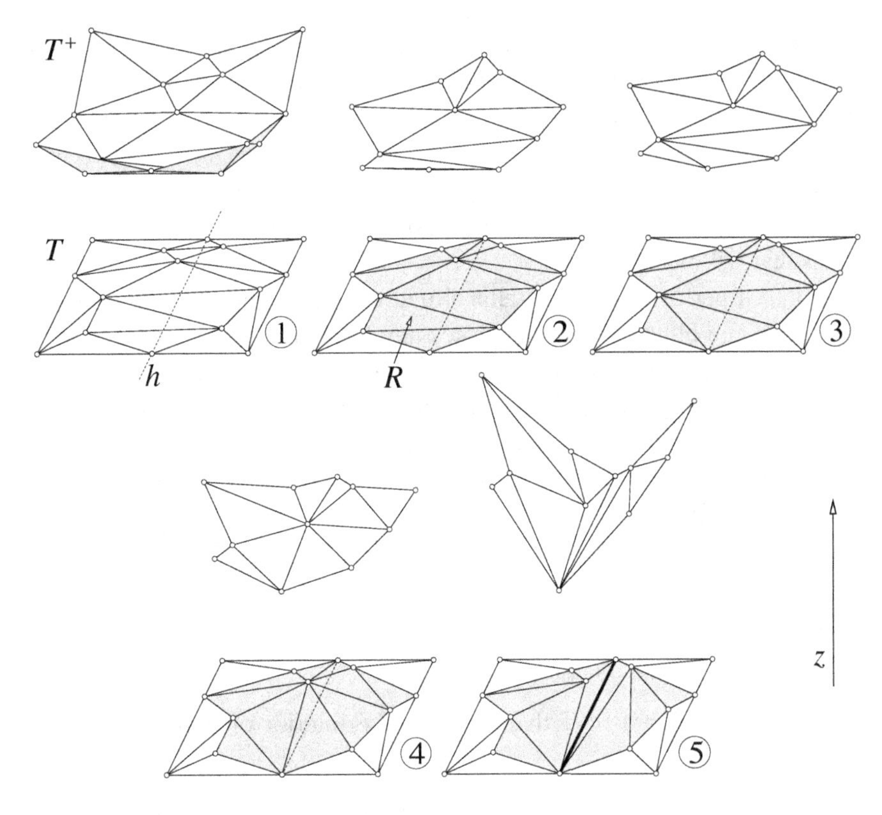

Figure 5.4: A two-dimensional example of inserting a segment into a CDT. The algorithm extends to any dimension.

When a set of vertices is transformed affinely, its convex hull undergoes no combinatorial change. Likewise, an affine transformation of the vertex heights in a lifted triangulation does not change which facets are locally weighted Delaunay. In the facet insertion algorithm, however, each half of space undergoes a different affine transformation, so the simplices that cross the plane h change as the time κ increases. Observe that an algorithm in which only the heights of the right vertices change (at twice the speed) is equivalent. For numerical reasons, it is better to raise only half the vertices.

Let $\mathcal{P}(\kappa)$ be a time-varying weighted PLC, which is identical to $\mathcal{P}$ except that each right vertex v is lifted to a height of $v_z(\kappa)$. As κ increases, the algorithm FLIPINSERTPOLYGON maintains a triangulation of R that is *weighted constrained Delaunay* with respect to $\mathcal{P}(\kappa)$, meaning that every triangular facet is locally weighted Delaunay except those included in a polygon.

Every simplex in the evolving triangulation that has no left vertex, or no right vertex, remains constrained Delaunay with respect to $\mathcal{P}(\kappa)$ as κ increases. The algorithm FLIPINSERTPOLYGON deletes only simplices that have both a left and a right vertex *and* pass through f. All simplices outside the region R, or strictly on the boundary of R, remain intact.

When κ is sufficiently large, the flip algorithm reaches a state where no simplex in the region R has both a left vertex and a right vertex, hence f is a union of faces of the triangulation. At this time, the triangulation is the CDT of $\mathcal{P}^f$, and the job is done.

Pseudocode for FLIPINSERTPOLYGON appears below. The loop (Step 4) dynamically maintains the triangulation $\mathcal{T}$ as κ increases from 0 to ∞ and the lifted companions of the right vertices move up. For certain values of κ, the following event occurs: some facet g in the region R is no longer locally weighted Delaunay after time κ, because the two lifted tetrahedra that include g^+ are cohyperplanar at time κ. Upon this event, an update operation replaces these and other simplices that will not be locally weighted Delaunay after time κ with simplices that will be.

To ensure that it performs each bistellar flip at the right time, the algorithm maintains a priority queue (e.g. a binary heap) that stores any flip that might occur. For each facet g that could be flipped at some time in the future, the procedure CERTIFY determines when g might be flipped and enqueues a flip event. The main loop of FLIPINSERTPOLYGON repeatedly removes the flip with the least time from the priority queue and performs a flip if the facet still exists, i.e. was not eliminated by other flips. When the queue is empty, $\mathcal{T}$ has transformed into the CDT of $\mathcal{P}^f$.

FLIPINSERTPOLYGON($\mathcal{T}, f$)

1. Find one tetrahedron in $\mathcal{T}$ that intersects the interior of f by a rotary search around an edge of f.
2. Find all tetrahedra in $\mathcal{T}$ that intersect the interior of f by a depth-first search.
3. Initialize Q to be an empty priority queue. For each facet g in $\mathcal{T}$ that intersects the interior of f and has a vertex on each side of f, call CERTIFY(g, Q).
4. While Q is not empty,
   (a) Remove (g', κ) with minimum κ from Q.
   (b) If g' is still a facet in $\mathcal{T}$:
      (i) Call FLIP($\mathcal{T}, g'$) to eliminate g'.
      (i) For each facet g that lies on the boundary of the cavity retriangulated by the flip and has a vertex on each side of f, call CERTIFY(g, Q).

CERTIFY(g, Q)

1. Let σ and τ be the tetrahedra that share the facet g.
2. If the interior of σ^+ will be below aff τ^+ at time ∞,
   (a) Compute the time κ at which σ^+ and τ^+ are cohyperplanar.
   (b) Insert (g, κ) into the priority queue Q.

Just as gift-wrapping can fail when $\mathcal{P}$ is not generic, FACETINSERTPOLYGON can fail when $\mathcal{P}$ is not generic or simultaneous events occur for other reasons. For an implementation to succeed in practice, Step 2 of CERTIFY should perturb the vertex weights as described in Section 2.9 and in Exercise 2 of Chapter 3. It is possible for several events that take place at the same time (or nearly the same time, if roundoff error occurs) to come off the priority queue in an unexecutable order—recall that there are several circumstances annotated in the FLIP pseudocode in which it might be impossible to perform a flip that eliminates a specified facet. In these circumstances, Step 4(a) of FLIPINSERTPOLYGON should dequeue an event with an admissible flip and hold back events with inadmissible flips until they become admissible.

The correctness proof for FLIPINSERTPOLYGON is omitted, but it relies on the Constrained Delaunay Lemma (Lemma 4.10). All the tetrahedra outside the region R remain constrained Delaunay, so their facets are constrained Delaunay, except those included in a polygon. When the algorithm is done, the original vertex heights are restored (returning them to the paraboloid). This is an affine transformation of the lifted right vertices, so the facets created by bistellar flips remain locally Delaunay, except the facets included in f. Therefore, every facet not included in a polygon is locally Delaunay. By the Constrained Delaunay Lemma, the final triangulation is a CDT of $\mathcal{P}^f$.

For an analysis of the running time, let m be the number of vertices in the region R, and let n be the number of vertices in $\mathcal{T}$.

Proposition 5.5. *Steps 3 and 4 of* FLIPINSERTPOLYGON *run in* $O(m^2 \log m)$ *time.*

PROOF. Every bistellar flip either deletes or creates an edge in the region R. Because the vertex weights vary linearly with time, an edge that loses the weighted constrained Delaunay property will never regain it; so once an edge is deleted, it is never created again. Therefore, FLIPINSERTPOLYGON deletes fewer than $m(m-1)/2$ edges and creates fewer than $m(m-1)/2$ edges over its lifetime, and thus performs fewer than m^2 flips. Each flip enqueues at most a constant number of events. Each event costs $O(\log m)$ time to enqueue and dequeue, yielding a total cost of $O(m^2 \log m)$ time. □

Miraculously, the worst-case running time for any sequence of FLIPINSERTPOLYGON operations is only a constant factor larger than the worst-case time for one operation.

Proposition 5.6. *The running time of any sequence of valid* FLIPINSERTPOLYGON *operations applied consecutively to a triangulation is* $O(n^2 \log n)$*, if Step 1 of* FLIPINSERTPOLYGON *is implemented efficiently enough.*

PROOF. A sequence of calls to FLIPINSERTPOLYGON, like a single call, has the property that every simplex deleted is never created again. (Every deleted simplex crosses a polygon and cannot return after the polygon is inserted.) Therefore, a sequence of calls deletes fewer than $n(n-1)/2$ edges, creates fewer than $n(n-1)/2$ edges, and performs fewer than n^2 flips. Each flip enqueues a constant number of events. Each event costs $O(\log n)$ time to enqueue and dequeue, summing to $O(n^2 \log n)$ time for all events.

The cost of Step 2 of FLIPINSERTPOLYGON (the depth-first search) is proportional to the number of tetrahedra that intersect the relative interior of the polygon f. Either these tetrahedra are deleted when f is inserted, or they intersect f's relative interior without crossing f. A tetrahedron can intersect the relative interiors of at most ten PLC polygons that it does not cross, so each tetrahedron is visited in at most eleven depth-first searches. At most $O(n^2)$ tetrahedra are deleted and created during a sequence of calls to FLIPINSERTPOLYGON. Therefore, the total cost of Step 2 over all calls to FLIPINSERTPOLYGON is $O(n^2)$.

The cost of Step 1 of FLIPINSERTPOLYGON (identifying a tetrahedron that intersects the relative interior of f) is $O(n)$. This is a pessimistic running time; the worst case is achieved only if the edge used for the rotary search is an edge of $\Theta(n)$ facets. If $\Omega(n \log n)$ polygons are inserted into a single triangulation (which is possible but unlikely in practice), we must reduce the cost of Step 1 below $O(n)$. This can be done by giving each segment a balanced search tree listing the adjoining facets in the triangulation, in rotary order around the segment. Then, Step 1 executes in $O(\log n)$ time. The balanced trees are updated in $O(\log n)$ time per facet created or deleted. □

Recall from Section 4.4 that a bistellar flip replaces the top faces of a 4-simplex with its bottom faces. Because no face reappears after it is deleted, the sequence of flips performed during incremental polygon insertion is structurally similar to a four-dimensional triangulation. It has often been observed that most practical point sets have linear-size Delaunay triangulations in three or higher dimensions, so it seems like a reasonable inference that for most practical PLCs, the sequence of flips should have linear length. For those PLCs, incremental CDT construction with FLIPINSERTPOLYGON runs in $\Theta(n \log n)$ time.

5.9 Notes and exercises

An incremental insertion algorithm that works in any dimension was discovered independently by Bowyer [33], Hermeline [111, 112], and Watson [222]. Bowyer and Watson submitted their articles to *Computer Journal* and found them published side by side in 1981.

Although there is rarely a reason to choose them over the Bowyer–Watson algorithm, there is a literature on vertex insertion algorithms that use bistellar flips. Joe [118] generalizes Lawson's flip-based vertex insertion algorithm [131] to three dimensions, Rajan [173] generalizes it to higher dimensions, Joe [119] improves the speed of Rajan's generalization, and Edelsbrunner and Shah [91] generalize Joe's latter algorithm to weighted Delaunay triangulations.

Clarkson and Shor [64] introduce conflict graphs and show that randomized incremental insertion with a conflict graph runs in expected $O(n^{\lceil d/2 \rceil})$ time for Delaunay triangulations in $d \geq 3$ dimensions. Amenta, Choi, and Rote [7] propose the idea of a biased randomized insertion order and give the analysis of the randomized incremental insertion algorithm in Section 5.4. The bound on the number of j-tetrahedra in Proposition 5.1 was itself a major breakthrough of Clarkson and Shor [64]. The simpler proof given here is due to Mulmuley [154]. See Seidel [194] for a backward analysis of the algorithm with a vertex permutation chosen uniformly at random, in the style of Section 3.6.

Guibas and Stolfi [106] give an algorithm for walking point location in a planar Delaunay triangulation, and Devillers, Pion, and Teillaud [72] compare walking point location algorithms in two and three dimensions. The discussion in Section 5.5 of combining a BRIO with walking point location relies on experiments reported by Amenta et al. [7].

The first gift-wrapping algorithm for constructing Delaunay triangulations in three or more dimensions appeared in 1979 when Brown [34] published his lifting map and observed that the 1970 gift-wrapping algorithm of Chand and Kapur [41] for computing general-dimensional convex hulls can construct Voronoi diagrams. The first gift-wrapping algorithm for constructing three-dimensional CDTs appears in a 1982 paper by Nguyen [159], but as with the two-dimensional CDT paper of Frederick, Wong, and Edge [97], the author does not appear to have been aware of Delaunay triangulations at all. There is a variant of the gift-wrapping algorithm for CDTs that, by constructing the tetrahedra in a disciplined order and using other tricks to avoid visibility computations [200], runs in $O(nt)$ worst-case time, where n is the number of vertices and t is the number of tetrahedra produced.

Devillers [71] and Shewchuk [200] give gift-wrapping algorithms for deleting a degree-k vertex from a Delaunay or constrained Delaunay triangulation in $O(t \log k)$ time, where t is the number of tetrahedra created by the vertex deletion operation and can be as small as $\Theta(k)$ or as large as $\Theta(k^2)$. In practice, most vertices have a small degree, and naive gift-wrapping is usually fast enough.

The polygon insertion algorithm of Section 5.8 is due to Shewchuk [203]. Grislain and Shewchuk [105] show that it is NP-hard to determine whether a non-generic PLC has a CDT.

Exercises

1. You are using the incremental Delaunay triangulation algorithm to triangulate a cubical $\sqrt[3]{n} \times \sqrt[3]{n} \times \sqrt[3]{n}$ grid of n vertices. The vertices are not inserted in random order; instead, an adversary chooses the order to make the algorithm as slow as possible. As an asymptotic function of n, what is the largest *total* number of changes that might be made to the mesh (i.e. tetrahedron creations and deletions, summed over all vertex insertions), and what insertion order produces that asymptotic worst case? Ignore the time spent doing point location.

2. Let S and T be two sets of points in $\mathbb{R}^3$, having s points and t points, respectively. Suppose $S \cup T$ is generic. Suppose we are given Del S, and our task is to incrementally insert the points in T and thus construct Del $(S \cup T)$.

   We use the following algorithm. First, for each point p in T, find a tetrahedron or ghost tetrahedron in Del S that contains p by brute force (checking each point against each tetrahedron), and thereby build a conflict graph in $O(ts^2)$ time. Second, incrementally insert the points in T in random order, with each permutation being equally likely.

   If we were constructing Del $(S \cup T)$ from scratch, with the insertion order wholly randomized, it would take at worst expected $O((s + t)^2) = O(s^2 + t^2)$ time. However, because S is *not* a random subset of $S \cup T$, the expected running time for this algorithm can be worse. An adversary could choose S so that inserting the points in T is slow.

   Prove that the expected time to incrementally insert the points in T is $O(t^2 + ts^2)$.

3. Extend the analysis of the randomized incremental insertion algorithm to point sets that are not generic. To accomplish that, we piggyback on the proof for generic point sets. Let S be a finite set of points in $\mathbb{R}^3$, not necessarily generic, and let $S[\omega]$ be

the same points with their weights perturbed as described in Section 2.9. By Theorem 5.4, the randomized incremental insertion algorithm constructs Del $S[\omega]$ in expected $O(n^2)$ time, and in expected $O(n \log n)$ time in the optimistic case where the expected complexity of the Delaunay triangulation of a random subset of the points is linear.

When the randomized incremental insertion algorithm constructs Del S, the tetrahedra and conflicts created and deleted are not necessarily the same as when it constructs Del $S[\omega]$, but we can argue that the numbers must be comparable.

(a) Show that immediately after a vertex v is inserted during the construction of Del S, all the edges that adjoin v are strongly Delaunay. Show that immediately after v is inserted during the construction of Del $S[\omega]$, the same edges adjoin v, and perhaps others do too.

(b) The boundary of the retriangulated cavity is planar. Use this fact to show that the number of tetrahedra created when v is inserted during the construction of Del S cannot exceed the number of tetrahedra created when v is inserted during the construction of Del $S[\omega]$.

(c) Recall that each polyhedral cell of the Delaunay subdivision of S has its vertices on a common sphere, and that the cell is subdivided into tetrahedra in Del $S[\omega]$. Show that the number of conflicts of a cell (vertices in its open circumball) in the Delaunay subdivision of S cannot exceed the number of conflicts of any of the corresponding tetrahedra in Del $S[\omega]$. Show that the number of conflicts created when v is inserted during the construction of Del S cannot exceed the number of conflicts created when v is inserted during the construction of Del $S[\omega]$.

4. Recall from Figure 3.14 that gift-wrapping can fail to construct the Delaunay triangulation of a point set that is not generic. One way to make the algorithm robust is to use symbolic weight perturbations. A different way is to identify groups of points that are cospherical during the gift-wrapping step and triangulate them all at once.

   Design a tetrahedral gift-wrapping algorithm that implements the second suggestion, without help from any kind of perturbations. Recall from Figure 4.4 that polyhedra of the Delaunay subdivision cannot be subdivided into tetrahedra independently of each other. How does your solution ensure that these subdivisions will be consistent with each other?

5. Give an example of a PLC $\mathcal{P}$ that has a CDT and a polygon f such that $\mathcal{P} \cup \{f\}$ is a valid PLC but does not have a CDT.

Chapter 6

Delaunay refinement in the plane

Delaunay refinement algorithms generate high-quality meshes by inserting vertices into a Delaunay or constrained Delaunay triangulation. The vertices are placed to ensure domain conformity and to eliminate elements that are badly shaped or too large. Delaunay refinement has many virtues: the optimality of Delaunay triangulations with respect to criteria related to interpolation and the smallest angles; our rich inheritance of theory and algorithms for Delaunay triangulations; the fast, local nature of vertex insertion; and most importantly, the guidance that the Delaunay triangulation provides in finding locations to place new vertices that are far from the other vertices, so that short edges do not appear.

This chapter describes an enormously influential Delaunay refinement algorithm for triangular mesh generation invented by Jim Ruppert—arguably the first provably good mesh generation algorithm to be truly satisfying in practice. Ruppert's algorithm was inspired by Chew's algorithm, discussed in Section 1.2, but it has the advantage that it constructs graded meshes and it offers mathematical guarantees on triangle grading and size optimality. Figure 6.1 shows a mesh produced by Ruppert's algorithm.

We use Ruppert's algorithm to introduce ideas that recur throughout the book. We begin by introducing a generic template that summarizes most Delaunay refinement algorithms. Then we describe Ruppert's algorithm and show that, under the right conditions, it is mathematically guaranteed to produce high-quality triangular meshes. Ruppert's work established the theoretical strength of the generic algorithm, which we extend to tetrahedral meshing in subsequent chapters.

One guarantee is that the triangles in these meshes have high quality. Another guarantee is that the edges are not unduly short: there is a lower bound on the edge lengths proportional to a space-varying function called the *local feature size*, which roughly quantifies the longest edge lengths possible in any high-quality mesh of a specified piecewise linear complex. A highlight of the theory of mesh generation is that one can estimate both the maximum edge lengths and the minimum number of triangles in a high-quality mesh. Ruppert's algorithm comes within a constant factor of optimizing both these quantities, and so we say that the meshes it produces are *size-optimal*, a notion we define formally in Section 6.5.

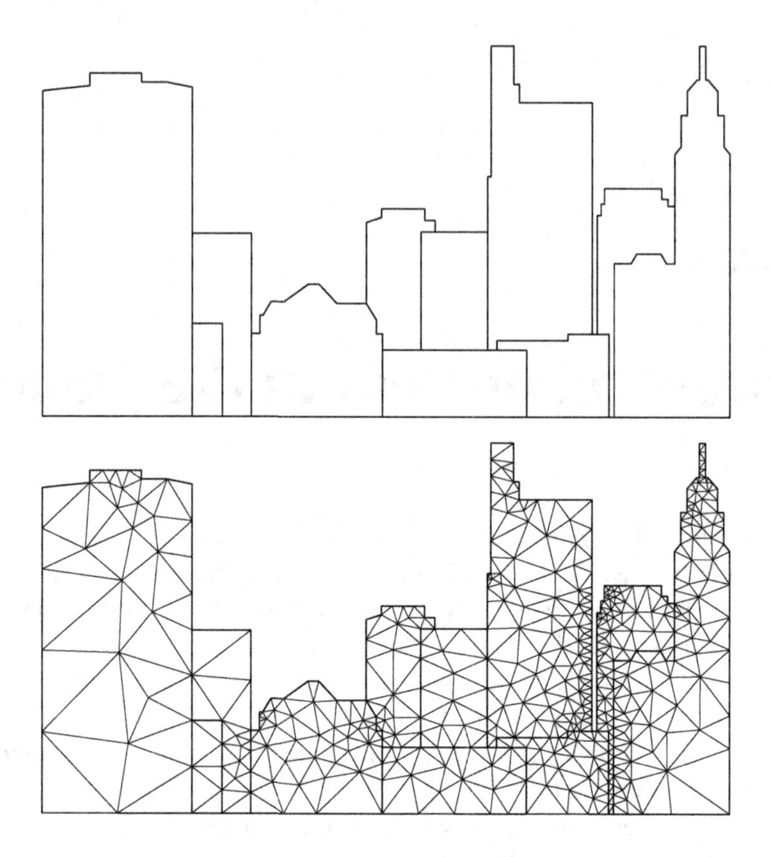

Figure 6.1: Meshing the Columbus skyline: a PLC and a Delaunay mesh thereof with no angle less than 28°, produced by Ruppert's Delaunay refinement algorithm.

6.1 A generic Delaunay refinement algorithm

The following generic Delaunay refinement method meshes a complex $\mathcal{P}$.

DelaunayRefinement($\mathcal{P}$)

1. Choose a vertex set $S \subset |\mathcal{P}|$.
2. Compute Del S.
3. If Del S fails to satisfy a property guaranteeing its geometric or topological conformity to $\mathcal{P}$, choose a point $c \in |\mathcal{P}|$ at or near the violation, insert c into S, update Del S, and repeat Step 3.
4. If an element $\tau \in$ Del S is poorly shaped or too large, choose a point $c \in |\mathcal{P}|$ in or near τ's circumball, insert c into S, update Del S, and go to Step 3.
5. Return the mesh $\{\sigma \in \text{Del } S : \sigma \subseteq |\mathcal{P}|\}$.

The first step depends on the type of complex. For a piecewise linear domain, S is the set of vertices in $\mathcal{P}$. For a smooth or piecewise smooth domain, points are chosen carefully from

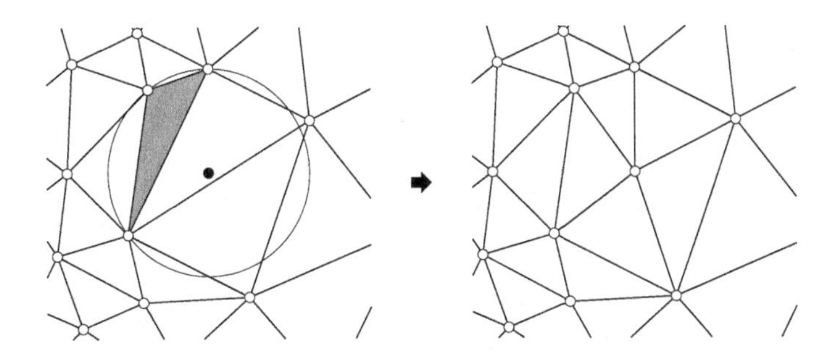

Figure 6.2: Inserting the circumcenter of a triangle whose radius-edge ratio exceeds 1. Every new edge adjoins the circumcenter and is at least as long as the circumradius of the skinny triangle, which is at least as long as the shortest edge of the skinny triangle.

the surfaces to form an initial vertex set. Step 2 constructs an initial Delaunay triangulation. Some algorithms maintain a constrained Delaunay triangulation instead; see Section 6.7. Steps 3 and 4 refine the triangulation by calling a vertex insertion algorithm such as the Bowyer–Watson algorithm. The properties enforced by Step 3 can encompass conforming to the domain boundary, capturing the domain topology, and approximating curved domain geometry with sufficient accuracy. Step 5 obtains a mesh of $\mathcal{P}$ by deleting the extraneous simplices that are not included in $|\mathcal{P}|$.

DelaunayRefinement terminates only when the mesh has the desired geometric properties and element shapes and sizes. The algorithm designer's burden is to choose the properties to enforce, choose vertices to insert, and prove that the mesh will eventually satisfy the properties, so the algorithm will terminate. Counterintuitive as it may seem, the proof usually proceeds by first showing that the algorithm must terminate, and hence the properties must be satisfied and the elements must have high quality.

The main insight behind all Delaunay refinement algorithms is that they constrain how close together two vertices can be, and thus constrain how short an edge can be. Consider inserting a new vertex at the circumcenter of an element whose radius-edge ratio is one or greater, as illustrated in Figure 6.2. Because a Delaunay simplex has an empty circumball, the distance from the new vertex to any other vertex is at least as great as the circumball's radius, which is at least as great as the simplex's shortest edge. Therefore, *the new vertex cannot participate in an edge shorter than the shortest edge already existing*. This explains why circumcenters are excellent places to put new vertices, why the radius-edge ratio is the quality measure naturally optimized by Delaunay refinement algorithms, and why these algorithms must terminate: they eventually run out of places to put new vertices. The following lemma formalizes the last idea.

Lemma 6.1 (Packing Lemma). *Let $D \subset \mathbb{R}^d$ be a bounded domain. Let $S \subset D$ be a point set and $\lambda > 0$ a scalar constant such that for every two distinct points u and v in S, $d(u, v) \geq \lambda$. Then there is a constant ξ depending solely on D and λ such that $|S| \leq \xi$.*

Proof. Consider the point set $D_{\lambda/2} = \{x \in \mathbb{R}^d : d(x, D) \leq \lambda/2\}$, known as the *Minkowski sum* of D with a ball of radius $\lambda/2$. $D_{\lambda/2}$ is bounded, as D is bounded. Every Euclidean d-ball having a center in S and radius $\lambda/2$ is included in $D_{\lambda/2}$. Because every pair of points in S is separated by a distance of at least λ, these balls have disjoint interiors, and we call this set of d-balls a *packing*. Therefore, $|S| \leq \text{volume}(D_{\lambda/2})/((\lambda/2)^d V_d)$, where V_d is the volume of a unit d-ball. □

The Delaunay refinement algorithms we study in this book place a lower bound λ on inter-vertex distances that is proportional to the aforementioned *local feature size* function, which locally estimates the longest edge lengths possible in a high-quality mesh. For a polyhedral domain, the local feature size is related to the distances between disjoint linear cells. For a domain whose boundary is a smooth surface, it is related to the distance from that boundary to its medial axis. We give formal definitions for the local feature size in the sections where they are first used.

6.2 Ruppert's Delaunay refinement algorithm

The input to Ruppert's algorithm is a piecewise linear complex $\mathcal{P}$ (recall Definition 2.8) and a positive constant $\bar{\rho}$ that specifies the maximum permitted radius-edge ratio for triangles in the output mesh. Recall from Section 1.7 that this equates to a minimum permitted angle of $\theta_{\min} = \arcsin \frac{1}{2\bar{\rho}}$ and a maximum permitted angle of $\theta_{\max} = 180° - 2\theta_{\min}$. The output is a mesh of $\mathcal{P}$—that is, a Steiner triangulation of $\mathcal{P}$ (recall Definition 2.12)—whose triangles all meet the standard of quality. The algorithm is guaranteed to terminate if $\bar{\rho} \geq \sqrt{2}$ and no two edges in $\mathcal{P}$ meet at an acute angle. (We remove the latter restriction in Section 6.6).

Ruppert's algorithm begins by setting S to be the set of vertices in $\mathcal{P}$ and constructing Del S. Let $\rho(\tau)$ denote the radius-edge ratio of a triangle τ. Suppose that Step 4 of DelaunayRefinement is elaborated as

> If there is a triangle $\tau \in$ Del S with $\rho(\tau) > \bar{\rho}$, insert the circumcenter c of τ into S, update Del S, and go to Step 3.

Let λ be the shortest distance between any two points in S before c is inserted. Let r be τ's circumradius, and let ℓ be the length of τ's shortest edge. Recall the main insight from the previous section: because τ's open circumdisk is empty, $d(c, S) \geq r = \rho(\tau)\ell > \bar{\rho}\lambda$. If we choose $\bar{\rho} \geq 1$, then Step 4 maintains a lower bound of λ on inter-point distances and we can apply the Packing Lemma (Lemma 6.1). If we choose $\bar{\rho} = 1$, every triangle whose smallest angle is less than 30° will be split by the algorithm. Therefore, if the algorithm terminates, every triangle has all its angles between 30° and 120°.

Unfortunately, this simple refinement rule has a serious flaw, because some of the new vertices might lie outside the domain $|\mathcal{P}|$. Moreover, recall from Figure 2.12 that Del S does not always respect the domain boundaries. We fix both problems by incorporating special treatment of the domain boundaries into Step 3 of the algorithm.

A goal of the algorithm is to force each *segment*—each edge in $\mathcal{P}$—to be represented by a sequence of edges in Del S. During refinement, the algorithm maintains a set E of *subsegments*, edges that should ideally appear in the mesh. Initially, E is the set of segments in $\mathcal{P}$; but as the algorithm adds new vertices to S, these vertices subdivide segments into subsegments, and subsegments into shorter subsegments. The algorithm explicitly maintains the identity of the subsegments, whether or not they appear as edges in Del S.

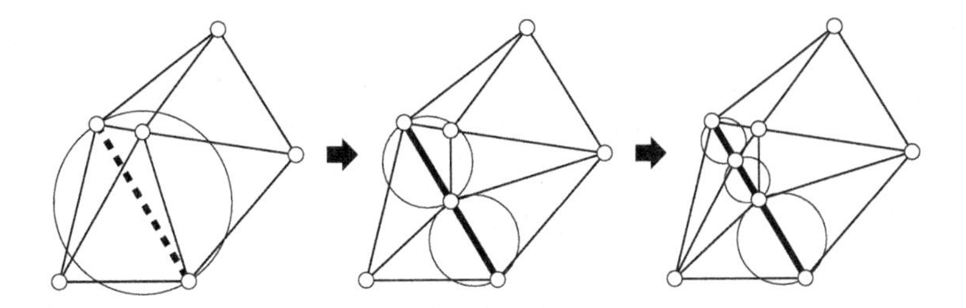

Figure 6.3: Segments (bold) are split until no subsegment is encroached.

Consider a subsegment e that is absent from Del S, as at upper right in Figure 2.12. By Proposition 2.10, Del S contains every strongly Delaunay edge, so e is not strongly Delaunay, and some vertex $v \in S$ must lie in e's closed diametric disk besides e's vertices.

Definition 6.1 (encroachment)**.** A vertex v that lies in the closed diametric ball of a subsegment e but is not a vertex of e is said to *encroach upon e*.

The following procedure SplitSubsegment treats an encroached subsegment by inserting a new vertex at its midpoint, thereby dividing it into two subsegments of half the length. All encroached subsegments are treated this way, whether they are present in Del S or not, as illustrated in Figure 6.3. The recovery of a boundary segment by repeatedly inserting vertices on its missing subsegments is sometimes called *stitching*.

SplitSubsegment(e, S, E)

1. Insert the midpoint of e into S.
2. Remove e from E and add its two halves to E.

We will see at the end of this section that if no subsegment is encroached, every triangle's circumcenter lies in the domain $|\mathcal{P}|$, so we can remove any skinny triangle by inserting a vertex at its circumcenter. However, this triangle refinement rule still has a serious flaw: a vertex inserted at a circumcenter can lie arbitrarily close to a segment, thereby forcing the Delaunay refinement algorithm to generate extremely small triangles in the region between the vertex and segment. We sidestep this hazard with a more complicated rule for treating skinny triangles that prevents vertices from getting dangerously close to segments, at the cost of weakening the angle bounds from 30° and 120° to approximately 20.7° and 138.6°. If a triangle τ has a radius-edge ratio greater than $\bar{\rho}$, we consider inserting its circumcenter c, as before. But if c encroaches upon a subsegment e, we reject c and split e instead.

SplitTriangle(τ, S, E)

1. Let c be the circumcenter of τ.
2. If c encroaches upon some subsegment $e \in E$, call SplitSubsegment(e, S, E). Otherwise, insert c into S.

If SPLITTRIANGLE elects not to insert c and instead calls SPLITSUBSEGMENT, we say that c is *rejected.* Although rejected vertices do not appear in the final mesh, they play a significant role in the algorithm's analysis. The new vertex inserted at the midpoint of e by SPLITSUBSEGMENT might or might not lie in τ's open circumdisk. If not, τ survives the call to SPLITTRIANGLE, but the Delaunay refinement algorithm can subsequently attempt to split τ again.

The following pseudocode implements Ruppert's Delaunay refinement algorithm.

DELTRIPLC($\mathcal{P}, \bar{\rho}$)

1. Let S be the set of vertices in $\mathcal{P}$. Let E be the set of edges in $\mathcal{P}$.
2. Compute Del S.
3. While some subsegment $e \in E$ is encroached upon by a vertex in S, call SPLITSUBSEGMENT(e, S, E), update Del S, and repeat Step 3.
4. If Del S contains a triangle $\tau \subseteq |\mathcal{P}|$ for which $\rho(\tau) > \bar{\rho}$, call SPLITTRIANGLE(τ, S, E), update Del S, and go to Step 3.
5. Return the mesh $\{\sigma \in \text{Del } S : \sigma \subseteq |\mathcal{P}|\}$.

Figure 6.4 illustrates the algorithm DELTRIPLC with a sequence of snapshots of the triangulation. The theoretical minimum angle of 20.7° is quite pessimistic compared to the algorithm's observed performance; it is almost always possible to obtain a minimum angle of 33° in practice.

Observe that Step 4 ignores skinny triangles that lie outside the domain. DELTRIPLC gives priority to encroached subsegments and splits a triangle only if no subsegment is encroached. This policy guarantees that every new vertex lies in the domain $|\mathcal{P}|$.

Proposition 6.2. *Let $\mathcal{T}$ be a Steiner triangulation of a PLC $\mathcal{P}$ in the plane. If no subsegment in $\mathcal{T}$ (i.e. edge in $\mathcal{T}$ included in a segment in $\mathcal{P}$) is encroached, then every triangle in $\mathcal{T}$ has its circumcenter in $|\mathcal{P}|$.*

PROOF. Suppose for the sake of contradiction that some triangle $\tau \in \mathcal{T}$ has a circumcenter $c \notin |\mathcal{P}|$. Let p be a point in the interior of τ. As Figure 6.5 shows, the line segment pc crosses from the interior of $|\mathcal{P}|$ to its exterior, and therefore must cross a subsegment e on the boundary of $|\mathcal{P}|$. We will show that e is encroached.

Let B_τ be the closed circumdisk of τ, let B_e be the closed diametric disk of e, and let H be the closed halfplane containing p whose boundary line is e's affine hull, so $B_e \cap H$ is a half-disk. The interior of B_τ must intersect e, as $pc \subset B_\tau$, but not e's vertices, as τ is Delaunay. Because the center c of B_τ lies outside H, it follows that $B_e \supset B_\tau \cap H \supset \tau$. Therefore, B_e contains all three vertices of τ. Two of τ's vertices might be vertices of e, but the third vertex encroaches on e. This contradicts the assumption that no subsegment is encroached. □

In Chapter 7, we describe a generalization of Proposition 6.2 called the Orthocenter Containment Lemma (Lemma 7.7), which provides an an alternative proof of the proposition.

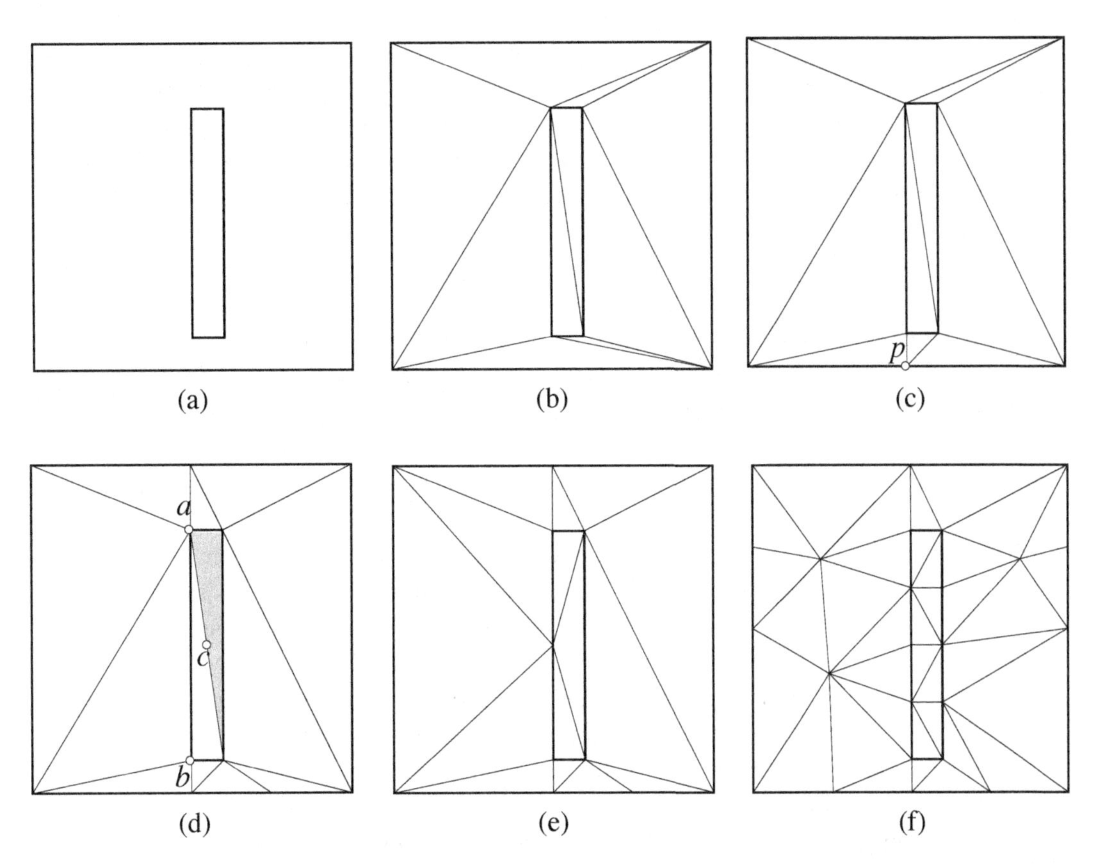

Figure 6.4: The algorithm DelTriPLC in action. (a) The input PLC. (b) The Delaunay triangulation of the input vertices. (c) The bottom edge is encroached, so it is split at its midpoint p. (d) After all the encroached subsegments are split, the shaded triangle has poor quality, so the algorithm considers inserting its circumcenter c. (e) Because c encroaches upon the segment ab, the algorithm does not insert c; instead it splits ab. (f) The final mesh.

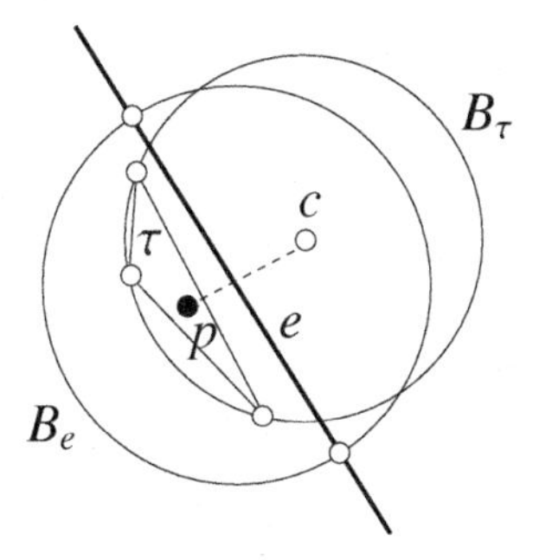

Figure 6.5: B_e includes the portion of τ's circumdisk B_τ on the same side of e as p.

6.3 Implementation and running time

We offer some recommendations on how to implement Ruppert's algorithm efficiently.

First, an implementation should maintain a queue of skinny and oversized triangles and a queue of encroached subsegments, which help Steps 3 and 4 to run quickly. Newly created triangles and subsegments are added to these queues if they ought to be split, but deleted triangles and subsegments are not removed from the queues until they reach the front of a queue and are discovered to be no longer in the mesh. Second, there is an easy, local, constant-time test for whether a subsegment e is encroached: either $e \notin \text{Del}\,S$, or $\text{Del}\,S$ contains a triangle that has e for an edge and an angle of 90° or greater opposite e—see Exercise 1. Therefore, after the initial Delaunay triangulation is computed, the queues can be initialized in time linear in the number of triangles.

Third, when a new vertex is inserted, only the newly created triangles need be checked to see if they are skinny, and only the edges of those triangles need be checked to see if they are encroached subsegments. Recall that the dictionary data structure of Section 3.2 maps each edge of the mesh to the two triangles that have it for an edge in expected constant time; the same data structure can record which edges of the mesh are subsegments. If the new vertex splits a subsegment, the two new subsegments must be tested; recall Figure 6.3.

Fourth, the easiest way to discover whether a triangle circumcenter encroaches upon a subsegment and should be rejected is to insert it, then check the edges opposite it. It is wise to store the deleted triangles so they can be rapidly restored if the circumcenter is rejected. A triangle that is deleted then restored should be in the queue of bad triangles if and only if it was in the queue before the vertex insertion, *including the triangle that the circumcenter was meant to split*—the rejected circumcenter might be successfully inserted later.

Fifth, if the bound $\bar{\rho}$ on triangle quality is strict, prioritizing the skinny triangles so that those with the largest radius-edge ratios (the smallest angles) are split first usually reduces the number of triangles in the final mesh by as much as 35%. (If the bound is not strict, ordering seems to make little difference.) Circumcenters of very skinny triangles tend to eliminate more skinny triangles than circumcenters of mildly skinny triangles. Although one can use a binary heap to prioritize triangles, experiments show that it is much faster, and equally effective, to maintain multiple queues representing different quality ranges.

Sixth, if the domain $|\mathcal{P}|$ is not convex, the triangulation $\text{Del}\,S$ contains triangles that lie outside the domain but in its convex hull. If a tiny angle forms between the boundary of $|\mathcal{P}|$ and the boundary of $\text{conv}\,|\mathcal{P}|$, DelTriPLC could perpetually add new vertices, trying in vain to get rid of the skinny triangle. To avoid this fate, Step 4 of the algorithm declines to split triangles that are not in $|\mathcal{P}|$. Unfortunately, identifying these triangles can be costly. A simple alternative—in fact, Ruppert's original algorithm—is to enclose $\mathcal{P}$ in a large square bounding box and mesh the entire box. The advantage is that no small angle can form between $\mathcal{P}$ and the bounding box. The disadvantages are that time is spent creating triangles that will be discarded in the end, and that small *exterior* angles of $|\mathcal{P}|$ can still put the algorithm into an infinite loop of refinement. Section 6.7 discusses another alternative that works better but requires an implementation of constrained Delaunay triangulations.

There is a notable discrepancy between the running time of Ruppert's algorithm in practice and its running time in the theoretical worst case. Let n be the number of vertices in the PLC $\mathcal{P}$, and let $N \geq n$ be the number of vertices in the final mesh. With a careful implementation, Ruppert's algorithm is observed to consistently take $O(n \log n + N)$ time in practice. The term $n \log n$ covers the construction of the initial triangulation. The term N covers the cost of refinement, because most vertex insertions during Delaunay refinement take constant time. There is little need for point location during the refinement stage, as almost every new vertex the Delaunay refinement algorithm generates is associated with a known subsegment that contains it or a known triangle whose open circumball contains it. The exception is when the algorithm inserts the midpoint of a subsegment that is missing from the mesh, in which case the algorithm must search the triangles adjoining a vertex of the subsegment to find one whose open circumball contains the midpoint. The use of a CDT eliminates this small cost.

However, recall from Section 3.5 that an unlucky vertex insertion can delete and create a linear number of triangles. PLCs are known for which refinement takes $\Theta(N^2)$ time, but such examples are contrived and do not arise in practice. Delaunay refinement tends to rapidly even out the distribution of vertices so that most new vertices have constant degree and take constant time to insert.

6.4 A proof of termination

DelTriPLC has two possible outcomes: either it will eventually delete the last skinny triangle or encroached subsegment and succeed, or it will run forever, creating new skinny triangles and encroached subsegments as fast as it deletes old ones. If we prove it does not run forever, we prove it succeeds.

If $\bar{\rho} > \sqrt{2}$ and no two edges in $\mathcal{P}$ meet at an acute angle, we can prove that DelTriPLC terminates by showing that it maintains a lower bound on the distances between mesh vertices and then invoking the Packing Lemma. The lengths of the edges in a mesh produced by Ruppert's algorithm are roughly proportional to a space-varying function that characterizes the local distances between the geometric features of a PLC.

Definition 6.2 (local feature size). The *local feature size* (LFS) of a PLC $\mathcal{P}$, where $|\mathcal{P}| \subset \mathbb{R}^d$, is a function $f : \mathbb{R}^d \to \mathbb{R}$ such that $f(x)$ is the radius of the smallest ball centered at x that intersects two disjoint linear cells in $\mathcal{P}$.

Figure 6.6 shows local feature sizes for a PLC at three points p, q, and r. A useful intuition is to imagine an inflating ball centered at a point; the ball stops growing when it first intersects two mutually disjoint features. For example, the ball growing outward from p in the figure does not stop growing when it first intersects two segments, because they adjoin each other; it continues to grow until it touches a vertex that is disjoint from the upper segment, as shown. Then $f(p)$ is the radius of the ball.

For every point $p \in \mathbb{R}^d$, there is a linear cell at a distance $f(p)$ from p. Moreover, there is a linear cell with dimension less than d at a distance $f(p)$, because the boundary of every d-cell in $\mathcal{P}$ is a union of lower-dimensional cells in $\mathcal{P}$. Omitting the d-cells from a PLC does not change f.

An important property of the local feature size is that it is 1-Lipschitz.

Definition 6.3 (k-Lipschitz). A real-valued function φ is k-Lipschitz if for any two points x and y,

$$\varphi(x) \leq \varphi(y) + k\, d(x, y).$$

The 1-Lipschitz property implies that f is continuous and its gradient, where it exists, has magnitude no greater than 1; but f is not differentiable everywhere.

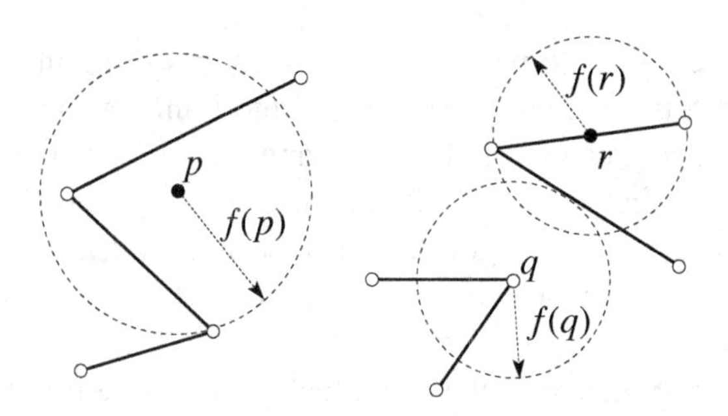

Figure 6.6: Examples of local features sizes in $\mathbb{R}^2$. At each point p, q, and r, the local feature size $f(p)$, $f(q)$, or $f(r)$ is the radius of the smallest disk that intersects two mutually disjoint linear cells.

Proposition 6.3. *The local feature size f is 1-Lipschitz—that is, $f(x) \leq f(y) + d(x, y)$ for any two points $x, y \in \mathbb{R}^2$.*

Proof. By the definition of local feature size, $B(y, f(y))$ intersects two mutually disjoint cells in $\mathcal{P}$. The ball $B(x, f(y) + d(x, y))$ includes $B(y, f(y))$ and, therefore, intersects the same two cells. It follows that $f(x) \leq f(y) + d(x, y)$. □

We assign every mesh vertex x an *insertion radius* $r_x = d(x, S)$, the distance from x to the nearest distinct vertex in the set S of mesh vertices the instant before x is added to S. Observe that immediately after x is added to S, r_x is the length of the shortest edge adjoining x in Del S. Every vertex that SplitTriangle rejects for encroaching upon a subsegment has an insertion radius too, equal to its distance to the nearest vertex in S at the moment it is considered and rejected. To show that DelTriPLC does not run forever, we will prove that the insertion radii do not become much smaller than the local feature size.

We say that a vertex is of *type i* if it lies on a linear i-cell in $\mathcal{P}$ but not on a lower-dimensional cell. Every vertex inserted or rejected by DelTetPLC, except vertices of type 0, has a *parent* vertex that, loosely speaking, is blamed for generating the child vertex. The parent might be in S, or it might be a rejected vertex.

- *Type 0 vertices* are input vertices in $\mathcal{P}$. A type 0 vertex has no parent. The insertion radius r_x of a type 0 vertex x is the distance from x to the nearest distinct vertex in $\mathcal{P}$.

- *Type 1 vertices* are those inserted at the midpoints of encroached subsegments by SplitSubsegment. For a type 1 vertex x generated on an encroached subsegment e, there are two possibilities. If the point encroaching upon e is a vertex in S, define x's parent p to be the vertex in S nearest x; then the insertion radius of x is $r_x = d(x, p)$. Otherwise, the encroaching point p is a rejected circumcenter, which we define to be x's parent, and x's insertion radius r_x is the radius of e's diametric ball. Because p is in that ball, $d(x, p) \leq r_x$.

- *Type 2 vertices* are those inserted or rejected at triangle circumcenters by SplitTriangle. For a type 2 vertex x generated at the circumcenter of a triangle τ, define x's parent p to be the most recently inserted vertex of the shortest edge of τ; if both ver-

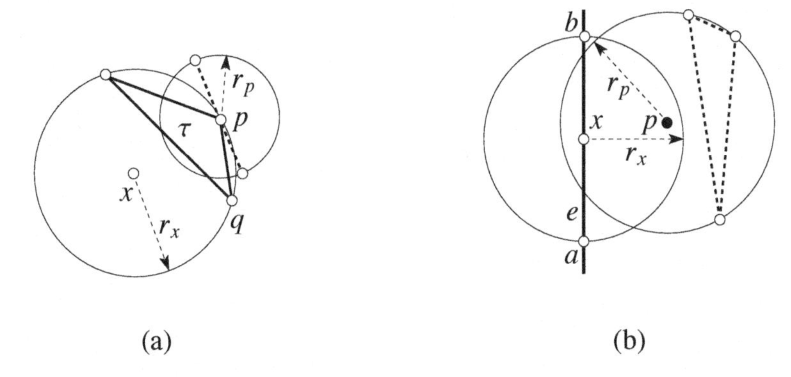

Figure 6.7: (a) The insertion radius r_x of the circumcenter x of a triangle τ is at least $\rho(\tau)$ times the insertion radius r_p of the most recently inserted vertex p of τ's shortest edge. (b) The insertion radius r_p of an encroaching circumcenter p of a Delaunay triangle is at most $\sqrt{2}$ times the insertion radius r_x of the midpoint x of the encroached subsegment.

tices are in $\mathcal{P}$, choose one arbitrarily. Because τ's circumdisk is empty, the insertion radius of x is $r_x = d(x, p)$, the circumradius of τ.

The following proposition proves a lower bound on the insertion radius of each vertex in terms of either its parent's insertion radius or the local feature size. These bounds will lead to lower bounds on the edge lengths in the final mesh.

Proposition 6.4. *Let $\mathcal{P}$ be a PLC in which no two edges meet at an acute angle. Let x be a vertex inserted into S or rejected by* DelTriPLC. *Let p be the parent of x, if one exists.*

(i) *If x is of type 0, then $r_x \geq f(x)$.*

(ii) *If x is of type 1, and its parent p is of type 0 or 1, then $r_x \geq f(x)$.*

(iii) *If x is of type 1, and its parent p is of type 2, then $r_x \geq r_p/\sqrt{2}$.*

(iv) *If x is of type 2, then $r_x > \bar{\rho} r_p$.*

Proof. If x is of type 0, then the disk $B(x, r_x)$ contains $x \in \mathcal{P}$ and another vertex in $\mathcal{P}$, hence $f(x) \leq r_x$ by the definition of local feature size.

If x is of type 2, then x is the circumcenter of a Delaunay triangle τ with $\rho(\tau) > \bar{\rho}$, as illustrated in Figure 6.7(a). The parent p is the most recently inserted vertex of τ's shortest edge pq, or both p and q are vertices in $\mathcal{P}$; it follows from the definition of insertion radius that $r_p \leq d(p, q)$. The insertion radius of x (and the circumradius of τ) is $r_x = d(x, p) = \rho(\tau)\, d(p, q) > \bar{\rho} r_p$ as claimed.

If x is of type 1, it is the midpoint of a subsegment e of a segment $s \in \mathcal{P}$, and x's parent p encroaches upon e. There are three cases to consider, depending on the type of p.

- If p is of type 0, then $p \in \mathcal{P}$, and the disk $B(x, d(x, p))$ intersects two disjoint linear cells in $\mathcal{P}$, namely, p and s. By the definition of local feature size, $f(x) \leq d(x, p) = r_x$.

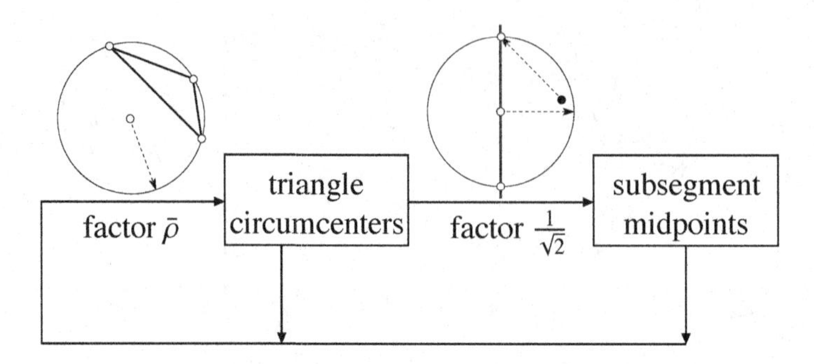

Figure 6.8: Flow graph illustrating the worst-case relation between a vertex's insertion radius and the insertion radii of the children it begets. If no cycle has a product smaller than one, Ruppert's algorithm will terminate.

- If p is of type 1, then p was previously inserted on some segment $s' \in \mathcal{P}$. The segments s and s' cannot share a vertex, because the presence of p in e's diametric disk would imply that s meets s' at an angle less than 90°. Thus $B(x, d(x, p))$ intersects two disjoint segments and $f(x) \leq d(x, p) = r_x$.

- If p is of type 2, then p is a circumcenter rejected for encroaching upon e, so it lies in the diametric disk of e, whose center is x, as illustrated in Figure 6.7(b). Let b be the vertex of e nearest p; then $\angle pxb \leq 90°$ and $d(p, b) \leq \sqrt{2}d(x, b)$. DELTRIPLC calls SPLITTRIANGLE only when no subsegment is encroached, so $r_x = d(x, b)$. The parent p is the center of the circumdisk of a Delaunay triangle, so r_p is the radius of the circumdisk. The triangle's open circumdisk does not contain b, so $r_p \leq d(p, b) \leq \sqrt{2}d(x, b) = \sqrt{2}r_x$, as claimed.

□

The flow graph in Figure 6.8 depicts the relationship between the insertion radius of a vertex and the smallest possible insertion radii of its children, from Proposition 6.4. The boxes represent type 2 and type 1 vertices, respectively. Type 0 vertices are omitted because they cannot contribute to cycles in the flow graph. We can prove that DELTRIPLC terminates by showing that it cannot produce sequences of vertices with ever-diminishing insertion radii—that is, there is no cycle in the flow graph whose product is less than one. This is true if we choose $\bar{\rho}$ to be at least $\sqrt{2}$. When a rejected circumcenter splits a subsegment, the newly created edges can be a factor of $\sqrt{2}$ shorter than the circumradius of the skinny triangle; we compensate for that by trying to split a triangle only if its circumradius is at least a factor of $\sqrt{2}$ greater than the length of its shortest edge. If $\bar{\rho} = \sqrt{2}$, the final mesh has no angle less than $\arcsin \frac{1}{2\sqrt{2}} \doteq 20.7°$.

From this reasoning, it follows that if $\bar{\rho} \geq \sqrt{2}$, DELTRIPLC creates no edge shorter than the shortest distance between two disjoint linear cells in $\mathcal{P}$. Proposition 6.5 below makes a stronger statement: DELTRIPLC spaces vertices proportionally to the local feature size. If a user chooses $\bar{\rho} < \sqrt{2}$, the algorithm will try to obtain the quality requested, but it might fail to terminate, or it might generate a mesh that is not properly graded.

Proposition 6.5. *Suppose that $\bar{\rho} > \sqrt{2}$ and no two segments in $\mathcal{P}$ meet at an angle less than $90°$. Define the constants*

$$C_S = \frac{\left(\sqrt{2}+1\right)\bar{\rho}}{\bar{\rho}-\sqrt{2}}, \qquad C_T = \frac{\bar{\rho}+1}{\bar{\rho}-\sqrt{2}}.$$

Let x be a vertex inserted or rejected by DelTriPLC($\mathcal{P}, \bar{\rho}$).

(i) *If x is of type 1, then $r_x > f(x)/C_S$.*

(ii) *If x is of type 2, then $r_x > f(x)/C_T$.*

Proof. The expressions for C_S and C_T above arise as the solution of the equations $C_S = \sqrt{2}C_T + 1$ and $C_T = 1 + C_S/\bar{\rho}$, which are both used below. Observe that $C_S > C_T > 1$. The proof proceeds by induction on the sequence of points that are inserted or rejected by DelTriPLC. Let x be one such point, and suppose for the sake of induction that the claim holds for every previously inserted or rejected point.

If x is of type 1 and its parent p is of type 0 or 1, then $r_x \geq f(x) > f(x)/C_S$ by Proposition 6.4, confirming property (i). If x is of type 1 and p is of type 2, then $r_x \geq r_p/\sqrt{2}$ by Proposition 6.4 and $d(x, p) \leq r_x$. Inductive application of property (ii) gives $r_p > f(p)/C_T$. By the Lipschitz property of f (Proposition 6.3),

$$f(x) \leq f(p) + d(x,p) < C_T r_p + r_x \leq \left(\sqrt{2}C_T + 1\right) r_x = C_S r_x,$$

confirming property (i).

If x is of type 2, then by the inductive hypothesis its parent p satisfies $r_p > f(p)/C_S$. By Proposition 6.4, $r_x > \bar{\rho} r_p$. The Lipschitz property implies $f(x) \leq f(p) + d(x, p) = f(p) + r_x$, so

$$r_x > \bar{\rho} r_p > \frac{\bar{\rho}}{C_S} f(p) \geq \frac{\bar{\rho}}{C_S}(f(x) - r_x).$$

Rearranging terms gives

$$r_x > \frac{f(x)}{1 + C_S/\bar{\rho}} = \frac{f(x)}{C_T},$$

confirming property (ii). □

Proposition 6.5 gives a lower bound on the distance between a newly inserted vertex and all the preceding vertices. We want a more general bound on the distance between a vertex and all the other vertices, including those that are inserted later. The Lipschitz property of f allows us to derive the latter bound from the former.

Proposition 6.6. *Suppose that $\bar{\rho} > \sqrt{2}$ and no two segments in $\mathcal{P}$ meet at an angle less than $90°$. For any two vertices p and q that* DelTriPLC *inserts, $d(p, q) \geq f(p)/(C_S + 1)$.*

Proof. If p is inserted after q, Proposition 6.5 states that $d(p, q) \geq f(p)/C_S$. If q is inserted after p, $d(p, q) \geq f(q)/C_S$. Because f is 1-Lipschitz, $f(p) \leq f(q) + d(p, q) \leq (C_S + 1)\, d(p, q)$. Either way, the result follows. □

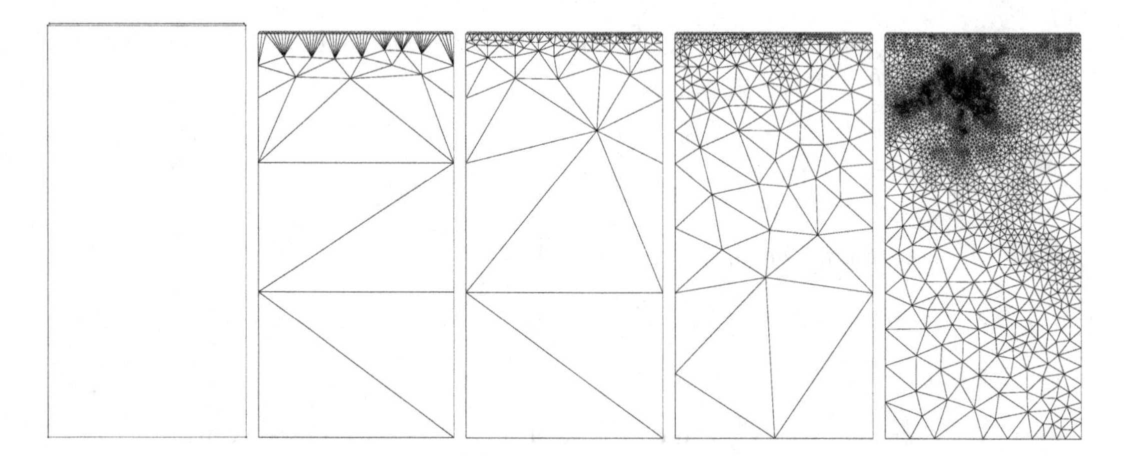

Figure 6.9: A domain with two polygons, the top one being extremely thin compared to the bottom one. Four meshes generated by Ruppert's algorithm, with no angle smaller than 5°, 20°, 30°, and 34.2°, respectively.

Proposition 6.6 establishes a lower bound on the distances between vertices proportional to the local feature size, which implies that DelTriPLC is guaranteed to produce graded meshes if the local feature size function is strongly graded, as Figure 6.9 illustrates. We call this guarantee *provably good grading*, because it implies that small geometric features of the domain do not cause the algorithm to produce unduly short edges far from those features.

To make the proposition concrete, consider choosing $\bar{\rho} = 1.93$ to guarantee that no angle is smaller than roughly 15°. Then $C_S \doteq 9.01$, so the spacing of vertices is at worst about ten times smaller than the local feature size. This worst theoretical outcome never occurs in practice; for example, the edges of the 20° mesh in Figure 6.9 are at least one third as long as their local feature sizes. The bound permits us to apply the Packing Lemma to prove that the algorithm does not run forever.

Theorem 6.7. *Suppose that $\bar{\rho} > \sqrt{2}$ and no two segments in $\mathcal{P}$ meet at an angle less than 90°. Then* DelTriPLC($\mathcal{P}, \bar{\rho}$) *terminates and returns a Steiner Delaunay triangulation of $\mathcal{P}$ whose triangles have radius-edge ratios at most $\bar{\rho}$.*

Proof. Let $f_{\min} = \min_{x \in |\mathcal{P}|} f(x)$. Because $\mathcal{P}$ is finite and any two disjoint linear cells in $\mathcal{P}$ are separated by a positive distance, $f_{\min} > 0$. By Proposition 6.6, DelTriPLC maintains an inter-vertex distance of least $f_{\min}/(C_S + 1) > 0$. By the Packing Lemma (Lemma 6.1), there is an upper bound on the number of vertices in the triangulation, so DelTriPLC must terminate. The algorithm terminates only if no subsegment is encroached and no skinny triangle lies in the domain, so it returns a high-quality mesh of $\mathcal{P}$ as stated. □

Figure 6.9 shows that the algorithm often succeeds for angle bounds well in excess of 20.7°, failing to terminate on the depicted domain only for angle bounds over 34.2°. The meshes illustrate the expected trade-off between mesh size and quality, albeit with longer edges than the lower bounds suggest.

For simplicity, we have not discussed the effects of refining triangles for being too large. In practice, it is usual for a user to specify a *size field* $\lambda : \mathbb{R}^2 \to \mathbb{R}$ that dictates space-varying upper bounds on the edge lengths or triangle circumradii in the mesh. Triangles that violate these bounds are split, just like skinny triangles. See Section 14.4 for an example of an analysis method that can extend the termination guarantee and derive lower bounds on the edge lengths in the mesh when refinement is driven by both a size field and the geometry of the domain.

6.5 A proof of size optimality and optimal grading

An algorithm is said to generate *size-optimal* meshes if the number of triangles in every mesh it produces is within a constant factor of the minimum possible number.

Definition 6.4 (size optimality). Let P be a class of piecewise linear complexes—that is, a set containing all PLCs that satisfy some criterion. For every PLC $\mathcal{P} \in P$, let $\mathcal{T}(\mathcal{P}, \bar{\rho})$ be the triangulation with the fewest triangles among all possible Steiner triangulations of $\mathcal{P}$ whose triangles' radius-edge ratios do not exceed $\bar{\rho}$. Let $\mathcal{M}(\mathcal{P}, \bar{\rho})$ be the Steiner triangulation of $\mathcal{P}$ generated by an algorithm that guarantees that no triangle in the mesh has a radius-edge ratio greater than $\bar{\rho}$. The triangulations this algorithm generates are *size-optimal* if for every $\mathcal{P} \in P$, the number of triangles in $\mathcal{M}(\mathcal{P}, \bar{\rho})$ is at most c times the number of triangles in $\mathcal{T}(\mathcal{P}, \bar{\rho})$, where c is a constant that depends solely on $\bar{\rho}$.

Ruppert's algorithm generates size-optimal meshes of the class of PLCs whose underlying spaces are convex and in which no two segments meet at an angle less than $90°$, for $\bar{\rho} \in (\sqrt{2}, \infty)$. The inclusion is strict: the constant c approaches infinity as $\bar{\rho}$ approaches $\sqrt{2}$ from above or infinity from below, so the guarantee is most meaningful for moderate demands on triangle quality.

Size optimality does not mean that we can find the perfectly optimal mesh $\mathcal{T}$—likely a futile quest—but we can still reason about its size and prove that the size of $\mathcal{M}$ is asymptotically optimal. The reader might ask, asymptotic in relation to what? One of the most interesting theoretical discoveries about mesh generation is that there is a natural measure of how many elements are required in any high-quality simplicial mesh: the integral over the domain of the inverse squared local feature size.

The following proposition shows that this integral is an asymptotic upper bound on the number of triangles in a mesh $\mathcal{M}$ generated by Ruppert's algorithm. Subsequent propositions show that it is an asymptotic lower bound on the number of triangles in any high-quality mesh, including $\mathcal{T}$, so $\mathcal{M}$ is size-optimal. Unfortunately, the lower bound is contingent on $|\mathcal{P}|$ being convex; in its original form, Ruppert's algorithm does not offer a size-optimality guarantee for nonconvex domains. In Section 6.7, we discuss how a variant of Ruppert's algorithm that uses constrained Delaunay triangulations does offer size-optimality for nonconvex domains, with the insight that the analysis method must redefine the local feature size function to use shortest distances in $|\mathcal{P}|$.

Proposition 6.8. *Let $\mathcal{P}$ be a PLC in the plane in which no two segments meet at an angle less than $90°$. Let $\mathcal{M}$ be a mesh of $\mathcal{P}$ generated by* DelTriPLC$(\mathcal{P}, \bar{\rho})$ *with $\bar{\rho} > \sqrt{2}$. Then the* *number of triangles in $\mathcal{M}$ is less than*

$$\frac{8(3 + 2C_S)^2}{\pi} \cdot \int_{|\mathcal{P}|} \frac{dx}{f(x)^2},$$

where C_S is a constant that depends solely on $\bar{\rho}$, defined in Proposition 6.5, and dx represents an infinitesimal measure of area in the plane.

Proof. Let S be the set of vertices in $\mathcal{M}$, and let $|S|$ denote the number of vertices in $\mathcal{M}$. For each vertex $v \in S$, consider the Euclidean disk $B_v = B(v, r_v)$ where $r_v = f(v)/(2 + 2C_S)$. The interiors of these disks are pairwise disjoint by Proposition 6.6. As no two segments in $\mathcal{P}$ meet each other at an acute angle, at least one quarter of each disk is included in $|\mathcal{P}|$. Therefore,

$$\begin{aligned}
\int_{|\mathcal{P}|} \frac{dx}{f(x)^2} &> \sum_{v \in S} \int_{B_v \cap |\mathcal{P}|} \frac{dx}{f(x)^2} \\
&> \sum_{v \in S} \int_{B_v \cap |\mathcal{P}|} \frac{dx}{(f(v) + r_v)^2} \\
&\geq \frac{1}{4} \sum_{v \in S} \int_{B_v} \frac{dx}{(f(v) + r_v)^2} \\
&= \frac{1}{4} \sum_{v \in S} \frac{\pi r_v^2}{(3 + 2C_S)^2 r_v^2} \\
&= \frac{\pi}{4(3 + 2C_S)^2} |S|.
\end{aligned}$$

Recall from Section 2.1 that an $|S|$-vertex triangulation has at most $2|S| - 5$ triangles. The result follows. □

The matching lower bound on the number of triangles in a high-quality mesh depends on several observations that seem unsurprising, but require care to prove: small domain features are surrounded by proportionally small triangles; triangles that adjoin each other cannot have arbitrarily different sizes; and triangles that are distant from each other cannot have a size difference far greater than the distance between them. Together, these observations imply that the local feature size function places an upper bound on the local edge lengths in a good mesh. None of these observations is true if arbitrarily skinny triangles are allowed; all of them depend on having an upper bound on the radius-edge ratio or, equivalently, a lower bound on the smallest angle. They also require that $|\mathcal{P}|$ be convex.

The following series of propositions formalizes these observations to prepare for the lower bound proof. A triangle τ has three altitudes—the distance from a vertex of τ to the affine hull of the opposite edge; let $h(\tau)$ denote its shortest altitude. The following proposition shows that between any two disjoint simplices in a triangulation, there is a triangle whose shortest altitude does not exceed the distance between the simplices. The proposition holds for any triangulation, of good quality or not. But note that good quality implies that the triangle with a bounded altitude also has bounded edge lengths.

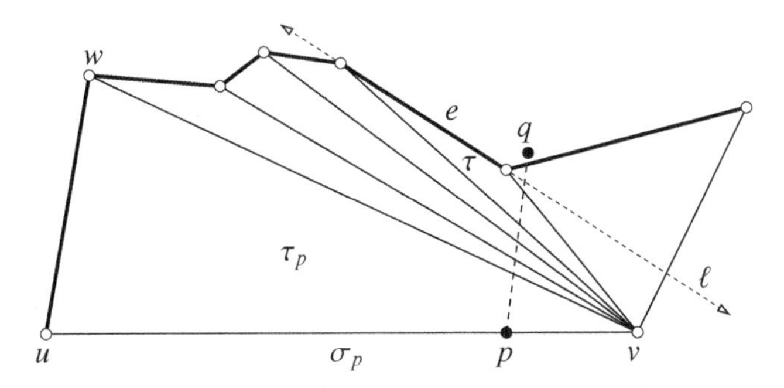

Figure 6.10: Between p and q, at least one triangle τ adjoining uv has an altitude of $d(p, q)$ or less.

Proposition 6.9. *Let $\mathcal{T}$ be a triangulation in the plane. Let p and q be two points such that $pq \in |\mathcal{T}|$. Let σ_p and σ_q be the unique simplices in $\mathcal{T}$ whose relative interiors contain p and q, respectively. If σ_p and σ_q are disjoint, there is a triangle $\tau \in \mathcal{T}$ that intersects both pq and σ_p such that $h(\tau) \leq d(p, q)$.*

Proof. Consider three cases: σ_p is a vertex, an edge, or a triangle.

If σ_p is a vertex, namely, the point p, let $\tau \in \mathcal{T}$ be a triangle adjoining p that intersects $pq \setminus \{p\}$—there are either one or two such triangles. Because p and σ_q are disjoint, the edge of τ opposite p intersects pq. Therefore, $h(\tau) \leq d(p, q)$ and the proposition holds.

If σ_p is an edge, consider several possibilities. If q is collinear with σ_p, replace p with the vertex of σ_p nearest q and apply the reasoning above. Otherwise, let $\tau_p \in \mathcal{T}$ be the triangle that has σ_p for an edge and intersects $pq \setminus \{p\}$. If $h(\tau_p) \leq d(p, q)$, the proposition holds. Otherwise, let u, v, and w be the vertices of τ_p, with u and v being the vertices of σ_p. Assume without loss of generality that the plane is rotated so that σ_p is horizontal with τ_p above it, that the plane is reflected so that w lies to the left of pq, and that the vertices are labeled so u lies to the left of v, as illustrated in Figure 6.10.

Consider the fan of triangles in $\mathcal{T}$ that adjoin v and have interiors that intersect pq, starting with τ_p and proceeding in clockwise order, as illustrated. None of these triangles' interiors can contain q, because then σ_q would not be disjoint from σ_p. Therefore, the chain of edges opposite v in the fan starts with the edge uw and ends with an edge that intersects pq; these edges are bold in Figure 6.10. Because $h(\tau_p) > d(p, q)$, the apex w is higher than the point q—that is, it is further above the affine hull of σ_p. But the last edge in the chain intersects pq and so must have at least one vertex as low as q or lower. Therefore, the chain includes at least one *down edge* whose clockwise vertex is lower than its counterclockwise vertex. Let e be the most clockwise down edge in the chain, let ℓ be e's affine hull, and let $\tau \in \mathcal{T}$ be the triangle joining edge e with vertex v. Because e is the last down edge, all subsequent edges and the point q must lie above or on ℓ. Because e is a down edge, p is further from ℓ than v. It follows that $h(\tau) \leq d(v, \ell) < d(p, \ell) \leq d(p, q)$.

In the third and final case, σ_p is a triangle. Let p' be the point where pq intersects the boundary of σ_p, and let σ'_p be the face of σ_p whose relative interior contains p'. Replace p with p', replace σ_p with σ'_p, and apply the reasoning above. □

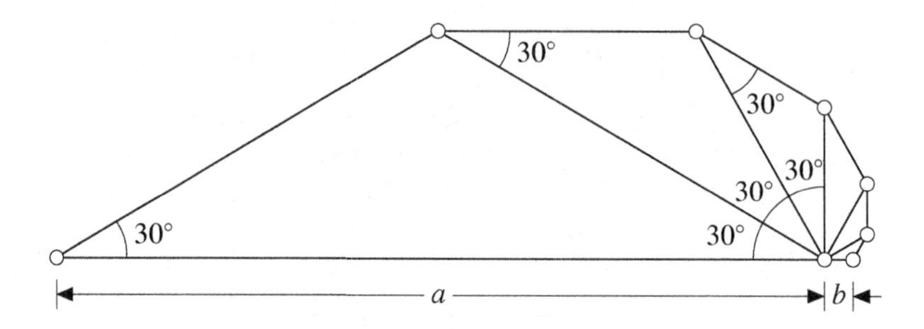

Figure 6.11: In a triangulation with no angle smaller than 30°, the ratio a/b cannot exceed 27.

Any bound on the smallest angle of a triangulation imposes a limit on the grading of triangle sizes. The next proposition bounds the difference in sizes between two triangles that share a vertex. In the following propositions, let $\ell_{\max}(\tau)$ denote the length of τ's longest edge.

Proposition 6.10. *Let $\mathcal{T}$ be a triangulation in the plane with $|\mathcal{T}|$ convex. Let $\bar{\rho} = \max_{\sigma \in \mathcal{T}} \rho(\sigma)$ be the maximum radius-edge ratio among the triangles in $\mathcal{T}$; thus, $\theta_{\min} = \arcsin \frac{1}{2\bar{\rho}}$ is the minimum angle. Let τ and τ' be two triangles in $\mathcal{T}$ that share a vertex v. Then $\ell_{\max}(\tau) \leq \eta\, h(\tau')$, where $\eta = (2 \cos \theta_{\min})^{1+180^\circ/\theta_{\min}} / \sin \theta_{\min} = 2\bar{\rho}(4 - 1/\bar{\rho}^2)^{0.5+90^\circ/\arcsin 1/(2\bar{\rho})}$.*

Proof. Let a be the length of the longest edge adjoining the vertex v, let b be the length of the shortest, and let $\phi \leq 180^\circ$ be the angle separating the two edges. We claim that the ratio a/b cannot exceed $(2 \cos \theta_{\min})^{\phi/\theta_{\min}}$. This bound is tight if $\phi/\theta_{\min}$ is an integer; Figure 6.11 offers an example where the bound is obtained.

We prove this claim by induction on the sequence of edges around v from the longest edge to the shortest. For the base case, suppose the longest and shortest edges belong to a common triangle. Let α and β be the angles opposite the edges of lengths a and b, respectively; then $\alpha + \beta + \phi = 180^\circ$ and $\sin \alpha = \sin(\beta + \phi) = \sin \beta \cos \phi + \sin \phi \cos \beta$. By the Law of Sines, $a/b = \sin \alpha / \sin \beta = \cos \phi + \sin \phi / \tan \beta$, so a/b is maximized when $\beta = \theta_{\min}$. Observe that if ϕ also equals $\theta_{\min}$, then $a/b = 2 \cos \theta_{\min}$. It is straightforward to verify that if $\phi > \theta_{\min}$, then $a/b < (2 \cos \theta_{\min})^{\phi/\theta_{\min}}$, because the former grows more slowly than the latter as ϕ increases above $\theta_{\min}$. This establishes the base case.

If the longest and shortest edges adjoining v are not edges of a common triangle, let c be the length of an intermediate edge adjoining v. Then by the inductive hypothesis, $a/b = (a/c)(c/b) \leq (2 \cos \theta_{\min})^{\angle ac/\theta_{\min}} (2 \cos \theta_{\min})^{\angle bc/\theta_{\min}} = (2 \cos \theta_{\min})^{\phi/\theta_{\min}}$, and the claim holds.

Because τ has two edges no longer than a and no angle smaller than $\theta_{\min}$, its longest edge satisfies $\ell_{\max}(\tau) \leq 2a \cos \theta_{\min}$. Because τ' has two edges no shorter than b and no angle smaller than $\theta_{\min}$, its shortest altitude satisfies $h(\tau') \geq b \sin \theta_{\min}$. The result follows by combining inequalities. □

The constant η in Proposition 6.10 can be improved; see Exercise 8.

The next proposition shows that in a high-quality triangulation, every triangle's longest edge has an upper bound proportional to the local feature size at any point in the triangle.Recall that Proposition 6.6 gives a proportional lower bound for the edges produced by DELTRIPLC. Together, the two propositions show that Ruppert's algorithm generates meshes whose edge lengths are within a constant factor of the longest possible. We call this guarantee *optimal grading*.

Proposition 6.11. *Let $\mathcal{P}$ be a PLC in the plane with $|\mathcal{P}|$ convex. Let $\mathcal{T}$ be a Steiner triangulation of $\mathcal{P}$. Let $\bar{\rho} = \max_{\sigma \in \mathcal{T}} \rho(\sigma)$. Let τ be a triangle in $\mathcal{T}$. Let x be a point in τ. Then $\ell_{\max}(\tau) \leq 2\eta f(x)$, where η is a constant that depends solely on $\bar{\rho}$, specified in Proposition 6.10.*

PROOF. By the definition of local feature size, the disk $B(x, f(x))$ intersects two disjoint linear cells in $\mathcal{P}$, each a vertex or edge, at two points p and q, respectively. Because p and q lie on disjoint edges or vertices in $\mathcal{P}$, they lie on disjoint edges or vertices in $\mathcal{T}$. Because $|\mathcal{P}|$ is convex, $pq \in |\mathcal{P}|$ and we can apply Proposition 6.9 to show there is a triangle $\tau' \in \mathcal{T}$ that intersects pq such that $h(\tau') \leq d(p, q)$.

If τ adjoins τ', then $\ell_{\max}(\tau) \leq \eta\, h(\tau')$ by Proposition 6.10. It follows that $\ell_{\max}(\tau) \leq \eta\, d(p, q) \leq 2\eta f(x)$, and the claim holds.

Otherwise, let u be a point in $\tau' \cap pq$. By a second application of Proposition 6.9 to the points x and u, there is a triangle $\tau'' \in \mathcal{T}$ that adjoins τ and satisfies $h(\tau'') \leq d(x, u)$. Because u lies on pq, it lies in $B(x, f(x))$ and $d(x, u) \leq f(x)$. Therefore, $\ell_{\max}(\tau) \leq \eta\, h(\tau'') \leq \eta\, d(x, u) \leq \eta f(x)$, and the claim holds. □

We can now prove a lower bound on the size of a high-quality mesh.

Proposition 6.12. *Let $\mathcal{P}$ be a PLC in the plane with $|\mathcal{P}|$ convex. Let $\mathcal{T}$ be a Steiner triangulation of $\mathcal{P}$. Let $\bar{\rho} = \max_{\sigma \in \mathcal{T}} \rho(\sigma)$. The number of triangles in $\mathcal{T}$ is at least*

$$\frac{1}{\sqrt{3}\eta^2} \cdot \int_{|\mathcal{P}|} \frac{dx}{f(x)^2},$$

where η is a constant that depends solely on $\bar{\rho}$, specified in Proposition 6.10.

PROOF. Let $\ell_{\max}(x)$ be a function that maps each point $x \in |\mathcal{P}|$ to the length of the longest edge of the triangle in $\mathcal{T}$ that contains x, taking the greatest value if more than one triangle contains x. By Proposition 6.11, $\ell_{\max}(x) \leq 2\eta f(x)$, so

$$\begin{aligned}
\int_{|\mathcal{P}|} \frac{dx}{f(x)^2} &\leq 4\eta^2 \int_{|\mathcal{P}|} \frac{dx}{\ell_{\max}(x)^2} \\
&= 4\eta^2 \sum_{\tau \in \mathcal{T}} \int_{\tau} \frac{dx}{\ell_{\max}(\tau)^2} \\
&= 4\eta^2 \sum_{\tau \in \mathcal{T}} \frac{\text{area}(\tau)}{\ell_{\max}(\tau)^2} \\
&\leq \sqrt{3}\eta^2 \sum_{\tau \in \mathcal{T}} 1,
\end{aligned}$$

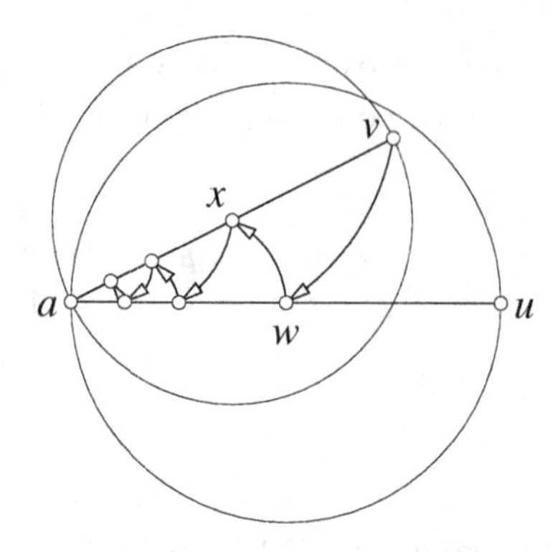

Figure 6.12: Ping-pong encroachment caused by a small input angle. Vertex v encroaches upon au, which is split at w. Vertex w encroaches upon av, which is split at x, which encroaches upon aw, and so on.

because $\text{area}(\tau)/\ell_{\max}(\tau)^2 = h(\tau)/(2\ell_{\max}(\tau))$ attains its maximum possible value of $\sqrt{3}/4$ for an equilateral triangle. The summation is the number of triangles in $\mathcal{T}$, so the claim follows. □

Propositions 6.8 and 6.12 together establish the size optimality of meshes produced by DelTriPLC, formally stated in the following theorem.

Theorem 6.13. *Let $\mathcal{P}$ be a PLC in the plane such that $|\mathcal{P}|$ is convex and no two segments in $\mathcal{P}$ meet at an angle less than* 90°. *Let $\bar{\rho}$ be a real number greater than* $\sqrt{2}$. DelTriPLC *produces a mesh $\mathcal{M}$ of $\mathcal{P}$ whose triangles' radius-edge ratios do not exceed $\bar{\rho}$, such that the number of triangles in $\mathcal{M}$ is at most a constant factor greater than the number of triangles in any other Steiner triangulation of $\mathcal{P}$ whose radius-edge ratios do not exceed $\bar{\rho}$.*

6.6 Meshing domains with small angles

Ruppert's algorithm requires that no two segments meet at an acute angle. This is a severe restriction. In practice, the algorithm often succeeds despite acute angles, but as domain angles drop below about 45°, it becomes increasingly likely to fail to terminate.

Figure 6.12 demonstrates one difficulty caused by small input angles. If two adjoining segments have unequal lengths, an endless cycle of mutual encroachment may produce ever shorter subsegments incident to the apex of the small angle. This phenomenon, sometimes called *ping-pong encroachment*, is observed only with angles of 45° or less.

Sometimes it is impossible to obtain good element quality. If two segments of a domain adjoin each other at a 1° angle, some triangle of the final mesh will have an angle of 1° or less. Moreover, a small domain angle sometimes necessitates generating elements with new small angles that are not inherited from the domain (recall Figure 1.5). Given a domain with small angles, a mesh generator must diagnose where it is necessary to give up and accept some poor-quality elements.

This section discusses two modifications to Ruppert's algorithm that extend it so it works remarkably well with domains that have small angles. The first modification was

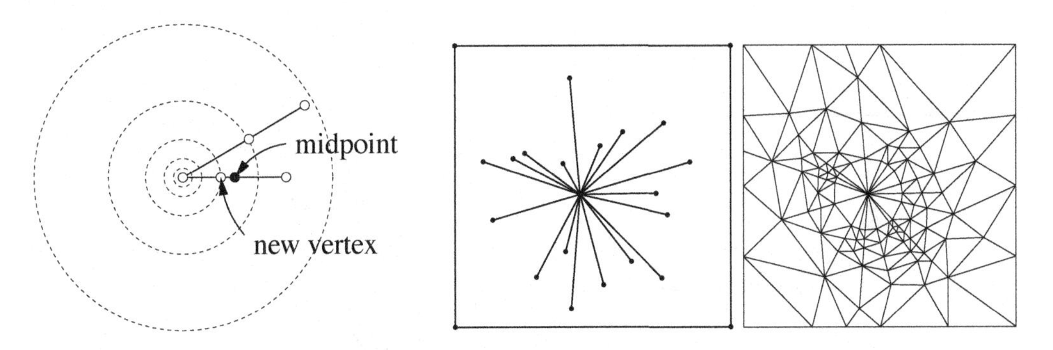

Figure 6.13: Concentric circular shells appear at left. If an encroached subsegment meets another segment at an acute angle, the subsegment is split at its intersection with a circular shell whose radius is 2^i for some integer i. The illustrations at right are a sample input and output of Ruppert's algorithm with concentric shell segment splitting.

proposed by Ruppert himself; he calls it "modified segment splitting using concentric circular shells." The second modification is a simple observation about which skinny triangles the mesh generator should not try to split. Together, these two modifications yield a variant of Ruppert's algorithm that always terminates and has some impressive properties. Most importantly, it can guarantee that no triangle has an angle greater than 138.6°. It also guarantees that skinny triangles appear only between segments separated by small angles. Recall that for many applications, bounding the largest angles is more important than bounding the smallest angles, because the former are related to the discretization and interpolation errors.

The first modification is to split some encroached subsegments off-center, rather than at their midpoints. Imagine that each input vertex is enclosed by concentric circles whose radii are all the powers of two—that is, 2^i for all integers i, as illustrated in Figure 6.13. When an encroached subsegment adjoins another segment at an angle less than 90°, split the subsegment not at its midpoint, but at one of the circular shells centered at the shared vertex, so that one of new subsegments has a power-of-two length. Choose the shell that gives the best-balanced split, so the two new subsegments produced by the split are between one third and two thirds the length of the split subsegment.

If both vertices of a segment adjoin other segments, the segment may undergo up to two unbalanced splits—one for each end. Choose one vertex arbitrarily, and split the segment so the subsegment adjoining that vertex has a power-of-two length between one quarter and one half the length of the split subsegment. The other subsegment produced by this split might undergo a subsequent off-center split, in which case all three subsegments will be at least one fifth the length of the original segment. All subsequent subsegment splits are bisections.

Concentric shell segment splitting prevents the runaway cycle of ever shorter subsegments portrayed in Figure 6.12, because adjoining subsegments of equal length do not encroach upon each other. Ruppert also suggests changing his algorithm so that it does not attempt to split a skinny triangle nestled in the corner of a small input angle. These changes are often effective, as the mesh at right in Figure 6.13 shows, and they always suffice for simple polygons with no internal boundaries.

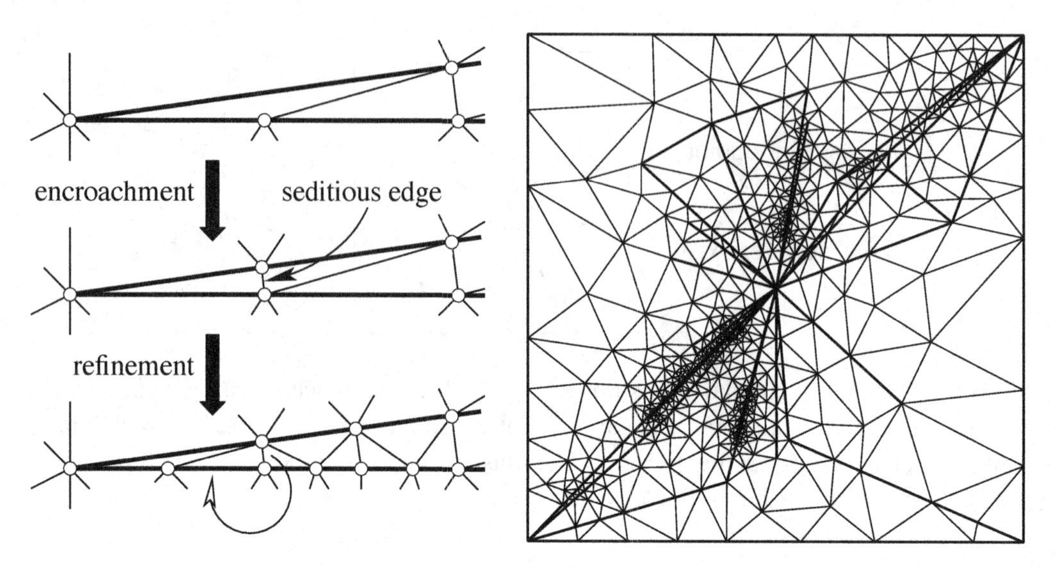

Figure 6.14: At left, a demonstration of how segments separated by small angles create short, seditious edges as they are split; the refinement of skinny triangles can cause the subsegments to be split again. At right, a mesh generated by Ruppert's algorithm with concentric shells when it declines to split triangles whose shortest edges are seditious. No angle in this mesh is greater than 127.1°, and no triangle has an angle less than 26.45° unless its shortest edge is seditious.

However, Figure 6.14 illustrates a more treacherous way by which small input angles and internal boundaries can cause Delaunay refinement to fail to terminate. Recall the key idea that Delaunay refinement should create no new edge that is shorter than the shortest edge previously existing. If two subsegments that adjoin each other at a very small angle are bisected, the new edge connecting their two midpoints can starkly violate this rule. The new, shorter edge can cause subsequent refinement as the algorithm removes skinny triangles, as illustrated, which can cause the subsegments to be split again, creating a yet shorter edge, and the cycle may continue forever.

An idea that breaks this cycle is to deny these new, unduly short edges the privilege of causing further refinement. Specifically, call an edge *seditious* if its vertices lie on two distinct segments that meet each other at an angle less than 60°, the two vertices lie on the same concentric shell, and the two vertices are true midpoints (not off-center splits), as illustrated in Figure 6.14.

The second modification is to simply decline to try to split any skinny triangle whose shortest edge is seditious. This precaution prevents the short lengths of seditious edges from propagating through the mesh. Triangles with small angles can survive, but only between segments adjoining each other at small angles. Figure 6.14 depicts a mesh generated by the modified algorithm for a PLC that requires both modifications to stop the algorithm from refining forever.

The observation behind why this modified algorithm terminates is that unduly short edges—edges shorter than those predicted by Proposition 6.4—can be created in only two circumstances. Off-center subsegment splits can create them, but only twice per PLC segment. Unduly short edges are also created by cascading bisections of adjoining segments, as illustrated in Figure 6.14, but these edges are all seditious, and are prevented from causing further refinement.

Proposition 6.14. *Let $\bar{\rho} > \sqrt{2}$ be the maximum permitted radius-edge ratio of a triangle whose shortest edge is not seditious. If Ruppert's algorithm is modified to use concentric shells for segment splitting and to decline to try to split any triangle whose shortest edge is seditious, it is guaranteed to terminate for any two-dimensional PLC $\mathcal{P}$, with no restrictions on the angles at which segments meet. Moreover, no triangle of the final mesh has an angle greater than $180° - 2\arcsin\frac{1}{2\bar{\rho}}$ nor an angle less than $\sin\phi_{\min}/\sqrt{5 - 4\cos\phi_{\min}}$, where $\phi_{\min}$ is the smallest angle separating two adjoining segments in $\mathcal{P}$.*

Proof. Let $\mathcal{P}'$ be a copy of the PLC $\mathcal{P}$ modified to include every off-center vertex the algorithm inserts on a segment; i.e. each vertex that is not the true midpoint of the subsegment being split. The segments in $\mathcal{P}'$ are subdivided accordingly. At most two off-center splits occur for each segment in $\mathcal{P}$, so $\mathcal{P}'$ has only finitely many extra vertices. Let $f(\cdot)$ denote the local feature size with respect to $\mathcal{P}'$, and let $f_{\min} = \min_{x \in |\mathcal{P}'|} f(x)$.

Consider a group of segments that meet at a common vertex z in $\mathcal{P}'$, with consecutive segments in the group separated by angles less than 60°. These segments have power-of-two lengths (possibly excepting some segments that will never be split, which we can ignore). When they are refined, they are split at their true midpoints, so their subsegments have power-of-two lengths. At any time during refinement, if the shortest subsegment of the segments in the group has length 2^i, then all the vertices on the segments lie on circular shells centered at z of radii $j \cdot 2^i$ for positive integers j, and two vertices can be separated by a distance less than 2^i only if they lie on the same shell. If a vertex v on one segment encroaches upon a subsegment e of another subsegment in the group, then e crosses the shell that v lies on, and the two subsegments created when e is split cannot be shorter than the two subsegments adjoining v. It follows that a cascading chain of mutual encroachments solely within the group does not create a subsegment shorter than the shortest subsegment already in the group.

Say that a pair of vertices (v, x) is *seditious* if they lie on two distinct segments in $\mathcal{P}'$ that meet each other at an angle less than 60° at some vertex z, and they lie on the same shell; that is, $d(z, v) = d(z, x)$. Thus, vx is a seditious edge if the mesh contains it. We claim that the modified algorithm never generates two vertices separated by a distance less than $f_{\min}$ unless they are a seditious pair. Suppose for the sake of contradiction that v is the first vertex inserted that breaks this invariant. Then there is a vertex x such that (v, x) is not seditious but $d(v, x) < f_{\min}$. Let w be the vertex nearest v at the moment v is inserted. Then $d(v, w) \leq d(v, x) < f_{\min}$. It is possible that w and x are the same vertex.

We claim that v is neither a circumcenter nor a type 1 vertex whose parent is a rejected circumcenter. If v is the circumcenter of a skinny triangle τ, then τ's shortest edge has length at least $f_{\min}$ because the algorithm does not split a triangle whose shortest edge is seditious, and by the inductive hypothesis, all the nonseditious edges had length $f_{\min}$ or greater before v was inserted. But τ's radius-edge ratio exceeds $\bar{\rho}$, so its circumradius is greater than $\bar{\rho} f_{\min} > \sqrt{2} f_{\min}$ and thus $d(v, w) > \sqrt{2} f_{\min}$, a contradiction. If v is a type 1 vertex whose parent p is a rejected circumcenter, then $r_p > \sqrt{2} f_{\min}$ by the same reasoning, so Proposition 6.4(iii) implies that $r_v \geq r_p/\sqrt{2} > f_{\min}$. Then $d(v, w) \geq r_v > f_{\min}$, a contradiction.

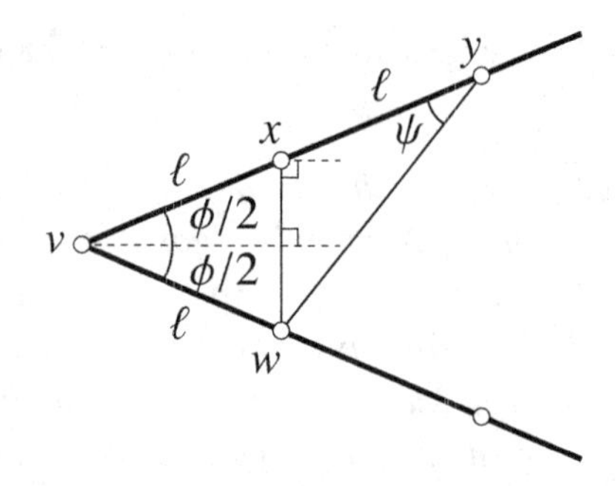

Figure 6.15: If a triangle's shortest edge wx is seditious and subtends an input angle ϕ, the triangle has no angle greater than $90^\circ + \phi/2$ nor less than ψ.

Therefore, v is a type 1 vertex inserted on an encroached subsegment e of a segment s, and the diametric disk of e contains some encroaching vertex and, hence, contains w. Thus, w is not a circumcenter (which would have been rejected). The fact that $d(v, w) < f_{\min}$ implies that w is not a vertex in $\mathcal{P}'$ and does not lie on a segment disjoint from s. The same is true for x. Therefore, w lies on a segment in $\mathcal{P}'$ that adjoins s at a shared vertex z. By our reasoning above, the two subsegments created when e is split are not shorter than the two subsegments adjoining w, which have lengths of at least $f_{\min}$ by the inductive hypothesis. But $d(v, x) < f_{\min}$, so x is in e's diametric disk. Thus x, like w, lies on a segment s' that adjoins s at z, and adjoins two subsegments whose lengths are at least $f_{\min}$. The fact that $d(v, x) < f_{\min}$ implies that v and x lie on a common circular shell. The radius of that shell is at least $f_{\min}$, so s and s' meet at an angle less than 60°. This contradicts our assumption that (v, x) is not seditious. It follows that only seditious pairs can be separated by a distance less than $f_{\min}$.

Because every subsegment has a length of at least $f_{\min}$, every seditious edge has a length of at least $2f_{\min} \sin \frac{\phi_{\min}}{2}$. It follows from the Packing Lemma (Lemma 6.1) that the modified algorithm terminates.

When the algorithm terminates, every triangle whose shortest edge is not seditious has no angle less than $\arcsin \frac{1}{2\bar{\rho}}$ and, thus, no angle greater than $180^\circ - 2 \arcsin \frac{1}{2\bar{\rho}}$. To bound the angles of the other triangles, consider the seditious edge wx in Figure 6.15. Its vertices lie on two distinct segments that meet at a vertex v at an angle $\phi < 60^\circ$, and the vertex x is a true midpoint of vy. If a triangle whose shortest edge is wx respects the segments, its largest angle cannot exceed $\angle wxy = 90^\circ + \phi/2 < 120^\circ$, which establishes our claim about the largest angles.

Let $\psi = \angle xyw$, and observe that $\psi < \phi$. No Delaunay triangle with shortest edge wx can have an angle less than ψ, because by the Inscribed Angle Theorem, any such triangle would have either y or v inside its circumdisk. By the Law of Sines, $\sin \psi / d(v, w) = \sin \angle vwy / d(v, y)$, hence

$$2 \sin \psi = \sin \angle vwy = \sin(180^\circ - \phi - \psi) = \sin(\phi + \psi) = \sin \phi \cos \psi + \cos \phi \sin \psi.$$

Therefore, $(2 - \cos \phi)^2 \sin^2 \psi = \sin^2 \phi \cos^2 \psi = \sin^2 \phi (1 - \sin^2 \psi)$. Rearranging terms gives $\sin \psi = \sin \phi / \sqrt{5 - 4 \cos \phi}$, which establishes our claim about the smallest angles. □

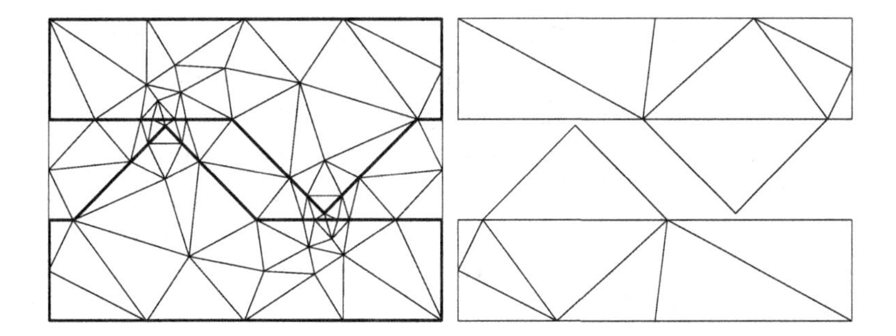

Figure 6.16: Two variations of Ruppert's Delaunay refinement algorithm with a 20° minimum angle. Left: Overrefinement with a Delaunay triangulation in a box. Right: Refinement with a constrained Delaunay triangulation.

Proposition 6.14 guarantees termination, but not good grading. It is possible to salvage a weakened proof of good grading; see the bibliographical notes for details.

In a practical implementation, it is wise to use an inter-segment angle smaller than 60° to define seditious edges, so that Delaunay refinement is less tolerant about leaving skinny triangles behind. This change breaks the termination proof, but in practice it threatens termination only if the angle threshold for seditious edges is substantially smaller than 20°. (The algorithm must still decline to try to split triangles that are right in the corners of small domain angles, of course, as these cannot be improved.)

6.7 Constrained Delaunay refinement

If software for constructing and updating constrained Delaunay triangulations is available, Ruppert's algorithm is easily modified to construct and maintain a CDT instead of Del S, and it enjoys several advantages by doing so. First, the algorithm stores no triangles outside the domain, even if $|\mathcal{P}|$ is not convex, and therefore saves the costs of maintaining them and checking which triangles are in the domain. Second, every subsegment is an edge of the CDT; whereas Ruppert's original algorithm sometimes must pay for point location to insert the midpoint of a subsegment that is absent from Del S, constrained Delaunay refinement requires no point location, because every newly inserted vertex is associated with a mesh edge or triangle. Third, and most important, CDTs prevent overrefinement that can occur where geometric features are separated by small distances exterior to the domain, as illustrated in Figure 6.16. As the mesh at left shows, Ruppert's original algorithm with a bounding box can refine a mesh much more than necessary, because of encroachments and skinny triangles exterior to the domain. A CDT prevents this overrefinement, as the mesh at right illustrates.

A nuisance in Section 6.5 is that the proof of size optimality holds only if $|\mathcal{P}|$ is convex. Figure 6.16 shows that this is not merely a technical flaw in the proofs; Ruppert's algorithm does not always generate size-optimal meshes of nonconvex domains. The local feature size does not distinguish exterior distances from interior distances, so it correctly predicts the behavior of Ruppert's algorithm, but it is not an accurate estimate of the longest possible edge lengths.

Once modified to maintain a CDT, Ruppert's algorithm is size-optimal even for nonconvex domains. We can prove this by replacing the Euclidean distance with the *intrinsic distance* between two points—the length of the shortest path connecting the points that lies entirely in $|\mathcal{P}|$. An intrinsic path must go around holes and concavities. Redefine the local feature size $f(x)$ at a point x to be the smallest value such that there are two disjoint linear cells in $\mathcal{P}$ within an intrinsic distance of $f(x)$ from x. The edge lengths in the mesh at right in Figure 6.16 are locally proportional to this modified local feature size. It is a tedious but straightforward exercise to show that the proofs in Sections 6.4 and 6.5 all hold for Ruppert's algorithm with a CDT and the intrinsic local feature size, without the assumption that $|\mathcal{P}|$ is convex.

Consider a different meshing problem: to produce a triangular mesh of a piecewise linear complex in three-dimensional space with no 3-cells, composed of polygons meeting at shared segments. The polygon triangulations must conform to each other—that is, match triangle edge to triangle edge—along their shared boundaries. This problem arises in boundary element methods for solving partial differential equations and in global illumination methods for computer graphics.

The constrained Delaunay refinement algorithm can solve this problem, with no conceptual changes, by meshing all the surfaces simultaneously. Again, the key is to define the local feature size in terms of intrinsic distances in the underlying space of the PLC. Where polygons meet at shared segments, features in one polygon may affect the local feature size in another, reflecting the fact that the refinement of one polygon can propagate into an adjoining polygon by splitting their shared segments.

6.8 Notes and exercises

For the sake of establishing precedence, we note that Ruppert's 1995 article [180] is his fullest presentation of his algorithm and its analysis, but earlier versions appeared in 1992 and 1993 [178, 179]. In 1993, Chew [61] independently discovered a very similar Delaunay refinement algorithm that guarantees a minimum angle of 30°. Unlike his 1989 algorithm described in Section 1.2, his 1993 algorithm offers optimal grading and size optimality for any angle bound less than 26.5°—compared to 20.7° for Ruppert's algorithm—although this property was proven not by Chew, but subsequently by Shewchuk [201]. The improved angle bounds are obtained by using a constrained Delaunay triangulation and a more conservative procedure for treating encroached subsegments. Miller, Pav, and Walkington [147] reanalyze Ruppert's original algorithm and extend its angle guarantee to 26.4°, with size optimality and optimal grading intact. The same techniques show that Chew's algorithm guarantees size optimality and optimal grading up to an angle guarantee of 28.6°. For angle bounds between 28.6° and 30°, Chew's algorithm is guaranteed to terminate but is not guaranteed to produce a graded or size-optimal mesh.

Most of the analysis in this chapter is taken from Ruppert's article, but the proof of Proposition 6.10 is adapted from Mitchell [150] and the proof of Proposition 6.9 is new. The idea to analyze Delaunay meshing algorithms in terms of the radius-edge ratio comes from Miller, Talmor, Teng, and Walkington [148]. The first size-optimality proof for a mesh generation algorithm was given by Bern, Eppstein, and Gilbert [18] for their provably good quadtree mesher, and their quadtree box sizes are a forerunner of Ruppert's local feature size. Mitchell [150] gives a stronger lower bound on the number of triangles in a high-quality triangulation that shrinks proportionally to $\theta_{\min}$ as $\theta_{\min}$ approaches zero, whereas the bound given here shrinks proportionally to $2^{-O(1/\theta_{\min})}$.

The idea to place new vertices at triangle circumcenters originates in a 1987 paper by William Frey [100], who appears to be the first to suggest using the Delaunay triangulation to guide vertex placement, rather than generating all the vertices before triangulating them. Circumcenters are not always the optimal locations to place new vertices. If a skinny triangle's circumcircle is substantially larger than its shortest edge, it is often better to place the new vertex closer to the short edge, so they form an acceptable new triangle. The effect is to make Delaunay refinement behave like an advancing front method. Frey [100] and Üngör [219] report that these *off-centers* give an excellent compromise between the quality and the number of triangles, and Üngör also shows that Ruppert's theoretical results remain true for properly chosen off-centers. A more aggressive algorithm of Erten and Üngör [94] optimizes the placement of a new vertex in a circumcircle; it often generates triangular meshes that have no angle smaller than 41°.

The idea to recover a boundary in a Delaunay triangulation by repeatedly inserting vertices at missing portions of the boundary originates in a 1988 paper by Schroeder and Shephard [188], who named this process *stitching*.

The suggestion to use concentric circular shells for segment splitting comes from Ruppert's original paper. The idea to decline to split triangles with seditious edges, as described in Section 6.6, is a slight variation of an algorithm of Miller, Pav, and Walkington [147]. Pav [167] proves that their algorithm offers good grading. The software TRIANGLE[1] implements Ruppert's algorithm, Chew's 1993 algorithm, Üngör's off-centers, and the modifications for domains with small angles discussed in Section 6.6. Chapters 9 and 15 address the difficulty of meshing three-dimensional domains with small angles.

Another important extension of Ruppert's algorithm is to domains with curved boundaries. The first such extension, by Boivin and Ollivier-Gooch [32], uses CDTs to aid the recovery of curved ridges in triangular meshes. A more recent algorithm by Pav and Walkington [166] can handle cusps—curved ridges that meet at an angle of zero. This algorithm maintains and returns a true Delaunay triangulation.

Yet another important extension is to generate anisotropic meshes. Practical Delaunay mesh generators adapt easily to anisotropy; for instance, George and Borouchaki [103] modify the Bowyer–Watson algorithm to use circumellipses instead of circumspheres. However, if the anisotropy field varies over space, the newly created elements might not have empty circumellipses, and the quality of the mesh cannot be guaranteed. Labelle and Shewchuk [127] propose a provably good algorithm for anisotropic triangular mesh generation that introduces anisotropic Voronoi diagrams to provide a foundation for the mathematical guarantees.

The fact that Ruppert's algorithm might perform work quadratic in the size of the final mesh was first noted by Ruppert in an unpublished manuscript. Barbič and Miller [15] work out an example in detail. Har-Peled and Üngör [109] describe a Delaunay refinement algorithm, essentially Ruppert's algorithm with off-centers, that uses a quadtree to help it

[1]http://www.cs.cmu.edu/~quake/triangle.html

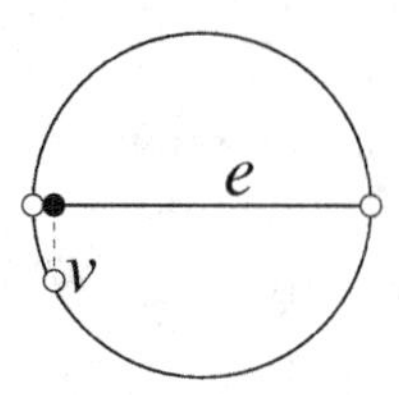

Figure 6.17: Projecting an encroaching vertex onto an encroached subsegment can create a dangerously short edge.

run in optimal $O(n \log n + N)$ time, where n is the number of input vertices and N is the number of output vertices. See also the discussion of *sparse Voronoi refinement* by Hudson, Miller, and Phillips [115] in the Chapter 8 notes, which achieves virtually the same running time without the need for a quadtree.

Exercises

1. Show that an edge $e \in \text{Del}\, S$ is encroached if and only if Del S contains a triangle that has e for an edge and a nonacute angle ($\geq 90^\circ$) opposite e. Therefore, a subsegment's encroachment can be diagnosed in $O(1)$ time.

2. Suppose DelTriPLC takes as input a PLC in which some segments meet each other at angles less than 90°, but never less than 60°. Then Proposition 6.4 no longer suffices, because a vertex inserted on one segment might encroach upon a subsegment on an adjoining segment. Show that nonetheless, DelTriPLC must terminate.

3. Suppose DelTriPLC enforces a stricter standard of quality for triangles that do not intersect the relative interior of a segment: a triangle that does not intersect a segment interior is split if its radius-edge ratio exceeds 1. A triangle that intersects a segment interior is split if its radius-edge ratio exceeds $\sqrt{2}$. Given a PLC in which no two segments meet at an angle less than 90°, show that DelTriPLC still must terminate.

4. Ruppert's algorithm splits encroached subsegments at their midpoints. We sometimes achieve smaller meshes if we use off-center splits when the encroaching vertex is not a rejected circumcenter. For example, if an input vertex v encroaches upon a subsegment e, and v is very close to e but not to e's midpoint, then splitting e off-center might reduce the number of triangles in the final mesh. One idea is to project v orthogonally onto e. Unfortunately, this idea might create an unreasonably short edge, as Figure 6.17 illustrates. Explain how to modify this idea so that Proposition 6.4 still holds, while still splitting subsegments as close to the projected point as possible.

5. If two segments meet each other at an angle of 45° or less, DelTriPLC may fail to terminate because of ping-pong encroachment, illustrated in Figure 6.12. However, suppose we modify DelTriPLC by eliminating Step 4 so skinny triangles are ignored and changing Step 3 so that only subsegments missing from Del S are considered encroached:

3. While some subsegment $e \in E$ is missing from Del S, call SPLITSUBSEGMENT(e, S, E), update Del S, and repeat Step 3.

Let $\mathcal{P}$ be a PLC in which no four segments meet at a single vertex. Show that this modified DELTRIPLC always produces a Steiner Delaunay triangulation of $\mathcal{P}$, no matter how many small angles it has.

6. Suppose that we modify Step 3 of DELTRIPLC as described in the previous exercise, but instead of eliminating Step 4, we replace it with the following.

   4. While Del S contains a triangle τ for which $\rho(\tau) > \bar{\rho}$ and the circumcenter c of τ does not encroach upon any subsegment in E, insert c into S, update Del S, and go to Step 3.

   In this modified algorithm, skinny triangles may survive, but only near the domain boundaries. Quantify the minimum quality of the triangles relative to their proximity to the domain segments in terms of their edge lengths and $\bar{\rho}$.

7. Show that if the local feature size function is modified to use intrinsic distances, it is still 1-Lipschitz, i.e. it still satisfies Proposition 6.3.

8. Show that if $\theta_{\min} \leq 30°$, the inequality in Proposition 6.10 can be improved so that $\eta = (2\cos\theta_{\min})^{180°/\theta_{\min}} / \sin\theta_{\min}$. Hint: This expression follows immediately if the longest edge of τ adjoins v. If τ's longest edge does not adjoin v, what is the angle separating the two edges of τ that do adjoin v?

9. Consider the problem of meshing the PLC $\mathcal{P}$ illustrated in Figure 1.5, which includes two adjoining segments that are separated by a very small angle ϕ. Clearly, it is impossible to avoid placing a triangle with angle ϕ or less at the small domain angle. But ideally, the mesh would have no other angle less than 30°. With help from Proposition 6.10, show that for sufficiently small ϕ, no such Steiner triangulation of $\mathcal{P}$ exists.

10. Prove that Ruppert's algorithm with "modified segment splitting using concentric circular shells" always terminates for domains that are simple polygons with no internal boundaries, even if it splits skinny triangles whose shortest edges are seditious, so long as it declines to split triangles nestled right in the corners of small domain angles.

11. Prove that DELTRIPLC maintains a quarantined complex (see Chapter 7) when it inserts a circumcenter of a triangle.

Chapter 7

Voronoi diagrams and weighted complexes

This chapter presents several complexes used in subsequent chapters. *Voronoi diagrams* are among the oldest geometric complexes studied; they are unbounded polyhedral complexes that associate each point in space with the closest site in a fixed set of sites—for instance, indicating for every point in town the nearest post office, as illustrated in Figure 7.1. Voronoi diagrams arise in the study of natural phenomena such as crystal formation, meteorology, computational chemistry, condensed matter physics, and epidemiology. Historically, they have been rediscovered several times and are also known as Dirichlet tessellations, Thiessen polygons, and Wigner–Seitz cells. They are the natural combinatorial duals of Delaunay subdivisions. In later chapters, they play important roles in ensuring the topological correctness of algorithms for meshing curved surfaces.

We also study *weighted* Voronoi diagrams, in which each site is equipped with a numerical weight; a larger weight allows a site to claim more territory. Weighted Voronoi diagrams are Voronoi diagrams in which the Euclidean distance is replaced with a nonmetric measure called the *power distance*. We have already introduced weighted Delaunay

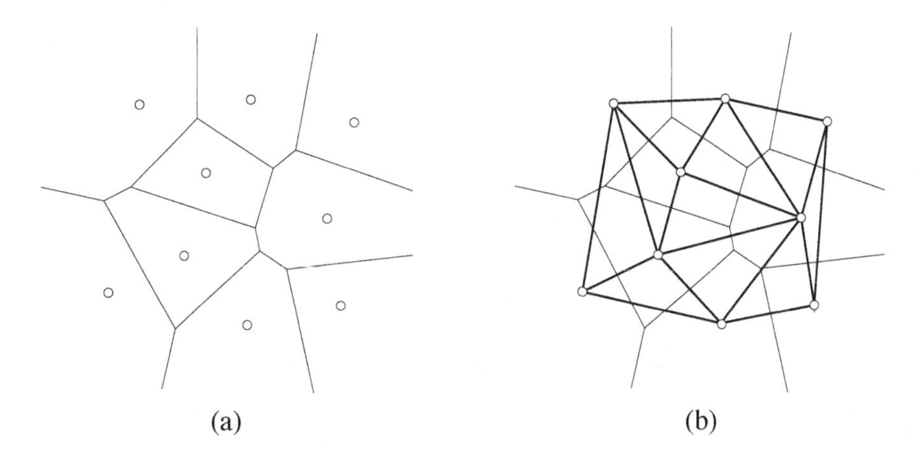

Figure 7.1: (a) The Voronoi diagram of a planar point set. (b) With the Delaunay triangulation.

triangulations, the duals of weighted Voronoi diagrams, in Section 2.8. Here we explore their properties in further detail.

Lastly, we present a kind of subcomplex of a weighted Delaunay triangulation that we call a *quarantined complex*, which satisfies conditions that guarantee that the circumcenters of the simplices lie in the triangulation. Quarantined complexes generalize Proposition 6.2 to higher dimensions and to weighted Delaunay triangulations. We prove results about quarantined complexes called the Monotone Power, Orthoball Cover, and Orthocenter Containment Lemmas, and apply them in Chapters 8, 9, and 11.

7.1 Voronoi diagrams

Let S be a finite set of points, called *sites*, in $\mathbb{R}^d$. The *Voronoi cell* V_u of a site $u \in S$ is the set of all points in $\mathbb{R}^d$ that are at least as close to u as to any other site in S. Formally,

$$V_u = \{ p \in \mathbb{R}^d : \forall w \in S,\ d(u, p) \leq d(w, p) \},$$

and u is said to be the *generating site* of V_u. Each Voronoi cell is a convex polyhedron, possibly an unbounded polyhedron. The faces of the Voronoi cells are called *Voronoi faces*. Voronoi faces of dimensions d, $d-1$, 2, 1, and 0 are called *Voronoi cells*, *Voronoi facets*, *Voronoi polygons*, *Voronoi edges*, and *Voronoi vertices*, respectively. All Voronoi faces are convex, and all except the vertices can be unbounded.

Definition 7.1 (Voronoi diagram)**.** The *Voronoi diagram* of S, denoted Vor S, is the polyhedral complex containing the Voronoi cell of every site $u \in S$ and all the faces of the Voronoi cells.

Observe from the Voronoi diagram in Figure 7.1(a) that sites on the boundary of conv S have unbounded Voronoi cells. In the plane, two Voronoi edges of each unbounded Voronoi cell are rays—unless the sites are collinear, in which case all the Voronoi edges are lines.

Voronoi faces can be characterized by the sites that generate them. Define V_{uw} to be the set of points in the plane that have both sites u and w as their nearest neighbor. Formally,

$$\begin{aligned}
V_{uw} &= V_u \cap V_w \\
&= \{ p \in \mathbb{R}^d : d(u, p) = d(w, p) \text{ and } \forall x \in S,\ d(u, p) \leq d(x, p) \}. \\
V_{u_1 \ldots u_j} &= V_{u_1} \cap \ldots \cap V_{u_j} \\
&= \{ p \in \mathbb{R}^d : d(u_1, p) = \cdots = d(u_j, p) \text{ and } \forall x \in S,\ d(u_1, p) \leq d(x, p) \}.
\end{aligned}$$

If V_{uw} is nonempty, it is a face of the Voronoi diagram and of both Voronoi cells V_u and V_w, and u and w are said to be *Voronoi neighbors* and the *generating sites* or simply *generators* of V_{uw}.

If S is generic (recall Definition 4.2), a Voronoi face generated by j sites has dimension $d+1-j$. For example, in the plane, V_{uw} is an edge of two cells and V_{uwx} is a vertex of three cells. If S is not generic, the dimensionality could be lower or higher; for example, if four vertices in the plane lie in the circular order u, v, w, x on the boundary of an empty open disk, then $V_{uw} = V_{uwv} = V_{uwvx}$ is a vertex. For $j \geq 3$, the face might have any dimension less than or equal to $d-2$; for $j=2$, any dimension less than or equal to $d-1$; for $j=1$, it is always a d-face.

Each Voronoi facet lies on a bisector between two sites. For two distinct points $u, w \in S$, the *bisector* of u and w is the set of points at equal distance from u and w, or equivalently, the hyperplane orthogonal to the segment uw that intersects the midpoint of uw. Clearly, V_{uw} is a subset of the bisector. The bisector cuts $\mathbb{R}^d$ into two halfspaces. Let H_{uw} be the closed halfspace containing u. Then

$$V_u = \bigcap_{w \in S \setminus \{u\}} H_{uw}.$$

Recall that Definition 1.5 defines a *convex polyhedron* to be the convex hull of a finite point set. For Voronoi diagrams, we define it more generally to be the intersection of a finite set of halfspaces, thereby permitting unbounded polyhedra.

For any finite point set S, the Voronoi diagram of S and the Delaunay subdivision of S defined in Section 4.2 are related to each other by a duality illustrated in Figure 7.1(b). A *duality* is a bijective map between the faces of one complex and the faces of another that reverses the property of inclusion; thus, if $\sigma \subset \tau$ are faces of the Delaunay subdivision of S, they map to dual faces $\sigma^* \supset \tau^*$ of the Voronoi diagram of S. Each k-face of the Delaunay subdivision is paired with a $(d-k)$-face of the Voronoi diagram. For example, each Delaunay vertex is paired with its Voronoi cell, and each Voronoi vertex is paired with a Delaunay d-face that is the union of all Delaunay d-simplices whose circumspheres are centered at that Voronoi vertex. The following theorem makes precise the nature of this duality.

Theorem 7.1. *Consider a site set $P = \{u_1, \ldots, u_j\} \subseteq S$. The polyhedron* conv P *is a k-face of the Delaunay subdivision of S if and only if $V_{u_1 \ldots u_j}$ is a $(d-k)$-face of* Vor S *and there is no site $u_{j+1} \in S \setminus P$ such that $V_{u_1 \ldots u_j} = V_{u_1 \ldots u_j u_{j+1}}$.*

Proof. If conv P is a face of the Delaunay subdivision, there exists a closed d-ball B that contains no site except $u_1, \ldots, u_j$, all of which lie on B's boundary. The Voronoi cell $V_{u_1 \ldots u_j}$ contains the center of B, so $V_{u_1 \ldots u_j}$ is nonempty and hence a face of Vor S. Because every site in $S \setminus P$ is further from the center of B than the radius of B, there is no site $u_{j+1} \in S \setminus P$ such that $V_{u_1 \ldots u_j} = V_{u_1 \ldots u_j u_{j+1}}$.

For the reverse implication, let B be a closed d-ball whose center is an arbitrary point in the interior of $V_{u_1 \ldots u_j}$ and whose boundary passes through $u_1, \ldots, u_j$. No other site is in B, so conv P is a face of the Delaunay subdivision.

If conv P has dimension k, let Π be the $(d-k)$-flat that contains B's center and is orthogonal to aff P. Every point on Π is equidistant from the sites in P, but no point in $\mathbb{R}^d \setminus \Pi$ is equidistant, so $V_{u_1 \ldots u_j} \subseteq \Pi$. Every point on Π whose distance from B's center is less than $d(B, S \setminus P)/2$ is closer to the sites in P than to any other site in S and is, therefore, in $V_{u_1 \ldots u_j}$. It follows that $V_{u_1 \ldots u_j}$ has the same dimension as Π, namely, $d-k$. □

To complete the duality, it is sometimes convenient to pair the unbounded outer face of the Delaunay triangulation with a "vertex at infinity" where every ray of the Voronoi diagram is imagined to terminate.

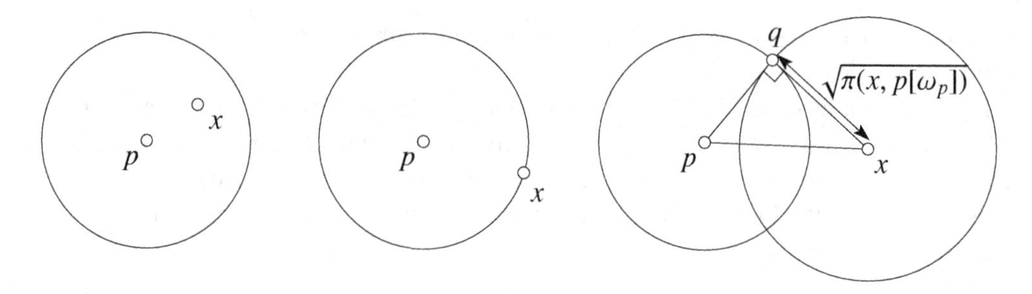

Figure 7.2: From left to right, the power distance $\pi(x, p[\omega_p])$ is negative, zero, and positive. The two balls at right are orthogonal, as $\pi(x[\omega_x], p[\omega_p]) = 0$.

Thanks to duality, the bounds on the complexity of a Delaunay triangulation given in Sections 2.1 and 4.1 apply to Voronoi diagrams. An n-site Voronoi diagram in the plane has at most $2n - 5$ vertices and $3n - 6$ edges. An n-site Voronoi diagram in $\mathbb{R}^3$ has at most $(n^2 - 3n - 2)/2$ vertices. Just as an n-vertex triangulation in $\mathbb{R}^d$ can have as many as $\Theta(n^{\lceil d/2 \rceil})$ d-simplices, an n-site Voronoi diagram can have up to $\Theta(n^{\lceil d/2 \rceil})$ Voronoi vertices.

To compute the Voronoi diagram of a set S of sites, first compute a Delaunay triangulation Del S; Chapters 3 and 5 describe algorithms for doing so. Then compute the circumcenter of each d-simplex in Del S. Where two or more adjoining d-simplices have the same circumcenter, fuse them into a polyhedron (having the same circumcenter), yielding the Delaunay subdivision. Finally, build the dual of the Delaunay subdivision. This last step might require no computation at all; a data structure that represents the Delaunay subdivision and the circumcenters implicitly represents the Voronoi diagram too.

7.2 Weighted Voronoi and weighted Delaunay

Just as Delaunay triangulations naturally generalize to weighted Delaunay triangulations, described in Section 2.8, Voronoi diagrams generalize to *weighted Voronoi diagrams*, the duals of weighted Delaunay subdivisions. Recall that a *weight assignment* $\omega : S \mapsto \mathbb{R}$ maps each site $u \in S$ to a weight ω_u. Negative weights are allowed. A *weighted point set* is written $S[\omega]$, and a *weighted point* is written $u[\omega_u]$.

Definition 7.2 (power distance). The *power distance* between two weighted points $p[\omega_p]$ and $q[\omega_q]$ is

$$\pi(p[\omega_p], q[\omega_q]) = d(p, q)^2 - \omega_p - \omega_q.$$

We sometimes apply the notation to unweighted points, which by default have weight zero.

A weighted point $p[\omega_p]$ can be interpreted as a d-ball $B_p = B(p, \sqrt{\omega_p})$. For an unweighted point x outside B_p, the power distance $\pi(x, p[\omega_p])$ is the square of the length of the line segment that extends from x to touch B_p tangentially, illustrated in Figure 7.2. The power distance is zero if x lies on the ball's boundary, and negative if x lies inside the ball. We sometimes write the power distance $\pi(x, p[\omega_p])$ as $\pi(x, B_p)$. If the weight ω_p is negative, the ball radius $\sqrt{\omega_p}$ is imaginary, and every point lies outside the ball.

A weighted Voronoi diagram is defined like a Voronoi diagram, except that the Euclidean distance is replaced by the power distance. Observe that the larger a point's weight, the closer it is to other points, so a larger weight enables a site to claim a larger cell. The *weighted Voronoi cell* of a site $u \in S[\omega]$ is

$$W_u = \{ p \in \mathbb{R}^d : \forall w \in S[\omega], \pi(p, u[\omega_u]) \leq \pi(p, w[\omega_w]) \}.$$

Counterintuitively, if u's weight is too small to compete with its neighbors, u can lie outside W_u, and W_u can even be empty (in which case u is submerged—not a vertex of the weighted Delaunay triangulation of S). Moreover, W_u is not necessarily a d-face; it can be of any dimension.

Definition 7.3 (weighted Voronoi diagram)**.** The *weighted Voronoi diagram* of $S[\omega]$, denoted Vor $S[\omega]$ and also known as the *power diagram*, is the polyhedral complex containing the weighted Voronoi cell of every site $u \in S$ and all the faces of the weighted Voronoi cells.

As a weighted point can be interpreted as a ball, we sometimes talk about the weighted Voronoi diagram of a set of balls. The faces of a weighted Voronoi cell are called *weighted Voronoi faces* and, like Voronoi faces, are characterized by their generating sites as

$$W_{u_1 \ldots u_j} = W_{u_1} \cap \ldots \cap W_{u_j}.$$

Unlike Voronoi faces, a weighted Voronoi face generated by two or more sites can have any dimension less than d.

Under the power distance, the bisector of $u[\omega_u]$ and $w[\omega_w]$ is still a hyperplane orthogonal to uw, but it intersects the midpoint of uw only if the points have equal weight; otherwise, it is further from the site with greater weight, and it might not intersect uw at all. As Figure 7.3 shows, if the balls centered at u and w intersect, the intersection of their boundaries lies on the bisector. The bisector cuts $\mathbb{R}^d$ into two closed halfspaces; let I_{uw} be the closed halfspace dominated by u. Then

$$W_u = \bigcap_{w \in S \setminus \{u\}} I_{uw}.$$

A fundamental property of the weighted Voronoi diagram is that there is a symmetric relationship between the sites and the vertices of the weighted Voronoi diagram, if we interpret both types of points as balls. Observe that a vertex v of the weighted Voronoi diagram is equidistant in power distance from the weighted points that generate it. Assign v a weight ϖ_v equal to the power distance from v to its generators. Then the power distance from $v[\varpi_v]$ to its generators is zero. Let B_v be the ball centered at v whose radius is $\sqrt{\varpi_v}$. We say that $v[\varpi_v]$ is *orthogonal* to its generators, and we call B_v the *orthoball* centered at v.

Definition 7.4 (orthogonal)**.** Two weighted points—equivalently, two balls—are *orthogonal* to each other if the power distance between them is zero. They are *farther than orthogonal* if the power distance is positive. They are *closer than orthogonal* if the power distance is negative.

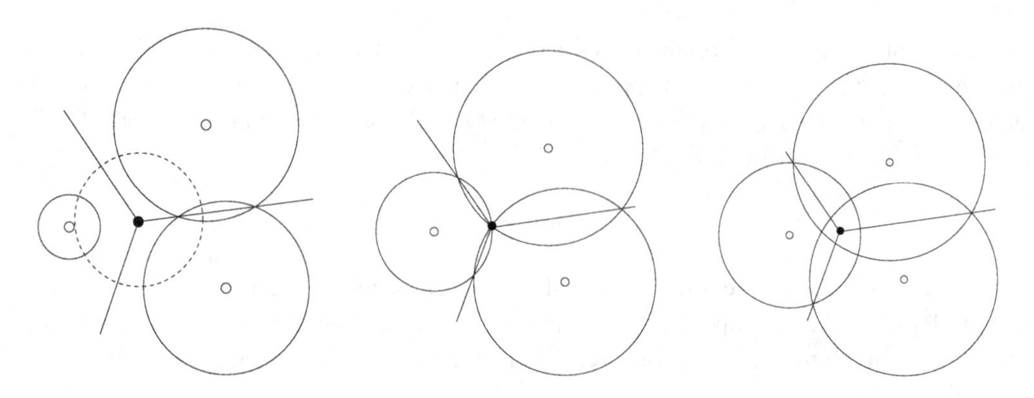

Figure 7.3: Three weighted point sets that have the same weighted Voronoi diagram, although the weights are different in each set. At left, the orthoball (dashed) centered at the weighted Voronoi vertex is orthogonal to all three weighted points. At center, the orthoball has radius zero. At right, the radius of the orthoball is imaginary.

If the weights are positive, orthogonality implies that the boundaries of the balls intersect each other at right angles, as illustrated at right in Figure 7.2. To see this, consider weighted points p and x with associated balls B_p and B_x, and let q be a point on the boundary of both balls. If the balls meet at right angles, $\angle pqx$ is a right angle, and by Pythagoras' Theorem $\omega_p + \omega_x = d(p, x)^2$, so the power distance between $p[\omega_p]$ and $x[\omega_x]$ is zero.

Orthoballs generalize the notion of circumballs to simplices with weighted vertices. For example, a d-simplex with weighted vertices has a unique orthoball, namely, the orthoball centered at the vertex of the weighted Voronoi diagram of the simplex's $d + 1$ vertices, as illustrated at left in Figure 7.3. More generally, an orthoball of a simplex σ can be understood in two ways: as a ball that is orthogonal to σ's weighted vertices, or as the intersection of a hyperplane that includes the lifted simplex σ^+ with the paraboloid induced by the parabolic lifting map.

Definition 7.5 (orthoball). Let σ be a k-simplex or a convex k-polyhedron with weighted vertices. An *orthoball* of σ is a ball (open or closed, as desired) that is orthogonal to σ's weighted vertices. In other words, the center c and radius $\sqrt{\varpi_c}$ of the orthoball satisfy $d(v, c)^2 - \omega_v - \varpi_c = 0$ for every vertex v of σ. Note that ϖ_c can be negative, in which case the radius is imaginary. An *orthosphere* of σ is the boundary of an orthoball of σ.

If σ has an orthoball (every simplex does, but not every polyhedron), there is one unique orthoball of σ whose center lies on the affine hull of σ, called the *diametric orthoball* of σ. The diametric orthoball's center is called the *orthocenter* of σ, its radius is the *orthoradius* of σ, and its intersection with aff σ is the *k-orthoball* of σ, which has the same center and radius as the diametric orthoball. An *orthodisk* is a 2-orthoball.

Analogous to circumballs, a k-simplex in $\mathbb{R}^d$ has exactly one k-orthoball, but if $k < d$ it has infinitely many d-orthoballs. The k-orthoball is a cross-section of every d-orthoball (see Lemma 7.4). Among all d-orthoballs, the diametric orthoball has minimum radius. Not every convex polyhedron has an orthoball, but every face of a weighted Delaunay subdivision does.

The alternative way to understand an orthosphere O of a simplex σ is to imagine lifting O to $\mathbb{R}^{d+1}$ with the parabolic lifting map, yielding the ellipsoid $O^+ = \{(p, \|p\|^2) : p \in O\}$, which is the intersection of a hyperplane with the paraboloid $p_{d+1} = \|p\|^2$. With this observation, every orthosphere induces a non-vertical hyperplane in $\mathbb{R}^{d+1}$ and vice versa. To see this, recall from the proof of Lemma 2.1 that if O has center o and radius $\sqrt{\varpi_o}$, the lifted ellipsoid O^+ lies on a hyperplane $h \subset \mathbb{R}^{d+1}$ whose formula is $p_{d+1} = 2o \cdot p - \|o\|^2 + \varpi_o$, where p varies over $\mathbb{R}^d$. By Definition 7.5, $\|v\|^2 - \omega_v = 2o \cdot v - \|o\|^2 + \varpi_o$ for every vertex v of σ. Recall that a weighted vertex v is lifted to a height of $\|v\|^2 - \omega_v$; it follows that $v^+ \in h$ for every vertex v of σ, so $\sigma^+ \subset h$.

This intuition fails for an orthoball of σ with imaginary radius, which induces a hyperplane that includes σ^+ but passes below the paraboloid and does not intersect it. But because imaginary radii are permitted, the one-to-one map from orthospheres to hyperplanes still holds. This map also provides an alternative way to understand the power distance; see Exercise 6.

Just as a simplex is Delaunay if it has an empty open circumball, a simplex is *weighted Delaunay* if it has an orthoball at a nonnegative power distance from every weighted vertex, or equivalently if it has a witness hyperplane (recall Definition 2.5). Observe that these are precisely the conditions in which its vertices $u_1, \ldots, u_j$ induce a face $W_{u_1 \ldots u_j}$ of the weighted Voronoi diagram.

Theorem 7.2. *Consider a weighted site set* $P[\omega] = \{u_1, \ldots, u_j\} \subseteq S[\omega]$. *The polyhedron* conv P *is a* k*-face of the weighted Delaunay subdivision of* S *if and only if* $W_{u_1 \ldots u_j}$ *is a* $(d-k)$*-face of* Vor $S[\omega]$ *and there is no site* $u_{j+1} \in S \setminus P$ *such that* $W_{u_1 \ldots u_j} = W_{u_1 \ldots u_j u_{j+1}}$.

Proof. Analogous to the proof of Theorem 7.1. □

This duality is a nearly symmetric relationship, because every weighted vertex of a weighted Delaunay face of any dimension is orthogonal to every weighted vertex of its dual face in the weighted Voronoi diagram. Therefore, every weighted Voronoi d-cell W_u has an orthoball, namely, the ball B_u induced by the weighted site $u[\omega_u]$. The only break in the symmetry arises from the unbounded cells of the weighted Voronoi diagram. If $U[\varpi]$ is the set of weighted vertices of the weighted Voronoi diagram Vor $S[\omega]$, then the weighted vertices of the weighted Voronoi diagram Vor $U[\varpi]$ include the original sites in $S[\omega]$ except those on the boundary of conv S.

A useful fact is that the orthocenter of a k-face σ of a weighted Delaunay subdivision is the point aff $\sigma \cap$ aff σ^*, where σ^* is σ's dual face in the weighted Voronoi diagram. Because every weighted Voronoi diagram is the dual of a weighted Delaunay subdivision, the complexity bounds given for Voronoi diagrams in Section 7.1 apply to weighted Voronoi diagrams as well.

7.2.1 Properties of orthoballs

This section proves some geometric properties of orthoballs, the power distance, and weighted Delaunay triangulations. The Delaunay triangulation is a special case of the weighted Delaunay triangulation, so the propositions that follow apply to ordinary Delaunay triangulations as well.

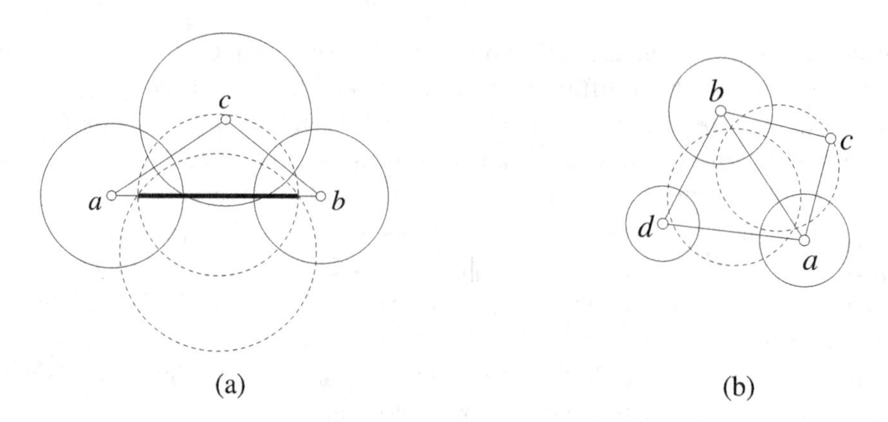

Figure 7.4: (a) The intersection of edge ab and the orthoball of abc (the bigger dashed circle) is an interval (bold). The smaller dashed circle indicates the diametric ball of this interval and the diametric orthoball of ab. (b) The bisector of the orthoballs of abc and abd is the line that includes the edge ab.

The *orthogonal projection* of a point p onto a flat or a linear cell $g \subseteq \mathbb{R}^3$ is the point $\tilde{p} \in \operatorname{aff} g$ such that the line segment $p\tilde{p}$ is perpendicular to the affine hull $\operatorname{aff} g$.

An interesting property of weighted Voronoi diagrams is that any cross-section of a weighted Voronoi diagram is a lower-dimensional weighted Voronoi diagram of a modified set of sites that lie in the cross-section. The modified sites are found by orthogonally projecting the sites onto the cross-sectional flat and adjusting their weights as described in the following proposition.

Proposition 7.3. *Let $\Pi \subset \mathbb{R}^d$ be a flat. Let $p \in \mathbb{R}^d$ be a point with weight ω_p. Let $\tilde{p}$ be the orthogonal projection of p onto Π, with weight $\omega_{\tilde{p}} = \omega_p - d(p, \tilde{p})^2$. For every point $x \in \Pi$, $\pi(x, p[\omega_p]) = \pi(x, \tilde{p}[\omega_{\tilde{p}}])$.*

Proof. $\pi(x, \tilde{p}[\omega_{\tilde{p}}]) = d(x, \tilde{p})^2 - \omega_{\tilde{p}} = d(x, \tilde{p})^2 + d(\tilde{p}, p)^2 - \omega_p = d(x, p)^2 - \omega_p = \pi(x, p[\omega_p])$. □

If $\omega_{\tilde{p}}$ is nonnegative, the projected site $\tilde{p}[\omega_{\tilde{p}}]$ can be interpreted as $B(\tilde{p}, \sqrt{\omega_{\tilde{p}}})$, the smallest d-ball that includes the cross-section $B(p, \sqrt{\omega_p}) \cap \Pi$. If $\omega_{\tilde{p}}$ is negative, Proposition 7.3 still applies, but the "cross-section" has imaginary radius. The same principle applies not only to weighted sites, but also to orthoballs. The following proposition shows that cross-sections of an orthoball of a simplex indicate the diametric orthoballs of its faces, as illustrated in Figure 7.4(a).

Lemma 7.4 (Orthoball Restriction Lemma). *Let τ be a simplex with weighted vertices in $\mathbb{R}^d$. Let $B(c, \sqrt{\varpi_c})$ be an orthoball of τ, and let σ be a face of τ. Let $\tilde{c}$ be the orthogonal projection of c onto $\operatorname{aff} \sigma$. Then the orthocenter and orthoradius of σ are $\tilde{c}$ and $\sqrt{\varpi_{\tilde{c}}}$, respectively, where $\varpi_{\tilde{c}} = \varpi_c - d(c, \tilde{c})^2$, and $B(\tilde{c}, \sqrt{\varpi_{\tilde{c}}})$ is the diametric orthoball of σ.*

Proof. By the definition of orthoball, $B(c, \sqrt{\varpi_c})$ is orthogonal to the weighted vertices of τ. By Proposition 7.3, $B_{\tilde{c}} = B(\tilde{c}, \sqrt{\varpi_{\tilde{c}}})$ also is orthogonal to the weighted vertices of σ and is, therefore, an orthoball of σ. Because $\tilde{c} \in \operatorname{aff} \sigma$, $B_{\tilde{c}}$ is the diametric orthoball. □

Just as every weighted Voronoi facet lies on a bisector of two weighted points, every weighted Delaunay facet not on the boundary of the triangulation lies on a bisector of two orthoballs of d-simplices, as illustrated in Figure 7.4(b). See the proof of Lemma 7.5 for confirmation.

7.3 Quarantined complexes

In our presentation of Ruppert's algorithm, Proposition 6.2 shows that when no subsegment is encroached, every triangle's circumcenter lies in the domain, so the algorithm can safely insert the circumcenter of any skinny triangle. In this section, we generalize the proposition to higher-dimensional triangulations and weighted Delaunay triangulations to support the Delaunay refinement algorithms in this book. We study triangulations called *quarantined complexes* that have the following desirable properties.

Definition 7.6 (quarantined complex). Let $\mathcal{P}$ be a PLC in $\mathbb{R}^d$, let $S \subset |\mathcal{P}|$ be a finite point set that includes the vertices of $\mathcal{P}$, and let ω be a weight assignment such that a subcomplex $\mathcal{Q}$ of Del $S[\omega]$ is a Steiner triangulation of $\mathcal{P}$. The *dimension* k of both $\mathcal{P}$ and $\mathcal{Q}$ is the dimension of their highest-dimensional cell, which is not necessarily d. A j-simplex in $\mathcal{Q}$ is called a *boundary simplex* of $\mathcal{Q}$ if $j < k$ and it is included in a linear j-cell in $\mathcal{P}$. We call $\mathcal{Q}$ a quarantined complex if it satisfies the following conditions.

(i) The dimension of aff $|\mathcal{Q}|$ is equal to the dimension of $\mathcal{Q}$.

(ii) Every vertex in $\mathcal{Q}$ has nonnegative weight.

(iii) The power distance between every pair of vertices in $\mathcal{Q}$ is nonnegative.

(iv) For every boundary simplex σ of $\mathcal{Q}$ and every vertex v in $\mathcal{Q}$, the power distance between $v[\omega_v]$ and the diametric ball B_σ of σ is nonnegative, i.e. $\pi(v[\omega_v], B_\sigma) \geq 0$.

Figure 7.5 illustrates a quarantined complex and some complexes that are not quarantined. Note that $\mathcal{P}$ is sometimes a subcomplex of a larger PLC; perhaps $\mathcal{P}$ contains a linear cell and its faces. Observe that if every vertex has weight zero, condition (iii) is trivially satisfied, as $\pi(u, v)^2 = d(u, v)^2 > 0$, and condition (iv) requires each boundary simplex to have an open diametric ball that contains no vertex. In Ruppert's algorithm, condition (iv) holds when no subsegment is encroached.

Definition 7.7 (balls in a quarantined complex). For any quarantined complex $\mathcal{Q}$ of dimension k in $\mathbb{R}^d$, let Ball($\mathcal{Q}$) denote the set of closed balls that contains $B(v, \sqrt{\omega_v})$ for every vertex v in $\mathcal{Q}$, the diametric orthoball of every k-simplex in $\mathcal{Q}$, and the diametric orthoball of every boundary simplex of $\mathcal{Q}$ that is not a vertex.

Figure 7.6 illustrates the balls in Ball($\mathcal{Q}$) for one quarantined complex. No vertex in $\mathcal{Q}$ is closer than orthogonal to any ball in Ball($\mathcal{Q}$) except that vertex's own ball: this is true for vertex balls by condition (iii), for diametric orthoballs of boundary simplices by condition (iv), and for diametric orthoballs of k-simplices because $\mathcal{Q}$ is a Steiner weighted Delaunay triangulation.

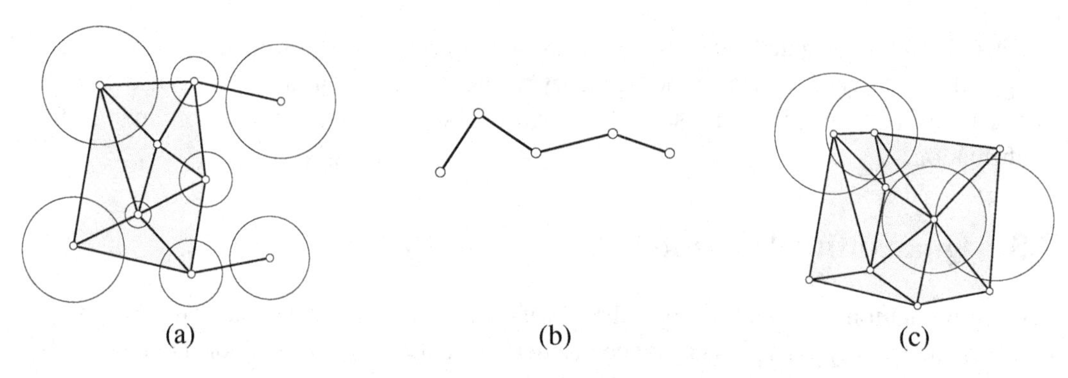

Figure 7.5: (a) A quarantined complex. The weighted vertices are represented by balls. (b) This one-dimensional complex is not a quarantined complex because it has dimension one and its affine hull has dimension two. (c) This two-dimensional complex is not a quarantined complex because two weighted vertices, represented by the balls at upper left, are closer than orthogonal, and the diametric orthoball of the boundary simplex at right is closer than orthogonal to a vertex.

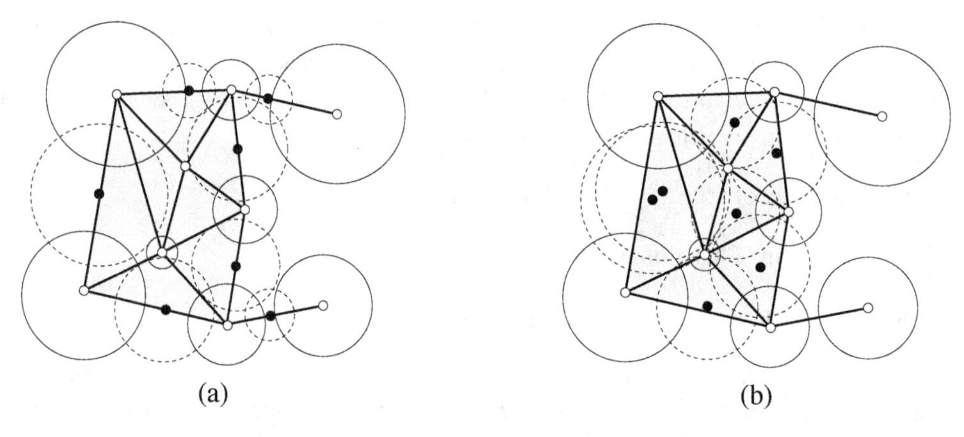

Figure 7.6: Balls in a quarantined complex. (a) The weighted vertices (solid circles) and the diametric orthoballs of the boundary edges (dashed). (b) Orthoballs of triangles (dashed).

7.3.1 The Monotone Power Lemma

The Monotone Power Lemma states that in a k-dimensional quarantined complex, the diametric orthoballs of the k-simplices and boundary simplices that intersect an arbitrary ray have power distances from the origin of the ray that increase monotonically along the ray, as illustrated in Figure 7.7. The result also holds for the power distances from a point whose orthogonal projection onto the affine hull of the quarantined complex is the origin of the ray. We use the lemma to prove the correctness of mesh generation algorithms in Chapters 8, 9, and 11. To apply the lemma, let $\mathcal{Q}$ be a quarantined complex in $\mathbb{R}^d$, let Π be the affine hull of $|\mathcal{Q}|$, and let σ and τ be simplices in $\mathcal{Q}$.

Lemma 7.5 (Monotone Power Lemma)**.** *Let Π be a flat in $\mathbb{R}^d$. Let $\vec{\gamma} \subset \Pi$ be a ray that intersects two simplices $\sigma, \tau \subset \Pi$ in that order—meaning that the ray strikes a point in σ then a distinct point in τ. (Either point may be in $\sigma \cap \tau$, and one simplex may be a face of*

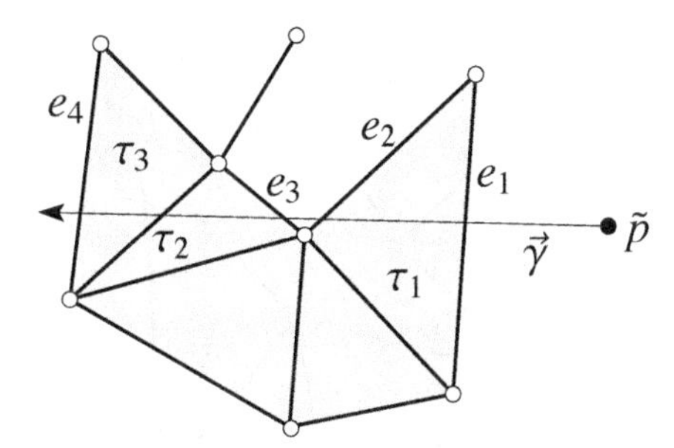

Figure 7.7: A demonstration of the Monotone Power Lemma. The ray $\vec{\gamma}$ crosses the simplices $e_1, \tau_1, e_2, e_3, \tau_2, \tau_3, e_4$ in that order. The power distances from $\tilde{p}$ to their diametric orthoballs increase monotonically in that order.

the other.) Let B_σ and B_τ be orthoballs of σ and τ whose centers lie on Π. Suppose that no vertex of σ is closer than orthogonal to B_τ and no vertex of τ is closer than orthogonal to B_σ. Let $p \in \mathbb{R}^d$ be an unweighted point whose orthogonal projection onto Π, written $\tilde{p}$, is the source of $\vec{\gamma}$. Then $\pi(p, B_\sigma) \leq \pi(p, B_\tau)$.

PROOF. If $B_\sigma = B_\tau$, then $\pi(p, B_\sigma) = \pi(p, B_\tau)$ and the result follows, so assume that $B_\sigma \neq B_\tau$. Because $p\tilde{p}$ is orthogonal to Π, $\pi(p, B_i) = d(p, \tilde{p})^2 + \pi(\tilde{p}, B_i)$ for $i \in \{\sigma, \tau\}$, and it suffices to prove that $\pi(\tilde{p}, B_\sigma) \leq \pi(\tilde{p}, B_\tau)$.

Let $H = \{x \in \mathbb{R}^d : \pi(x, B_\sigma) = \pi(x, B_\tau)\}$ be the bisector of B_σ and B_τ with respect to the power distance. Every vertex of σ is orthogonal to B_σ and orthogonal or farther than orthogonal to B_τ, and therefore lies in the closed halfspace $\{x \in \mathbb{R}^d : \pi(x, B_\sigma) \leq \pi(x, B_\tau)\}$ bounded by H. It follows that every point in σ lies in the same halfspace. Symmetrically, every point in τ lies in the closed halfspace $\{x \in \mathbb{R}^d : \pi(x, B_\sigma) \geq \pi(x, B_\tau)\}$. Because $\vec{\gamma}$ intersects a point in σ then a point in τ, its origin $\tilde{p}$ is in the former halfspace—that is, $\pi(\tilde{p}, B_\sigma) \leq \pi(\tilde{p}, B_\tau)$. □

7.3.2 The Orthoball Cover Lemma

The balls of a quarantined complex $\mathcal{Q}$ cover its underlying space—every point in $|\mathcal{Q}|$ is in some ball in $\text{Ball}(\mathcal{Q})$. More generally, suppose that a weighted point $x[\varpi_x]$ in $|\mathcal{Q}|$ has nonnegative power distances from the weighted vertices of a quarantined complex $\mathcal{Q}$. Then for any point p in space, there is a ball in $\text{Ball}(\mathcal{Q})$ at least as close to p in power distance as $x[\varpi_x]$ is, as illustrated in Figure 7.8. (See Exercise 10 for the relationship between these two claims.)

The Orthoball Cover Lemma below generalizes this claim further. It is a crucial step to proving the Orthocenter Containment Lemma, and it is also the key to guaranteeing boundary conformity for a postprocessing step in Chapter 11 that modifies vertex weights to eliminate sliver tetrahedra from a mesh.

Lemma 7.6 (Orthoball Cover Lemma). *Let $\mathcal{Q}$ be a quarantined complex in $\mathbb{R}^d$, with a weight assignment ω to the vertices of $\mathcal{Q}$. Let $x[\varpi_x]$ be a weighted point in $\text{aff}\,|\mathcal{Q}|$ such that $\pi(v[\omega_v], x[\varpi_x]) \geq 0$ for every vertex v in $\mathcal{Q}$. Let p be a point in $\mathbb{R}^d$, and let $\tilde{p}$ be the*

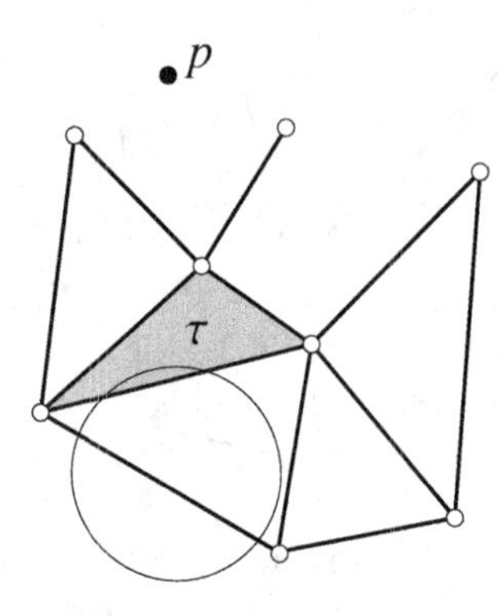

Figure 7.8: All vertices are unweighted. A weighted point $x[\varpi_x]$ is drawn as a circle, at a positive power distance from every vertex. The power distance from p to the circumball of the triangle τ is no greater than the power distance from p to $x[\varpi_x]$.

orthogonal projection of p onto aff $|\mathcal{Q}|$. *If $x\tilde{p}$ intersects $|\mathcal{Q}|$, then there is a ball $B \in \text{Ball}(\mathcal{Q})$ such that $\pi(p, B) \leq \pi(p, x[\varpi_x])$.*

Proof. If p coincides with a vertex $v \in \mathcal{Q}$, the ball associated with $v[\omega_v]$ satisfies the claim because $\pi(p, v[\omega_v]) = -\omega_v$, which is nonpositive because $\mathcal{Q}$ is a quarantined complex, and $\pi(p, x[\varpi_x]) = \pi(v[\omega_v], x[\varpi_x]) + \omega_v \geq \omega_v$, which is nonnegative.

Consider a point $p \in |\mathcal{Q}|$ that is not a vertex. Let σ be the highest-dimensional simplex in $\mathcal{Q}$ that contains p. Either σ has the same dimensionality as $\mathcal{Q}$ or σ is a boundary simplex. In either case, the diametric orthoball B_σ of σ is in Ball($\mathcal{Q}$). We will see that B_σ satisfies the claim.

Following the proof of Lemma 2.1, associate the weighted point $x[\varpi_x]$ with the hyperplane $h \subset \mathbb{R}^{d+1}$ whose formula is $q_{d+1} = 2x \cdot q - \|x\|^2 + \varpi_x$, where q varies over $\mathbb{R}^d$. The power distance $\pi(v[\omega_v], x[\varpi_x])$ is the signed vertical distance of v^+ above h (see Exercise 6), and the requirement that $\pi(v[\omega_v], x[\varpi_x]) \geq 0$ for every vertex v in $\mathcal{Q}$ is equivalent to requiring that every lifted vertex $v^+ \in \mathbb{R}^{d+1}$ lie on or above h. Therefore, the entire lifted complex $\mathcal{Q}^+ = \{\sigma^+ : \sigma \in \mathcal{Q}\}$ lies on or above h.

Likewise, $\pi(p, B_\sigma)$ is the signed vertical distance of p^+ above σ^+. As σ^+ lies on or above h, $\pi(p, B_\sigma) \leq \pi(p, x[\varpi_x])$ and B_σ satisfies the claim.

Consider a point $p \notin |\mathcal{Q}|$. If $\tilde{p} \in |\mathcal{Q}|$, the reasoning above establishes that there is a ball $B \in \text{Ball}(\mathcal{Q})$ such that $\pi(\tilde{p}, B) \leq \pi(\tilde{p}, x[\varpi_x])$. Because $p\tilde{p}$ is orthogonal to aff $|\mathcal{Q}|$ and B is centered on aff $|\mathcal{Q}|$, $\pi(p, B) = d(p, \tilde{p})^2 + \pi(\tilde{p}, B) \leq \pi(p, x[\varpi_x])$ and B satisfies the claim.

If $p \notin |\mathcal{Q}|$ and $\tilde{p} \notin |\mathcal{Q}|$, we prove the lemma by induction on the dimension k of $\mathcal{Q}$. In the base case, $k = 0$ and $\tilde{p}$ must be in $|\mathcal{Q}|$, so the lemma holds. If $k > 0$, assume for the inductive step that the lemma holds for quarantined complexes of dimension less than k. Because $x\tilde{p}$ intersects $|\mathcal{Q}|$, a walk from x to $\tilde{p}$ leaves $|\mathcal{Q}|$ at a point y on the boundary of $|\mathcal{Q}|$. Let $\mathcal{P}$ be the PLC that $\mathcal{Q}$ triangulates, let $g \in \mathcal{P}$ be the linear cell of least dimension that contains y, as illustrated in Figure 7.9, and observe that $\tilde{p} \notin \text{aff}\, g$. Let $\mathcal{Q}|_g = \{\sigma \in \mathcal{Q} : \sigma \subseteq g\}$ be the subset of simplices that triangulate g, and observe that $\mathcal{Q}|_g$ is a quarantined complex of dimension less than k.

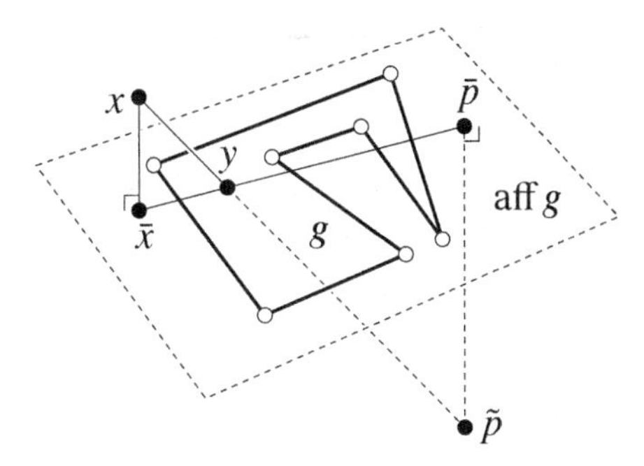

Figure 7.9: The linear cell g of least dimension through which $x\tilde{p}$ exits $|\mathcal{Q}|$.

Let $\bar{x}$ be the projection of x onto aff g, and let $\varpi_{\bar{x}} = \varpi_x - d(x, \bar{x})^2$. By Proposition 7.3, the weighted point $\bar{x}[\varpi_{\bar{x}}]$ is at nonnegative power distances from the weighted vertices of g because $x[\varpi_x]$ is. Observe that the projection $\bar{p}$ of p onto aff g is also the projection of $\tilde{p}$ onto aff g, and that $\bar{x}\bar{p}$ intersects g because g, $x\tilde{p}$, and $\bar{x}\bar{p}$ all contain y, as illustrated. By the inductive assumption, there is a ball $B \in \text{Ball}(\mathcal{Q}|_g)$ such that $\pi(p, B) \leq \pi(p, \bar{x}[\varpi_{\bar{x}}])$.

The ball B satisfies the claim by the following reasoning. Because $x\tilde{p}$ intersects g but $\tilde{p} \notin$ aff g, either x lies on g or $x\tilde{p}$ crosses aff g. In the former case, $\bar{x} = x$; in the latter case, $\angle x\bar{x}\tilde{p} > 90°$ and therefore $\angle x\bar{x}p > 90°$. In either case, $d(p, \bar{x})^2 + d(\bar{x}, x)^2 \leq d(p, x)^2$. It follows that $\pi(p, B) \leq d(p, \bar{x})^2 - \varpi_{\bar{x}} = d(p, \bar{x})^2 + d(\bar{x}, x)^2 - \varpi_x \leq d(p, x)^2 - \varpi_x = \pi(p, x[\varpi_x])$. □

The proof of the Orthoball Cover Lemma uses only the first two properties of a quarantined complex, and it does not use the weighted Delaunay property. Hence, it holds for a wide class of PLC triangulations besides quarantined complexes.

7.3.3 The Orthocenter Containment Lemma

The Orthocenter Containment Lemma states that all the orthocenters of the k-simplices and boundary simplices in a quarantined complex lie in its underlying space. Delaunay refinement algorithms insert orthocenters of simplices to obtain domain conformity and high element quality, so it is important to ensure that the orthocenters lie in the domain.

Lemma 7.7 (Orthocenter Containment Lemma). *Let $\mathcal{Q}$ be a quarantined complex of dimension k in $\mathbb{R}^d$. The orthocenter of every k-simplex lies in $|\mathcal{Q}|$, as does the orthocenter of every boundary simplex of $\mathcal{Q}$. More specifically, let $\mathcal{P}$ be the PLC that $\mathcal{Q}$ triangulates. If an i-simplex $\tau \in \mathcal{Q}$ is included in a linear i-cell $g \in \mathcal{P}$, then the orthocenter of τ lies in g.*

Proof. Suppose for the sake of contradiction that some i-simplex $\tau \in \mathcal{Q}$ is included in a linear i-cell $g \in \mathcal{P}$ but has an orthocenter $c \notin g$. Let p be a point in the interior of τ. The line segment from p to c leaves g at a point y on g's boundary, illustrated in Figure 7.10. Let $f \in \mathcal{P}$ be the face of g of least dimension that contains y, and observe that $c \notin$ aff f and $p \notin$ aff f. Let $\tilde{c}$ be the orthogonal projection of c onto aff f. Because pc crosses f, $\angle p\tilde{c}c > 90°$.

Let $\sqrt{\varpi_c}$ be the orthoradius of τ, and let $\varpi_{\tilde{c}} = \varpi_c - d(c, \tilde{c})^2$. By Proposition 7.3, the weighted point $\tilde{c}[\varpi_{\tilde{c}}]$ has nonnegative power distances from the weighted vertices of

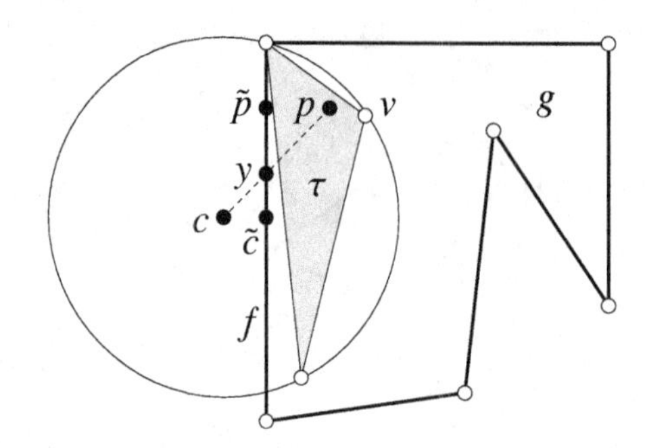

Figure 7.10: A face f of g separates the simplex $\tau \subseteq g$ from its orthocenter c.

f because $c[\varpi_c]$ does. Let $\mathcal{Q}|_f = \{\sigma \in \mathcal{Q} : \sigma \subseteq f\}$ be the subset of simplices that triangulate f. Observe that $\mathcal{Q}|_f$ is a quarantined complex. Let $\tilde{p}$ be the orthogonal projection of p onto aff f, and observe that $\tilde{p}\tilde{c}$ contains y and therefore intersects f, as illustrated. By the Orthoball Cover Lemma (Lemma 7.6), there is a ball $B \in \mathrm{Ball}(\mathcal{Q}|_f)$ such that $\pi(p, B) \leq \pi(p, \tilde{c}[\varpi_{\tilde{c}}])$. From this and the fact that $\angle p\tilde{c}c > 90°$, we have $\pi(p, B) \leq d(p, \tilde{c})^2 - \varpi_{\tilde{c}} = d(p, \tilde{c})^2 + d(\tilde{c}, c)^2 - \varpi_c < d(p, c)^2 - \varpi_c = \pi(p, c[\varpi_c])$. Because $p \in \tau$ and $\pi(p, B) - \pi(p, c[\varpi_c])$ is a linear function of p, there is a vertex v of τ for which $\pi(v, B) < \pi(v, c[\varpi_c])$. Because $c[\varpi_c]$ is orthogonal to $v[\omega_v]$, we conclude that $\pi(v[\omega_v], B) < \pi(v[\omega_v], c[\varpi_c]) = 0$. But this implies that v is closer than orthogonal to a ball in $\mathrm{Ball}(\mathcal{Q})$, which is not permitted in a quarantined complex. □

The proof can be interpreted as saying that if the orthocenter of τ lies outside g, with a face f of g separating τ from its orthocenter, then some vertex of τ encroaches upon a boundary simplex included in f or is too close to a vertex of f for $\mathcal{Q}$ to be a quarantined complex. The encroaching vertex is the vertex of τ that lies farthest toward B's side of the bisector between B and the orthocenter $c[\varpi_c]$.

7.4 Notes and exercises

Gustav Dirichlet [81] first used two-dimensional and three-dimensional Voronoi diagrams in the study of quadratic forms in 1850. The general Voronoi diagram in $\mathbb{R}^d$ is named after Georgy Feodosevich Voronoi [220] (sometimes written Voronoy) who defined it in 1908. Voronoi was one of Boris Delaunay's two doctoral advisors.

The informal use of the Voronoi diagram dates back to Descartes's *Principia Philosophiae* [70] of 1644. In an illustration, he decomposes space into convex Voronoi-like regions that contain one star each. Each star seems to exert more influence in its region than the other stars. The concept of Voronoi diagrams has been conceived in several fields and given different names, such as *domains of action* in crystallography [160], *Wigner–Seitz zones* in metallurgy [224], and *Thiessen polygons* in geography and meteorology [114, 214].

The power distance as a generalized distance function was known to Voronoi. The weighted Voronoi diagram, also known as the *power diagram*, has been used in problems in packings and coverings of spheres [177] and number theory [135]. In computational geometry, Aurenhammer [9] pioneered the study of the weighted Voronoi diagram. For in-depth surveys of Voronoi diagrams, see Aurenhammer [10] and Aurenhammer and Klein [11].

The Monotone Power Lemma generalizes a result of Edelsbrunner [85], who uses it to show that if the vertices of a weighted Delaunay triangulation are generic, then from the perspective of any fixed point in $\mathbb{R}^d$, it is impossible to have a cycle of d-simplices in which each simplex overlaps the next simplex in the cycle. The insight is that any ray shot from the viewpoint intersects d-simplices in increasing order according to the power distances of their orthoballs from the viewpoint. The result does not hold if the vertices are not generic, because it is possible to have a cycle of overlapping simplices that have the same orthoball.

The Orthocenter Containment Lemma began with Ruppert's proof of Proposition 6.2 for unweighted points in the plane [180], was partly extended to three dimensions by Shewchuk [198], and was extended to weighted Delaunay triangulations by Cheng and Dey [48].

Exercises

1. Draw the Voronoi diagram and Delaunay triangulation of the following point set.

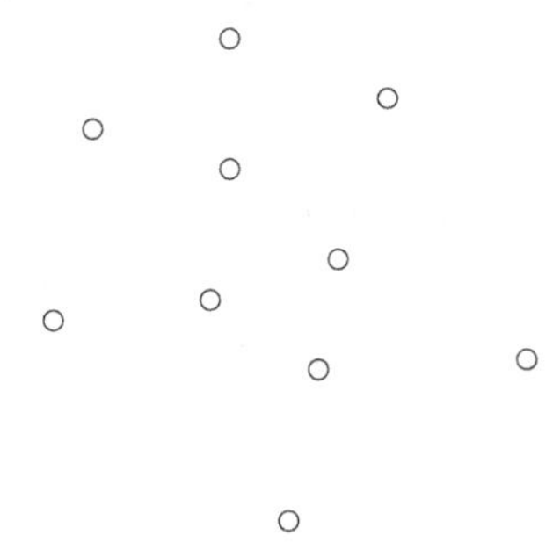

2. Let Y and Z be two point sets in the plane, together having n points. Consider overlaying their Voronoi diagrams $\text{Vor}\, Y$ and $\text{Vor}\, Z$: that is, draw the edges of both Voronoi diagrams and add a new vertex wherever an edge of $\text{Vor}\, Y$ intersects an edge of $\text{Vor}\, Z$, thereby creating a polyhedral complex called the *overlay* of $\text{Vor}\, Y$ and $\text{Vor}\, Z$. Formally, for each face $f \in \text{Vor}\, Y$ and each face $g \in \text{Vor}\, Z$, the overlay contains the face $f \cap g$ if it is nonempty.

   (a) Give an example showing how to choose sets Y and Z so the overlay has $\Theta(n^2)$ faces. Make sure it is clear how to generalize your example as $n \to \infty$.

   (b) Prove that for any two point sets Y and Z, the overlay of their Voronoi diagrams has only $O(n)$ *unbounded* faces.

3. The worst-case complexity of a Voronoi diagram of n sites is $\Theta(n^2)$ in three dimensions and $\Theta(n^{\lceil d/2 \rceil})$ in $\mathbb{R}^d$.

   (a) Prove that a planar cross-section of a d-dimensional Voronoi diagram—in other words, the polygonal complex formed by intersecting a 2-flat with a Voronoi diagram—has complexity no greater than $O(n)$.

(b) Does a planar cross-section of a d-dimensional Delaunay triangulation also always have a complexity of $O(n)$? Explain.

4. Show that the weighted Voronoi bisector of two weighted points is a hyperplane orthogonal to the line through the two points. Then show that if the points have nonnegative weights and are farther than orthogonal to each other, the bisector lies between them.

5. Show that at most two triangular faces of a tetrahedron can lie between the tetrahedron's interior and its circumcenter, but if its vertices have nonzero weights, three triangular faces can lie between its interior and its orthocenter.

6. Let $p[\omega_p]$ and $q[\omega_q]$ be two weighted points in $\mathbb{R}^d$, each representing either a weighted vertex or an orthoball. Lemma 2.1 associates each ball B_p in $\mathbb{R}^d$ with a hyperplane h_p in $\mathbb{R}^{d+1}$, and thereby associates each weighted point $p[\omega_p]$ with a hyperplane h_p. Show that the power distance $\pi(p[\omega_p], q[\omega_q])$ is equal to the vertical distance of p^+ above h_q, or symmetrically, the vertical distance of q^+ above h_p.

7. Let $\mathcal{T}$ be a triangulation of a point set in the plane, and let x be an arbitrary point in the plane. Say that a triangle $t_2 \in \mathcal{T}$ is behind a triangle $t_1 \in \mathcal{T}$ from x's perspective if one of the following conditions is satisfied.

   - Some ray shot from x intersects first t_1 and then t_2.
   - There is a triangle $t \in \mathcal{T}$ such that t is behind t_1 and t_2 is behind t from x's perspective. (Note that there need not be a single ray originating at x that intersects both t_1 and t_2, and that this definition is recursive.)

   Draw a triangulation $\mathcal{T}$ in the plane and a viewpoint x such that some triangle in $\mathcal{T}$ is behind itself. In other words, there is a cyclical list of triangles such that each triangle is behind the preceding triangle in the cycle.

8. Consider a triangulation comprising just one triangle abc and its faces. Even for this simple triangulation, the consequence of the Monotone Power Lemma can fail to hold if $c[\omega_c]$ is closer than orthogonal to the diametric orthoball of ab. Design a counterexample that demonstrates this failure.

9. Consider a triangulation comprising just one edge ab and its vertices. The consequence of the Orthoball Cover Lemma can fail to hold if the condition that $\pi(a[\omega_a], x[\varpi_x]) \geq 0$ is not satisfied. Design a counterexample that demonstrates this failure.

10. Let $\mathcal{Q}$ be a quarantined complex with a weight assignment ω. Let $x[\varpi_x]$ be a weighted point such that $x \in |\mathcal{Q}|$ and $\pi(x[\varpi_x], v[\omega_v]) \geq 0$ for every vertex v in $\mathcal{Q}$. Show that $B(x, \sqrt{\varpi_x})$ is included in the union of the balls in $\mathrm{Ball}(\mathcal{Q})$.

11. Consider a triangulation comprising just one triangle τ and its faces. Suppose that every vertex weight is zero and τ has an obtuse angle. Show that the consequence of the Orthocenter Containment Lemma fails to hold, and specify which property of a quarantined complex the triangle violates.

Chapter 8

Tetrahedral meshing of PLCs

This chapter extends Ruppert's Delaunay refinement algorithm to tetrahedral meshing of three-dimensional piecewise linear complexes. The only part of the extension that is not straightforward is the procedure for enforcing conformity of the mesh to the domain polygons. Given a PLC with no acute angles, the algorithm described here generates a mesh whose tetrahedra all have small radius-edge ratios, as Figure 8.1 illustrates. Chapter 9 extends Delaunay refinement to polyhedra with small angles.

A small radius-edge ratio is a weaker shape guarantee than a small aspect ratio. Recall from Section 1.7 that there is one type of bad tetrahedron, called a sliver, that can have a very good radius-edge ratio and arbitrarily bad dihedral angles. Delaunay refinement is guaranteed to eliminate most skinny tetrahedra, but not slivers. It can split slivers, just like any other tetrahedra, but there is no guarantee that new slivers will not appear as quickly as old ones are split, so Delaunay refinement tends to overrefine the mesh where it attacks slivers. An alternative is to remove slivers through means other than refinement. Meshes with small radius-edge ratios are a good starting point for mesh improvement methods that eliminate the remaining skinny tetrahedra, such as the sliver exudation algorithm described in Chapter 10.

There are applications for which triangulations with small radius-edge ratios suffice. For example, some finite volume methods use the Voronoi dual of a Delaunay mesh and require only that the Voronoi cells have small aspect ratios, a requirement that is always met if the primal Delaunay tetrahedra have small radius-edge ratios.

Like Ruppert's algorithm, tetrahedral Delaunay refinement offers provably good grading—all the edges in the final mesh have lengths proportional to the local feature size. Unlike Ruppert's algorithm, the tetrahedral algorithm does not always produce size-optimal meshes, for technical reasons related to slivers.

Many methods of mesh generation, including most advancing front methods, triangulate the domain boundary as a separate step prior to tetrahedralizing the interior. A shortfall of that approach is that it is difficult to predict how small the triangles must be to facilitate a high-quality mesh of the domain geometry without unnecessarily small elements. By contrast, Delaunay refinement generates the boundary triangulation as a byproduct of the tetrahedral mesh, which is not beholden to poor decisions made earlier.

Figure 8.1: Three PLCs and their meshes generated by DelTetPLC.

8.1 A tetrahedral Delaunay refinement algorithm

The input to the tetrahedral Delaunay refinement algorithm is a piecewise linear complex $\mathcal{P}$ (recall Definition 4.3) and a positive constant $\bar{\rho}$ that specifies the maximum permitted radius-edge ratio for tetrahedra in the output mesh. The algorithm is guaranteed to terminate if $\bar{\rho} \geq 2$ and $\mathcal{P}$ satisfies a condition governing the angles at which its linear cells meet, described in Section 8.3.

The tetrahedral algorithm, like Ruppert's algorithm, maintains the Delaunay triangulation Del S of a growing vertex set S, and begins by setting S to be the set of vertices in $\mathcal{P}$. The first priority is to recover the segments of $\mathcal{P}$, so that each segment is a union of edges in Del S. The second priority is to recover the polygons of $\mathcal{P}$, so that each polygon is a union of triangles in Del S. When the mesh conforms to $\mathcal{P}$, the third priority is to improve its quality by splitting tetrahedra whose radius-edge ratios exceed $\bar{\rho}$. When the algorithm terminates, it writes as output the subset of the Delaunay simplices included in $|\mathcal{P}|$.

As in Ruppert's algorithm, the segments in $\mathcal{P}$ are subdivided into *subsegments* by the vertices in S. A vertex $v \in S$ that lies in the closed diametric ball of a subsegment e but is not a vertex of e is said to *encroach upon* e. By Proposition 4.3, Del S contains every strongly Delaunay edge, so every subsegment absent from Del S is encroached. Encroached subsegments, whether absent or present in Del S, are split by the procedure SplitSubsegment, which is unchanged from Ruppert's algorithm except that it operates in three dimensions; see Figure 8.2(a).

SplitSubsegment(e, S, E)

1. Insert the midpoint of e into S.
2. Remove e from E and add its two halves to E.

Recall from Chapter 6 that there are two reasons to split encroached subsegments: to enforce domain conformity by stitching missing segments into the mesh, and to prevent circumcenters of bad elements from being inserted outside the domain or too close to a domain boundary. For the same reasons, we subdivide each polygon in $\mathcal{P}$ into triangular *subpolygons* that we want to appear in Del S, and split them when they are encroached.

The main difficulty is that if a polygon does not appear in Del S as a union of triangles, it is not obvious how the polygon should be subdivided into subpolygons. Proposition 4.11 provides an answer by stating that if a polygon $g \in \mathcal{P}$ appears as a union of simplices in Del S, then those simplices are a two-dimensional Steiner Delaunay triangulation of g. Proposition 4.11 is technically about CDTs, but it is applicable here because we invoke it only when no subsegment is encroached, so Del $(S \cap g)$ includes a Steiner Delaunay triangulation of g, which is a Steiner CDT of g as well.

The algorithm exploits this insight by explicitly maintaining for each polygon g a two-dimensional Delaunay triangulation $\mathcal{T}_g$ of the vertices lying on g. These triangulations are updated independently from Del S, and they are compared against Del S to identify missing subpolygons.

Definition 8.1 (subpolygon). A *subsegment* of a polygon $g \in \mathcal{P}$ is a subsegment of a segment of g. For each polygon $g \in \mathcal{P}$, consider the two-dimensional Delaunay triangulation

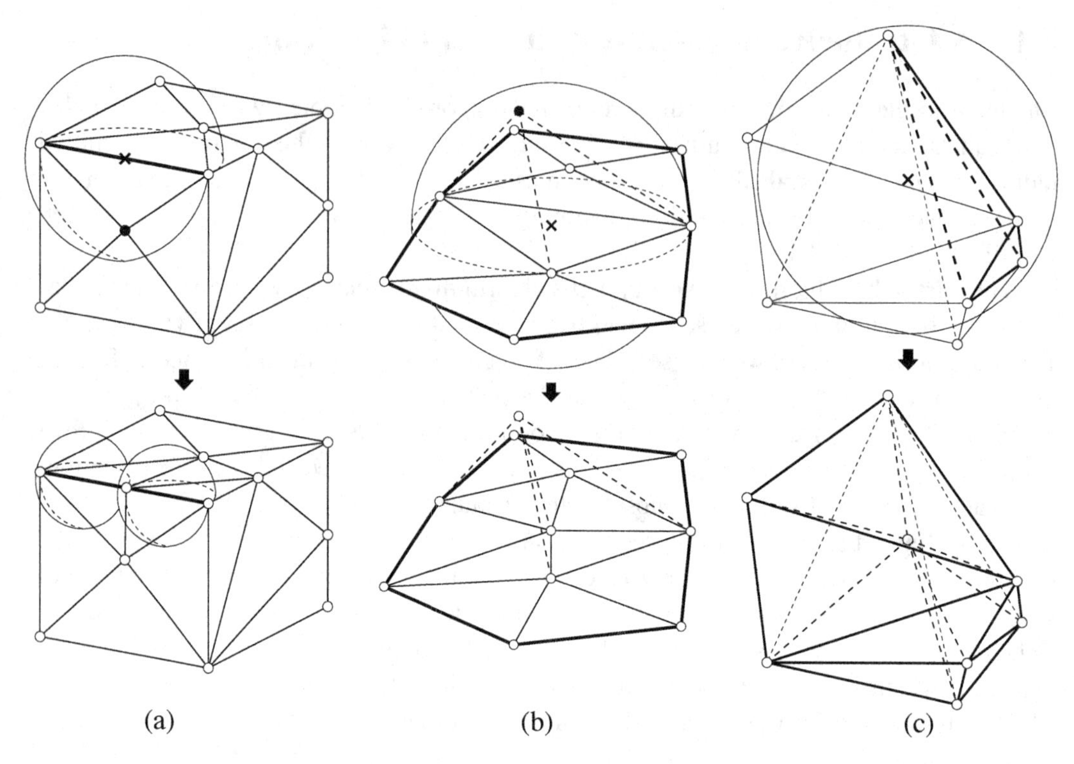

Figure 8.2: Three operations for tetrahedral Delaunay refinement. (a) Splitting an encroached subsegment. The black vertex is in the diametric ball of the bold subsegment. (b) Splitting an encroached subpolygon. The black vertex is in the diametric ball illustrated. The triangular faces shown are subpolygons of a single polygon; the tetrahedra atop them are not shown. (c) Splitting a skinny tetrahedron.

Del $(S \cap g)$ of the vertices in S that lie on g. If no subsegment of g is encroached, then Del $(S \cap g)$ contains every subsegment of g's segments, and hence $\mathcal{T}_g = \{\sigma \in \text{Del}\,(S \cap g) : \sigma \subseteq g\}$ is a Steiner Delaunay triangulation of g. We call the triangles in $\mathcal{T}_g$ the *subpolygons* of g, whether they appear in Del S or not.

All of the triangular faces visible in the middle row of Figure 8.1 are subpolygons, but most of the triangular faces in the interiors of the meshes are not. Subpolygons are encroached under similar circumstances as subsegments.

Definition 8.2 (subpolygon encroachment)**.** A vertex v that lies in the closed diametric ball of a subpolygon σ but does not lie on σ's circumcircle is said to *encroach upon* σ.

Recall from Definition 1.17 that σ's circumcircle is the unique circle passing through its three vertices; it lies in σ's affine hull and is an equator of its diametric ball. Vertices on σ's circumcircle are considered not to encroach upon σ for two reasons. One is so that σ's own vertices do not encroach upon σ. The second is because if other vertices lie on σ's circumcircle, they probably also lie on the same polygon g of which σ is a subpolygon, and it is not desirable that vertices on g encroach upon subpolygons of g. Observe that in this case, $S \cap g$ has multiple two-dimensional Delaunay triangulations, not all of which contain σ.

By Proposition 4.3, Del S contains every strongly Delaunay triangle, so if a subpolygon $\sigma \in \text{Del}(S \cap g)$ is absent from Del S, either σ is encroached or $S \cap g$ has multiple Delaunay triangulations and Del S patches the hole with triangles from a different Delaunay triangulation of $S \cap g$. In theory this mismatch could be prevented by the symbolic weight perturbation method of Section 2.9, but in practice mismatches occur anyway because the coordinates of the vertices in $S \cap g$ are represented with finite precision and are not always exactly coplanar. A practical solution is for the Delaunay algorithm to check whether σ is "encroached upon" solely by vertices on g; if so, the algorithm should locally correct $\text{Del}(S \cap g)$ to match Del S instead of splitting σ. Hereafter, we assume that for every polygon g, $\text{Del}(S \cap g)$ is matched as closely as possible with Del S whenever S is updated.

Encroached subsegments always have priority over encroached subpolygons. But when no subsegment is encroached, the procedure SPLITSUBPOLYGON splits encroached subpolygons, whether absent or present in Del S, by inserting new vertices at their circumcenters, as illustrated in Figure 8.2(b). However, if a subpolygon circumcenter encroaches upon one or more subsegments, it is rejected, and one of the subsegments it encroaches upon is split instead. This policy prevents the dangers of a circumcenter being inserted outside the polygon or very close to a segment.

SPLITSUBPOLYGON(σ, S, E)

1. Let c be the circumcenter of σ.
2. If c encroaches upon a subsegment $e \in E$, call SPLITSUBSEGMENT(e, S, E). Otherwise, insert c into S.

When no subsegment is encroached—which is always true when SPLITSUBPOLYGON is called—each polygon triangulation $\mathcal{T}_g$ is a quarantined complex. The Orthocenter Containment Lemma (Lemma 7.7) assures us that the circumcenter of every subpolygon in $\mathcal{T}_g$ lies on g, as Proposition 6.2 did for Ruppert's algorithm in the plane.

Figure 8.3 shows how the algorithm stitches a missing polygon into a mesh by splitting encroached subpolygons. Shaded triangular subpolygons in the polygon triangulation at top center are missing from the tetrahedralization at bottom center; the bold dashed line represents a tetrahedralization edge that passes through the polygon. SPLITSUBPOLYGON patches a hole in the polygon by creating a vertex at the circumcenter of a missing subpolygon and inserting it independently into both triangulations.

The order in which encroached subpolygons are split is important. The proposition below shows that if a point c encroaches upon a subpolygon σ of a polygon g, then the orthogonal projection of c onto g lies in a subpolygon of g that c also encroaches upon. The algorithm produces meshes with better radius-edge ratios if it splits the latter subpolygon first.

Proposition 8.1. *Let g be a polygon in a three-dimensional PLC $\mathcal{P}$. Let S be a vertex set that includes the vertices in $\mathcal{P}$. Suppose that no vertex in S encroaches upon any subsegment of g's segments; hence $\mathcal{T}_g = \{\sigma \in \text{Del}(S \cap g) : \sigma \subseteq g\}$ is a Steiner Delaunay triangulation of g that contains those subsegments. Let $c \in \mathbb{R}^3$ be a point that encroaches upon a subpolygon $\sigma \in \mathcal{T}_g$ but does not encroach upon any subsegment. Then the orthogonal projection of c*

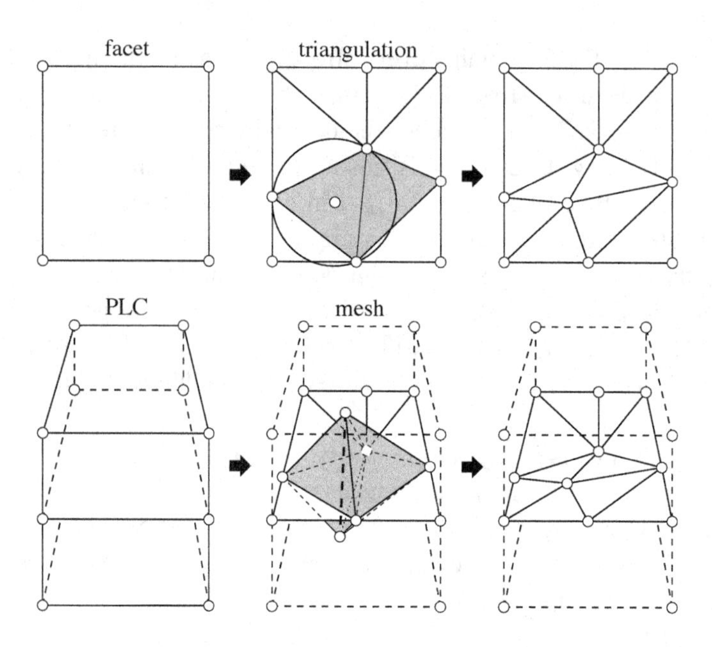

Figure 8.3: The top row depicts a rectangular polygon and its Delaunay triangulation. The bottom row depicts the polygon's position as an internal boundary of a domain and its subdivision as it is recovered in a Delaunay tetrahedralization. Most of the vertices and tetrahedra of the mesh are omitted for clarity.

onto g's affine hull lies in g's interior but not on a face (internal or external) of g, and c encroaches upon the subpolygon in $\mathcal{T}_g$ containing c's projection.

Proof. Because no subsegment of g is encroached, $\mathcal{T}_g$ is a quarantined complex. Let $\tilde{c}$ be the projection of c onto g's affine hull. Because c encroaches upon σ, $\tilde{c}$ cannot lie on a vertex of g. Suppose for the sake of contradiction that $\tilde{c}$ lies outside g or on a segment of g. Then the line segment connecting $\tilde{c}$ to a point in σ's interior intersects a subsegment e of g, and the Monotone Power Lemma (Lemma 7.5) states that $\pi(c, B_e) \leq \pi(c, B_\sigma)$, where B_e and B_σ are the diametric balls of e and σ. Because c encroaches upon σ but not e, $0 < \pi(c, B_e) \leq \pi(c, B_\sigma) \leq 0$, a contradiction. Therefore, $\tilde{c}$ lies in g's interior but not on a face of g, and some subpolygon $\tau \in \mathcal{T}_g$ contains $\tilde{c}$. By the Monotone Power Lemma, $\pi(c, B_\tau) \leq \pi(c, B_\sigma) \leq 0$, so c encroaches upon τ too. □

Encroached subsegments and subpolygons always have priority over skinny tetrahedra. When none is encroached, the algorithm attacks tetrahedra with large radius-edge ratios. In this circumstance, the tetrahedral subcomplex $\{\sigma \in \text{Del}\, S : \sigma \subseteq |\mathcal{P}|\}$ is a quarantined complex and a Steiner triangulation of $\mathcal{P}$. The Orthocenter Containment Lemma (Lemma 7.7) guarantees that every tetrahedron's circumcenter lies in the domain $|\mathcal{P}|$.

The procedure SplitTetrahedron splits tetrahedra whose radius-edge ratios exceed $\bar{\rho}$ by inserting new vertices at their circumcenters, as illustrated in Figure 8.2(c). However, if a tetrahedron circumcenter encroaches upon a subsegment or subpolygon, SplitTetrahedron rejects it and splits an encroached subsegment or subpolygon instead.

SPLITTETRAHEDRON(τ, S, E)

1. Let c be the circumcenter of τ.
2. If c encroaches upon a subsegment $e \in E$, call SPLITSUBSEGMENT(e, S, E) and return.
3. If c encroaches upon a subpolygon of some polygon $g \in \mathcal{P}$, let σ be a subpolygon of g that contains the orthogonal projection of c onto g. Call the procedure SPLITSUBPOLYGON(σ, S, E) and return.
4. Insert c into S.

Figure 8.4 illustrates the effect of SPLITTETRAHEDRON. If c is rejected and a subsegment or subpolygon is split instead, the new vertex might or might not eliminate τ from Del S. If τ survives, the algorithm can subsequently try to split τ again.

Pseudocode for the tetrahedral Delaunay refinement algorithm follows.

DELTETPLC($\mathcal{P}, \bar{\rho}$)

1. Let S be the set of vertices in $\mathcal{P}$. Let E be the set of edges in $\mathcal{P}$.
2. Compute the three-dimensional triangulation Del S. For each polygon $g \in \mathcal{P}$, compute the two-dimensional triangulation Del $(S \cap g)$.
3. While some subsegment $e \in E$ is encroached upon by a vertex in S, call SPLITSUBSEGMENT(e, S, E), update Del S and the two-dimensional triangulation Del $(S \cap g)$ for every polygon g that includes e, and repeat Step 3.
4. If some subpolygon of some polygon $g \in \mathcal{P}$ is encroached upon by a vertex $v \in S$, let σ be a subpolygon of g that contains the orthogonal projection of v onto g. Call SPLITSUBPOLYGON(σ, S, E); update Del S, Del $(S \cap g)$, and the other two-dimensional triangulations; and go to Step 3.
5. If Del S contains a tetrahedron $\tau \subseteq |\mathcal{P}|$ for which $\rho(\tau) > \bar{\rho}$, call SPLITTETRAHEDRON(τ, S, E), update Del S and the two-dimensional triangulations, and go to Step 3.
6. Return the mesh $\{\sigma \in \text{Del}\, S : \sigma \subseteq |\mathcal{P}|\}$.

Because every output tetrahedron has a radius-edge ratio at most $\bar{\rho}$, every triangular face of every output tetrahedron also has a radius-edge ratio at most $\bar{\rho}$. For full generality, it is sometimes desirable for $\mathcal{P}$ to include *dangling simplices* that are not faces of any linear 3-cell. Dangling vertices and edges present no difficulty, but DELTETPLC might triangulate dangling polygons with arbitrarily skinny triangles. If these are undesirable, modify the algorithm so that it splits subpolygons whose radius-edge ratios exceed $\bar{\rho}$. Encroached subsegments must take priority over skinny subpolygons.

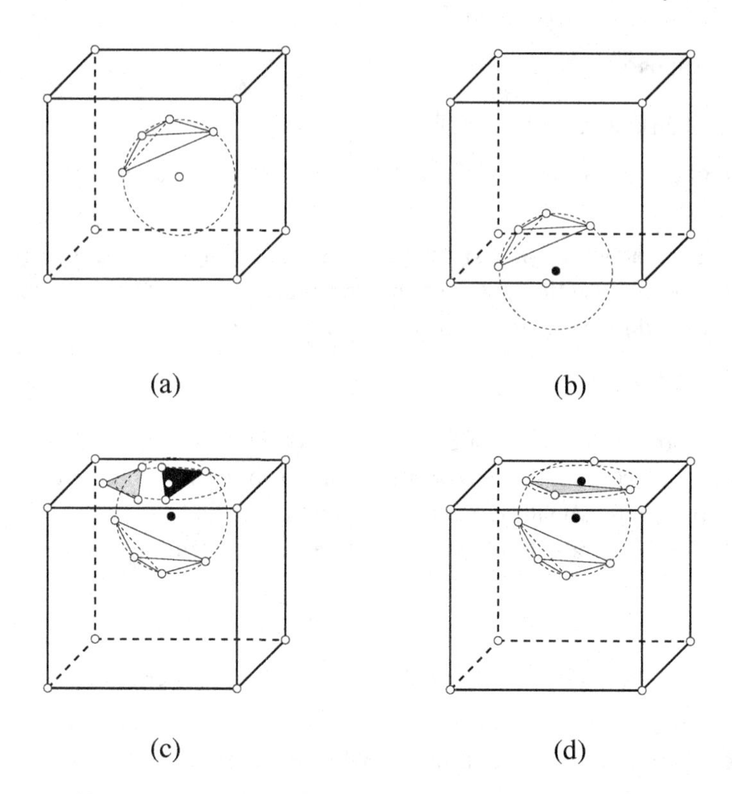

Figure 8.4: (a) A skinny tetrahedron is split by the insertion of its circumcenter. (b) The circumcenter of a tetrahedron is rejected for encroaching upon a subsegment, whose midpoint is inserted instead. (c) The circumcenter of a tetrahedron is rejected for encroaching upon the shaded subpolygon. The circumcenter of the black subpolygon is inserted instead. (d) The circumcenter of the shaded subpolygon is rejected for encroaching upon a subsegment, whose midpoint is inserted instead.

8.2 Implementation and running time

Much of our implementation advice from Section 6.3 applies to the tetrahedral algorithm as well. The dictionary data structure of Section 3.2 can store the polygon triangulations, and the similar data structure of Section 5.1 can store the mesh. An implementation should also maintain a queue of encroached subsegments, a queue of encroached subpolygons, and a queue of skinny and oversized tetrahedra so that Steps 3–5 run quickly. Items that reach the front of a queue must be checked to ensure they still exist.

There is a constant-time test for whether a subpolygon σ is encroached: either $\sigma \notin \text{Del}\, S$ or $\text{Del}\, S$ contains a tetrahedron that has σ for a face and whose fourth vertex encroaches upon σ—see Exercise 1. The test for whether a subsegment e is encroached is essentially the same as the test in the plane: e is encroached if $e \notin \text{Del}\, S$ or $\text{Del}\, S$ contains a triangle that has e for an edge and an angle of 90° or greater opposite e. However, in the plane there are at most two triangles to test, whereas in three dimensions a subsegment might be an edge of arbitrarily many triangles. Nevertheless, after the initial Delaunay triangulation is computed, the queues can be initialized in time linear in the number of tetrahedra.

When a new vertex is inserted during refinement, the queues are updated by checking only the newly created tetrahedra and their faces. It is necessary to be able to rapidly determine whether a face of the mesh is a subsegment or subpolygon. The mesh data structure maps triangles to tetrahedra in expected constant time and can also record which triangular faces of the mesh are subpolygons. Subsegments can be recorded in a separate hash table or in the mesh data structure by using a dummy third vertex. Overall, queue management and encroachment identification should take time linear in the number of tetrahedra created.

The easiest way to discover whether a circumcenter encroaches upon a subsegment or subpolygon and should be rejected is to insert it and check the triangles and edges of the cavity evacuated by the Bowyer–Watson algorithm. If the circumcenter is rejected, reverse the changes and ensure that the encroached subpolygon or skinny tetrahedron that generated the circumcenter remains in the appropriate queue so the algorithm can try to insert the same circumcenter again if necessary. See also Exercise 8.

As with Ruppert's algorithm, we observe that a careful implementation of DelTetPLC runs in $O(n \log n + N)$ time in practice on most PLCs, where n is the number of vertices in $\mathcal{P}$ and $N \geq n$ is the number of vertices in the final mesh. However, recall from Section 4.1 that there exist sets of n input vertices whose Delaunay triangulations have $\Theta(n^2)$ complexity. Section 5.4 shows that those triangulations can be constructed in expected $O(n^2)$ time. It is uncommon but plausible for a real-world input to include vertex configurations like that illustrated in Figure 4.2.

After the construction of the initial triangulation, the remaining $N - n$ vertex insertions take at most $O(N^2)$ time by the argument in Section 5.2. Point location is needed only to insert midpoints of missing subsegments and circumcenters of missing subpolygons. All costs considered, the Delaunay refinement algorithm takes $O(N^2)$ time in the worst case. As in two dimensions, Delaunay refinement rapidly evens out the distribution of vertices, so most new vertices take constant time to insert, and the refinement stage normally takes $O(N)$ time in practice.

8.3 A proof of termination and good grading

DelTetPLC terminates only if no subsegment or subpolygon is encroached—hence Del S conforms to $\mathcal{P}$—and no tetrahedron in the domain has a radius-edge ratio greater than $\bar{\rho}$. Here we prove that the tetrahedral Delaunay refinement algorithm terminates and, therefore, generates a high-quality mesh of $\mathcal{P}$, if $\bar{\rho} > 2$ and the input PLC $\mathcal{P}$ satisfies a *projection condition* that rules out most acute angles.

Definition 8.3 (projection condition). Two adjoining linear cells f and g in a PLC $\mathcal{P}$ are said to satisfy the *projection condition* if the orthogonal projection of f onto g's affine hull does not intersect $g \setminus f$, and the orthogonal projection of g onto f's affine hull does not intersect $f \setminus g$. A PLC $\mathcal{P}$ satisfies the projection condition if every pair of adjoining linear cells in $\mathcal{P}$ does.

Three-dimensional PLCs have several kinds of angles, illustrated in Figure 8.5. Two edges adjoining at a shared vertex induce an angle called a *plane angle*. The projection condition prohibits edges from meeting at angles less than 90°. A polygon g and an edge

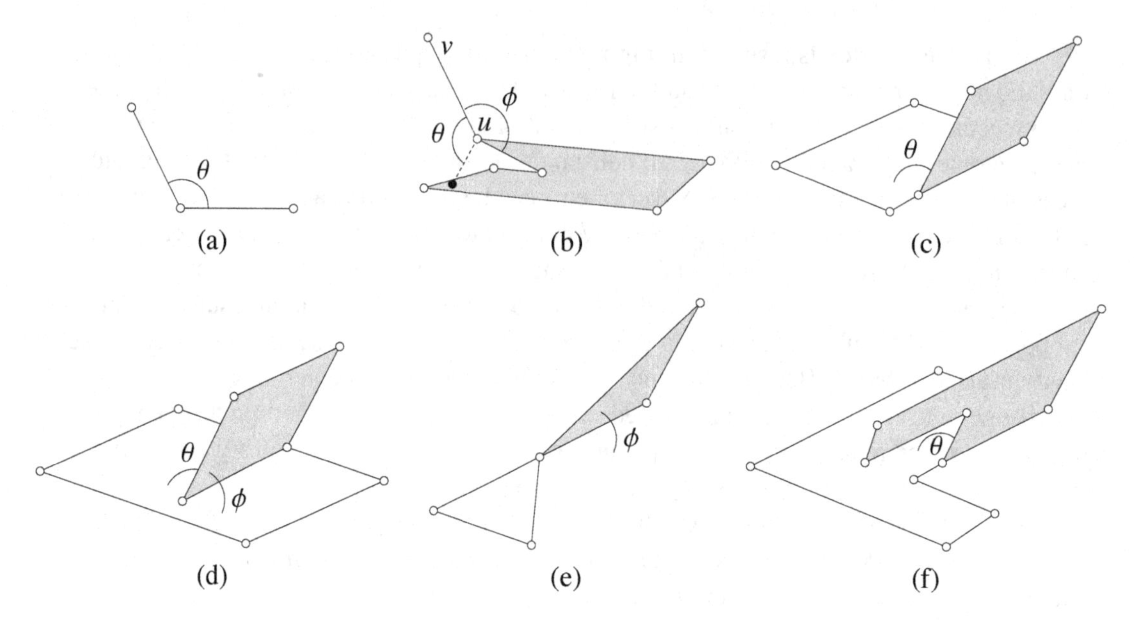

Figure 8.5: (a) The plane angle between two edges. (b) Plane angles between an edge and a polygon. (c) A dihedral angle between two polygons. (d) Two dihedral angles. (e) An acute dihedral angle that is not forbidden. (f) Forbidden by the projection condition.

uv adjoining at a shared vertex u with $v \notin g$ can induce two kinds of plane angles, illustrated in Figure 8.5(b): $\theta = \min\{\, \angle(\overrightarrow{up}, \overrightarrow{uv}) : p \in g \setminus \{u\}\}$ and $\phi = \min\{\, \angle(\overrightarrow{up}, \overrightarrow{uv}) : up \subset g$ and $p \neq u\}$. The projection condition prohibits acute angles of both types—the former type because of the projection of the polygon onto the edge. However, see the notes at the end of the chapter for why acute angles of the former type can be permitted.

The projection condition rules out most, but not all, acute dihedral angles. For example, the two polygons meeting at an obtuse dihedral angle θ in Figure 8.5(c) are permitted, but the two polygons in Figure 8.5(d) are forbidden because they meet along a shared edge at an acute angle ϕ. Sometimes the projection condition is more subtle. The polygons in Figure 8.5(e) are permitted, even though they nominally meet at an acute dihedral angle. The polygons in Figure 8.5(f) are forbidden even though they nominally meet only at an obtuse dihedral angle, but see the notes for how to remove this restriction.

The following two propositions say that if two adjoining linear cells satisfy the projection condition, and neither is a face of the other, then a point on one cannot encroach upon a subsegment of the other, nor can it encroach upon a subpolygon of the other if the other has no encroached subsegments.

Proposition 8.2. *Let s be a segment in a PLC $\mathcal{P}$, and let $g \not\supseteq s$ be a linear cell in $\mathcal{P}$ that adjoins s. Suppose s and g satisfy the projection condition. No point on g ever encroaches upon a subsegment of s.*

Proof. By the projection condition, the orthogonal projection of a point $p \in g$ onto the affine hull of s cannot lie in the interior of s. Therefore, p cannot lie in the diametric ball of any subsegment of s—unless p is a vertex of s, which cannot encroach on a subsegment of s. □

Proposition 8.3. *Let g be a polygon in a PLC $\mathcal{P}$, and let $s \not\supseteq g$ be a linear cell in $\mathcal{P}$ that adjoins g. Suppose s and g satisfy the projection condition. At a given time during the execution of* DelTetPLC, *suppose no subsegment of g's segments is encroached. If a point $p \in s$ does not encroach upon any subsegment of g, then p does not encroach upon any subpolygon of g.*

Proof. Let $\tilde{p}$ be the orthogonal projection of p onto the affine hull of g. Because s and g satisfy the projection condition, $\tilde{p}$ does not lie on $g \setminus s$, so it lies either off of g or on a face that g shares with s. Therefore, by Proposition 8.1, $\tilde{p}$ cannot encroach upon a subpolygon of g. □

As with Ruppert's algorithm, we prove that DelTetPLC terminates by demonstrating that there is a lower bound on the distances between any two mesh vertices and applying the Packing Lemma (Lemma 6.1). We reuse some terms introduced in Section 6.4: the *insertion radius* r_x of a vertex x is the distance $d(x, S)$ at the instant just before x is inserted into S, or when x is rejected for encroaching upon a subsegment or subpolygon. A vertex is of *type i* if it lies on a linear i-cell in $\mathcal{P}$ but not on a lower-dimensional cell—although vertices rejected by SplitSubpolygon have type 2 and vertices rejected by SplitTetrahedron have type 3, regardless of whether they happen to fall on a lower-dimensional cell. Every vertex inserted or rejected by DelTetPLC, but not of type 0, has a *parent* vertex, which also may have been inserted or rejected.

- *Type 0 vertices* are input vertices in $\mathcal{P}$; they have no parents. The insertion radius r_x of a type 0 vertex x is the distance from x to the nearest distinct vertex in $\mathcal{P}$.

- *Type 1 vertices* are those inserted at the midpoints of encroached subsegments by SplitSubsegment, and *type 2 vertices* are those inserted or rejected at the circumcenters of encroached subpolygons by SplitSubpolygon. For a type 1 or type 2 vertex x generated on an encroached simplex σ, there are two possibilities, illustrated in Figure 8.6(a, b, c). If the point encroaching upon σ is in S, x's parent p is the vertex in S nearest x, and the insertion radius of x is $r_x = d(x, p)$. Otherwise, the encroaching point p is a rejected circumcenter, which is x's parent, and x's insertion radius r_x is the radius of σ's diametric ball. As p is in that ball, $d(x, p) \leq r_x$.

- *Type 3 vertices* are those inserted or rejected at tetrahedron circumcenters by SplitTetrahedron. For a type 3 vertex x generated at the circumcenter of a tetrahedron τ, x's parent p is the most recently inserted vertex of the shortest edge of τ; if both vertices are in $\mathcal{P}$, choose one arbitrarily. As τ's circumball is empty, the insertion radius of x is $r_x = d(x, p)$, the circumradius of τ, as illustrated in Figure 8.6(d).

The following proposition proves a lower bound on the insertion radius of each vertex in the style of Proposition 6.4. Recall from Definition 6.2 that the local feature size $f(x)$ is the radius of the smallest ball centered at x that intersects two disjoint linear cells.

Proposition 8.4. *Let $\mathcal{P}$ be a PLC that satisfies the projection condition. Let x be a vertex inserted into S or rejected by* DelTetPLC. *Let p be the parent of x, if one exists.*

(i) *If x is of type 0, then $r_x \geq f(x)$.*

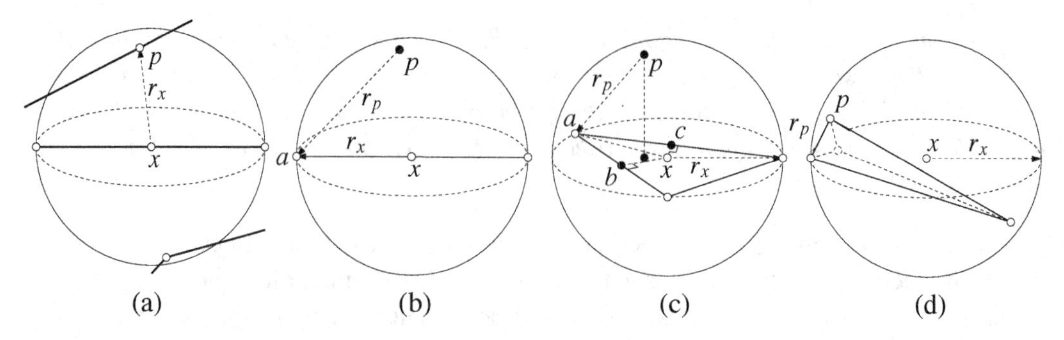

Figure 8.6: Four examples of a new vertex x with parent p. (a) A vertex p encroaches upon a subsegment, causing a vertex to be inserted at its midpoint x with insertion radius $r_x = d(x, p)$. (b) A circumcenter p encroaches upon a subsegment, causing p to be rejected and a vertex to be inserted at the midpoint x with insertion radius $r_x = d(x, a)$. The parent's insertion radius r_p is at most $\sqrt{2}r_x$. (c) A circumcenter p encroaches upon a subpolygon, causing p to be rejected and a vertex to be inserted at the circumcenter x. Similar to case (b). (d) A vertex is inserted at a tetrahedron circumcenter x with insertion radius $r_x = d(x, p)$, where p is the most recently inserted vertex of the tetrahedron's shortest edge.

(ii) *If x is of type 1 or 2, and the type of p is less than or equal to that of x, then $r_x \geq f(x)$.*

(iii) *If x is of type 1 or 2, and the type of p is greater than that of x, then $r_x \geq r_p / \sqrt{2}$.*

(iv) *If x is of type 3, then $r_x > \bar{\rho} r_p$.*

Proof. If x is of type 0, $x \in \mathcal{P}$ and hence $f(x) \leq r_x$ by the definition of local feature size.

If x is of type 3, let τ be the tetrahedron with circumcenter x, illustrated in Figure 8.6(d). Because SplitTetrahedron tried to split τ, $\rho(\tau) > \bar{\rho}$. The parent p is the most recently inserted vertex of τ's shortest edge pq, or both p and q are vertices in $\mathcal{P}$, so it follows from the definition of insertion radius that $r_p \leq d(p, q)$. The insertion radius of x (and the circumradius of τ) is $r_x = d(x, p) = \rho(\tau)\, d(p, q) > \bar{\rho} r_p$ as claimed.

If x is of type 1, it is the midpoint of a subsegment e of a segment $s \in \mathcal{P}$, and x's parent p encroaches upon e. If p is of type $i \leq 1$, as illustrated in Figure 8.6(a), p lies on some linear i-cell in $\mathcal{P}$ that is disjoint from s by Proposition 8.2, so $r_x = d(x, p) \geq f(x)$. Otherwise, p is a type 2 or 3 circumcenter rejected for encroaching upon e, as illustrated in Figure 8.6(b). Let a be the vertex of e nearest p. Then $\angle pxa \leq 90^\circ$ and $r_p \leq d(p, a) \leq \sqrt{2}\, d(x, a) = \sqrt{2} r_x$, as claimed.

If x is of type 2, it is the circumcenter of a subpolygon σ of a polygon $g \in \mathcal{P}$, and x's parent p encroaches upon σ. If p is of type $i \leq 2$, p lies on some linear i-cell in $\mathcal{P}$ that is disjoint from g by Proposition 8.3, so $r_x = d(x, p) \geq f(x)$ as claimed. Otherwise, p is a type 3 tetrahedron circumcenter rejected for encroaching upon σ, as illustrated in Figure 8.6(c). Let $\tilde{p}$ be the orthogonal projection of p onto the affine hull of g. Recall that DelTetPLC chooses σ so that $\tilde{p} \in \sigma$. Let a be the vertex of σ nearest $\tilde{p}$, and let b and c be the midpoints of the edges of σ adjoining a. The projection $\tilde{p}$ lies on at least one of the right triangles abx or acx, both of which have the hypotenuse ax. Therefore, $d(\tilde{p}, a) \leq d(x, a)$. Because p is in the diametric ball of σ, $d(p, \tilde{p})$ cannot exceed the radius $d(x, a)$ of that ball. The line segments $p\tilde{p}$ and $\tilde{p}a$ are orthogonal, so by Pythagoras' Theorem, $d(p, a) \leq \sqrt{2}\, d(x, a)$.

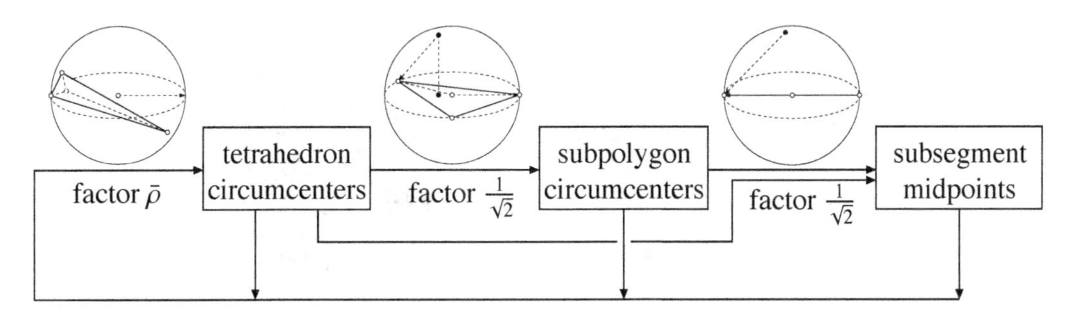

Figure 8.7: Flow graph illustrating the worst-case relation between a vertex's insertion radius and the insertion radii of the children it begets. If no cycle has a product smaller than one, the tetrahedral Delaunay refinement algorithm will terminate.

Hence, $r_p \leq d(p, a) \leq \sqrt{2}\, d(x, a) = \sqrt{2} r_x$, and the result follows. □

The flow graph in Figure 8.7 depicts the relationship between the insertion radius of a vertex and the smallest possible insertion radii of its children, from Proposition 8.4. The boxes represent type 3, 2, and 1 vertices, respectively. DelTetPLC terminates if it cannot produce vertices with ever-diminishing insertion radii—that is, there is no cycle in the flow graph whose product is less than one. This is true if we choose $\bar{\rho}$ to be at least two.

From this reasoning, it follows that if $\bar{\rho} \geq 2$, DelTetPLC creates no edge shorter than the shortest distance between two disjoint linear cells in $\mathcal{P}$. We leave that termination proof for Exercise 4 and give here an alternative proof that establishes the provably good grading of the mesh. Proposition 8.5 below establishes that if $\bar{\rho} > 2$, the algorithm does not place a vertex much closer to other vertices than the local feature size at the vertex. If a user chooses $\bar{\rho} < 2$, the algorithm will try to achieve the specified radius-edge ratio, but it might fail to terminate.

Proposition 8.5. *Suppose that $\bar{\rho} > 2$ and $\mathcal{P}$ satisfies the projection condition. Define the constants*

$$C_0 = 1, \quad C_1 = \frac{(3 + \sqrt{2})\bar{\rho}}{\bar{\rho} - 2}, \quad C_2 = \frac{(\sqrt{2} + 1)\bar{\rho} + \sqrt{2}}{\bar{\rho} - 2}, \quad C_3 = \frac{\bar{\rho} + \sqrt{2} + 1}{\bar{\rho} - 2}.$$

Let x be a vertex of type i inserted or rejected by DelTetPLC($\mathcal{P}, \bar{\rho}$). *Then $r_x \geq f(x)/C_i$.*

Proof. The proof proceeds by induction on the order by which DelTetPLC inserts or rejects vertices. Type 0 vertices are the inductive basis: for every vertex x in $\mathcal{P}$, $r_x \geq f(x)$ by Proposition 8.4(i).

Consider any vertex x inserted or rejected by DelTetPLC. For the inductive step, assume the proposition holds for x's parent p. By the Lipschitz property,

$$f(x) \leq f(p) + d(p, x) \leq f(p) + r_x. \tag{8.1}$$

If x is of type 3, then $r_x \geq \bar{\rho} r_p$ by Proposition 8.4(iv). Observe that $C_1 > C_2 > C_3 > C_0 = 1$. By the induction assumption, $f(p) \leq C_1 r_p$ irrespective of the type of p. Inequality (8.1) gives

$$f(x) \leq C_1 r_p + r_x \leq \left(\frac{C_1}{\bar{\rho}} + 1\right) r_x = C_3 r_x.$$

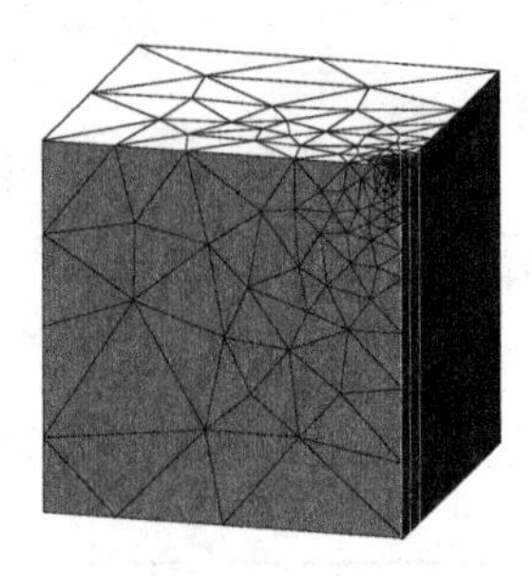
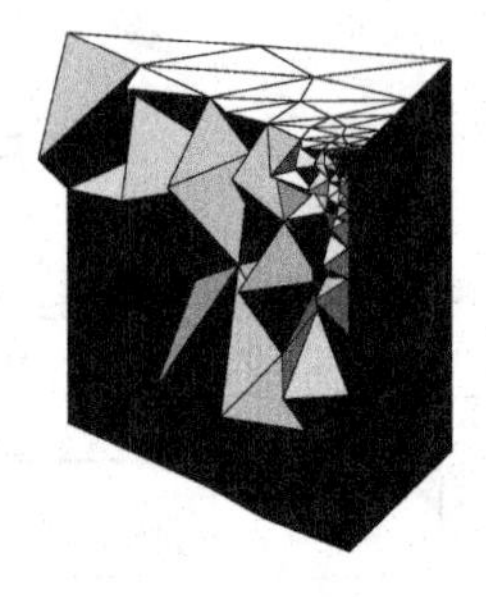

Figure 8.8: A mesh of a truncated cube and a diagonal cross-section thereof.

Suppose that x is of type $i \in \{1,2\}$. If p is of type $j \leq i$, by Proposition 8.4(ii), $r_x \geq f(x) \geq f(x)/C_i$ and the result holds. If p is of type $j > i$, then $r_x \geq r_p/\sqrt{2}$ by Proposition 8.4(iii). By the inductive assumption, $f(p) \leq C_j r_p \leq C_{i+1} r_p$. Inequality (8.1) gives

$$f(x) \leq C_{i+1} r_p + r_x \leq \left(\sqrt{2}C_{i+1} + 1\right) r_x = C_i r_x.$$

□

With Proposition 8.5, we can show that the final mesh has provably good grading, in the sense that there is a lower bound on its edge lengths proportional to the local feature size.

Proposition 8.6. *Suppose that $\bar{\rho} > 2$ and $\mathcal{P}$ satisfies the projection condition. For any two vertices p and q that* DelTetPLC *inserts, $d(p,q) \geq f(p)/(C_1 + 1)$.*

Proof. If p is inserted after q, Proposition 8.5 states that $d(p,q) \geq f(p)/C_1$, and the result follows. If q is inserted after p, the proof is identical to that of Proposition 6.6. □

To make the proposition concrete, consider choosing $\bar{\rho} = 2.5$. Then $C_1 \doteq 22.1$, so the spacing of vertices is at worst about 23 times smaller than the local feature size. As with Ruppert's algorithm, the outcome in practice is invariably much better than the worst case in theory. In Figure 8.8, a cube has been truncated at one corner, cutting off a portion whose width is one-millionth that of the cube. No tetrahedron in the mesh has radius-edge ratio greater than $\bar{\rho} = 1.2$, for which the algorithm is not guaranteed to terminate, but good grading is apparent. The worst edge is 73 times shorter than the local feature size at one of its vertices. With a bound of $\bar{\rho} = 2.5$, the worst edge is 9 (rather than 23) times shorter than the local feature size at one of its vertices.

The lower bound on edge lengths implies that the algorithm terminates and succeeds.

Theorem 8.7. *Suppose that $\bar{\rho} > 2$ and $\mathcal{P}$ satisfies the projection condition. Then* DelTetPLC$(\mathcal{P}, \bar{\rho})$ *terminates and returns a Steiner Delaunay triangulation of $\mathcal{P}$ whose tetrahedra have radius-edge ratios at most $\bar{\rho}$.*

Proof. Let $f_{\min} = \min_{x \in |\mathcal{P}|} f(x)$. By Proposition 8.6, DelTetPLC maintains an inter-vertex distance of at least $f_{\min}/(C_1 + 1) > 0$. By the Packing Lemma (Lemma 6.1), there is an upper bound on the number of vertices in the triangulation, so DelTetPLC must terminate.

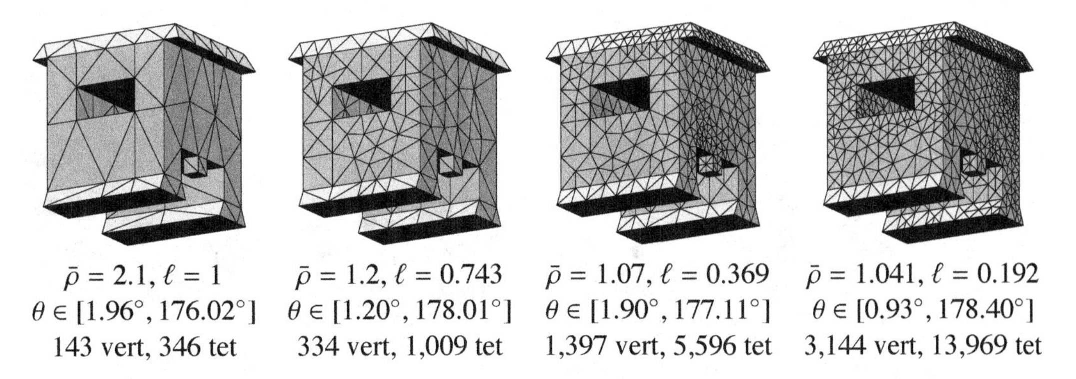

Figure 8.9: Meshes of a PLC. For each mesh, $\bar{\rho}$ is the bound on radius-edge ratio provided as an input to DelTetPLC, ℓ is the length of the shortest edge (compare with the PLC dimensions of $10 \times 10 \times 10$), θ indicates the range of dihedral angles in the mesh tetrahedra, vert denotes the number of vertices, and tet denotes the number of tetrahedra.

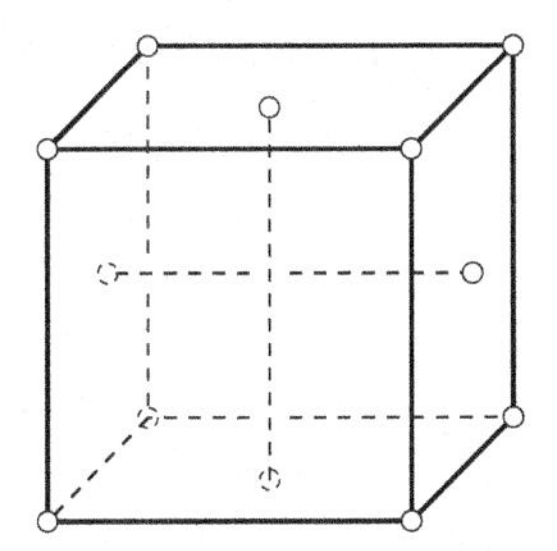

Figure 8.10: A PLC demonstrating that tetrahedral Delaunay refinement is not size-optimal.

The algorithm terminates only if no subsegment or subpolygon is encroached and no skinny tetrahedron lies in the domain, so it returns a high-quality mesh of $\mathcal{P}$ as stated. □

Figure 8.9 shows the trade-off between mesh quality and size. The algorithm often succeeds for $\bar{\rho}$ substantially less than 2, failing to terminate on this PLC only for $\bar{\rho} = 1.04$ or less. Although excellent radius-edge ratios can be obtained, a few poor dihedral angles make these meshes unsuitable for many applications.

Unfortunately, DelTetPLC does not inherit size-optimality nor optimal grading from Ruppert's algorithm. Figure 8.10 gives a counterexample, a PLC inside which two segments pass very close to each other without intersecting. This PLC has a CDT comprising 19 tetrahedra with good radius-edge ratios, but one of them is a sliver tetrahedron whose vertices are the endpoints of the two interior segments. However, the best that DelTetPLC can promise near the center of the domain is to generate edges whose lengths are proportional to the distance between the two segments, which can be arbitrarily small. The only known size-optimal tetrahedral mesh generation algorithms guarantee a bound on the aspect ratios of the tetrahedra, not merely the radius-edge ratios, and therefore prohibit slivers. These algorithms are size-optimal not because they produce coarser meshes, but because the lower bounds on the number of tetrahedra are substantially higher when slivers are outlawed.

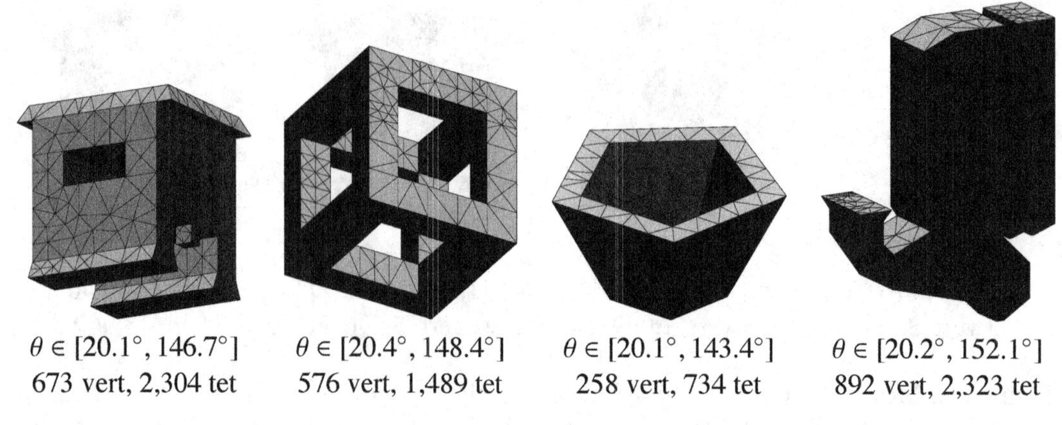

Figure 8.11: Meshes created by Delaunay refinement with a 20° bound on the smallest dihedral angle. For each mesh, θ indicates the range of dihedral angles in the mesh tetrahedra, vert denotes the number of vertices, and tet denotes the number of tetrahedra. Although the radius-edge ratios are not explicitly constrained, none is greater than 1.94. For comparison with Figure 8.9, the shortest edge in the leftmost mesh has length 0.346.

8.4 Refining slivers away

DelTetPLC is easily modified to split slivers as well as tetrahedra with small radius-edge ratios—but there is no guarantee that it will terminate. Nevertheless, it is natural to ask whether the algorithm terminates in practice.

As Figure 8.11 demonstrates, Delaunay refinement often succeeds for useful dihedral angle bounds. To create these meshes, we modified DelTetPLC to split a tetrahedron if and only if its smallest dihedral angle is less than 20°. Although the radius-edge ratio was not a criterion for determining which tetrahedra to split, we used it to determine the order in which skinny tetrahedra are split: tetrahedra with the largest radius-edge ratios are split first, and slivers with good radius-edge ratios are split last. In our experience, this ordering helps to reduce the number of tetrahedra.

The meshes in the figure have dihedral angles between 20° and 153°. Experiments with large meshes suggest that a minimum angle of 19° can be obtained reliably. However, the distribution of vertices is uneven in comparison with meshes generated using the radius-edge ratio as the sole splitting criterion, because excessive refinement occurs in locations that have bad luck with slivers. For this reason, it is attractive to use alternative methods of removing slivers (see Exercise 6 and Chapter 10) or to smooth the vertices of the over-refined mesh.

8.5 Constrained Delaunay tetrahedral refinement

The tetrahedral Delaunay refinement algorithm can be modified to maintain a CDT instead of an ordinary Delaunay triangulation and thereby gain some, but not all, of the benefits enumerated in Section 6.7 for Ruppert's algorithm.

Recall from Section 4.5 that the input PLC $\mathcal{P}$ might not have a CDT. The constrained Delaunay refinement algorithm begins by setting S to be the vertices in $\mathcal{P}$, constructing

Del S, and splitting the subset of encroached subsegments that are not strongly Delaunay, adding the new midpoint vertices to S. Let $\mathcal{P}'$ be $\mathcal{P}$ with its edges split by the extra vertices in S. When every subsegment is strongly Delaunay, by the CDT Theorem (Theorem 4.9), $\mathcal{P}'$ has a CDT. At that point, the algorithm constructs a CDT of $\mathcal{P}'$, which is a Steiner CDT of $\mathcal{P}$. Thereafter, it updates the CDT as new vertices are inserted into S. (There is also a lazy man's version of the algorithm, which splits encroached subsegments and missing subpolygons until Del S conforms to $\mathcal{P}$; then the algorithm discards the simplices outside $|\mathcal{P}|$ and thereafter maintains a CDT. This version requires no CDT construction code.)

Because of the need to split subsegments before constructing a CDT, the algorithm is only partly successful in preventing overrefinement due to geometric features separated by small distances exterior to the domain. However, once domain conformity has been achieved, the algorithm enjoys the other advantages: it saves the cost of maintaining tetrahedra outside the domain and testing which tetrahedra are in the domain; and every subsegment and subpolygon is a face of the CDT, so there is no need for point location when splitting an encroached one.

The algorithm also maintains a triangulation for each polygon g, and realizes the same advantages as Ruppert's algorithm if it maintains CDT $(S \cap g)$ instead of Del $(S \cap g)$.

8.6 Notes and exercises

The tetrahedral Delaunay refinement algorithm presented in this chapter is by Shewchuk [198]. It builds on ideas from the first provably good tetrahedral Delaunay refinement algorithm, by Dey, Bajaj, and Sugihara [74], which works only for convex polyhedra and generates tetrahedra of uniform size. Si [206] developed a constrained Delaunay version of the algorithm, available as the software TetGen. For the sake of simplifying the analysis, we outlaw some PLCs for which the algorithm is guaranteed to work. It is possible to prove that the polygons in Figure 8.5(f) do not prevent the Delaunay refinement algorithm from terminating, nor do angles like θ in Figure 8.5(b), but the proof requires redefining the notion of local feature size. See Shewchuk's Ph.D. dissertation [197] for details.

Miller, Talmor, Teng, Walkington, and Wang [149] show that Delaunay tetrahedralizations with bounded radius-edge ratios have Voronoi duals whose cells have bounded aspect ratios, and that some finite volume methods are guaranteed to have low discretization errors when they employ these dual Voronoi cells as control volumes.

The counterexample to the size-optimality of the tetrahedral Delaunay refinement algorithm in Figure 8.10 comes from an unpublished manuscript by Gary Miller and Dafna Talmor. Size-optimality has been proven for the octree algorithms of Mitchell and Vavasis [151, 152].

Because the number of tetrahedra in a Delaunay tetrahedralization can be quadratic in the number of vertices, it appears that Delaunay refinement on some domains must take at least quadratic time. However, Miller, Talmor, Teng, and Walkington [148] prove that a high-quality tetrahedralization has only a linear number of vertices. Hudson, Miller, and Phillips [115] give an alternative to Delaunay refinement called *sparse Voronoi refinement* that interleaves the creation of new vertices with the insertion of input vertices, always maintaining elements with bounded radius-edge ratios, rather than triangulating all the input vertices from the start. They call it "sparse" because throughout the refinement process, no vertex's degree ever exceeds a constant. Sparse Voronoi refinement is guaranteed to run in $O(n \log \Delta + N)$ time, where n is the input PLC complexity, N is the number of output vertices, and Δ is the *spread*, defined to be the diameter of the domain divided by the smallest distance between two disjoint linear cells in the input. If the spread is polynomial in n, the running time is $O(n \log n + N)$. It is an open problem, albeit solely of theoretical interest, to achieve an $O(n \log n + N)$ running time for domains with a superpolynomial spread.

Nave, Chrisochoides, and Chew [157] present an algorithm and implementation for parallel tetrahedral Delaunay refinement, demonstrating that it is possible to perform vertex insertions in parallel by having processors lock individual tetrahedra without suffering an unreasonable loss of speed. Spielman, Teng, and Üngör [208] give alternative parallel algorithms for both triangular and tetrahedral Delaunay refinement. They prove that there is always a large number of vertices that can be inserted simultaneously, yielding polylogarithmic guarantees on parallel running time.

Exercises

1. Show that a triangle $\sigma \in \mathrm{Del}\, S$ is encroached if and only if $\mathrm{Del}\, S$ contains a tetrahedron that has σ for a face and whose fourth vertex lies in σ's diametric sphere. Therefore, a subpolygon's encroachment can be diagnosed in $O(1)$ time.

2. Let S be a point set in $\mathbb{R}^3$ whose Delaunay triangulation $\mathrm{Del}\, S$ has no tetrahedron whose radius-edge ratio exceeds $\bar{\rho}$.

   (a) Let V_x be the Voronoi cell of a vertex $x \in S$. Let R be the radius of the smallest ball centered at x that contains the vertices of V_x (not counting a "vertex at infinity"). Let r be the radius of the largest ball centered at x and included in V_x. Prove that there is a constant depending solely on $\bar{\rho}$ that the ratio R/r cannot exceed.

   (b) Use (a) to show that there is a constant upper bound on the degree of vertices in $\mathrm{Del}\, S$, and hence $\mathrm{Del}\, S$ has $O(|S|)$ tetrahedra.

3. DelTetPLC splits one encroached subsegment, encroached subpolygon, or skinny tetrahedron at a time. Assuming that the input PLC satisfies the projection condition, is the algorithm still correct if we split every encroached subsegment at once? That is, if we make a list of the midpoints of every encroached subsegment and then insert them all (sequentially) into the triangulation? Explain. What if we split every encroached subpolygon at once? What if we split every skinny tetrahedron at once?

4. Let $\mathcal{P}$ be a three-dimensional PLC that satisfies the projection condition, and let ℓ be the shortest distance between two disjoint linear cells in $\mathcal{P}$. Use Proposition 8.4 and the idea in Figure 8.7 to prove that if $\bar{\rho} \geq 2$, DelTetPLC($\mathcal{P}, \bar{\rho}$) generates a mesh with no edge shorter than ℓ. Note that this guarantee is stronger than Proposition 8.5 in some part of the domain and weaker in other parts, as it does not guarantee a graded mesh.

5. Let $\mathcal{P}$ be a three-dimensional PLC that satisfies the projection condition. Let $\bar{r}$ be an input constant such that $\bar{r} \leq f(x)$ for all $x \in |\mathcal{P}|$. Suppose we modify Step 5 of DelTetPLC so that it is oblivious to radius-edge ratios, and instead splits tetrahedra whose circumradii exceed $\bar{r}$.

   5. If Del S contains a tetrahedron $\tau \subseteq |\mathcal{P}|$ whose circumradius exceeds $\bar{r}$, call SplitTetrahedron(τ, S, E), update Del S and the two-dimensional triangulations, and go to Step 3.

   Prove that the modified algorithm terminates, with the distance between every pair of vertices no less than $c \cdot \bar{r}$ for some positive constant c.

6. Chew [62] suggests the following idea for reducing the number of slivers created during Delaunay refinement. When the algorithm wishes to split a tetrahedron τ whose radius-edge ratio exceeds $\bar{\rho}$, do not insert a new vertex precisely at τ's circumcenter. Instead, let B be a ball with the same center as τ's circumball, but half its radius. Generate a bunch of random vertices in $B(t)$ and determine which of them will maximize the quality of the worst new tetrahedron, as determined by a quality measure of your choice, if it is inserted. Insert the winning vertex—unless it encroaches upon a subsegment or subpolygon, in which case the algorithm follows the usual procedure (reject the vertex, split a subsegment or subpolygon instead).

   For this modified algorithm, what is the best upper bound $\bar{\rho}$ on radius-edge ratio that can (without extraordinary efforts) be guaranteed with the same proof techniques employed in this chapter? Explain your answer.

7. To guarantee that no tetrahedron has a radius-edge ratio greater than 2, DelTetPLC requires that encroached subpolygons not be split in an arbitrary order; rather, a subpolygon may be split only if it contains the orthogonal projection of an encroaching vertex. If encroached subpolygons *are* split in an arbitrary order, what is the best upper bound $\bar{\rho}$ on radius-edge ratio that can (without extraordinary efforts) be guaranteed with the same proof techniques employed in this chapter? Explain your answer.

8. In the algorithm DelTetPLC, the circumcenter c of a subpolygon of a polygon g is rejected if it encroaches upon a subsegment. But if no subsegment that c encroaches upon is included in g, it is okay to insert c anyway, in addition to splitting all the subsegments it encroaches upon. Explain why this change does not affect the termination guarantee.

Chapter 9

Weighted Delaunay refinement for PLCs with small angles

The Delaunay refinement algorithm DelTetPLC in the previous chapter requires that the input PLC satisfy the projection condition, which rules out linear cells adjoining each other at dihedral angles or plane angles less than 90°. For many engineering applications, this restriction is not acceptable. Unfortunately, DelTetPLC, like Ruppert's original algorithm, can fail to terminate because of ping-pong encroachment (recall Figure 6.12). Ruppert's "modified segment splitting using concentric circular shells," described in Section 6.6, extends easily to spherical shells, and it copes reasonably well with segments that meet at small plane angles in tetrahedral meshes. It is not a complete solution, because of the seditious edges discussed in Section 6.6.

Three-dimensional domains introduce the additional danger of ping-pong encroachment among adjoining polygons, which is substantially harder to control than mutual encroachment among segments, especially if many polygons share a single segment. Efforts have been made to adapt spherical or cylindrical shells for the protection of segments where such polygons meet at small dihedral angles. However, algorithms for constructing Steiner Delaunay triangulations of three-dimensional PLCs often place undesirably many vertices in the vicinity of small domain angles and create undesirably short edges—even when they are not concerned with the quality of the tetrahedra.

This chapter offers an alternative approach to small domain angles that takes advantage of the properties of the weighted Delaunay triangulation, and does not overrefine as algorithms that use shells often do. We present here a new Delaunay refinement algorithm (it has not previously appeared in the literature) that generates a graded mesh in which no tetrahedron has a radius-edge ratio greater than 2, except possibly tetrahedra that adjoin a vertex or segment where two linear cells in the PLC meet at an acute angle. The tactic is to place positively weighted vertices at those meeting points. A weighted Delaunay triangulation tends to preferentially create edges and triangles adjoining more strongly weighted vertices, thereby helping to enforce domain conformity at the cost of locally sacrificing quality.

Because we use weighted vertices and a weighted Delaunay triangulation, we insert new vertices at orthocenters of simplices instead of their circumcenters. Orthocenters are further from the highly weighted vertices and closer to the unweighted vertices than circumcenters are. The weighted vertices act as protecting balls that prevent new vertices from being inserted too close to an apex of a small angle. The cycle of mutual encroachment in Figure 6.12 cannot occur because the apex is protected by a ball in which no new vertex can appear. Although we must compromise on tetrahedron quality near the apex, we guarantee that no tetrahedron has an orthoradius greater than twice the length of its shortest edge. The value of this guarantee is weakest where the vertex weights are greatest. Tetrahedra that do not adjoin small domain angles do not have weighted vertices, so their radius-edge ratios are at most 2, as usual.

9.1 The ideas behind weighted Delaunay refinement

Recall that each weighted vertex $v[\omega_v]$ is represented by a ball $B_v = B(v, \sqrt{\omega_v})$. We call these balls *protecting balls*. The weighted Delaunay refinement algorithm creates a weighted vertex with the intention that no vertex shall ever be inserted inside its protecting ball. It never assigns a vertex a negative weight. Most of the vertices that the algorithm creates are *unweighted*, meaning they have weight zero; we conceive of their protecting balls as being single points.

Weighted Delaunay refinement is similar to ordinary Delaunay refinement, but encroachment is defined in terms of orthospheres instead of circumspheres, and new vertices are inserted at orthocenters instead of circumcenters. Roughly speaking—we will be more precise later—a simplex σ with vertices from a weighted point set $S[\omega]$ is encroached if some other vertex in $S[\omega]$ is orthogonal or closer than orthogonal to the diametric orthoball of σ. In particular, an unweighted vertex must lie in σ's diametric orthoball to encroach upon σ. An encroached simplex is split with a new vertex at its orthocenter. The new vertex is not too close to any other vertex, because no vertex lies in the interior of that simplex's diametric orthoball, as no vertex has negative weight.

Several other observations are crucial to understanding the algorithm. First, if the intersection of the interiors of the protecting balls centered at σ's vertices is nonempty, then σ has an imaginary orthoradius. It follows that no unweighted vertex can encroach upon σ. The refinement stage of the algorithm inserts only unweighted vertices, so simplices with imaginary orthoradii are invulnerable to encroachment or to removal from the weighted Delaunay triangulation. In terms of the parabolic lifting map, the entirety of σ^+ is below the paraboloid, so no vertex inserted on the paraboloid can remove σ^+ from the convex hull. Our algorithm begins with a protection stage that identifies segments where polygons meet at acute dihedral angles and covers them with overlapping protecting balls, thereby ensuring that their subsegments cannot be split.

Second and conversely, if the intersection of the interiors of the protecting balls centered at σ's vertices is empty, then σ's orthocenter does not lie inside its vertices' protecting balls. If in addition σ is weighted Delaunay, then σ's orthocenter does not lie inside any protecting ball whatsoever. Therefore, splitting an encroached simplex never entails the risk of inserting a new vertex inside a protecting ball. Neither does splitting a tetrahedron with positive orthoradius.

Third, a tetrahedron is split if its *orthoradius-edge ratio*, its orthoradius divided by the length of its shortest edge, exceeds a threshold $\bar{\rho}$. If a tetrahedron has no weighted vertex, its orthoradius is its circumradius. Therefore, most tetrahedra in the final mesh have good circumradius-edge ratios, but tetrahedra that adjoin the apex of an acute domain angle might not. This is a reasonable compromise, as it is sometimes impossible to achieve high quality near small domain angles.

The weighted Delaunay refinement algorithm, called DelTetAcutePLC, takes as input a function $\lambda : |\mathcal{P}| \to \mathbb{R}$, called a *size field*, that reflects the user's locally desired spacing of vertices in the domain. The algorithm might be forced to generate much shorter edges in the vicinity of small angles and small geometric features, but it will not leave behind edges that are substantially longer than the user requests. A user desiring the sparsest possible mesh can simply set $\lambda \equiv \infty$. We require that $\inf_{x \in |\mathcal{P}|} \lambda(x) > 0$.

DelTetAcutePLC begins with a protection stage called Protect that centers protecting balls on vertices and segments in $\mathcal{P}$ where linear cells meet at acute angles. A protected segment is covered by a union of protecting balls. The initial vertex set $S[\omega]$ contains a weighted vertex $v[r^2]$ for each protecting ball $B(v, r)$ and an unweighted vertex for each vertex in $\mathcal{P}$ that is not protected. When the protection stage is done, DelTetAcutePLC constructs the weighted Delaunay triangulation Del $S[\omega]$ and executes a refinement stage called Refine, which inserts unweighted vertices into Del $S[\omega]$ to enforce domain conformity and eliminate poor-quality tetrahedra.

DelTetAcutePLC($\mathcal{P}, \lambda, \bar{\rho}$)

1. $S[\omega] \leftarrow$ Protect($\mathcal{P}, \lambda$).
2. Construct Del $S[\omega]$.
3. $S[\omega] \leftarrow$ Refine($\mathcal{P}, S[\omega], \lambda, \bar{\rho}$).
4. Return the mesh $\{\sigma \in \text{Del}\, S[\omega] : \sigma \subseteq |\mathcal{P}|\}$.

9.2 Protecting vertices and segments

The protection stage, implemented by the Protect pseudocode below, first identifies every vertex v in $\mathcal{P}$ that adjoins two linear cells that fail the projection condition (Definition 8.3), and for each such vertex constructs a protecting ball $B_v = B(v, \min\{\lambda(v), f(v)/(2\sqrt{2})\})$, where f is the local feature size function. Then, it identifies every segment in $\mathcal{P}$ that is an edge of two polygons in $\mathcal{P}$ that fail the projection condition, and covers each such segment with protecting balls (with the subroutine Cover). We call these specially treated vertices and segments *acute*. Clearly, both vertices of an acute segment are acute.

Protect($\mathcal{P}, \lambda$)

1. Let $S[\omega]$ be an empty point set.
2. For each vertex v in $\mathcal{P}$, determine whether v is acute. If so, add the weighted point $v[\omega_v]$ to $S[\omega]$ to represent the protecting ball $B_v = B(v, \sqrt{\omega_v})$, where $\sqrt{\omega_v} = \min\{\lambda(v), f(v)/(2\sqrt{2})\}$. Otherwise, add the unweighted point v to $S[\omega]$.

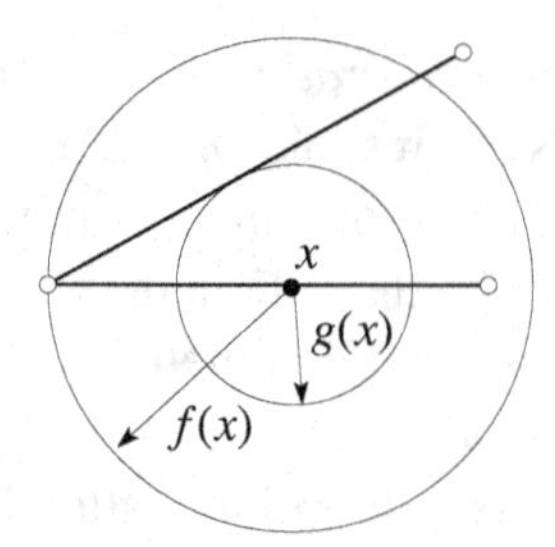

Figure 9.1: Local gap size $g(x)$ versus local feature size $f(x)$.

3. For each segment uv in $\mathcal{P}$, if uv is acute, call COVER($S[\omega], u, v$).
4. Return the weighted point set $S[\omega]$.

To set the radii of the protecting balls on segments, we use the notion of the local gap size.

Definition 9.1 (local gap size)**.** For every $x \in |\mathcal{P}|$, *the local gap size* $g(x)$ is the smallest radius $r > 0$ such that $B(x, r)$ intersects two linear cells in $\mathcal{P}$, one of which does not contain x.

Figure 9.1 illustrates the difference between the local gap size and the local feature size. At a vertex v in $\mathcal{P}$, the two notions coincide: $g(v) = f(v)$. Observe that g is not continuous: it approaches zero as x approaches a vertex, but is nonzero at the vertex. However, g is 1-Lipschitz in the interior of a segment.

We require the radius of every protecting ball with center x to be less than or equal to $g(x)/2$. Hence, no protecting ball centered on the interior of a segment s may reach more than halfway to a segment other than s, nor halfway to a polygon that does not include s. This restriction limits the ability of protecting balls to encroach upon subsegments and subpolygons, and ensures that no three protecting balls have interiors with a common intersection.

A recursive procedure COVER covers each acute segment in $\mathcal{P}$ with protecting balls such that any two consecutive balls are orthogonal. Let B_x and B_y be two balls that have already been placed on a segment. COVER($S[\omega], x, y$) begins by computing the unique ball on xy orthogonal to both B_x and B_y. If that ball is not too big, it is accepted to cover the remainder of xy. Otherwise, COVER centers a smaller ball at a point z in the middle of the gap between B_x and B_y and recursively calls itself to cover xz and zy. To ensure that it will not produce an unduly small ball, COVER maintains the invariant that the gap between nonorthogonal balls is always at least $2\sqrt{2} - 2 \doteq 0.828$ times the radius of the smaller of the two balls that bookend the gap. Figure 9.2 illustrates its workings.

COVER($S[\omega], x, y$)

1. Compute the center z on xy of the ball orthogonal to B_x and B_y with the relation
$$d(x, z) = \frac{d(x, y)^2 + \text{radius}(B_x)^2 - \text{radius}(B_y)^2}{2d(x, y)}.$$
Compute its radius $Z = \sqrt{d(x, z)^2 - \text{radius}(B_x)^2}$.

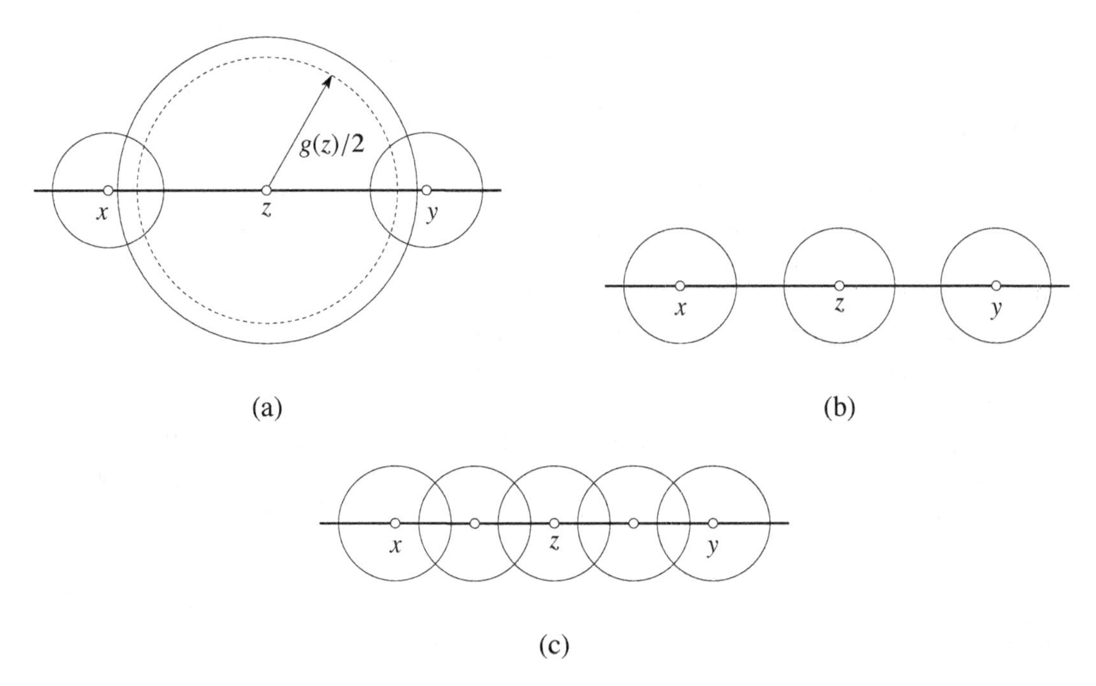

Figure 9.2: (a) The middle solid ball is orthogonal to B_x and B_y. Because its radius exceeds $g(z)/2$, we (b) use a ball of radius proportional to the gap size and (c) recursively cover xz and zy.

2. If $Z \leq \min\{\lambda(z), g(z)/2\}$, add the weighted point $z[Z^2]$ to $S[\omega]$ to represent the protecting ball $B(z, Z)$ and return.
3. Let $G = d(x, y) - \text{radius}(B_x) - \text{radius}(B_y)$ be the length of the gap between B_x and B_y. Let w be the point in the middle of the gap; i.e. $d(x, w) = \text{radius}(B_x) + G/2$. Let $W = \min\{\lambda(w), g(w)/2, G/(4\sqrt{2} - 2)\}$.
4. Add the weighted point $w[W^2]$ to $S[\omega]$ to represent the protecting ball $B(w, W)$, call Cover($S[\omega], x, w$), and call Cover($S[\omega], w, y$).

Cover ensures that protecting balls with consecutive centers on a segment are orthogonal. With this property, the subset of $\mathcal{P}$ in the union of the protecting balls can be triangulated with Delaunay tetrahedra whose aspect ratios have an upper bound that depends on the domain angles—that is, so that the protecting balls do not include arbitrarily thin slivers (see the notes at the end of the chapter). For simplicity, we will not make use of this property.

Protect and Cover must compute local gap sizes at points on acute vertices and segments. These can be determined by computing the distances from a point to all the linear cells in $\mathcal{P}$ that do not contain it and taking the minimum.

We show that Protect terminates and does not create unduly short edges by deriving a lower bound on the radius of every protecting ball. The following proposition is a preliminary step to that bound.

Proposition 9.1. Protect *and* Cover *maintain the invariant that if two balls have consecutive centers on a segment but are not orthogonal, there is a gap between them whose length* *is at least* $2\sqrt{2} - 2 \doteq 0.828$ *times the radius of the smaller ball. The gap can accommodate a ball that is at least as large as the smaller neighbor and not closer than orthogonal to either neighbor.*

Proof. For an acute segment uv, Step 2 of Protect creates a ball B_u whose radius is at most $f(u)/(2\sqrt{2}) \leq d(u,v)/(2\sqrt{2})$ and a ball B_v whose radius is also at most $d(u,v)/(2\sqrt{2})$. The gap between the two balls has length at least $(2\sqrt{2}-2)\, d(u,v)/(2\sqrt{2})$, so the invariant holds before Cover$(S[\omega], u, v)$ is called.

Step 3 of Cover places a ball of radius at most $G/(4\sqrt{2} - 2)$ in the center of a gap of length G, leaving smaller gaps of length at least $(2\sqrt{2} - 2)\, G/(4\sqrt{2} - 2)$ on either side of it. Again the invariant holds.

If we place at the center of the gap a ball with the same radius as the smaller ball adjoining the gap, the distance from its center to the center of the smaller ball is at least $\sqrt{2}$ times their radius. Therefore, it is not closer than orthogonal to either neighbor. □

Proposition 9.2. *Let uv be an acute segment in $\mathcal{P}$. The radius of every protecting ball B_z centered on uv is between* $c \cdot \inf_{x \in (uv \setminus (B_u \cup B_v)) \cup \{u,v\}} \min\{\lambda(x), g(x)/2\}$ *and* $\min\{\lambda(z), g(z)/2\}$, *where* $c = \left(4\sqrt{2} - 2\right)^{-1} \left(1 - 1/\sqrt{2}\right)^{1/2} > 1/7$. *If* $\inf_{x \in uv} \lambda(x)$ *is positive,* Cover$(S[\omega], u, v)$ *terminates, whereupon uv is covered by protecting balls such that any two balls with consecutive centers are orthogonal.*

Proof. Every ball placed by Step 2 of Protect or Step 2 or 4 of Cover satisfies the upper bound by construction. The radii of B_u and B_v, placed by Step 2 of Protect, also satisfy the lower bound by construction.

Any ball placed by Step 4 of Cover is centered at a point $w \in uv \setminus (B_u \cup B_v)$ at the midpoint of a gap of width G, and has radius either $W = \min\{\lambda(w), g(w)/2\}$, satisfying the lower bound, or $W = G/(4\sqrt{2} - 2)$. Let us find a lower bound for W in the latter case. By Proposition 9.1, the gap is next to a ball B_y of radius at most $G/(2\sqrt{2} - 2)$; let y be the center of B_y. The fact that Cover reached Step 4 means that a ball B_z constructed by Step 1 was rejected for being too large; specifically, $\text{radius}(B_z) > \min\{\lambda(z), g(z)/2\}$. As the center z of B_z lies in the gap, $d(y,z) \leq G + \text{radius}(B_y)$. As B_y and B_z are orthogonal, $\text{radius}(B_z)^2 = d(y,z)^2 - \text{radius}(B_y)^2 \leq G^2 + 2G \cdot \text{radius}(B_y) \leq G^2 + G^2/(\sqrt{2} - 1) = \sqrt{2}G^2/(\sqrt{2} - 1)$. Combining inequalities yields $W \geq c \cdot \text{radius}(B_z) > c \cdot \min\{\lambda(z), g(z)/2\}$.

Every ball placed on uv by Step 2 of Cover is at least as large as one of its two neighbors by Proposition 9.1. It satisfies the lower bound because its neighbor does.

The infimum $\inf_{x \in (uv \setminus (B_u \cup B_v)) \cup \{u,v\}} g(x)$ is positive for every PLC. If $\inf_{x \in uv} \lambda(x)$ is positive, there is a constant lower bound on the size of every protecting ball, so Cover terminates. By design, it can terminate only when any two balls with consecutive centers are orthogonal. □

See the end of Section 9.4 for a discussion of the mysterious expression $\inf_{x \in (uv \setminus (B_u \cup B_v)) \cup \{u,v\}} g(x)$ and its relationship to the small angles of $\mathcal{P}$.

In practice, it may be wise to modify Step 2 of Cover to make Z somewhat smaller than $g(z)/2$ (perhaps $g(z)/3$) to make room for more Steiner points to be inserted between two segments meeting at a small angle, and thereby make room to create tetrahedra with good radius-edge ratios in the gap.

9.3 The refinement stage

After the protection stage ends, the refinement stage triangulates $\mathcal{P}$. Each protecting ball B_x is represented by a weighted point $x[\omega_x]$ in $S[\omega]$, where $\omega_x = \text{radius}(B_x)^2$. Unprotected vertices in $\mathcal{P}$ are represented by unweighted points in $S[\omega]$. DelTetAcutePLC constructs the weighted Delaunay triangulation $\text{Del}\, S[\omega]$, and the procedure Refine refines it by inserting new vertices, all unweighted. We recall the notions of subsegments, subpolygons, and encroachment from Chapter 8, but we adjust their definitions to treat weighted vertices.

Definition 9.2 (subsegment encroachment). A vertex v *encroaches upon* a subsegment s if v is not a vertex of s and v has a nonpositive power distance from the diametric orthoball of s.

Definition 9.3 (subpolygon encroachment). For a polygon $h \in \mathcal{P}$, if no subsegment of h is encroached, then the two-dimensional weighted Delaunay triangulation $\text{Del}\,(S[\omega] \cap h)$ includes a Steiner triangulation of h. The *subpolygons* of h are the triangles in this Steiner triangulation, whether they appear in $\text{Del}\, S[\omega]$ or not. A vertex $v \in S[\omega]$ *encroaches upon* a subpolygon σ of h if the power distance between $v[\omega_v]$ and the diametric orthoball of σ is nonpositive, unless v lies on $\text{aff}\, h$ and the power distance is zero. (The exception is analogous to the exception for cocircular vertices in Definition 8.2. If the weighted Delaunay subdivision of $S[\omega] \cap h$ contains a polygon that is not a triangle, each weighted Delaunay triangulation subdivides the polygon into triangles, and those triangles should not be encroached upon by each other's vertices.)

There are three differences between the Refine stage and the Delaunay refinement algorithm in Chapter 8. First, encroachment is determined by orthoballs; second, new vertices are inserted at orthocenters. Even subsegments are split at their orthocenters rather than their midpoints.

SplitWeightedSubsegment($e, S[\omega]$)

Insert the orthocenter of e into $S[\omega]$.

Observe that every subsegment of an acute segment has a diametric orthoball of imaginary radius and is never encroached, so no such subsegment is ever passed to SplitWeightedSubsegment. Observe also that some segments are not acute but have one or even two vertices that are acute, and therefore weighted; it is these segments that necessitate inserting subsegment orthocenters rather than midpoints.

SplitWeightedSubpolygon($\sigma, S[\omega]$)

1. Let c be the orthocenter of σ.
2. If c encroaches upon a subsegment e, call SplitWeightedSubsegment($e, S[\omega]$) and return. Otherwise, insert c into $S[\omega]$.

The procedure for splitting encroached subpolygons is similar to that of Chapter 8: if no subsegment is encroached and a vertex v encroaches upon some subpolygon of a polygon h, the Monotone Power Lemma (Lemma 7.5) in Chapter 7 shows that v also encroaches upon the subpolygon σ of h that contains the orthogonal projection of v onto h. We split σ by inserting a new vertex at σ's orthocenter. Splitting σ rather than an arbitrary encroached subpolygon makes it possible to guarantee a better bound on the quality of the tetrahedra.

The third difference is that a tetrahedron is split if its orthoradius-edge ratio exceeds $\bar{\rho}$.

SplitWeightedTetrahedron($\tau, S[\omega]$)

1. Let c be the orthocenter of τ.
2. If c encroaches upon a subsegment e, call SplitWeightedSubsegment($e, S[\omega]$) and return.
3. If c encroaches upon a subpolygon of some polygon $h \in \mathcal{P}$, let σ be a subpolygon of h that contains the orthogonal projection of c onto h. Call SplitWeightedSubpolygon($\sigma, S[\omega]$) and return.
4. Insert c into $S[\omega]$.

The procedure Refine is the entry point for the refinement stage. Termination is guaranteed only if the bound $\bar{\rho}$ on the largest permissible orthoradius-edge ratio is 2 or greater. For simplicity, we omit from the pseudocode the necessary record-keeping of the subsegments, the subpolygons, and the mesh, addressed by the pseudocode in Chapter 8.

Refine($\mathcal{P}, S[\omega], \lambda, \bar{\rho}$)

1. While some vertex v in $S[\omega]$ encroaches upon a subsegment e, call SplitWeightedSubsegment($e, S[\omega]$) and repeat Step 1.
2. If some vertex v in $S[\omega]$ encroaches upon a subpolygon of a polygon $h \in \mathcal{P}$, let σ be a subpolygon of h that contains the orthogonal projection of v onto h. Call SplitWeightedSubpolygon($\sigma, S[\omega]$) and go to Step 1.
3. If Del $S[\omega]$ contains a tetrahedron $\tau \subseteq |\mathcal{P}|$ whose orthoradius-edge ratio exceeds $\bar{\rho}$ or whose orthoradius exceeds $\lambda(c)$ where c is the orthocenter of τ, then call SplitWeightedTetrahedron($\tau, S[\omega]$) and go to Step 1.
4. Return $S[\omega]$.

SplitWeightedSubpolygon is invoked only when there is no encroached subsegment, in which case for every polygon h in $\mathcal{P}$, the subcomplex $\{\sigma \in \text{Del}\, S[\omega] : \sigma \subseteq h\}$ triangulates h and is a quarantined complex. By the Orthocenter Containment Lemma (Lemma 7.7), h contains the orthocenters of all its subpolygons. Therefore, when SplitWeightedSubpolygon inserts the orthocenter of a subpolygon $\sigma \subseteq h$, the orthocenter refines h.

Likewise, SplitWeightedTetrahedron is invoked only when there is no encroached subsegment or subpolygon, in which case the subcomplex $\{\tau \in \text{Del}\, S[\omega] : \tau \subseteq |\mathcal{P}|\}$ is a quarantined complex. By the Orthocenter Containment Lemma, $|\mathcal{P}|$ contains the orthocenter of every tetrahedron in it, so it is safe to insert the orthocenters.

9.4 A proof of termination and good grading

In this section, we prove that if $\bar{\rho} \geq 2$, then, DELTETACUTEPLC terminates and returns a mesh of $\mathcal{P}$ in which no tetrahedron has an orthoradius-edge ratio greater than $\bar{\rho}$, hence no tetrahedron with unweighted vertices has a circumradius-edge ratio exceeding $\bar{\rho}$. We derive a lower bound on the distances between points in S, thereby showing that DELTETACUTEPLC terminates and produces nicely graded meshes.

The edge lengths in the final mesh are proportional to the local feature size function for a modified version of $\mathcal{P}$ we call a *eunuch structure*. Let C be the set of vertices at centers of protecting balls, and let U be the union of the interiors of all the protecting balls. Let $\mathcal{E} = \{g \setminus U : g \in \mathcal{P}\} \cup C$ be a set of cells defined by removing from each linear cell every protecting ball, then adding the protecting ball centers as vertices. $\mathcal{E}$ is not generally a complex, as the boundary of a cell in $\mathcal{E}$ is not necessarily a union of cells in $\mathcal{E}$. It is important to keep it this way. Observe that $\mathcal{E}$ satisfies the projection condition, because if two adjoining cells in $\mathcal{P}$ fail the projection condition, their counterparts in $\mathcal{E}$ do not adjoin each other at all.

The cells in $\mathcal{E}$ are not necessarily polyhedra, but $\mathcal{E}$ has a well-defined local feature size function, denoted $f_{\mathcal{E}}$. Multiplied by the right constant, this function is a lower bound on the edge lengths in the mesh generated by DELTETACUTEPLC (see Theorem 9.4). All vertices in that mesh lie on cells in $\mathcal{E}$. Thus $f_{\mathcal{E}}$ gives useful intuition on how mesh vertices are distributed.

The analysis follows the pattern established in Chapters 6 and 8. A vertex is of type i if it lies on a linear i-cell in $\mathcal{P}$ but not on a lower-dimensional cell. Every vertex in the mesh has a type, and so does every rejected orthocenter that SPLITWEIGHTEDSUBPOLYGON or SPLITWEIGHTEDTETRAHEDRON declines to insert because of encroachment. Vertices in $\mathcal{P}$ are of type 0. Other vertices lying on segments are of type 1, including vertices placed on segments by COVER and vertices inserted by SPLITWEIGHTEDSUBSEGMENT. Vertices inserted or rejected by SPLITWEIGHTEDSUBPOLYGON are of type 2. Vertices inserted or rejected by SPLITWEIGHTEDTETRAHEDRON are of type 3. Protecting balls are centered on some vertices of types 0 and 1.

We take the notion of insertion radius from Chapters 6 and 8, but change the definition. The *insertion radius* r_x of a vertex x of type 0 or a weighted vertex (of type 0 or type 1), created during the protection stage, is the Euclidean distance from x to the nearest other vertex created during the protection stage. For a vertex x inserted during the refinement stage—every unweighted vertex not of type 0—the power distance replaces the Euclidean distance: r_x is the square root of the power distance from x to the vertex in S (weighted or unweighted) that is nearest to x by power distance, at the instant before x is inserted into S or rejected. In other words, $r_x^2 = \min_{y \in S} \pi(x, y) = \min_{y \in S} \left(d(x, y)^2 - \omega_y\right)$. When a vertex is inserted at the center of an orthoball of a weighted Delaunay simplex, the insertion radius of the vertex is equal to the radius of the orthoball.

Each unweighted vertex inserted during the refinement stage has a *parent* vertex, whose insertion radius helps to place a lower bound on the child's. Centers of protecting balls do not have parents. The parent of an unweighted type 1 or 2 vertex x, inserted at the orthocenter of an encroached subsegment or subpolygon, is the closest encroaching vertex by power distance; this vertex might be in S or might be a rejected circumcenter. The parent of a type 3 vertex x, inserted at the orthocenter of a tetrahedron τ, is the most recently inserted vertex of τ's shortest edge. If both vertices were present from the start, choose one arbitrarily.

The following proposition places a lower bound on the insertion radius r_x of a vertex x in terms of $f_{\mathcal{E}}(x)$, $\lambda(x)$, and the insertion radius r_p of x's parent p. Recall that the relationship between r_x and r_p is the crux of the proof by induction of Proposition 8.5.

Proposition 9.3. *Let x be a vertex. Let p be its parent, if x has one.*

(i) *If x is a type 0 vertex or a weighted vertex, then $r_x \geq f_{\mathcal{E}}(x)$.*

(ii) *If x is an unweighted vertex of type $i \in \{1, 2\}$ and p is of type $j \leq i$, then $r_x \geq \sqrt{3} f_{\mathcal{E}}(x)/2$.*

(iii) *If x is an unweighted vertex of type $i \in \{1, 2\}$ and p is of type $j > i$, then $r_x \geq r_p/\sqrt{2}$.*

(iv) *If x is of type 3, then $r_x > \bar{\rho} r_p$ or $r_x > \lambda(x)$.*

Proof. If x is a type 0 vertex or a weighted vertex, $r_x \geq f_{\mathcal{E}}(x)$ because at the end of the stage Protect, r_x is the Euclidean distance from x to the nearest weighted vertex, and $f_{\mathcal{E}}(x)$ cannot exceed that value as $\mathcal{E}$ contains every weighted vertex.

If x is of type 3, it is the orthocenter of a tetrahedron τ whose orthoradius-edge ratio $\rho(\tau)$ exceeds $\bar{\rho}$ or whose orthoradius exceeds $\lambda(x)$. The insertion radius of x is the radius $r_x = \sqrt{\pi(x, p)}$ of τ's orthoball because the vertices of τ are orthogonal to the orthoball but, as τ is weighted Delaunay, no mesh vertex is closer than orthogonal to the orthoball. In the case where τ is split because its orthoradius exceeds $\lambda(x)$, clearly $r_x > \lambda(x)$. In the case where $\rho(\tau) > \bar{\rho}$, the parent p of x is the most recently inserted endpoint of τ's shortest edge. As no vertex in $S[\omega]$ is negatively weighted, r_p is no greater than the length ℓ of τ's shortest edge. Therefore, $r_x = \rho(\tau)\ell > \bar{\rho} r_p$.

If x is an unweighted vertex of type $i \in \{1, 2\}$, it is the orthocenter of an encroached subsegment or subpolygon σ. Let X be its orthoradius. If p is of type $j > i$, then p is a simplex orthocenter rejected for encroaching upon σ. Because σ was not encroached before p was rejected, $r_x = X$. The simplex σ contains the orthogonal projection of p onto aff σ by the algorithm's design. Therefore, σ has a vertex a such that $\angle pxa \leq 90°$, which implies that

$$d(p, a)^2 \leq d(p, x)^2 + d(x, a)^2 \leq X^2 + d(x, a)^2.$$

It follows that

$$r_p^2 \leq \pi(p, a) = d(p, a)^2 - \omega_a \leq X^2 + d(x, a)^2 - \omega_a = X^2 + \pi(x, a) = 2X^2 = 2r_x^2,$$

so $r_x \geq r_p/\sqrt{2}$ as claimed.

If x is an unweighted vertex of type $i \in \{1, 2\}$ and p is of type $j \leq i$, then let h be the linear cell of least dimension in $\mathcal{P}$ that includes the simplex σ split by x, and let h' be the linear cell of least dimension in $\mathcal{P}$ that contains p. Whether or not h and h' are disjoint, their counterparts $h \setminus U, h' \setminus U \in \mathcal{E}$ are disjoint, where U is the union of the interiors of the protecting balls. To see this, observe that if h' adjoins h, the encroachment implies that together they do not satisfy the projection condition, so $h \cap h'$ is covered by U.

Unweighted vertices are never inserted in protecting balls, so $x \in h \setminus U$, and if p is unweighted, $p \in h' \setminus U$. If p is weighted, it is a vertex in $\mathcal{E}$ with a protecting ball whose radius does not exceed half the local gap size, so $\sqrt{\omega_p} \leq g(p)/2 \leq d(x,p)/2$. In either case, by the definition of local feature size, $f_{\mathcal{E}}(x) \leq d(x,p)$, so

$$r_x^2 = \pi(x,p) = d(x,p)^2 - \omega_p \geq d(x,p)^2 - \left(\frac{d(x,p)}{2}\right)^2 \geq \frac{3}{4} f_{\mathcal{E}}(x)^2,$$

thus $r_x \geq \sqrt{3} f_{\mathcal{E}}(x)/2$ as claimed. □

We use Proposition 9.3 to show that the algorithm generates graded meshes with strong lower bounds on the edge lengths. We wish to incorporate the effect of the size field in addition to the local feature size. Define the field

$$\mu(x) = \min\left\{ f_{\mathcal{E}}(x), \inf_{y \in |\mathcal{P}|} (C_3 \lambda(y) + d(x,y)) \right\},$$

where $C_3 = (\bar{\rho} + \sqrt{2} + 1)/(\bar{\rho} - 2)$ (recall Proposition 8.5). Observe that the second expression in the braces is 1-Lipschitz—by design, it is the largest 1-Lipschitz function that is nowhere greater than $C_3 \lambda$. The field μ is 1-Lipschitz because it is a minimum of two 1-Lipschitz functions. It captures the combined influence of the local feature size and the size field on the edge lengths in the mesh.

Theorem 9.4. *Let $\mathcal{P}$ be a PLC embedded in $\mathbb{R}^3$. If $\bar{\rho} \geq 2$ and $\inf_{x \in |\mathcal{P}|} \lambda(x) > 0$, then* DelTetAcutePLC($\mathcal{P}, \bar{\rho}$) *terminates and returns a Steiner weighted Delaunay triangulation of $\mathcal{P}$ in which no tetrahedron has an orthoradius-edge ratio greater than $\bar{\rho}$. Therefore, tetrahedra with no weighted vertices have circumradius-edge ratios no greater than $\bar{\rho}$. For any two vertices p and q in the mesh, $d(p,q) \geq \mu(p)/(C_1 + 1)$ where $C_1 = (3 + \sqrt{2})\bar{\rho}/(\bar{\rho} - 2)$.*

Proof. By Proposition 9.2, the stage Protect terminates.

The inequalities stated in Proposition 9.3 are nearly the same as in Proposition 8.4, so we can reprise the proofs of Propositions 8.5 and 8.6 with μ replacing the local feature size f. Proposition 8.5 states that the insertion radius r_x of every vertex x of type i satisfies $r_x \geq \mu(x)/C_i$, where C_0, C_1, C_2, and C_3 are specified in the statement of Proposition 8.5. There are two changes to the algorithm for which we must verify that these invariants still hold. First, if a vertex x is inserted at the orthocenter of a tetrahedron because its orthoradius exceeds $\lambda(x)$, then $r_x > \lambda(x) \geq \mu(x)/C_3$ as stated. Second, the inequality $r_x \geq \sqrt{3} f_{\mathcal{E}}(x)/2$ from Proposition 9.3 is looser than its counterpart from Proposition 8.4, but it is tight enough to imply that $r_x > \mu(x)/C_2 > \mu(x)/C_1$.

By Proposition 8.6, for any two vertices in the mesh, $d(p,q) \geq \mu(p)/(C_1 + 1)$. By the Packing Lemma (Lemma 6.1), Refine terminates.

When DelTetAcutePLC terminates, no subsegment or subpolygon is encroached, hence $\{\sigma \in \text{Del}\, S[\omega] : \sigma \subseteq |\mathcal{P}|\}$ is a Steiner triangulation of $\mathcal{P}$. Moreover, because DelTetAcutePLC terminates, no tetrahedron's orthoradius-edge ratio exceeds $\bar{\rho}$. □

The merit of the eunuch structure $\mathcal{E}$ is that it gives us intuition about how small angles affect the edge lengths in the mesh. Theorem 9.4 shows that DelTetAcutePLC returns a

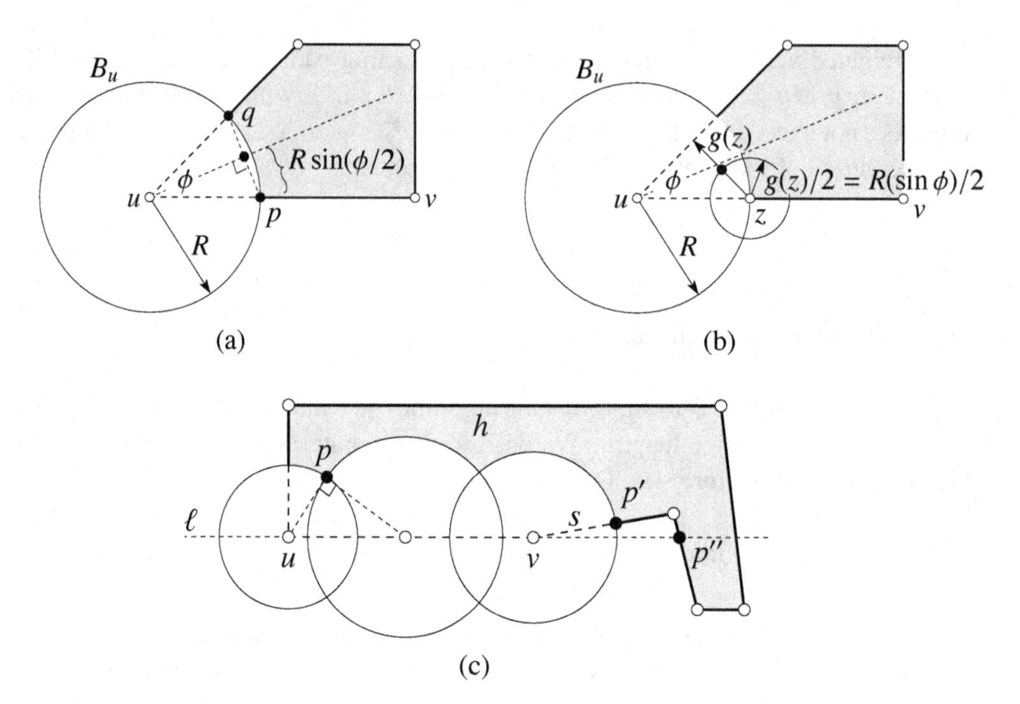

Figure 9.3: (a) The complex $\mathcal{E}$, shown as bold edges, has its minimum local feature size of $R\sin(\phi/2)$ at the midpoint of pq. (b) A protecting ball centered at z can have radius as great as half the gap size, as shown, or as small as nearly seven times smaller. (c) The polygon h meets another polygon h' (not shown) at a small dihedral angle along an edge uv. After the protecting balls are removed, the distance between the polygons depends partly on how close they are to the meeting line ℓ. A point p whose projection onto ℓ lies on $h \cap h'$ is kept at a distance by the orthogonal protecting balls. A point whose projection does not lie on $h \cap h'$, like p' and p'', can lie much closer to ℓ or even on ℓ.

mesh with edge lengths not much smaller than the size field λ or the local feature size $f_{\mathcal{E}}$ of the eunuch structure, whichever is smaller. There are two reasons why $f_{\mathcal{E}}(x)$ could be smaller than $f(x)$ at a point x: there are two adjoining linear cells in $\mathcal{P}$ whose counterparts in $\mathcal{E}$ are disjoint and whose distance from x is less than $f(x)$; or a new vertex was added to $\mathcal{E}$ at a protecting ball center near x.

Let us examine how small $f_{\mathcal{E}}$ can be in terms of the angles in $\mathcal{P}$. Recall from Chapter 8 that there are several notions of the angle at which two linear cells adjoin each other. For simplicity, we use the linear cells' affine hulls as proxies: if two segments or polygons (possibly one of each) h and h' adjoin each other and fail the projection condition, we take the smallest angle at which aff h meets aff h', and we say that h and h' meet at that angle.

Every acute vertex u is protected by a ball of radius $R = \min\{\lambda(u), f(u)/(2\sqrt{2})\}$. Consider a segment $uv \in \mathcal{P}$ and another segment or polygon $h \in \mathcal{P}$ that meet at u at an angle of ϕ. Their counterparts in $\mathcal{E}$, namely, $uv \setminus \text{Int}\, B(u, R)$ and $h \setminus \text{Int}\, B(u, R)$, can jointly generate a local feature size $f_{\mathcal{E}}$ as small as $R\sin(\phi/2)$, but not smaller (ignoring contributions from other linear cells). As Figure 9.3(a) shows, this value is achieved at the midpoint of a line segment pq where $p = uv \cap \text{Bd}\, B(u, R)$ and $q \in (\text{aff}\, h) \cap \text{Bd}\, B(u, R)$ is chosen so that $\angle qup = \phi$.

The success of Protect depends on the local gap size having a positive lower bound wherever protecting balls can be centered. If ϕ is the smallest angle at which an edge uv meets any other edge or polygon, then the local gap size $g(z)$ for a point z in the interior of uv is at least $\min\{f(z), d(u,z)\sin\phi, d(v,z)\sin\phi\}$; see Figure 9.3(b). Recall from Proposition 9.2 that every protecting ball centered on uv has radius greater than $\inf_{x\in(uv\setminus(B_u\cup B_v))\cup\{u,v\}} \min\{\lambda(x)/7, g(x)/14\} \geq \inf_{x\in(uv\setminus(B_u\cup B_v))\cup\{u,v\}} \min\{\lambda(x)/7, f(x)/14, d(u,x)\sin\phi/14, d(v,x)\sin\phi/14\}$. As the protecting balls at u and v do not contain the centers of the other protecting ball centered on uv, both $d(u,x)$ and $d(v,x)$ are greater than the radius R of the smaller of the protecting balls centered at u and v. Thus, the other balls centered on uv have radii at least $\inf_{x\in uv} \min\{\lambda(x)/7, f(x)/14, R(\sin\phi)/14\}$. A very small angle ϕ can yield very small protecting balls and a proportionally small local feature size $f_{\mathcal{E}}$ near the protecting balls.

Consider two polygons h and h' that meet at a small dihedral angle θ along a shared edge uv (or perhaps at a single vertex). How close can their counterparts in $\mathcal{E}$ be to each other? Let U be the union of the interiors of the protecting balls placed on $h \cap h'$. Let p be a point in $h \setminus U$ and q be a point in $h' \setminus U$ such that $d(p,q)$ is globally minimized. There are two cases. If the projection of p onto the line $\ell = \operatorname{aff} h \cap \operatorname{aff} h'$ lies on $h \cap h'$, as at left in Figure 9.3(c), then p cannot be closer to the line than $R'/\sqrt{2}$, where R' is the radius of the smallest protecting ball on uv. This follows because consecutive balls are orthogonal and the distance is minimized where the boundaries of two radius-R' balls meet. In this case, $d(p,q)$ can be as small as $\sqrt{2}R'\sin(\theta/2)$ if p and q are on the boundaries of the same two protecting balls, forcing the local feature size $f_{\mathcal{E}}$ to be as small as $R'\sin(\theta/2)/\sqrt{2}$ at the midpoint of pq. (Recall Figure 9.3(a), but imagine a dihedral angle.)

If p's projection does not lie on $h \cap h'$, as at the points p' and p'' in Figure 9.3(c), it is possible for $d(p,q)$ to be much smaller. Because $d(p,q)$ is minimized, at least one of p or q lies on an edge of its polygon; suppose p lies on an edge s of h. If s adjoins h' (see p' in the figure), let $\phi \leq \theta$ be the an angle at which they meet, let R be the radius of the protecting ball at the vertex where s and h' meet, and recall that their counterparts in $\mathcal{E}$ can be as close as $2R\sin(\phi/2)$; this is the minimum for $h \setminus U$ and $h' \setminus U$ as well. Therefore, they can generate a local feature size $f_{\mathcal{E}}$ as small as $R\sin(\phi/2)$, but not smaller.

Conversely, if s and h' are disjoint (see p'' in the figure), it is possible that h and h' make $f_{\mathcal{E}}$ smaller than f at some points, but h and h' cannot make $\min_{x\in\mathbb{R}^3} f_{\mathcal{E}}(x)$ smaller than $\min_{x\in\mathbb{R}^3} f(x)$ because s and h' are disjoint, and the distance between h and h' is minimized by $p \in s$ and $q \in h'$.

The worst circumstance is when an edge uv participates in both a very small plane angle ϕ that induces a tiny local gap size and forces Protect to place tiny protecting balls on uv, and a very small dihedral angle θ that forces Refine to produce tinier edges just outside the protecting balls. The lengths of these edges are $\Theta(\phi\theta)$ as ϕ and θ approach zero.

9.5 Notes and exercises

If a PLC in $\mathbb{R}^3$ has segments or polygons that adjoin each other at small angles, it can be quite difficult to triangulate it with Delaunay simplices, even if there is no constraint on the quality of the tetrahedra. Most algorithms for generating Steiner Delaunay triangulations of three-dimensional PLCs use the idea to protect PLC vertices and segments with balls inside which vertices may not be inserted. These algorithms do not use weighted vertices; rather, they prevent the insertions of vertices into the protecting balls by placing vertices on the boundary of the union of the protecting balls. Different protection methods have been proposed by Murphy, Mount, and Gable [156], Cohen-Steiner, Colin de Verdière, and Yvinec [65], Cheng and Poon [57], Cheng, Dey, Ramos, and Ray [52], and Pav and Walkington [165].

Murphy et al. [156] protect PLC vertices with balls whose radii are uniform—a fraction of the minimum local feature size among the vertices—and protect the segments with cylinders whose radii are uniform—a fraction of the minimum local gap size among the segments. The balls and cylinders are protected by vertices placed at points where PLC segments and polygons meet ball boundaries and cylinder boundaries. Finally, they triangulate the polygons by calling Chew's algorithm, described in Section 1.2. The triangles have uniform sizes, and their circumradii are a fraction of the minimum local gap size among the polygons. There are no restrictions on the shapes of the tetrahedra generated. This algorithm has the distinction of being the first proof that every three-dimensional PLC has a Steiner Delaunay triangulation, but it generates far more vertices than necessary.

Cohen-Steiner et al. [65] protect PLC vertices with balls whose radii are proportional to the local feature size and protect the segments with balls whose radii are proportional to the local gap size—much as we do in this chapter, except that two adjacent protecting balls may have a very small overlap, whereas the algorithm in this chapter places them so they are orthogonal. The balls are protected by vertices placed where ball boundaries intersect PLC polygons, as determined by special encroachment rules. These rules also recover the polygons. The algorithm produces graded triangulations in practice, but unduly short edges can appear.

The two algorithms above do not attempt to control the quality of the tetrahedra. The first Delaunay refinement algorithm to claim some theoretical guarantees on tetrahedron quality for domains with small angles is by Shewchuk [199]. It uses constrained Delaunay triangulations, the CDT Theorem (Theorem 4.9), and concentric spherical shell segment splitting to guarantee domain conformity. The final mesh is a graded Steiner CDT of the input PLC. The paper proves that for some PLCs, no algorithm can fix every skinny element, or even every skinny element that does not immediately adjoin a small domain angle: inherently, part of the problem of meshing domains with small angles is to decide when and where to leave skinny elements alone. The algorithm decides which skinny tetrahedra not to try to split by explicitly computing insertion radii and declining some of the vertex insertions that violate the consequences of Proposition 8.4, even if it means leaving a skinny tetrahedron intact.

Cheng and Poon [57] describe an algorithm that produces well-graded, high-quality Steiner Delaunay triangulations without the need for CDTs. They place protecting balls of graded sizes on vertices and segments, using a precursor of the scheme described in this chapter; then they run Delaunay refinement outside the protecting balls while placing vertices on the protecting ball boundaries according to encroachment rules. The union of the protecting balls is filled with Delaunay tetrahedra that are compatible with the triangulation outside the union. The algorithm guarantees an upper bound on the aspect ratios of the tetrahedra inside the protecting balls, ruling out the worst slivers. The bound depends on the domain angles, and degrades as the domain angles do. All the other tetrahedra in the domain have bounded radius-edge ratios.

The Delaunay meshing algorithm of Cheng, Dey, Ramos, and Ray [52] also protects vertices and segments with balls of graded sizes. The radii of the protecting balls centered at input vertices are chosen by computing local feature sizes, but the protecting balls on acute segments are chosen and refined adaptively by a specialized Delaunay refinement method. Once the algorithm has computed a Steiner Delaunay triangulation, the protecting balls are frozen. Then skinny tetrahedra are refined as described in Chapter 8, but tetrahedron circumcenters that fall in protecting balls are discarded. The algorithm is guaranteed to work only for polyhedra, i.e. domains in which there are no internal boundaries and no segment is a face of more than two polygons. This restriction makes it easier to produce a conforming Delaunay triangulation and obviates the need to compute ball-polygon intersections. The meshes in Figure 9.4 are generated by this algorithm.

Very shortly thereafter, Pav and Walkington [165] proposed a similar meshing algorithm that generates graded Steiner Delaunay triangulations of general PLCs. An advantage of the algorithm is related to local feature size computations: the algorithms of Cohen-Steiner et al. [65] and Cheng and Poon [57] require expensive explicit computations of local feature sizes at protecting ball centers; Cheng et al. [52] require them only at PLC vertices; Pav and Walkington [165] (and Shewchuk [199]) infer the local feature sizes inexpensively from the process of meshing itself, as do the algorithms DelTriPLC and DelTetPLC.

The constrained Delaunay refinement algorithm of Si [206] copes with small domain angles by maintaining a CDT and declining to insert a new vertex v if its insertion will create an edge shorter than $b \cdot \lambda(v)$, where b is a user-specified constant and λ is the user-specified size field. The algorithm takes advantage of the CDT Theorem (Theorem 4.9) to guarantee that a CDT exists; hence, some vertices may be inserted on segments despite creating short edges.

The new algorithm presented in this chapter adapts the segment protection scheme of Cheng and Poon [57], but it uses a Steiner weighted Delaunay triangulation both inside and outside the protecting balls. The ideas to turn protecting balls into weighted points and to refine by inserting orthocenters come from Cheng, Dey, and Ramos [51], whose algorithm for meshing piecewise smooth complexes appears in Chapter 15.

Exercises

1. Let s be a linear cell in a PLC, and let $\mathring{s}$ be the union of its proper faces in the PLC. Prove that the local gap size function g is 1-Lipschitz on the domain $s \setminus \mathring{s}$; that is, for any two points x and y in $s \setminus \mathring{s}$, $g(x) \leq g(y) + d(x, y)$.

2. (a) Derive the expressions in Step 1 of Cover for the position and radius of a ball orthogonal to two other balls with collinear centers.

   (b) Step 3 of Cover can be made more aggressive. Suppose that we modify it to compute three new balls having equal radii that fill the gap between B_x and B_y tightly, so that in the sequence of five balls from B_x to B_y, each consecutive pair of balls is orthogonal. Then the center ball is shrunk if necessary to accommo-

Figure 9.4: Three polyhedra that have small dihedral angles, and their tetrahedral meshes generated by a precursor of the algorithm described here.

date the size field and the local gap size, and the corresponding weighted point is inserted into $S[\omega]$.

Derive expressions for the radius and positions of the three balls. They need not be closed-form expressions, as the radius is the root of a quartic polynomial.

3. Let pqr be a triangle in $\mathbb{R}^2$. Suppose that we protect the segments of pqr as described in this chapter. Prove that for every protecting ball center x, $g(x) = \phi \cdot \Theta(f(x))$, where ϕ is the smallest angle of the triangle. Can you extend your proof to a polygon with polygonal holes?

4. Rather than have separate protection and refinement stages, we could change the algorithm so that it protects segments lazily during refinement, deciding whether to place a protecting ball on a segment only when one of its subsegments is encroached.

   (a) Write pseudocode for this algorithm.

   (b) When your algorithm decides to place a new protecting ball, can this ball contain an existing vertex? Why or why not?

   (c) Describe an advantage of the modified algorithm.

   (d) Prove that your algorithm produces a Steiner triangulation of the input PLC.

5. Derive formulae for computing the coordinates of the orthocenter of a triangle, given the coordinates of its vertices.

6. DELTETACUTEPLC does not protect nonacute segments. The advantage is that every mesh tetrahedron that does not adjoin an acute segment or vertex is guaranteed to have a good radius-edge ratio. However, if a Delaunay refinement algorithm protects every segment, acute and nonacute, then $\bar{\rho}$ can be reduced to less than 2, and REFINE can still be guaranteed to terminate and work correctly. For this modified algorithm, what is the best upper bound $\bar{\rho}$ on the orthoradius-edge ratio that can (without extraordinary efforts) be guaranteed with the same proof techniques employed in this chapter? Explain your answer.

Chapter 10

Sliver exudation

The quality measure naturally optimized by Delaunay refinement algorithms is the radius-edge ratio. Although it identifies all skinny triangles, the radius-edge ratio fails to screen out some sliver tetrahedra, so standard Delaunay refinement methods guarantee that they can remove most kinds of bad tetrahedra, but not all. Slivers have dihedral angles arbitrarily close to both 180° and 0°, and thereby poison both the discretization error and the conditioning of the stiffness matrix in the finite element method. Some of the quality measures we describe in Section 1.7—namely, the aspect ratio and the volume-length measure—give poor scores to all skinny tetrahedra. Unfortunately, these measures are not easy to optimize. It is notoriously difficult to devise algorithms that offer a mathematical guarantee that slivers will not survive.

This chapter describes such an algorithm. *Sliver exudation* takes as its input a Delaunay triangulation whose tetrahedra have good radius-edge ratios and returns a weighted Delaunay triangulation having the same vertices but somewhat different tetrahedra. Sliver exudation guarantees that the output mesh will not have extremely bad slivers; there is a fixed lower bound on a quality measure called the *sliver measure* of every tetrahedron. Unfortunately, that lower bound is too small to be a useful guarantee. It remains an open problem to design a tetrahedral meshing algorithm for PLCs that offers meaningful mathematical bounds on the dihedral angles. Fortunately, like many provably good mesh generation algorithms, sliver exudation works far better in practice than its theoretical bounds suggest; it removes most, but not all, skinny tetrahedra in practice.

Recall from Section 8.4 that slivers can be removed by refinement, although there is no mathematical guarantee. However, refinement to remove slivers often overrefines, creating smaller tetrahedra than the user desires. By contrast, sliver exudation has a mathematical guarantee and never overrefines—it never refines at all—but the guarantee is weak and it does not remove every unwanted tetrahedron in practice. Best results are obtained by combining the two—using sliver exudation to remove most of the slivers, then eliminating the few survivors with the weighted Delaunay refinement techniques of Chapter 9, perhaps interleaved with additional rounds of sliver exudation.

Because the analysis of sliver exudation is complicated, we defer a discussion of how to handle the domain boundaries to Chapter 11. To keep this chapter simple, assume that the input is a Delaunay triangulation of $\mathbb{R}^3$ with infinitely many tetrahedra or a triangulation of a periodic space.

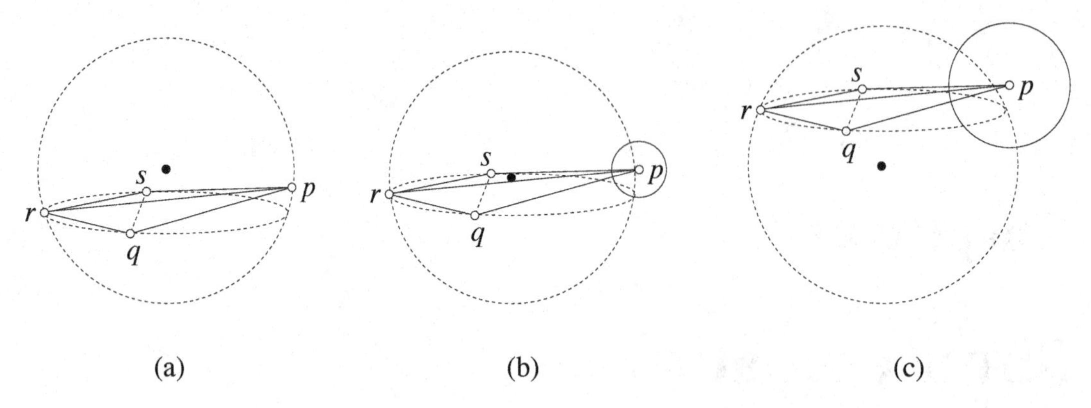

Figure 10.1: As the weight of the vertex p increases, the orthoball of the sliver $pqrs$ might initially shrink but soon expands rapidly.

10.1 The main idea and the algorithm

Let S be a countably infinite point set whose Delaunay triangulation fills space, thus $\text{conv}\, S = \mathbb{R}^3$. Suppose that there is a finite upper bound on the radius-edge ratios of the tetrahedra in $\text{Del}\, S$. We wish to assign appropriate weights to points in S and return a weighted Delaunay triangulation of the weighted points. We require that the balls that represent the weighted points do not intersect each other, to ensure that every weighted Delaunay edge, triangle, and tetrahedron has positive orthoradius. Therefore, the weight we assign to each point is nonnegative and less than one quarter of the square of the distance to its nearest neighbor in S.

Definition 10.1 (neighbor distance function). For every point $p \in \mathbb{R}^3$, let $N(p)$ be the distance between p and the second-closest point in S. We call N the *neighbor distance function.*

In the special case where $p \in S$, $N(p)$ is the distance from p to its nearest neighbor in S. Because $\text{Del}\, S$ connects every vertex to its nearest neighbor, it is easy to compute $N(p)$ over S. The function N is 1-Lipschitz: $N(x) \leq N(y) + d(x, y)$ for all points $x, y \in \mathbb{R}^3$.

Figure 10.1 gives some intuition about how a smart weight assignment can eliminate slivers. In Figure 10.1(a), the sliver $pqrs$ is Delaunay, so no vertex lies in its open circumball. As the vertices are initially unweighted, the circumball of $pqrs$ is its orthoball. If we continuously increase the weight of p, while keeping q, r, and s unweighted, the orthoball of $pqrs$ moves away from p as illustrated in Figures 10.1(b) and 10.1(c), though its boundary still passes through the other three vertices. Although the orthoball might initially shrink, it soon increases in size until it contains some other vertex in S, whereupon $pqrs$ is no longer weighted Delaunay and disappears from the weighted Delaunay triangulation. The closer the vertices of $pqrs$ are to being coplanar, the more unstable the orthoball is, and the faster it expands.

Several technical hurdles must be overcome to turn this idea into a proof. First, we quantify the relationship between a change in the weight of p and the change in the size of the orthoball. We cannot increase the weight of p to more than $N(p)^2/4$, so we must argue that a small increase in the weight of a sliver vertex induces a sufficiently large increase in the size of its orthoball. Second, increasing the weight of a second vertex might counteract the effect of the first one. For instance, if every vertex is assigned the same weight, the weighted Delaunay triangulation is identical to the Delaunay triangulation, and all the slivers survive! We must show that weights can be assigned so that slivers are not resurrected, nor are new slivers created. Most of this chapter is a sequence of derivations culminating in the Sliver Theorem, which shows that it is possible to eliminate the very skinniest slivers.

For the analysis of sliver exudation, it is convenient to define a quality measure similar to the volume-length measure, but with the root-mean-squared edge length replaced with the length of the longest edge. The volume-length measure is better for numerical optimization because it is a smooth function of the vertex positions, whereas the following sliver measure is easier to reason about in a discrete setting.

Definition 10.2 (sliver measure)**.** The *sliver measure* $\psi(\tau)$ of a tetrahedron τ is its volume divided by the cube of the length of its longest edge.

A positive lower bound on the sliver measure of a tetrahedron implies bounds on its aspect ratio and its smallest and largest plane and dihedral angles. A lower bound on the sliver measure *is* a lower bound on the volume-length measure.

The following procedure Exude takes a vertex set S and a parameter $\alpha \in (0, 1/2)$, and performs sliver exudation. The parameter α is best chosen by experimentation. The quality guarantee disappears at either extreme, as α approach zero or $1/2$.

Exude(S, α)

1. Let $\omega \equiv 0$ be a weight assignment; i.e. $\omega_p = 0$ for every $p \in S$. Compute Del S.
2. For each vertex $p \in S$:
   (a) Continuously vary the weight of p from zero to $\alpha^2 N(p)^2$ and keep track of the minimum sliver measure among the tetrahedra adjoining p whenever the weighted Delaunay triangulation changes.
   (b) Choose a weight ω_p for p that maximizes the minimum sliver measure among the tetrahedra that adjoin p. Update Del $S[\omega]$.
3. Return Del $S[\omega]$.

Exude offers a guarantee on the quality of the output mesh only if there is an upper bound on the radius-edge ratio of every tetrahedron in the input mesh Del S. The guarantee follows from the Sliver Theorem, which shows that Step 2 always finds a weight $\omega_p \in [0, \alpha^2 N(p)^2]$ such that every tetrahedron adjoining p in Del $S[\omega]$ has a sliver measure that exceeds some positive constant $\psi_{\min}$. The algorithm does not need to compute or know $\psi_{\min}$. The claim holds regardless of what weights have been assigned to the other vertices, and there is no need to revisit ω_p after assigning weights to subsequent vertices, although revisiting ω_p can be effective in practice. A crucial observation is that when ω_p increases, every new tetrahedron thus created adjoins p, so no new sliver is created.

Although the Sliver Theorem requires only that the tetrahedra in Del S have bounded radius-edge ratios, the quality guarantee $\psi_{\min}$ improves dramatically if no two adjoining

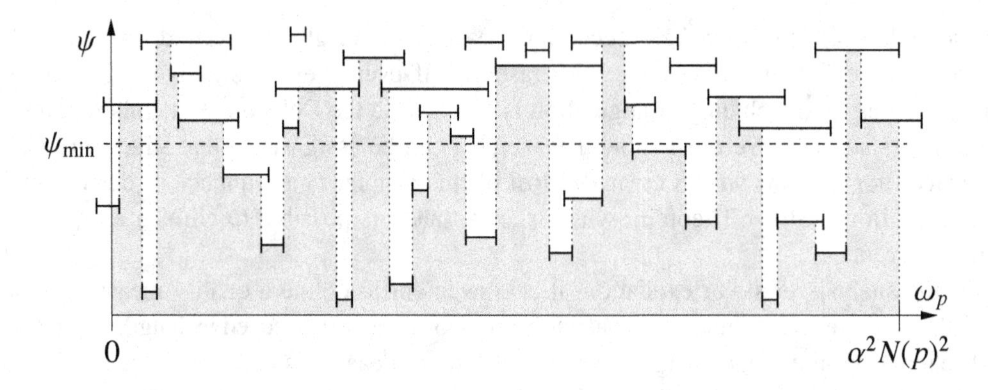

Figure 10.2: Each interval represents a tetrahedron that adjoins a vertex p and exists for an interval of values of the weight ω_p. A tetrahedron with a larger sliver measure ψ has the potential to survive over a wider range of weights. The shaded region graphs the minimum of the sliver measures. Sliver exudation chooses a weight that maximizes this minimum.

edges in Del S differ too much in length. We recommend modifying the Delaunay refinement algorithm to explicitly enforce this constraint. See Exercise 7 for details.

If S is infinite, a sequential Exude will run forever; but it can create an ever-expanding zone of tetrahedra with bounded quality. In principle, a parallel implementation with infinitely many processors could finish quickly because, as we shall see, each mesh vertex can interact with only a constant number of other vertices. The reasons why Exude does not work for a finite S are that Del $S[\omega]$ might not conform to the input domain and, more subtly, slivers near the domain boundary can survive if their orthocenters move outside the domain. Chapter 11 describes a Delaunay refinement algorithm that removes these obstacles.

10.2 Implementing sliver exudation

In principle, we find a weight ω_p for a vertex p by plotting the minimum sliver measure among the tetrahedra adjoining p as ω_p varies from zero to $\alpha^2 N(p)^2$, then choosing a value of ω_p that maximizes the minimum sliver measure, as illustrated in Figure 10.2. To turn this principle into an algorithm, we subdivide the domain of ω_p into discrete intervals in which Del $S[\omega]$ is constant. As we have seen in Section 5.8, when the weight of a vertex varies smoothly, the Delaunay Lemma helps us to predict the weights at which the weighted Delaunay triangulation will change. An implementation can maintain those weights in a priority queue.

Specifically, the algorithm examines every tetrahedron adjoining p, checks whether the triangular face opposite p in the tetrahedron is a face of a second tetrahedron, and if so, computes the weight ω_p above which the triangular face will no longer be locally Delaunay. That weight is the power distance from p to the orthoball of the second tetrahedron. If the weight does not exceed $\alpha^2 N(p)^2$, the priority queue stores a record of the triangular face with the power distance serving as its numerical key for the priority queue.

The algorithm simulates a continuously increasing vertex weight much like the procedure FlipInsertPolygon from Section 5.8: it repeatedly removes an event from the priority queue, checks whether the event is still relevant (the triangular face still exists), updates the weighted Delaunay triangulation with a bistellar flip or a variant of the Bowyer–Watson algorithm (see Exercise 1), and checks the newly created tetrahedra to see if new events should be enqueued.

After each update, the algorithm records the sliver measure of the worst tetrahedron adjoining p. When the priority queue is empty, the algorithm reverses the sequence of updates until it reaches the configuration that maximizes the minimum sliver measure. We will see that only a constant number of vertices are involved in the sequence of flips (Lemma 10.6), so finding the optimal weight for a vertex takes constant time. Sliver exudation thus runs in linear time.

An experimental evaluation of sliver exudation by Herbert Edelsbrunner and Damrong Guoy demonstrates that its performance in practice is better than the theory guarantees. Their implementation begins with tetrahedral Delaunay refinement, which produces a mesh in which a few percent of the tetrahedra have poor quality. Sliver exudation reduces this proportion, typically by a factor of about 50. Nevertheless, tetrahedra with dihedral angles less than 1° sometimes survive, and in most of their examples a few dihedral angles less than 5° survive. Most of the skinny tetrahedra that survive sliver exudation adjoin the boundary. This is not surprising, as slivers with orthocenters outside the domain can survive, and the implementation constrains the weight assignment so that surface triangles are never eliminated. Nevertheless, the method eliminates most skinny boundary tetrahedra, and the next chapter describes how to make sliver exudation perform as well at the boundary as it does in the interior.

The sliver measure in Step 2(b) can be replaced with any other quality measure; Edelsbrunner and Guoy use each tetrahedron's minimum dihedral angle.

Because the optimal weight of a vertex depends on the weights of its neighbors, a second and third pass of sliver exudation might further improve the mesh, although there is no improvement in the theoretical guarantee. However, if subsequent passes are permitted to reduce vertex weights, it is important to modify Step 2 to recognize that reducing the weight of p may create sliver tetrahedra that do not adjoin p, and the sliver measures of these tetrahedra must be considered in choosing the optimal weight. See Exercise 8 for a related suggestion that might further improve the algorithm's performance.

10.3 The union of weighted Delaunay triangulations

To ensure that the triangles in $\mathrm{Del}\, S[\omega]$ have good radius-edge ratios, we constrain the weight assignment ω so that the weight of each vertex p is in the range $[0, \alpha^2 N(p)^2]$, where α is a value chosen from $(0, 1/2)$. Let $\mathrm{W}(\alpha)$ be the set of all weight assignments ω such that $\omega_p \in [0, \alpha^2 N(p)^2]$ for every $p \in S$.

Different weight assignments yield different weighted Delaunay triangulations, motivating us to study the union of all weighted Delaunay triangulations. Define the set of simplices

$$K(S, \alpha) = \bigcup_{\omega \in \mathrm{W}(\alpha)} \mathrm{Del}\, S[\omega].$$

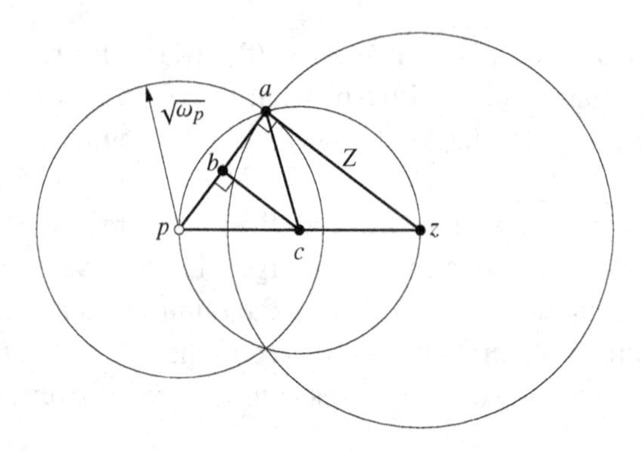

Figure 10.3: Because $B(z, Z)$ and $B(p, \sqrt{\omega_p})$ are orthogonal, every point a where their boundaries intersect forms a right triangle paz. Construct a similar triangle pbc of half the dimensions, and observe that abc is a reflection of pbc. Therefore, the ball $B(c, d(p,z)/2)$ has p, z, and a on its boundary and is included in $B(z, Z) \cup B(p, \sqrt{\omega_p})$.

In this section we prove several properties of $K(S, \alpha)$: the orthoradius-edge ratios of the tetrahedra do not exceed a constant; the triangles have circumradii at most a constant times their orthoradii; adjoining edges differ in length by most a constant factor; and the largest vertex degree does not exceed a constant.

10.3.1 Orthoradius-edge ratios of tetrahedra in $K(S, \alpha)$

The Orthoradius-Edge Lemma below shows that an upper bound on the radius-edge ratios of the tetrahedra in a space-covering Del S implies an upper bound on the orthoradius-edge ratios of the tetrahedra in $K(S, \alpha)$. Together with the Triangle Radius Lemma in the next section, this fact guarantees that the triangles in $K(S, \alpha)$ have good quality, although the tetrahedra might not. In Section 11.5, we show that the following lemma also holds for finite triangulations that contain their tetrahedra's orthocenters.

Lemma 10.1 (Orthoradius-Edge Lemma). *Let S be a point set in $\mathbb{R}^3$ such that* conv $S = \mathbb{R}^3$ *and no tetrahedron in* Del S *has a radius-edge ratio greater than a constant $\bar{\rho}$. If $\alpha \in (0, 1/2)$, no tetrahedron in $K(S, \alpha)$ has an orthoradius-edge ratio greater than $\mathring{\rho} = 2\bar{\rho}$.*

Proof. Let τ be a tetrahedron in $K(S, \alpha)$. Let $\omega \in \mathrm{W}(\alpha)$ be a weight assignment such that $\tau \in$ Del $S[\omega]$. Let z and Z be the orthocenter and orthoradius of τ for the weight assignment ω. Let p be a vertex of τ. By the definition of orthoball, the balls $B(z, Z)$ and $B(p, \sqrt{\omega_p})$ are orthogonal to each other, so $d(z, p)^2 = Z^2 + \omega_p$. No vertex in S lies inside $B(z, Z)$ because every vertex in S has nonnegative weight. No vertex in $S \setminus \{p\}$ lies in $B(p, 2\sqrt{\omega_p})$ because $\omega_p < N(p)^2/4$.

Let c be the midpoint of pz. The ball $B(c, d(p,z)/2)$ has p on its boundary and is included in the union $B(z, Z) \cup B(p, \sqrt{\omega_p})$; see Figure 10.3 and its caption for a demonstration of the latter claim. Therefore, $B(c, d(p,z)/2)$ contains no vertex in $S \setminus \{p\}$, and its center c is in p's Voronoi cell V_p.

This reasoning holds true for every vertex of τ, so let p and q be the vertices of τ's shortest edge, and let $c_p \in V_p$ and $c_q \in V_q$ be the midpoints of pz and qz, respectively. The

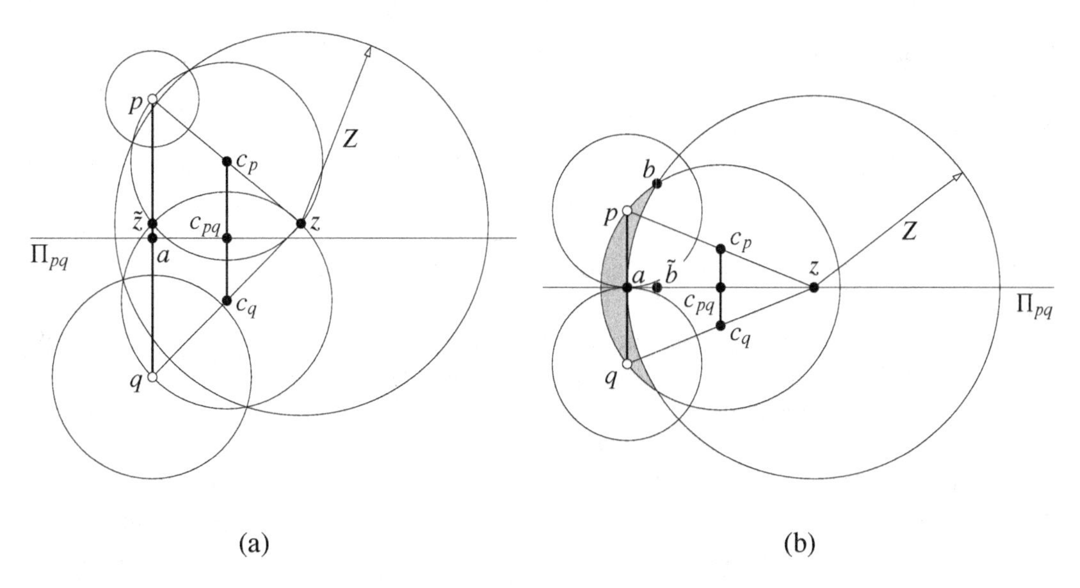

Figure 10.4: (a) The point c_p lies in p's Voronoi cell and c_q lies in q's. The line segment c_pc_q intersects at least one Voronoi polygon, which might or might not be $V_{pq} \subset \Pi_{pq}$. (b) The bulge $B(c_{pq}, d(c_{pq}, p)) \setminus \text{Int}\, B(z, Z)$ (shaded) is largest when p and q have maximum weight.

line segment c_pc_q is parallel to pq, has half the length of pq, and intersects the bisector plane Π_{pq} of pq at a point c_{pq}, as illustrated in Figure 10.4(a). The orthocenter z lies on the *weighted* bisector of pq, which intersects pq for every choice of weights in $\text{W}(\alpha)$. Hence, $\angle pc_pc_q$ is a nonacute angle, and among all the points on the line segment c_pc_q, c_p is closest to p. Therefore, $d(c_{pq}, p) \geq d(c_p, p) = d(z, p)/2 \geq Z/2$.

If c_{pq} lies on the Voronoi polygon V_{pq}, then pq is a Delaunay edge dual to V_{pq}. As V_{pq} is bounded and convex, it has a Voronoi vertex w such that $d(w, p) \geq d(c_{pq}, p) \geq Z/2$. There is a Delaunay tetrahedron $\sigma \in \text{Del}\, S$ dual to w that has pq for an edge and a circumradius of $d(w, p) \geq Z/2$. By assumption, the radius-edge ratio of σ is at most $\bar{\rho}$, so the length of pq is at least $Z/(2\bar{\rho})$, the orthoradius-edge ratio of τ is at most $2\bar{\rho}$, and the lemma holds.

Conversely, if c_{pq} does not lie on V_{pq} or (because S is not a generic vertex set) $pq \notin \text{Del}\, S$, the line segment c_pc_q intersects at least one other Voronoi cell besides V_p and V_q. For this case, we will show that there is a Delaunay tetrahedron with circumradius at least $Z/2$ and an edge even shorter than pq, then repeat the reasoning above. Assume that $Z \geq \sqrt{6}\, d(p, q)/2$; otherwise, τ satisfies the lemma immediately because $\bar{\rho}$ is at least $\sqrt{6}/4$, the radius-edge ratio of an equilateral tetrahedron.

Let $p, p', \ldots, q', q$ be the sequence of vertices whose Voronoi cells intersect c_pc_q, in order on a walk from c_p to c_q. The projections of these vertices onto aff c_pc_q occur in the same order; for example, it is not possible for V_p and $V_{p'}$ to intersect c_pc_q in one order if the projections of p and p' onto aff c_pc_q are in the opposite order. Therefore, although p' could be on the opposite side of the bisector Π_{pq} from p or q' could be on the opposite side of Π_{pq} from q, both cannot be true simultaneously. Therefore, assume without loss of generality that p' is not on the opposite side of Π_{pq} from p, so $d(p, p') \leq d(q, p')$.

Let $c_{pp'}$ be the point where the line segment c_pc_q intersects the Voronoi polygon $V_{pp'}$.

Echoing the argument above, $V_{pp'}$ has a Voronoi vertex w such that $d(w, p) \geq d(c_{pp'}, p) \geq d(c_p, p) \geq Z/2$, and the dual of w is a Delaunay tetrahedron $\sigma \in \text{Del}\, S$ that has pp' for an edge and a circumradius of at least $Z/2$; thus by assumption the length of pp' is at least $Z/(2\bar{\rho})$. We will show that pq is longer than pp', so the orthoradius-edge ratio of τ is less than $2\bar{\rho}$.

As $c_{pp'}$ lies on the Voronoi facet $V_{pp'}$, the ball $B(c_{pp'}, d(c_{pp'}, p))$ has p and p' on its boundary, and does not have q in its interior. The projection of p' onto aff c_pc_q lies between the projections of p and q, so p' lies between the two planes parallel to Π_{pq} that pass through p and q. Therefore, p' must lie in the ball $B(c_{pq}, d(c_{pq}, p))$, which is a superset of the portion of $B(c_{pp'}, d(c_{pp'}, p))$ between the two planes. No vertex lies in the interior of $B(z, Z)$, so p' lies in the *bulge* $B(c_{pq}, d(c_{pq}, p)) \setminus \text{Int}\, B(z, Z)$, illustrated in Figure 10.4(b). We examine the size of the bulge, which will prove to be small.

To bound its size, we argue that for a fixed edge length $d(p, q)$ and a fixed orthoradius Z, the *outer rim* of the bulge—the circle where the boundaries of $B(c_{pq}, d(c_{pq}, p))$ and $B(z, Z)$ intersect—achieves its maximum radius when the vertex weights assume their maximum values of $\omega_p = \omega_q = d(p, q)^2/4$, as illustrated. Hence, an upper bound on the radius of the outer rim that holds when p and q take their maximum weights holds for smaller weights as well.

First, we show that the *bulge ball* $B(c_{pq}, d(c_{pq}, p))$ grows as the weight of p grows. Consider the triangle pqz oriented as in Figure 10.4(a). Suppose p and q are fixed in place, and the weight ω_p is increased while the orthoradius Z stays fixed. Then the length of the edge pz increases and the orthocenter z moves while the edge pq remains stationary and the edge qz revolves about q with its length fixed. For any weight assignment in $W(\alpha)$, $\angle pqz$ is acute and increasing, so the orthocenter z moves further from aff pq. As pz grows, its midpoint c_p and the midpoint c_q also move further from aff pq, and therefore c_{pq} slides along the bisector Π_{pq}, maintaining the invariant $d(c_p, \text{aff}\, pq) = d(c_q, \text{aff}\, pq) = d(c_{pq}, \text{aff}\, pq) = d(z, \text{aff}\, pq)/2$. Therefore, the radius of the bulge ball $B(c_{pq}, d(c_{pq}, p))$ increases monotonically with ω_p.

Second, we show that, given our assumption that $Z \geq \sqrt{6}\, d(p, q)/2$, which implies that $d(p, q)^2 \leq 2Z^2/3$, less than half of the bulge ball's surface area is exposed outside the orthoball. Let a be the midpoint of pq, and observe that the orthoball $B(z, Z)$ must contain a for any valid weight assignment. By construction, $d(a, c_{pq}) = d(z, \text{aff}\, c_pc_q) \leq d(z, a)/2 \leq Z/2$. By Pythagoras' Theorem, the squared radius of the bulge ball is $d(c_{pq}, p)^2 = d(p, q)^2/4 + d(a, c_{pq})^2 \leq 5Z^2/12$. Let b be any point on the outer rim of the bulge, as illustrated in Figure 10.4(b), and observe that $d(z, b) = Z$. As both b and z lie in the bulge ball, $d(z, c_{pq})^2 + d(c_{pq}, b)^2 \leq 5Z^2/6 < Z^2$, implying that the angle $\angle zc_{pq}b$ is greater than 90°. This is true for every point b on the bulge's outer rim, so less than half the bulge ball's surface is exposed.

Third, we show that the distance $d(z, c_{pq})$ between the orthocenter and the center of the bulge ball increases monotonically with ω_p. To see this, let $\tilde{z}$ be the orthogonal projection of z onto the edge pq, and let $\gamma = d(p, z)^2$. By Pythagoras' Theorem,

$$d(z, \tilde{z})^2 = d(p, z)^2 - (d(p, q) - d(\tilde{z}, q))^2 = d(q, z)^2 - d(\tilde{z}, q)^2,$$

$$\text{therefore } d(\tilde{z}, q) = \frac{d(q, z)^2 - \gamma + d(p, q)^2}{2\, d(p, q)}.$$

The squared distance from the orthocenter to the center of the bulge ball is

$$\begin{aligned}
d(z, c_{pq})^2 &= \left(d(\tilde{z}, q) - \frac{d(p,q)}{2}\right)^2 + \left(\frac{d(z,\tilde{z})}{2}\right)^2 \\
&= d(\tilde{z}, q)^2 - d(\tilde{z}, q)\, d(p,q) + \frac{d(p,q)^2}{4} + \frac{d(q,z)^2 - d(\tilde{z}, q)^2}{4} \\
&= \frac{3\, d(\tilde{z}, q)^2}{4} - \frac{d(q,z)^2 - \gamma + d(p,q)^2}{2} + \frac{d(p,q)^2}{4} + \frac{d(q,z)^2}{4}.
\end{aligned}$$

The derivative of this value with respect to γ is

$$\frac{d}{d\gamma} d(z, c_{pq})^2 = \frac{3\, d(\tilde{z}, q)}{2} \frac{d}{d\gamma} d(\tilde{z}, q) + \frac{1}{2} = -3 \frac{d(q,z)^2 - d(p,z)^2 + d(p,q)^2}{8\, d(p,q)^2} + \frac{1}{2}.$$

For any weight assignment in $W(\alpha)$, $d(q,z)^2 - d(p,z)^2 = Z^2 + \omega_q - Z^2 - \omega_p \leq d(p,q)^2/4$, so the derivative above is at least $1/32$. Hence, $d(z, c_{pq})$ grows as the weight ω_p grows and pz lengthens.

Thus we have established that as the weight of p grows, both the radius of the bulge ball and its center's distance from the orthocenter grow, but not so much that half the bulge ball's surface area is exposed. It follows that the radius of the outer rim of the bulge also grows monotonically with ω_p. Symmetrically, the same is true as the weight of q grows.

Let us bound the radius of the outer rim of the bulge when the vertex weights assume their maximum values of $\omega_p = \omega_q = d(p,q)^2/4$, as illustrated in Figure 10.4(b). Then the balls $B(p, d(p,q)/2)$ and $B(q, d(p,q)/2)$ are tangent to each other and to the bisector Π_{pq} at the midpoint a of pq, where they meet $B(z, Z)$ orthogonally. The orthocenter z and the bulge center c_{pq} lie on Π_{pq}, with c_{pq} at the midpoint of za, so $d(c_{pq}, z) = d(c_{pq}, a) = Z/2$.

Let b be any point on the outer circular rim of the bulge, and let $\tilde{b}$ be the orthogonal projection of b onto za. The bulge ball's boundary also passes through p, so $d(b, c_{pq})^2 = d(p, c_{pq})^2 = d(p,a)^2 + Z^2/4$. By Pythagoras' Theorem,

$$\begin{aligned}
Z^2 &= d(b,z)^2 \\
&= d(b,\tilde{b})^2 + d(\tilde{b}, z)^2 \\
&= d(b,\tilde{b})^2 + (d(\tilde{b}, c_{pq}) + Z/2)^2 \\
&= d(b, c_{pq})^2 + Z\, d(\tilde{b}, c_{pq}) + Z^2/4 \\
&= d(p,a)^2 + Z^2/2 + Z\, d(\tilde{b}, c_{pq}), \\
\text{hence } d(\tilde{b}, c_{pq}) &= Z/2 - d(p,a)^2/Z.
\end{aligned}$$

The bulge rim has a squared radius of

$$\begin{aligned}
d(b,\tilde{b})^2 &= d(b, c_{pq})^2 - d(\tilde{b}, c_{pq})^2 \\
&= d(p,a)^2 + Z^2/4 - Z^2/4 + d(p,a)^2 - d(p,a)^4/Z^2 \\
&< 2d(p,a)^2 \\
&= d(p,q)^2/2.
\end{aligned}$$

We return now to the general case where the vertex weights are not necessarily at their maximum, nor equal to each other, and find an upper bound for $d(p,p')$. The plane aff pqz

bisects the bulge into two symmetric halves, both containing p and q. The plane through the ball centers z and c_{pq} and orthogonal to aff zpq divides these halves into symmetric quarters. This second cutting plane is Π_{pq} if p and q have equal weights; otherwise, it is slightly tilted, because z does not lie on Π_{pq}. Let x be the point at the center of the circular outer rim of the bulge. Both cutting planes contain x.

The bound on the bulge rim diameter implies that the bulge is included in a ball centered at x with radius less than $d(p,q)/\sqrt{2}$, and that this ball can be subdivided into eight identical octants by the same two cutting planes and a third orthogonal plane through x, the last being the affine hull of the circular rim. Each bulge quarter lies in a sphere octant. Any two points in a sphere octant are separated by a distance less than $d(p,q)$. The point p' shares a sphere octant with one of p or q, so either $d(p,p') < d(p,q)$ or $d(q,p') < d(p,q)$. In the latter case, recall that because p and p' are on the same side of the bisector Π_{pq}, $d(p,p') \leq d(q,p')$, which implies that $d(p,p') < d(p,q)$ as claimed. □

10.3.2 Circumradii of triangles in $K(S,\alpha)$

The circumradius of every triangle in $K(S,\alpha)$ is at most a constant times its orthoradius, as the following lemma shows. Together with the Orthoradius-Edge Lemma, this result implies a constant upper bound on the radius-edge ratio, and thus a constant lower bound on the minimum plane angle, of every triangle in $K(S,\alpha)$. The result holds only for triangles: a bound on the orthoradii of the tetrahedra in $K(S,\alpha)$ does not imply a bound on their circumradii.

Lemma 10.2 (Triangle Radius Lemma). *Let pqr be a triangle with circumradius X and orthoradius Z, where each vertex v of pqr has a weight $\omega_v \in [0,\alpha^2 N(v)^2]$ for $\alpha \in [0,1/2)$. Then $X \leq c_{\text{rad}}Z$, where $c_{\text{rad}} = 1/\sqrt{1-4\alpha^2}$.*

Proof. Suppose for the sake of contradiction that $Z < \sqrt{1-4\alpha^2}X$. Let pqr have circumcenter x and orthocenter z. For each vertex v of pqr, $d(z,v)^2 = Z^2 + \omega_v \leq Z^2 + \alpha^2 N(v)^2 < (1-4\alpha^2)X^2 + 4\alpha^2 X^2 = X^2 = d(x,v)^2$. Therefore, the orthocenter is closer to every vertex than the circumcenter is. Clearly, the orthocenter z of pqr lies in the intersection $B(p,d(z,p)) \cap B(q,d(z,q)) \cap B(r,d(z,r))$, but none of these three balls contains the circumcenter x. Every point in pqr besides z is closer to at least one vertex than z is, so every point in pqr is in at least one of the balls. If $x \in pqr$, the contradiction is immediate, so the lemma holds for triangles that contain their circumcenters.

Suppose $x \notin pqr$. Suppose without loss of generality that pq is the shortest edge and pr is the longest, as illustrated in Figure 10.5. Let $\ell = d(p,q)$ be the length of the shortest edge. The long edge pr separates the triangle from its circumcenter x, so the angle at q is obtuse. The short edge pq spans less than one-quarter turn on the circumcircle of pqr, so $\ell < \sqrt{2}X$.

The range $\omega_v \in [0,\alpha^2 N(v)^2]$ implies that $d(z,p)^2 \leq Z^2 + \alpha^2\ell^2$ and $d(z,r)^2 \leq Z^2 + \alpha^2 d(q,r)^2$, but $d(z,q) \geq Z$. Therefore, the orthocenter z must lie in $B\left(p, \sqrt{Z^2+\alpha^2\ell^2}\right) \cap B\left(r, \sqrt{Z^2+\alpha^2 d(q,r)^2}\right) \setminus \text{Int}\, B(q,Z)$. We claim that this set is empty because, loosely speaking, xq separates the first two balls from each other except perhaps in the interior of the third ball, so the lemma holds by contradiction.

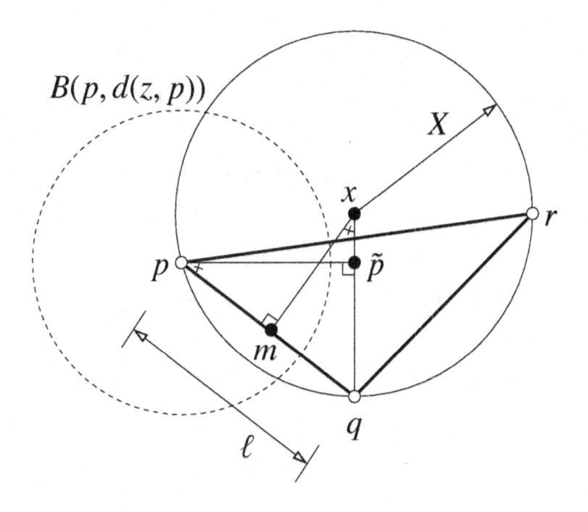

Figure 10.5: The orthoradius z is in the ball $B(p, d(z, p))$, which does not intersect xq.

To see this, let $\tilde{p}$ be the orthogonal projection of p onto xq, and let m be the midpoint of pq. The right triangle $p\tilde{p}q$ is similar to the right triangle xmq, so

$$\frac{d(q, \tilde{p})}{\ell} = \frac{\ell}{2X}. \tag{10.1}$$

The ball $B(p, d(z, p))$ intersects xq if and only if it contains $\tilde{p}$, or equivalently $d(z, p) \geq d(p, \tilde{p})$. However, if the ball $B(q, Z)$ also contains $\tilde{p}$, it is possible that $B(p, d(z, p))$ intersects xq but $B(p, d(z, p)) \setminus \operatorname{Int} B(q, Z)$ does not, which suffices for our proof. $B(q, Z)$ contains $\tilde{p}$ if and only if $d(q, \tilde{p}) \leq Z$, or equivalently

$$\ell^2 \overset{(10.1)}{=} 2X\, d(q, \tilde{p}) \leq 2XZ. \tag{10.2}$$

Therefore, we consider two cases: ℓ is in the range $(0, \sqrt{2XZ}]$ and thus $\tilde{p} \in B(q, Z)$, or ℓ is in the range $[\sqrt{2XZ}, \sqrt{2}X)$.

If $\ell \in [\sqrt{2XZ}, \sqrt{2}X)$, we will show that as Figure 10.5 depicts, $B(p, d(z, p))$ does not intersect xq. Suppose for the sake of contradiction that $d(p, \tilde{p}) \leq d(z, p)$. By Pythagoras' Theorem,

$$d(p, \tilde{p})^2 = d(p, q)^2 - d(q, \tilde{p})^2 \overset{(10.1)}{=} \ell^2 - \frac{\ell^4}{4X^2}.$$

As $p[\omega_p]$ and $B(z, Z)$ are orthogonal,

$$Z^2 + \alpha^2\ell^2 \geq Z^2 + \omega_p = d(z, p)^2 \geq d(p, \tilde{p})^2 = \ell^2 - \frac{\ell^4}{4X^2}.$$

Dividing this inequality by ℓ^2 gives

$$\frac{Z^2}{\ell^2} + \frac{\ell^2}{4X^2} \geq 1 - \alpha^2.$$

The left-hand side of this inequality approaches infinity as ℓ approaches either zero or infinity, and has one inflection point in between, so its maximum in the range $\ell \in [\sqrt{2XZ}, \sqrt{2}X]$ is at one of the two ends of the range. At the lower end of the range, the inequality reduces to $Z \geq (1-\alpha^2)X$, contradicting the assumption that $Z < \sqrt{1-4\alpha^2}X$. At the upper end, it reduces to $Z \geq \sqrt{1-2\alpha^2}X$, also contradicting the assumption. Because the inequality is false at both ends of the range, where the left-hand size is maximized, it is false for all values in the range. It follows that $B(p, d(z, p))$ lies strictly on one side of aff xq if $\ell \in [\sqrt{2XZ}, \sqrt{2}X)$.

If $\ell \in (0, \sqrt{2XZ}]$, $B(p, d(z, p))$ might intersect xq, but we will show that $B(p, d(z, p)) \setminus \text{Int}\, B(q, Z)$ does not. Let u be the point on xq for which $d(q, u) = Z$. Recall that $B(q, Z)$ contains $\tilde{p}$ because Inequality (10.2) holds, so the line segment uq also contains $\tilde{p}$. It follows that $B(p, d(z, p)) \setminus \text{Int}\, B(q, Z)$ intersects xq if and only if $B(p, d(z, p))$ contains u, or equivalently $d(p, u) \leq d(z, p)$. By the Law of Cosines, this would imply that

$$\begin{aligned} d(z, p)^2 &\geq d(p, u)^2 \\ &= d(q, u)^2 + d(p, q)^2 - 2d(q, u)\, d(p, q) \cos \angle pqu, \\ \text{therefore } Z^2 + \alpha^2 \ell^2 &\geq Z^2 + \ell^2 - \frac{Z\ell^2}{X} \\ \text{and } Z &\geq (1 - \alpha^2)X, \end{aligned}$$

contradicting the assumption that $Z < \sqrt{1-4\alpha^2}X$. Thus $u \notin B(p, d(z, p))$.

In either case, $B(p, d(z, p)) \setminus \text{Int}\, B(q, Z)$ lies strictly on one side of aff xq. If the edge rq, like pq, has length $\sqrt{2}X$ or less, it follows by symmetry that $B(r, d(z, r)) \setminus \text{Int}\, B(q, Z)$ lies strictly on the other side of aff xq, hence $B(p, d(z, p)) \cap B(r, d(z, r)) \setminus B(q, Z)$ is empty, contradicting the fact that every triangle has an orthocenter, so the lemma follows.

If the edge qr is longer than $\sqrt{2}X$, placing r in the upper right quadrant of Figure 10.5, then the fact that $B(r, d(z, r))$ does not intersect the circumcenter x implies that $B(r, d(z, r))$ does not intersect xq or the lower left quadrant of the figure. Symmetrically, $B(p, d(z, p))$ does not intersect the upper right quadrant. As the edge pr crosses xq and neither ball contains x, the balls $B(p, d(z, p))$ and $B(r, d(z, r))$ cannot intersect each other in the upper left quadrant. We have seen that $B(p, d(z, p)) \setminus \text{Int}\, B(q, Z)$ does not intersect the lower right quadrant. Again $B(p, d(z, p)) \cap B(r, d(z, r)) \setminus \text{Int}\, B(q, Z)$ is empty, and the lemma follows. □

We conjecture that the bound can be improved to $c_{\text{rad}} = 1/\sqrt{1-3\alpha^2}$. If so, this bound would be tight, as it is achieved by an equilateral triangle whose vertices have the maximum permitted weight.

The Orthoradius-Edge Lemma and the Triangle Radius Lemma show that the triangles in a mesh produced by sliver exudation have reasonable quality.

Lemma 10.3 (Triangle Quality Lemma). *Let S be a countably infinite point set in $\mathbb{R}^3$ such that* conv $S = \mathbb{R}^3$ *and no tetrahedron in* Del S *has a radius-edge ratio greater than a constant $\bar{\rho}$. If $\alpha \in (0, 1/2)$, for every weight assignment $\omega \in \text{W}(\alpha)$, every triangle $\sigma \in$* Del $S[\omega]$ *satisfies $\rho(\sigma) \leq c_{\text{rad}}\mathring{\rho}$, and thus has no angle less than* $\arcsin 1/(2c_{\text{rad}}\mathring{\rho})$.

Proof. By the Orthoradius-Edge Lemma (Lemma 10.1), every tetrahedron in Del $S[\omega]$ has an orthoradius-edge ratio of at most $\mathring{\rho}$. Every triangle in Del $S[\omega]$ is a face of a tetrahedron, and each triangle's orthoball is a cross-section of the tetrahedron's orthoball by the Orthoball Restriction Lemma (Lemma 7.4), so every triangle has an orthoradius-edge ratio of at most $\mathring{\rho}$. By the Triangle Radius Lemma (Lemma 10.2), every triangle in Del $S[\omega]$ has a radius-edge ratio of at most $c_{\text{rad}}\mathring{\rho}$ for every $\omega \in \text{W}(\alpha)$. □

10.3.3 The variation of edge lengths in Del S and $K(S, \alpha)$

If a mesh vertex adjoins both a very short edge and a very long one, sliver exudation is handicapped because the short edge restricts the weight of the vertex to a narrow range, but a large weight change might be necessary to eliminate a large sliver.

An upper bound $\bar{\rho}$ on the radius-edge ratio of a tetrahedron implies that its edge lengths differ by at most a factor of $2\bar{\rho}$. The following lemma establishes a more general fact: an upper bound on the radius-edge ratios of the tetrahedra in a Delaunay triangulation implies that the lengths of any two adjoining edges differ by at most a constant factor, even if they are not edges of a common tetrahedron. This lemma, called the Delaunay Length Lemma, generalizes Proposition 6.10 to three dimensions—or more; it holds equally well for higher-dimensional Delaunay triangulations. A second lemma, called the Edge Length Lemma, extends the result to weighted Delaunay tetrahedralizations and to $K(S, \alpha)$. These lemmas hold for space-covering triangulations; in Section 11.5, we extend the Delaunay Length Lemma to finite triangulations, albeit with a weaker constant.

Unfortunately, the constant factors are large. We hasten to add that there is a much more effective way to guarantee that adjoining edges differ by a small constant factor: change the Delaunay refinement algorithm so that it explicitly refines every edge that adjoins a much shorter edge. By this means, the edge ratio constant κ_{len} defined below can be made as small as 4. See Exercise 7 for details. In practice, large disparities in adjoining edge lengths are undesirable for most applications of meshes, and users typically specify a size field that prevents their formation.

Lemma 10.4 (Delaunay Length Lemma). *Let S be a point set in $\mathbb{R}^d$, and suppose that no d-simplex in* Del S *has a radius-edge ratio greater than a constant $\bar{\rho}$. Let p be a vertex in S that is not on the boundary of* conv S. *For any two adjoining edges $pq, pr \in$* Del S, *$d(p,q) \leq \kappa_{\text{len}} d(p,r)$, where $\kappa_{\text{len}} = 2\bar{\rho} e^{\sqrt{4\bar{\rho}^2-1}(\pi/2+\arcsin 1/(2\bar{\rho}))}$ and the arcsine is given in radians.*

Proof. Let V_p be p's Voronoi cell, and let x be a point on its boundary. As p is not on the boundary of conv S, V_p is bounded. We claim that px cannot meet a facet of V_p at an angle less than $\arcsin 1/(2\bar{\rho})$. To see this, let $q \in S$ be a vertex for which $V_{pq} = V_p \cap V_q$ is a facet of V_p that contains x, and observe that the dual of V_{pq} is the Delaunay edge pq, whose bisector includes V_{pq}. The distance from p to that bisector is $d(p,q)/2$. Every vertex of V_{pq} dualizes to a Delaunay d-simplex whose radius-edge ratio is at most $\bar{\rho}$, so $d(p,v) \leq \bar{\rho}\, d(p,q)$ for every vertex v of V_{pq} and $d(p,x) \leq \bar{\rho}\, d(p,q)$ for every point x on V_{pq}. The claim follows.

Let u be the nearest neighbor of p in S. The midpoint m of pu is the point nearest p on the boundary of V_p, as illustrated in Figure 10.6. Let v be the vertex of V_p farthest from p. No Delaunay edge adjoining p can have length greater than $2\,d(p,v)$ or less than $2\,d(p,m)$, so we wish to bound the ratio $d(p,v)/d(p,m)$.

Let Π be a plane that contains p, v, and m, and let $\Pi_p = \Pi \cap V_p$ be a two-dimensional cross-section of p's Voronoi cell; Π_p is a convex polygon. Define a polar coordinate system centered at p and oriented so that m lies at the angle $\theta = 0$ and v at an angle no greater than π, as illustrated.

Let $r(\theta)$ be a continuous function equal to the distance from p to the point on the boundary of Π_p at angle θ. Although $dr/d\theta$ is discontinuous at the vertices of Π_p, it is

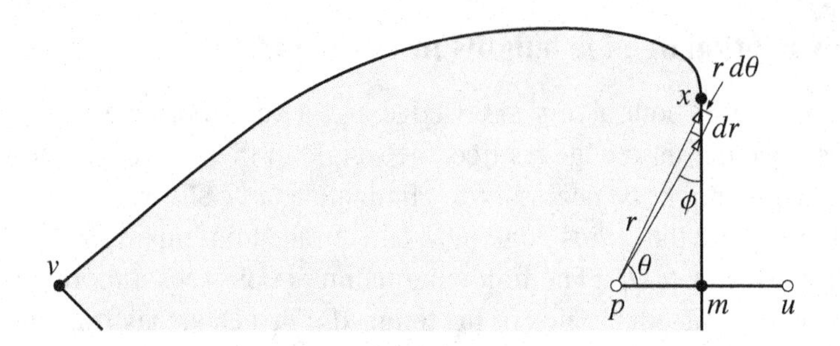

Figure 10.6: The boundary of p's Voronoi cell V_p parametrized with respect to a polar coordinate system centered at p, with p's nearest neighbor u at $\theta = 0$. Observe that $dr/d\theta = r \cot \phi$, where ϕ is the angle at which the radial ray meets the boundary. The spiral curve, which is not drawn to scale, represents the limiting outline of a sequence of Voronoi edges.

defined everywhere else on the boundary. If x is a point on an edge of Π_p and px meets the edge at an angle of ϕ, then $dr/d\theta = r \cot \phi$, as illustrated. We have seen that $\sin \phi \geq 1/(2\bar{\rho})$, so $\cot \phi \leq \sqrt{4\bar{\rho}^2 - 1}$.

Thus, $r(\theta)$ satisfies the differential inequality

$$\frac{dr}{d\theta} \leq \sqrt{4\bar{\rho}^2 - 1}\, r.$$

With the initial condition $r(0) = d(p, m)$ and θ expressed in radians, the solution is

$$r(\theta) \leq d(p, m)\, e^{\sqrt{4\bar{\rho}^2 - 1}\,\theta},$$

which describes a spiral. It follows that $d(p, v) \leq e^{\sqrt{4\bar{\rho}^2-1}\,\pi}\, d(p, m)$.

We improve this bound by observing that at $\theta = 0$, the line segment pm is orthogonal to the boundary and $dr/d\theta = 0$; it takes some distance for $r(\theta)$ to get up to speed, as the illustration shows. The function $r(\theta)$ is constrained by the bisector of pu for angles up to $\theta = \pi/2 - \arcsin 1/(2\bar{\rho})$, at which point $r \leq \bar{\rho}\, d(p, u) = 2\bar{\rho}\, d(p, m)$. The differential inequality bounds r for greater values of θ. Therefore, $d(p, v) \leq 2\bar{\rho} e^{\sqrt{4\bar{\rho}^2-1}(\pi/2 + \arcsin 1/(2\bar{\rho}))}\, d(p, m)$. □

Lemma 10.5 (Edge Length Lemma). *Let S be a countably infinite point set in $\mathbb{R}^3$ such that* conv *$S = \mathbb{R}^3$ and no tetrahedron in* Del *S has a radius-edge ratio greater than a constant $\bar{\rho}$. Let κ_{len} be the maximum ratio between the lengths of any two adjoining edges in* Del *S. For any two adjoining edges $pq, pr \in K(S, \alpha)$, $d(p, q) \leq c_{\text{len}} d(p, r)$, where*

$$c_{\text{len}} = \sqrt{4\bar{\rho}^2 \kappa_{\text{len}}^2 + \alpha^2 + 2\bar{\rho}\kappa_{\text{len}} \frac{1 + \alpha^2}{1 - \alpha^2}}.$$

Proof. As pq is in $K(S, \alpha)$, it is an edge in Del $S[\omega]$ for some $\omega \in \mathrm{W}(\alpha)$. Reducing the weight of any vertex besides p and q does not stop pq from being weighted Delaunay, so we can choose a weight assignment ω such that $\omega_v = 0$ for every $v \in S \setminus \{p, q\}$. Let $pquv \in$ Del $S[\omega]$ be a tetrahedron with edge pq. Let $B(z, Z)$ be the orthoball of $pquv$.

We will show that Z is at most proportional to $d(p, r)$, and that the distances $d(p, z)$ and $d(z, q)$ are at most proportional to Z as Figure 10.7 depicts, thereby establishing the lemma.

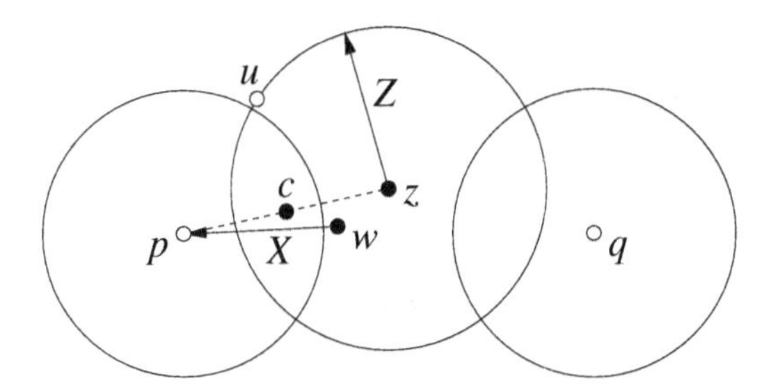

Figure 10.7: The length of a weighted Delaunay edge pq is not much larger than the diameter $2Z$ of an orthoball associated with pq.

Recall from the proof of the Orthoradius-Edge Lemma (Lemma 10.1) that the midpoint c of pz is in p's Voronoi cell V_p. Therefore, V_p has a vertex w such that $d(p, w) \geq d(p, c) \geq Z/2$, which dualizes to a Delaunay tetrahedron τ adjoining p with circumradius $X = d(p, w)$. Let ℓ be the length of the shortest edge of τ that adjoins p. The Delaunay triangulation connects p to its nearest neighbor, so $\ell \leq \kappa_{\text{len}}\, d(p, r)$ by assumption and

$$Z \leq 2\, d(p, w) = 2X \leq 2\bar{\rho}\ell \leq 2\bar{\rho}\kappa_{\text{len}}\, d(p, r).$$

The distance from p to the orthocenter z satisfies

$$d(p, z)^2 = Z^2 + \omega_p \leq 4\bar{\rho}^2\kappa_{\text{len}}^2\, d(p, r)^2 + \alpha^2 d(p, r)^2.$$

To bound the distance from q to the orthocenter z, observe that u has weight zero and lies on the boundary of the orthoball.

$$d(q, z)^2 = Z^2 + \omega_q \leq Z^2 + \alpha^2 d(q, u)^2 \leq Z^2 + \alpha^2 (d(q, z) + Z)^2.$$

Expanding yields a quadratic inequality whose positive solution is

$$d(q, z) \leq \frac{1 + \alpha^2}{1 - \alpha^2} Z.$$

The result follows by substituting these inequalities into $d(p, q) \leq d(p, z) + d(q, z)$. □

10.3.4 The degrees of vertices in $K(S, \alpha)$

The degree of a vertex in $K(S, \alpha)$ cannot exceed a constant that depends solely on $\bar{\rho}$ and α. This fact follows from the Edge Length Lemma (Lemma 10.5), which shows that the neighbors of a vertex cannot be too close to each other, hence only a constant number of neighbors can be packed within reach of a vertex. An important consequence is that sliver exudation runs in linear time.

Lemma 10.6 (Vertex Degree Lemma). *Let S be a countably infinite point set in $\mathbb{R}^3$ such that* conv S = $\mathbb{R}^3$ *and no tetrahedron in* Del S *has a radius-edge ratio greater than a constant $\bar{\rho}$. Every vertex in $K(S, \alpha)$ has degree at most* $c_{\text{deg}} = 3(2c_{\text{len}} + 1)^3 \ln(c_{\text{len}} + 2)$.

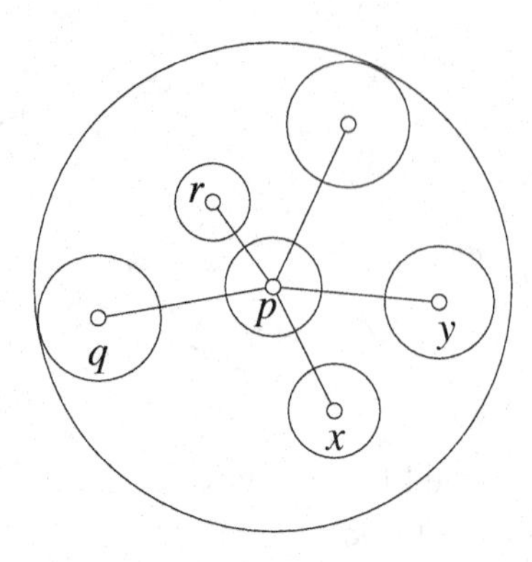

Figure 10.8: The small balls are centered on neighbors of p in $K(S,\alpha)$ and have radii proportional to their distances from p. They are disjoint and fit inside the large ball, which has room for no more than $O(c_{\mathrm{len}}^3 \log c_{\mathrm{len}})$ of them.

Proof. Let I be the set of neighbors in $K(S,\alpha)$ of a vertex $p \in S$, and consider the degree $|I|$ of p. Let $u \in I$ be a neighbor of p and let w be u's nearest neighbor in S; then $uw \in \operatorname{Del} S \subseteq K(S,\alpha)$. The Edge Length Lemma (Lemma 10.5) implies that $d(u,w) \geq d(p,u)/c_{\mathrm{len}}$. Therefore, if we center an open ball of radius $d(p,u)/(2c_{\mathrm{len}})$ at each vertex $u \in I$, as illustrated in Figure 10.8, the balls are pairwise disjoint. They are also disjoint from $B(p, N(p) - N(p)/(2c_{\mathrm{len}}))$.

Let pq be the longest edge adjoining p in $K(S,\alpha)$. The disjoint open balls are included in the ball $B(p,R)$ where $R = d(p,q) + d(p,q)/(2c_{\mathrm{len}})$. Therefore,

$$\int_{B(p,R)\setminus B\left(p,N(p)-\frac{N(p)}{2c_{\mathrm{len}}}\right)} \frac{1}{d(p,x)^3}\, dx > \sum_{u\in I} \int_{B\left(u,\frac{d(p,u)}{2c_{\mathrm{len}}}\right)} \frac{1}{d(p,x)^3}\, dx,$$

where dx represents an infinitesimal measure of volume. By the Edge Length Lemma, $d(p,q) \leq c_{\mathrm{len}} N(p)$, so $R \leq (c_{\mathrm{len}} + 1/2)\, N(p)$. We express the left integral in polar coordinates, and bound the right integral by taking the minimum value of $1/d(p,x)^2$ over each ball.

$$\begin{aligned}
\int_{N(p)-\frac{N(p)}{2c_{\mathrm{len}}}}^{R} \frac{4\pi r^2}{r^3}\, dr &> \sum_{u\in I} \int_{B\left(u,\frac{d(p,u)}{2c_{\mathrm{len}}}\right)} \frac{1}{\left(d(p,u)+\frac{d(p,u)}{2c_{\mathrm{len}}}\right)^3}\, dx. \\
\text{Therefore,}\quad [4\pi \ln r]_{N(p)-N(p)/(2c_{\mathrm{len}})}^{(c_{\mathrm{len}}+1/2)\,N(p)} &> \sum_{u\in I} \frac{1}{\left(d(p,u)+\frac{d(p,u)}{2c_{\mathrm{len}}}\right)^3}\, \frac{4\pi\, d(p,u)^3}{3(2c_{\mathrm{len}})^3}, \\
4\pi \ln\left(c_{\mathrm{len}} + \frac{2c_{\mathrm{len}}}{2c_{\mathrm{len}}-1}\right) &> \sum_{u\in I} \frac{4\pi}{3(2c_{\mathrm{len}}+1)^3}, \\
\text{and}\quad 3(2c_{\mathrm{len}}+1)^3 \ln(c_{\mathrm{len}}+2) &> |I|.
\end{aligned}$$

□

Observe that the Vertex Degree Lemma applies to ordinary Delaunay triangulations with unweighted vertices too, with c_{len} replaced by the smaller κ_{len} from the Delaunay Length Lemma.

10.4 The Sliver Theorem

Here we use the properties of $K(S,\alpha)$ established in Section 10.3 to prove the main result: if the tetrahedra in $\mathrm{Del}\, S$ have good radius-edge ratios, then there is a weight assignment $\omega \in \mathrm{W}(\alpha)$ such that $\mathrm{Del}\, S[\omega]$ is free of the skinniest slivers. The proof starts with the argument that no sliver can survive a small increase in the weight of one of its vertices. Then it uses the bounded degree of vertices in $K(S,\alpha)$ to show that EXUDE can assign any vertex a weight that guarantees that it does not adjoin a sliver.

Let $pqrs$ be a tetrahedron in $K(S,\alpha)$. Let $D_{pqrs}(\xi)$ be the signed distance from the orthocenter of $pqrs$ to $\mathrm{aff}\, qrs$ as a function of the weight ξ of p. The sign of $D_{pqrs}(\xi)$ is positive if the orthocenter of $pqrs$ lies on the same side of $\mathrm{aff}\, qrs$ as p, and negative otherwise. The following proposition shows that the orthocenter of $pqrs$ is close to $\mathrm{aff}\, qrs$.

Proposition 10.7. *Let S be a countably infinite point set in $\mathbb{R}^3$ such that $\mathrm{conv}\, S = \mathbb{R}^3$ and no tetrahedron in $\mathrm{Del}\, S$ has a radius-edge ratio greater than a constant $\bar{\rho}$. Let $pqrs$ be a tetrahedron in $\mathrm{Del}\, S[\omega]$ for some $\omega \in \mathrm{W}(\alpha)$. Let L be the length of the shortest edge of $pqrs$. Then $D_{pqrs}(\omega_p) \in [-c_{\mathrm{int}}L, c_{\mathrm{int}}L]$, where $c_{\mathrm{int}} = \sqrt{\mathring{\rho}^2 - 1/(3c_{\mathrm{rad}}^2)}$ and $\mathring{\rho}$ and c_{rad} are defined in Lemmas 10.1 and 10.2, respectively.*

PROOF. Let Y be the orthoradius of the triangle qrs, and let Z be the orthoradius of $pqrs$. By the Orthoball Restriction Lemma (Lemma 7.4), the orthoball of qrs is the intersection of $\mathrm{aff}\, qrs$ with the orthoball of $pqrs$, so

$$\begin{aligned} D_{pqrs}(\omega_p)^2 &= Z^2 - Y^2 \\ &\overset{\text{Lemma 10.1}}{\leq} \mathring{\rho}^2 L^2 - Y^2 \end{aligned}$$

The circumradius of qrs is at least $L/\sqrt{3}$, so the Triangle Radius Lemma (Lemma 10.2) implies that

$$Y \geq \frac{L}{\sqrt{3}c_{\mathrm{rad}}},$$

hence $D_{pqrs}(\omega_p)^2 \leq \mathring{\rho}^2 L^2 - L^2/\left(3c_{\mathrm{rad}}^2\right) = c_{\mathrm{int}}^2 L^2$. □

Next, we derive an expression for $D_{pqrs}(\xi)$ in terms of p's weight ξ and the altitude h of p from $\mathrm{aff}\, qrs$ and, thereby, show that $D_{pqrs}(\xi)$ is proportional to $-\xi/h$. A sliver has negligible volume, so h is negligible and ξ/h is large even for a small positive weight ξ. Thus, a small increase in p's weight can push $D_{pqrs}(\xi)$ outside the permissible interval derived in Proposition 10.7 and thereby eliminate the sliver $pqrs$. In other words, the orthoball is numerically unstable if the vertices of $pqrs$ are nearly coplanar.

Proposition 10.8. *$D_{pqrs}(\xi) = D_{pqrs}(0) - \xi/(2h)$, where $h = d(p, \mathrm{aff}\, qrs)$.*

PROOF. Let z and Z be the orthocenter and orthoradius of $pqrs$, respectively. Let ℓ be the distance from p to the line through z perpendicular to $\mathrm{aff}\, qrs$, illustrated in Figure 10.9. As $p[\xi]$ is orthogonal to the orthoball of $pqrs$,

$$\begin{aligned} \xi + Z^2 &= d(p,z)^2 \\ &= \ell^2 + (D_{pqrs}(\xi) - h)^2. \end{aligned} \tag{10.3}$$

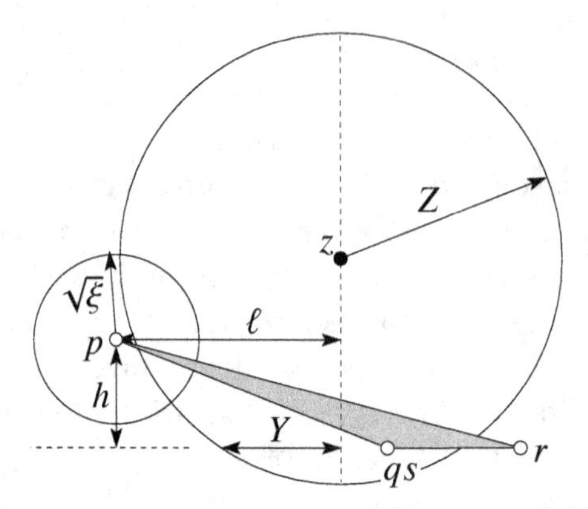

Figure 10.9: The left ball represents the weighted vertex p, and the right ball is the orthoball of the tetrahedron $pqrs$. The balls are viewed from a point in the line aff qs.

Let Y be the orthoradius of the triangle qrs. Recall from the proof of Proposition 10.7 that

$$Z^2 = Y^2 + D_{pqrs}(\xi)^2. \tag{10.4}$$

Substituting (10.4) into (10.3) and rearranging terms gives

$$D_{pqrs}(\xi) = \frac{h^2 + \ell^2 - Y^2}{2h} - \frac{\xi}{2h},$$

hence $D_{pqrs}(\xi) = D_{pqrs}(0) - \xi/(2h)$. □

The next proposition investigates the range of weights a vertex p can assume for which an adjoining sliver tetrahedron $pqrs$ in $K(S, \alpha)$ is weighted Delaunay, given that all the other vertex weights are fixed. The width of this range of weights is at worst proportional to the sliver measure $\psi(pqrs)$; recall Definition 10.2. Therefore, the thinner a sliver is, the easier it is to eliminate it by perturbing the weight of one of its vertices. In the following, let $\omega|p \mapsto \xi$ denote a weight assignment in which p has weight ξ rather than ω_p, but every other vertex in S has the weight specified by ω.

Proposition 10.9. *Let S be a countably infinite point set in $\mathbb{R}^3$ such that* conv $S = \mathbb{R}^3$ *and no tetrahedron in* Del S *has a radius-edge ratio greater than a constant $\bar{\rho}$. Let $pqrs$ be a tetrahedron in $K(S, \alpha)$, and let $\omega \in \mathrm{W}(\alpha)$ be a weight assignment such that $pqrs \in$* Del $S[\omega]$. *Let $\xi_{\min}$ and $\xi_{\max}$ be the minimum and maximum weights, respectively, in the set $\{\xi : pqrs \in$ Del $S[\omega|p \mapsto \xi]\}$. Then $\xi_{\max} - \xi_{\min} \leq c_{\mathrm{wt}} L_{\min} L_{\max}\, \psi(pqrs)$, where $c_{\mathrm{wt}} = 192 c_{\mathrm{int}} c_{\mathrm{rad}}^3 \mathring{\rho}^3$ and $L_{\min}$ and $L_{\max}$ are the lengths of the shortest and longest edges of $pqrs$, respectively.*

Proof. By Proposition 10.7, if $pqrs \in$ Del $S[\omega|p \mapsto \xi]$ for some weight ξ, then $D_{pqrs}(\xi)$ lies in the range $[-c_{\mathrm{int}} L_{\min}, c_{\mathrm{int}} L_{\min}]$. By Proposition 10.8, ξ lies in the range $[2hD_{pqrs}(0) - 2hc_{\mathrm{int}} L_{\min}, 2hD_{pqrs}(0) + 2hc_{\mathrm{int}} L_{\min}]$, where $h = d(p, \text{aff } qrs)$. Hence the interval $[\xi_{\min}, \xi_{\max}]$ has width at most $4hc_{\mathrm{int}} L_{\min}$.

We now show that h is at most proportional to $L_{\max}\, \psi(pqrs)$. By the Triangle Quality Lemma (Lemma 10.3), the triangular faces of $pqrs$ have radius-edge ratios at most $c_{\mathrm{rad}} \mathring{\rho}$.

Hence any two adjoining edges of $pqrs$ differ in length by a factor of at most $2c_{\text{rad}}\mathring{\rho}$, and the smallest angle of qrs is at least $\arcsin \frac{1}{2c_{\text{rad}}\mathring{\rho}}$. Let $\ell_{\max}$ and ℓ_{med} be the lengths of the longest and median edges of qrs. The shortest altitude of qrs is at least $\ell_{\text{med}}/(2c_{\text{rad}}\mathring{\rho})$, and

$$\text{area}(qrs) \geq \frac{\ell_{\text{med}}\ell_{\max}}{4c_{\text{rad}}\mathring{\rho}} \geq \frac{L_{\max}^2}{16c_{\text{rad}}^3\mathring{\rho}^3}. \tag{10.5}$$

It follows that

$$\begin{aligned} h &< 16c_{\text{rad}}^3\mathring{\rho}^3 L_{\max} \cdot \frac{h \cdot \text{area}(qrs)}{L_{\max}^3} \\ &= 48c_{\text{rad}}^3\mathring{\rho}^3 L_{\max}\,\psi(pqrs), \end{aligned}$$

and the difference $\xi_{\max} - \xi_{\min}$ is at most $4hc_{\text{int}}L_{\min} < 192c_{\text{int}}c_{\text{rad}}^3\mathring{\rho}^3 L_{\min}L_{\max}\,\psi(pqrs)$. □

Proposition 10.9 shows that if a sliver is skinny enough, we can eliminate it by perturbing the weight of one of its vertices. Because a vertex can adjoin only a constant number of tetrahedra in $K(S, \alpha)$, a proper choice of weight can eliminate *every* sufficiently skinny tetrahedron adjoining the vertex. Thus, we arrive at the main result of this chapter, the Sliver Theorem.

Theorem 10.10 (Sliver Theorem). *Let S be a countably infinite point set in $\mathbb{R}^3$ such that* $\text{conv}\, S = \mathbb{R}^3$ *and no tetrahedron in* $\text{Del}\, S$ *has a radius-edge ratio greater than a constant $\bar{\rho}$. Let α be a constant in $(0, 1/2)$. There is a weight assignment $\omega \in W(\alpha)$ such that every tetrahedron $\tau \in \text{Del}\, S[\omega]$ satisfies $\psi(\tau) > \psi_{\min}$ and every triangle $\sigma \in \text{Del}\, S[\omega]$ satisfies $\rho(\sigma) \leq 2\bar{\rho}/\sqrt{1 - 4\alpha^2}$, where $\psi_{\min} = \alpha^2/(2c_{\text{deg}}^2 c_{\text{len}}^2 c_{\text{wt}})$ is a constant that depends solely on α and $\bar{\rho}$. Moreover, the procedure* Exude *finds such a weight assignment.*

Proof. The quality of the triangles is guaranteed by the Triangle Quality Lemma (Lemma 10.3).

Let $pqrs$ be a tetrahedron in $K(S, \alpha)$. By the Edge Length Lemma (Lemma 10.5), no edge adjoining p has length greater than $c_{\text{len}}N(p)$, so no edge of $pqrs$ has length greater than $2c_{\text{len}}N(p)$. By Proposition 10.9, as the weight of p varies, $pqrs$ is weighted Delaunay only within a weight interval of width at most $2c_{\text{len}}^2 c_{\text{wt}}\,\psi(pqrs)\,N(p)^2$.

By the Vertex Degree Lemma (Lemma 10.6), p has degree at most c_{deg} in $K(S, \alpha)$. When the weight of p changes in the range $[0, \alpha^2 N(p)^2]$, only the tetrahedra whose vertices are all neighbors of p in $K(S, \alpha)$ (including p itself) change. Color p and the neighbors of p in $K(s, \alpha)$ blue. Consider the moment just before the algorithm begins to change p's weight. The tetrahedra in $\text{Del}\, S[\omega]$ whose vertices are all blue form a subcomplex $\mathcal{C}$ of $\text{Del}\, S[\omega]$ with at most $c_{\text{deg}} + 1$ vertices. The Upper Bound Theorem for Polytopes says that $\mathcal{C}$ contains at most $(c_{\text{deg}} + 1)(c_{\text{deg}} - 2)$ triangles. As the algorithm increases the weight of p from zero, every tetrahedron that ever adjoins p is the convex hull of p and some triangle in $\mathcal{C}$. Hence, fewer than $(c_{\text{deg}} + 1)(c_{\text{deg}} - 2) \leq c_{\text{deg}}^2$ tetrahedra ever adjoin p while the algorithm is changing p's weight. Call a tetrahedron a *sliver* if its sliver measure is less than $\psi_{\min}$; then the total width of the weight intervals such that p is a vertex of some weighted Delaunay sliver is less than $2c_{\text{deg}}^2 c_{\text{len}}^2 c_{\text{wt}}\psi_{\min}N(p)^2 = \alpha^2 N(p)^2$.

Therefore, p can be assigned a weight in the range $[0, \alpha^2 N(p)^2]$ such that no weighted Delaunay tetrahedron adjoining p has a sliver measure less than $\psi_{\min}$, regardless of the weights of the other vertices in S. If the weight of p is increased from zero, every new tetrahedron thus created adjoins p, so no new sliver is created. By initializing every vertex weight to zero then sequentially choosing a weight for each vertex in S, we obtain a weight assignment $\omega \in \mathrm{W}(\alpha)$ such that $\psi(\tau) > \psi_{\min}$ for every tetrahedron $\tau \in \mathrm{Del}\, S[\omega]$. □

There are many reasons to believe that the bound $\psi_{\min}$ is unrealistically pessimistic in practice: the weight intervals for the slivers are likely to overlap; the number of slivers adjoining p in $K(S, \alpha)$ is typically much smaller than c_{deg}^2; and every sliver has four vertices, meaning four opportunities to eliminate it. Exercise 8 describes a way to modify the algorithm that improves $\psi_{\min}$ by a factor of $c_{\mathrm{rad}}^2 \mathring{\rho}^2$ in theory. Although the modified algorithm is impractical, it suggests other changes that might improve sliver exudation's performance in practice.

10.5 Notes and exercises

The sliver exudation algorithm and its analysis are from a paper by Cheng, Dey, Edelsbrunner, Facello, and Teng [49]. Note that this chapter uses a different sliver measure than the original paper. New proofs are given here for the Orthoradius-Edge Lemma, the Triangle Radius Lemma, the Delaunay Length Lemma, the Edge Length Lemma, and the Vertex Degree Lemma, all with constants better than in the original paper. The experimental evaluation of sliver exudation discussed in Section 10.2 was performed by Edelsbrunner and Guoy [87, 107].

Chew [62] proposes an alternative provably good method for sliver elimination, which we have described briefly in Exercise 6 of Chapter 8. He removes slivers by refinement, as described in Section 8.4, but instead of always inserting a new vertex at a sliver's circumcenter, he tries multiple random locations near the circumcenter. The problem is that a newly inserted vertex v might form a new sliver tetrahedron $vwxy$. Chew observes that this can happen only if v falls in a narrow tubular *disallowed region* enclosing the circumcircle of wxy; his algorithm tests random new vertices until it finds one that is not in the disallowed region of any nearby triangle. If the mesh tetrahedra have good radius-edge ratios, the number of triangles that can form a sliver in conjunction with the new vertex v is bounded by a constant (recall the Vertex Degree Lemma, Lemma 10.6). If the disallowed regions are sufficiently narrow, there must be a point near the sliver's circumcenter that does not lie in any nearby triangle's disallowed region.

Like sliver exudation, Chew's method provably eliminates the very worst slivers, but the guaranteed bounds on quality are minuscule. Chew's method has the advantage of producing a truly Delaunay mesh and not requiring an algorithm for weighted Delaunay triangulations, but sliver exudation has the advantage that it refines the mesh less.

Another alternative to constructing a weighted Delaunay triangulation is to perturb the vertex positions, rather than the vertex weights. Edelsbrunner, Li, Miller, Stathopoulos, Talmor, Teng, Üngör, and Walkington [88] adapt the analysis of sliver exudation to show that spatial perturbations can achieve similar results. Li's doctoral dissertation [136] gives additional details and shows how to extend the algorithm to domains with boundaries.

Tournois, Srinivasan, and Alliez [218] experiment with sliver removal by random vertex perturbations. They report that it can produce reasonably good dihedral angles, but it is slow. The bottleneck is that many random perturbations of a vertex must be tried to find a good one. As a remedy, they propose and experiment with a deterministic point perturbation method, which gives minimum dihedral angles ranging from 15.51° to 28.55° in their experiments, and is only twice as slow as sliver exudation. Slivers adjoining the boundary sometimes survive because there is little freedom to perturb boundary vertices.

Labelle [126] proposes a method for refining an infinite mesh or a mesh of a periodic space that performs strikingly better than sliver exudation, vertex perturbation, or random refinement: every tetrahedron is guaranteed to have dihedral angles between 30° and 135° and a radius-edge ratio no greater than 1.368. The method refines the mesh with vertices chosen from a nested sequence of body centered cubic lattices, whose structure helps to ensure high quality. Unlike the other methods, however, nobody knows a practical way to extend the algorithm to domains with boundaries. The algorithm tends to overrefine. Nevertheless, the idea of choosing new vertices on a fixed lattice or a nested sequence of lattices seems promising in practice. A sliver cannot have all four of its vertices on the lattice; although a lattice vertex can participate in a sliver that adjoins at least one input vertex or boundary vertex, it has fewer opportunities to participate in slivers, so sliver exudation should be particularly effective on such meshes.

The Vertex Degree Lemma states that an upper bound on the radius-edge ratios of the tetrahedra in a three-dimensional Delaunay triangulation implies a constant upper bound on the maximum degree of a vertex. This fact was first proved by Miller, Talmor, Teng, and Walkington [148].

Exercises

1. (a) Describe a variant of the tetrahedral Bowyer–Watson algorithm that updates a weighted Delaunay triangulation when the weight of a vertex increases.

   (b) Prove that the two-dimensional variant of your algorithm is correct.

2. Consider an alternative measure $\chi(\tau)$ of the quality of a tetrahedron τ, defined to be the minimum altitude of τ divided by the length of τ's longest edge. Show that if there is a constant upper bound on the radius-edge ratios of τ's triangular faces, then $\chi(\tau)$ differs from the sliver measure $\psi(\tau)$ by at most a constant factor.

3. Consider an alternative definition of sliver. For any $\zeta \in (0, 1)$, a j-simplex is a ζ-sliver if its volume is less than $\zeta^j \ell^j / j!$, where ℓ is the length of the shortest edge of the j-simplex. What shapes of triangles can be ζ-slivers? What shapes of tetrahedra can be ζ-slivers? Describe the differences, if any, between the ζ-slivers and the tetrahedra having small sliver measures (Definition 10.2).

4. Show that a variant of the Sliver Theorem can be proven if we replace the sliver measure ψ with the definition of ζ-sliver in Exercise 3.

5. Call a d-simplex in $\mathbb{R}^d$ a *sliver* if it has poor dihedral angles despite having a good radius-edge ratio. With this definition, how successful do you expect sliver exudation would be in 10-dimensional space compared to 3-dimensional space, both in theory and in practice? In particular, how common do you think slivers are in 10 dimensions compared to 3 dimensions?

6. Give pseudocode for the weight finding algorithm described in Section 10.2, following the style of FLIPINSERTPOLYGON in Section 5.8. Your code should include a power distance computation and an explicit call to the subroutine FLIP from Section 4.4.

7. Suppose we modify the tetrahedral Delaunay refinement algorithm DELTETPLC by adding the following step.

   5b. If Del S contains an edge e that shares a vertex with another edge that is shorter by a factor of at least

   $$\kappa_{\text{len}} = \frac{(8 + 2\sqrt{2})\bar{\rho} - 4}{3 + \sqrt{2}},$$

   then let $\tau \in \text{Del}\, S$ be the tetrahedron with greatest radius-edge ratio that has e for an edge, call SPLITTETRAHEDRON(τ, S, E), update Del S and the two-dimensional triangulations, and go to Step 3.

   Show that Proposition 8.5 still holds for this modified algorithm. Hence, the modified DELTETPLC generates a mesh with the same guarantees on quality and grading, in which no two adjoining edges differ in length by a factor greater than κ_{len}. Thus, when such a mesh is treated by sliver exudation, the Sliver Theorem holds for much stronger values of κ_{len}, c_{len}, c_{deg}, and $\psi_{\min}$.

8. Proposition 10.8 implies that to exude a sliver, the most effective weight to change is that of the vertex having least altitude in the sliver, which is the vertex opposite the triangular face with greatest area. Because sliver exudation sometimes eliminates a sliver by changing the weight of the vertex opposite the triangular face with least area, the bound $\psi_{\min}$ is weakened by a factor of $2c_{\text{rad}}^2\mathring{\rho}^2$, lost in inequality (10.5). Here we consider a randomized algorithm that recovers most of this loss, and thus offers substantially better quality in theory. The idea is that a vertex v is not responsible for every sliver that adjoins it in $K(S, \alpha)$, but only those slivers in which it has least altitude.

   (a) Suppose we are given an infinite input mesh, but we are concerned only with the quality of the tetrahedra that intersect a bounded region of interest. Show that the following randomized replacement for EXUDE spends expected $O(1)$ time on each vertex and ultimately produces a mesh in which every tetrahedron τ in the region of interest satisfies $\psi(\tau) \geq \psi_{\min}/2$.

   While some tetrahedron τ in the region of interest has $\psi(\tau) < \psi_{\min}/2$, let v be an arbitrary vertex of τ, set its weight ω_v to a value chosen uniformly at random from $[0, \alpha^2 N(v)^2]$, and update Del $S[\omega]$.

(b) Show that the following randomized replacement for EXUDE spends expected $O(1)$ time on each vertex, and ultimately produces a mesh in which every tetrahedron τ in the region of interest satisfies $\psi(\tau) \geq c_{\text{rad}}^2 \mathring{\rho}^2 \psi_{\min}$.

> While some tetrahedron τ in the region of interest has $\psi(\tau) < c_{\text{rad}}^2 \mathring{\rho}^2 \psi_{\min}$, let v be the vertex of least altitude in τ, set its weight ω_v to a value chosen uniformly at random from $[0, \alpha^2 N(v)^2]$, and update Del $S[\omega]$.

Comment: Despite its theoretical superiority, the second algorithm is unlikely to beat EXUDE in practice. What might work well in practice is to perform several passes of weight optimization in which each vertex's weight is chosen to maximize the minimum quality of the adjoining tetrahedra in which the vertex has minimum altitude (ignoring the quality of the other adjoining tetrahedra), followed by a pass of standard sliver exudation in which each vertex's weight is chosen to maximize the minimum quality of all the tetrahedra that adjoin the vertex or are created by reducing its weight.

9. With weighted vertices, construct a tetrahedron $pqrs$ that has an arbitrarily large radius-edge ratio but a small orthoradius-edge ratio, thereby establishing that the Triangle Radius Lemma (Lemma 10.2) cannot be generalized to tetrahedra.

10. The proof of the Orthoradius-Edge Lemma shows that if a tetrahedron τ has orthocenter z, nonnegative orthoradius Z, and a nonnegatively weighted vertex p, and c is the midpoint of pz, then the ball $B(c, d(p,z)/2)$ has p on its boundary and is included in the union $B(z, Z) \cup B(p, \sqrt{\omega_p})$. If no vertex in S lies in the interior of the latter two balls, no vertex lies in the interior of $B(c, d(p,z)/2)$.

    In sliver exudation, a larger ball centered on p, namely $B(p, 2\sqrt{\omega_p})$, also has no vertex in S in its interior, so there exists a larger empty ball that touches p. Let u be a point that lies on pz one fifth of the way from z to p. Show that $B(u, d(u,p)) \subset B(z, Z) \cup B(p, 2\sqrt{\omega_p})$, so $B(u, d(u,p))$ also has no vertex in its interior. (Unfortunately, this ball does not yield a better constant in the Orthoradius-Edge Lemma, because the bulge grows proportionally, making room for longer edges.)

11. Identify the results in this chapter that can fail if S is finite—thus Del S has a boundary—and explain how they can fail.

Chapter 11

Refinement for sliver exudation

The Sliver Theorem states that, given a Delaunay mesh covering $\mathbb{R}^3$ whose tetrahedra have good radius-edge ratios, we can eliminate the worst slivers by assigning appropriate weights to the vertices. There are two hurdles in applying sliver exudation to finite domains. The first is domain conformity: if we increase the weight of a vertex to remove a sliver, some nearby subsegment or subpolygon may be removed as well. Loosely speaking, strongly weighted vertices can punch holes in the domain boundaries. The second is that a finite domain undercuts one of the premises of sliver exudation: that a sliver cannot be weighted Delaunay if its orthoball grows too large. If a sliver rests against the boundary of the domain's convex hull, its orthoball can grow arbitrarily large without ever enclosing a vertex.

In this chapter, we describe how a Delaunay refinement algorithm can mesh a PLC so that subsequent sliver exudation is effective and chances no risk of losing domain conformity. The idea is to add a step to the refinement algorithm that temporarily assigns a positive weight to each vertex, one vertex at a time. The increased weight may cause the vertex to encroach upon subsegments and subpolygons of the mesh—recall from Chapter 9 that a weighted vertex encroaches upon a subsegment or subpolygon if the former is orthogonal or closer than orthogonal to the latter's diametric orthoball. Only one vertex is weighted, so the diametric orthoballs are simply diametric balls. Subsegments and subpolygons encroached this way are split to ensure that they cannot become encroached during the subsequent sliver exudation stage.

When Delaunay refinement is finished, sliver exudation can proceed as usual with no danger that any subsegment or subpolygon will be encroached. Thus, every subsegment and subpolygon is strongly weighted Delaunay and is guaranteed to be in the weighted Delaunay triangulation.

Our algorithm takes as input a PLC $\mathcal{P}$, a parameter $\bar{\rho}$ that specifies the maximum permitted radius-edge ratio among the tetrahedra in the mesh prior to sliver exudation, and the sliver exudation parameter $\alpha \in (0, 1/2)$. The algorithm is guaranteed to terminate and produce a Steiner Delaunay triangulation of $\mathcal{P}$ if $\bar{\rho} > 2/(1 - \alpha^2)$ and $\mathcal{P}$ satisfies the projection condition (Definition 8.3). After sliver exudation, the algorithm returns a mesh in which every tetrahedron τ satisfies $\psi(\tau) > \psi_{\min}$ and every triangle σ satisfies $\rho(\sigma) \leq 2\bar{\rho}/\sqrt{1 - 4\alpha^2}$, where $\psi_{\min}$ is specified in Theorem 11.9, and is slightly weaker than in Chapter 10. Sliver exudation trades away a bound on the radius-edge ratios of the tetrahedra for a bound on their sliver measures.

11.1 Domain conformity with uncertain vertex weights

We combat both the problem of domain conformity and the problem of large orthoballs by modifying the Delaunay refinement algorithm to generate a mesh in which modest increases of the vertex weights cannot cause either problem. The idea is to treat each vertex by temporarily assigning it a positive weight—the maximum possible weight that sliver exudation might assign it—and split any subsegment or subpolygon that it thereby encroaches upon. The weight assignments are temporary because their sole purpose is to check whether a weight increase might threaten domain conformity or move a tetrahedron's orthocenter outside the domain. The refinement stage terminates and the sliver exudation stage begins when every tetrahedron has a good radius-edge ratio and no subsegment or subpolygon can become encroached if a vertex is assigned a weight within the range employed by sliver exudation.

The success of both stages depends on the *vertex weight invariants*.

Definition 11.1 (vertex weight invariants). Let S be the set of mesh vertices at a selected time during execution, including every vertex in $\mathcal{P}$ and those that have been inserted.

1. At every point in time during Delaunay refinement, every vertex $p \in S$ has a weight of either 0 or $\alpha^2 N(p)^2$, with the neighbor distance function N defined with respect to the current vertex set S.

2. At every point in time during sliver exudation, every vertex $p \in S$ has a weight in the range $[0, \alpha^2 N(p)^2]$.

Invariant 1 is maintained by our method of temporary weight assignment. Invariant 2 is maintained by the sliver exudation algorithm. As $\alpha < 1/2$, both invariants imply that for any two weighted vertices $p, q \in S$, the balls that represent them have disjoint interiors. Observe that $N(p)$ tends to shrink during refinement, but it is fixed during sliver exudation.

The mesh refinement stage maintains an unweighted vertex set S and its Delaunay triangulation Del S. When the algorithm temporarily assigns a vertex the maximum possible weight, it does not actually modify the triangulation; it merely checks whether the vertex encroaches upon any subsegment or subpolygon whose vertices are unweighted, employing the weighted notion of encroachment from Definitions 9.2 and 9.3. Because the triangulation is not modified, it is possible for the vertex to "encroach" upon a subpolygon of a polygon that contains the vertex. The algorithm ignores these spurious encroachments, because a change in the vertex's weight during sliver exudation will simply change how the polygon is triangulated.

The procedure TestConformity performs a temporary weight assignment. It calls the subroutine SplitSubsegment from Chapter 8 and a modified SplitSubpolygon that appears in Section 11.2. TestConformity is called only when no subsegment or subpolygon is encroached, so the subpolygons are well defined and present in Del S.

TestConformity(v, S, E, α)

1. Compute $N(v) = d(v, S)$. Let $\xi = \alpha^2 N(v)^2$. Determine whether the weighted vertex $v[\xi]$ encroaches upon any subsegment or subpolygon.

2. If $v[\xi]$ encroaches upon a subsegment $e \in E$ that does not adjoin v, call SplitSubsegment(e, S, E) and return.
3. If $v[\xi]$ encroaches upon a subpolygon of a polygon $g \in \mathcal{P}$ such that $v \notin g$, call SplitSubpolygon(g, v, S, E, α).

Step 1 can be implemented as a local test, thereby avoiding the expense of exhaustively checking every subsegment and subpolygon. Recall from Section 10.2 that the Bowyer–Watson algorithm can be modified to update a weighted Delaunay triangulation when a vertex weight increases. Likewise, an implementation of Step 1 of TestConformity can use the depth-first search step of the Bowyer–Watson algorithm to identify the cavity that will be retriangulated if the sliver exudation algorithm assigns the weight ξ to the vertex v. Every subsegment or subpolygon included in the boundary of this cavity must be tested to see if it is encroached. Every subsegment or subpolygon that intersects the interior of the cavity is encroached—no further testing is necessary. Subsegments and subpolygons not included in the cavity are not encroached; see Exercise 1.

11.2 A refinement algorithm for sliver exudation

Our Delaunay refinement algorithm uses temporary weight assignments not only for vertices already present in the mesh, but also for points being considered for insertion at the circumcenters of skinny tetrahedra and encroached subpolygons. If a circumcenter encroaches upon a subsegment when it is assigned its maximum permitted weight or a tetrahedron circumcenter at maximum weight encroaches upon a subpolygon, it is rejected and not inserted. Although this stronger rejection rule is not necessary, it improves the quality of the tetrahedra that the algorithm generates. We modify SplitSubpolygon and SplitTetrahedron accordingly.

SplitSubpolygon(g, p, S, E, α)

1. Let σ be a subpolygon of g that contains the orthogonal projection of p onto g. Let c be the circumcenter of σ. Compute $d(c, S)$. Let $\xi = \alpha^2 d(c, S)^2$.
2. If $c[\xi]$ encroaches upon a subsegment $e \in E$, call SplitSubsegment(e, S, E). Otherwise, insert c, unweighted, into S.

Step 1 of SplitSubpolygon relies on the fact that if a point $p[\omega_p]$ encroaches upon a subpolygon of a polygon g whose subsegments are not encroached, then $p[\omega_p]$ also encroaches upon the subpolygon of g that contains the orthogonal projection of p onto aff g. This fact follows from the Monotone Power Lemma (Lemma 7.5); its proof is an easy extension of the proof of Proposition 8.1 (see Exercise 2).

SplitTetrahedron(τ, S, E, α)

1. Let c be the circumcenter of τ. Let $\xi = \alpha^2 d(c, v)^2$ where v is any vertex of τ.

2. If $c[\xi]$ encroaches upon a subsegment $e \in E$, call SPLITSUBSEGMENT(e, S, E) and return.
3. If $c[\xi]$ encroaches upon a subpolygon of a polygon $g \in \mathcal{P}$, call the procedure SPLITSUBPOLYGON(g, c, S, E, α) and return.
4. Insert c, unweighted, into S.

The procedure DELTETEXUDEPLC below takes a PLC $\mathcal{P}$, uses Delaunay refinement to mesh it and prepare it for sliver exudation, and performs sliver exudation on the mesh. It also takes as input the maximum permitted radius-edge ratio $\bar{\rho}$ for tetrahedra *before* sliver exudation (a bound that sliver exudation does not necessarily preserve) and the sliver exudation parameter $\alpha \in (0, 1/2)$. It is in most respects the same as the procedure DELTETPLC in Chapter 8, but it adds calls to TESTCONFORMITY and a final call to EXUDE from Chapter 10.

If a call to TESTCONFORMITY on a vertex v does not cause a subsegment or subpolygon to become encroached, then v will never encroach upon a subsegment or subpolygon again; see the forthcoming Proposition 11.1. For the sake of speed, we label each vertex *clean* or *dirty*. Every vertex is initially dirty except for tetrahedron circumcenters, which are always labeled *clean* when they are inserted. This labeling convention is not explicitly stated in the pseudocode, but the reader should assume it is in force. A vertex is relabeled clean when a call to TESTCONFORMITY verifies that the vertex cannot encroach upon anything. Thereafter, the vertex is never tested again.

For theoretical reasons, DELTETEXUDEPLC begins by enclosing the input PLC $\mathcal{P}$ in a cubical bounding box; let $\mathcal{P}^{\Box}$ be the augmented PLC. DELTETEXUDEPLC meshes the entire cube. The bounding box is not at all necessary in practice, but without a bounding box, Del S can have arbitrarily bad tetrahedra that are outside the domain but in its convex hull, and it is necessary to modify both the refinement and sliver exudation stages so they do not measure the quality of tetrahedra that are not in $|\mathcal{P}|$. The bounding box is obligatory in theory because the Delaunay Length Lemma II (Lemma 11.8) requires every tetrahedron in Del S to have a good radius-edge ratio, which is not always possible if $|\mathcal{P}|$ is nonconvex. In practice, there are better ways to prevent large disparities in the lengths of the edges adjoining a vertex, discussed in Section 10.3.3.

DELTETEXUDEPLC$(\mathcal{P}, \bar{\rho}, \alpha)$

1. Let $\mathcal{P}^{\Box}$ be the PLC obtained by adding a large cubical bounding box around $\mathcal{P}$. Let S be the set of vertices in $\mathcal{P}^{\Box}$. Let E be the set of edges in $\mathcal{P}^{\Box}$.
2. Compute the three-dimensional triangulation Del S. For each polygon $g \in \mathcal{P}^{\Box}$, compute the two-dimensional triangulation Del $(S \cap g)$.
3. While some subsegment $e \in E$ is encroached upon by a vertex in S, call SPLITSUBSEGMENT(e, S, E), update Del S and the two-dimensional triangulation Del $(S \cap g)$ for every polygon g that includes e, and repeat Step 3.
4. If some subpolygon of some polygon $g \in \mathcal{P}^{\Box}$ is encroached upon by a vertex $v \in S$, call SPLITSUBPOLYGON(g, v, S, E, α), update Del S, Del $(S \cap g)$, and the other two-dimensional triangulations, and go to Step 3.

5. If some vertex $v \in S$ is dirty, call TestConformity(v, S, E, α). If TestConformity inserted a new vertex, go to Step 3. Otherwise, label v as clean and repeat Step 5.
6. If some tetrahedron $\tau \in \text{Del}\, S$ has $\rho(\tau) > \bar{\rho}$, call SplitTetrahedron(τ, S, E, α), update Del S and the two-dimensional triangulations, and go to Step 3.
7. Let $\mathcal{T} = \text{Exude}(S, \alpha)$.
8. Return the mesh $\{\sigma \in \mathcal{T} : \sigma \subseteq |\mathcal{P}|\}$.

As discussed in Chapter 8, the Orthocenter Containment Lemma (Lemma 7.7) guarantees that every vertex that SplitSubsegment, SplitSubpolygon, or SplitTetrahedron inserts lies inside the bounding box, and that every vertex that SplitSubpolygon inserts lies in the corresponding polygon.

11.3 A guarantee of domain conformity during sliver exudation

Here we verify one of the premises of the algorithm DelTetExudePLC: that if a vertex at maximum weight does not encroach upon any subsegment or subpolygon, it will never encroach upon anything again, even after the mesh is refined further and the sliver exudation stage assigns weights to the vertices.

This premise has two consequences. First, when the refinement stage terminates, Del $S[\omega]$ conforms to $\mathcal{P}$—that is, $\{\sigma \in \text{Del}\, S[\omega] : \sigma \subseteq |\mathcal{P}|\}$ is a Steiner triangulation of $\mathcal{P}$—for every weight assignment $\omega \in \text{W}(\alpha)$. Second, once a vertex is labeled *clean*, there is no need for the algorithm ever to check again whether it encroaches upon a subsegment or subpolygon.

The claim might seem obvious. If a subsegment is not encroached upon by a vertex at maximum weight, it will not be encroached upon by the vertex at a lesser weight. If a subsegment's own vertices are assigned positive weights, its diametric orthoball shrinks, making it harder to encroach. Therefore, if a vertex with maximum weight does not encroach upon a subsegment with unweighted vertices, it will not encroach for any weight assignment in $\text{W}(\alpha)$.

Unfortunately, subpolygons are more problematic than subsegments. Changing the weights of the vertices changes a polygon's subpolygons, and so does refinement. If a vertex encroaches upon no subpolygon now, will that still be true if the polygon is refined further? Does the fact that no subpolygon in Del S is encroached mean that no subpolygon in Del $S[\omega]$ will be encroached? Fortunately, we have a tool for guaranteeing that these subpolygons are not encroached: the Orthoball Cover Lemma (Lemma 7.6).

In the following proposition, we must distinguish between the subsegments and subpolygons of two different triangulations of the same complex. For a segment $s \in \mathcal{P}$, a subsegment of s *with respect to* $S[\omega]$ is an edge in $\text{Del}\,(S \cap s)[\omega]$. For a polygon $g \in \mathcal{P}$, a subpolygon of g *with respect to* $S[\omega]$ is a triangle in $\{\sigma \in \text{Del}\,(S \cap g)[\omega] : \sigma \subseteq g\}$. We say "subsegment or subpolygon of $\mathcal{P}$" to refer to a subsegment or subpolygon of any segment or polygon in $\mathcal{P}$.

Proposition 11.1. *Let $\mathring{S} \subseteq S \subset |\mathcal{P}|$ be two finite vertex sets that include the vertices in a PLC $\mathcal{P}$. Let $v \in \mathring{S}$ be a vertex and ξ a weight. Suppose that no vertex in $\mathring{S}$ (assigned weight* *zero) encroaches upon any subsegment of* $\mathcal{P}$ *with respect to* $\mathring{S}$. *Suppose also that* $v[\xi]$ *does not encroach upon any subsegment or subpolygon of* $\mathcal{P}$ *with respect to* $\mathring{S}$. *Then for every weight assignment* $\omega \in W(\alpha)$ *with* $\alpha \in [0, 1/2)$ *and* $\omega_v \leq \xi$, $v[\omega_v]$ *does not encroach upon any subsegment of* $\mathcal{P}$ *with respect to* $S[\omega]$, *nor does it encroach upon any subpolygon of* $\mathcal{P}$ *with respect to* $S[\omega]$ *whose orthocenter lies on the corresponding polygon.*

Proof. Suppose for the sake of contradiction that $v[\omega_v]$ encroaches upon a subsegment σ of some segment $g \in \mathcal{P}$ with respect to $S[\omega]$, or $v[\omega_v]$ encroaches upon a subpolygon σ of some polygon $g \in \mathcal{P}$ with respect to $S[\omega]$. In the latter case, suppose also that g contains σ's orthocenter. Observe that $v \notin g$, because a vertex of $S \cap g$ cannot encroach upon $\sigma \in \mathrm{Del}\,(S \cap g)[\omega]$.

Let B_σ be the diametric orthoball of σ; then $v[\omega_v]$ is orthogonal or closer than orthogonal to B_σ. The fact that σ is a simplex in $\mathrm{Del}\,(S \cap g)[\omega]$ implies that no weighted vertex in $S \cap g$ is closer than orthogonal to B_σ, so no unweighted vertex in $\mathring{S} \cap g$ is closer than orthogonal to B_σ.

Let $\mathcal{Q} = \{\tau \in \mathrm{Del}\,(\mathring{S} \cap g) : \tau \subseteq g\}$ be the triangulation of g induced by $\mathring{S}$. By assumption, no vertex in $\mathring{S}$ encroaches upon any subsegment with respect to $\mathring{S}$, so $\mathcal{Q}$ is a quarantined complex (recall Definition 7.6). As $|\mathcal{Q}|$ contains the center of B_σ, the Orthoball Cover Lemma (Lemma 7.6) states that there is a ball $B \in \mathrm{Ball}(\mathcal{Q})$ such that $\pi(v, B) \leq \pi(v, B_\sigma)$; hence, $\pi(v, B) - \xi \leq \pi(v, B) - \omega_v \leq \pi(v, B_\sigma) - \omega_v \leq 0$; hence, both $v[\xi]$ and $v[\omega_v]$ are orthogonal or closer than orthogonal to B. The ball B cannot represent a vertex of $\mathcal{Q}$ because no weight assignment in $W(\alpha)$ permits $v[\omega_v]$ to be orthogonal or closer than orthogonal to another vertex in S. Therefore, B is the diametric ball of a subsegment or subpolygon τ in $\mathcal{Q}$, and $v[\xi]$ encroaches upon τ, contradicting the assumption that $v[\xi]$ does not encroach upon any subsegment or subpolygon of g with respect to $\mathring{S}$. □

Proposition 11.2. *If* DelTetExudePLC($\mathcal{P}, \bar{\rho}, \alpha$) *terminates, it returns a Steiner triangulation of* $\mathcal{P}$.

Proof. The refinement phase of DelTetExudePLC terminates only if no subsegment or subpolygon is encroached, even by a vertex assigned its maximum possible weight under the vertex weight invariants. After the sliver exudation phase assigns weights to the vertices, by Proposition 11.1, no vertex in $S[\omega]$ encroaches upon any subsegment of $\mathrm{Del}\,S[\omega]$. Therefore, every subsegment is strongly weighted Delaunay and, by a weighted variant of Proposition 4.3, in $\mathrm{Del}\,S[\omega]$. The fact that no subsegment is encroached implies that every polygon in $\mathcal{P}$ is triangulated by a well-defined set of subpolygons whose orthocenters lie in the polygon by the Orthocenter Containment Lemma (Lemma 7.7). By Proposition 11.1, the subpolygons are not encroached, so they also are strongly weighted Delaunay and in $\mathrm{Del}\,S[\omega]$. The result follows. □

11.4 A proof of termination, good quality, and good grading

Here, we adapt the analysis of DelTetPLC from Chapter 8 to the more aggressive encroachment rules of DelTetExudePLC. The following proposition shows that if a PLC $\mathcal{P}$ satisfies the projection condition, a vertex on one segment or polygon in $\mathcal{P}$ cannot encroach upon a subsegment or subpolygon in an adjoining segment or polygon in $\mathcal{P}$, unless the latter is a face of the former.

Proposition 11.3. *Let $\mathcal{P}$ be a PLC that satisfies the projection condition. Let $S \subset |\mathcal{P}|$ be a finite vertex set that includes the vertices in $\mathcal{P}$. Let ω be a weight assignment for S that satisfies the vertex weight invariants. Let g and h be two adjoining linear cells in $\mathcal{P}$ such that neither is a face of the other; i.e. $g \not\subset h$ and $h \not\subset g$. Consider a vertex $v \in h \cap S$.*

(i) *If g is a segment, then $v[\omega_v]$ does not encroach upon any subsegment of g.*

(ii) *If g is a polygon and no vertex in $S[\omega]$ encroaches upon any subsegment of g, then $v[\omega_v]$ does not encroach upon any subpolygon of g.*

Proof. As g and h satisfy the projection condition, the orthogonal projection of v onto aff g is not in $g \setminus h$, and the orthogonal projection of h onto aff g does not intersect $g \setminus h$.

If g is a segment, the assumption that $g \not\subset h$ implies that g can intersect h only at g's vertices, so the projection condition implies that they adjoin only at a single vertex x. Suppose for the sake of contradiction that $v[\omega_v]$ encroaches upon a subsegment e of g with diametric orthoball B_e, thus $\pi(v[\omega_v], B_e) \leq 0$. Counterintuitive as it may seem, every weighted vertex $B(x, \sqrt{\omega_x})$ has an orthoball $B_x = B(x, \sqrt{-\omega_x})$ by Definition 7.5. The Monotone Power Lemma (Lemma 7.5) implies that $\pi(v[\omega_v], B_x) \leq \pi(v[\omega_v], B_e) \leq 0$ and, therefore, $d(v, x)^2 - \omega_v + \omega_x \leq 0$. But this is impossible because $\omega_v < d(v, x)^2/4$ and $\omega_x \geq 0$ by the vertex weight invariants.

If g is a polygon whose subsegments are not encroached, suppose for the sake of contradiction that $v[\omega_v]$ encroaches upon a subpolygon τ of g; thus, $\pi(v[\omega_v], B_\tau) \leq 0$. The projection of v onto aff g must lie outside g, on an edge of g, or on a vertex of g. By the Monotone Power Lemma, either some subsegment e of g satisfies $\pi(v[\omega_v], B_e) \leq \pi(v[\omega_v], B_\tau) \leq 0$, or some vertex x of g satisfies $\pi(v[\omega_v], B_x) \leq 0$. In the former case, $v[\omega_v]$ encroaches upon e, a contradiction. The latter case is impossible for the reasons stated in the previous paragraph. □

Recall from Section 8.3 that the *insertion radius* r_x of a vertex x that is inserted into S or rejected by DelTetExudePLC is the distance from x to the nearest distinct vertex in S at the moment when x is inserted or rejected. (We will not need the modified definition of insertion radius from Section 9.4, because when DelTetExudePLC inserts a vertex, all the vertices are unweighted.) Recall also that a vertex is of *type i* if it lies on a linear i-cell in $\mathcal{P}$ but not on a lower-dimensional cell—although vertices rejected by SplitSubpolygon have type 2 and vertices rejected by SplitTetrahedron have type 3, regardless of whether they happen to fall on a lower-dimensional cell. Finally, recall that every vertex not of type 0, including rejected vertices, has a *parent*, defined in Section 8.3.

The following proposition adapts Proposition 8.4 for the algorithm DelTetExudePLC, thereby giving a lower bound on the insertion radius of each vertex the algorithm inserts or rejects. The bounds are somewhat weaker than for DelTetPLC, because when encroachment causes the insertion of a vertex, its parent is not necessarily in the diametric ball of the encroached subsegment or subpolygon. Recall that $f(p)$ is the local feature size at a point p (Definition 6.2).

Proposition 11.4. *Let $\mathcal{P}$ be a PLC that satisfies the projection condition. Let x be a vertex placed in S or rejected by* DelTetExudePLC. *Suppose that vertex weight invariant 1 holds just before x is inserted or rejected. Let p be the parent of x, if one exists.*

(i) *If x is of type 0, then $r_x \geq f(x)$.*

(ii) *If x is of type 1 or 2, and the type of p is less than or equal to that of x, then $r_x \geq \sqrt{(1-\alpha^2)/(1+\alpha^2)} f(x)$.*

(iii) *If x is of type 1 or 2, and the type of p is greater than that of x, then $r_x \geq \sqrt{(1-\alpha^2)/2}\, r_p$.*

(iv) *If x is of type 3, then $r_x > \bar{\rho} r_p$.*

Proof. If x is of type 0 or 3, the proof from Proposition 8.4 remains valid without change.

If x is of type 1, its parent p encroaches upon a subsegment σ of a segment g, and x is the midpoint of σ. If x is of type 2, its parent p encroaches upon a subpolygon σ of a polygon g, and x is the circumcenter of σ. In either case, by the definition of encroachment, the weighted point $p[\omega_p]$ is orthogonal or closer than orthogonal to the diametric ball B_σ of σ. The radius of B_σ is $d(x, a)$, where a is the vertex of σ nearest p. By vertex weight invariant 1, $\omega_p \leq \alpha^2 N(p)^2$. Therefore,

$$\begin{aligned} d(p,x)^2 &\leq d(x,a)^2 + \omega_p \\ &\leq d(x,a)^2 + \alpha^2 N(p)^2 \\ &\leq d(x,a)^2 + \alpha^2 d(p,a)^2. \end{aligned}$$

Recall that DelTetExudePLC chooses σ so that it contains the orthogonal projection of p onto aff g. This fact and the fact that a is the vertex of σ nearest p imply that $\angle pxa \leq 90°$, as justified in the proof of Proposition 8.4. By Pythagoras' Theorem,

$$\begin{aligned} d(p,a)^2 &\leq d(p,x)^2 + d(x,a)^2 \\ &\leq 2d(x,a)^2 + \alpha^2 d(p,a)^2, \\ \text{therefore, } d(p,a)^2 &\leq \frac{2}{1-\alpha^2} d(x,a)^2 \\ \text{and } d(p,x)^2 &\leq \frac{1+\alpha^2}{1-\alpha^2} d(x,a)^2. \end{aligned}$$

If the type i of p is less than or equal to that of x, then p lies on some linear i-cell in $\mathcal{P}$ that is disjoint from g by Proposition 11.3, so $f(x) \leq d(p, x)$ by the definition of local feature size. If $p \in B_\sigma$, then $r_x = d(p, x) \geq f(x)$; otherwise, no vertex lies in the interior of B_σ, and $r_x = d(x, a) \geq \sqrt{(1-\alpha^2)/(1+\alpha^2)}\, d(p, x) \geq \sqrt{(1-\alpha^2)/(1+\alpha^2)}\, f(x)$, as claimed.

If the type of p is greater than that of x, then σ was not encroached before DelTetExudePLC tried to insert p, and p was rejected, so no vertex lies in the interior of B_σ and $r_x = d(x, a) \geq \sqrt{(1-\alpha^2)/2}\, d(p, a) \geq \sqrt{(1-\alpha^2)/2}\, r_p$, as claimed. □

The next proposition establishes the provably good grading of the mesh by showing that DelTetExudePLC does not place a vertex much closer to other vertices than the local feature size at the vertex. The result adapts Proposition 8.5.

Proposition 11.5. *Suppose that $\bar{\rho} > 2/(1-\alpha^2)$ and $\mathcal{P}$ satisfies the projection condition. Define the constants*

$$a = \sqrt{\frac{2}{1-\alpha^2}}, \qquad b = \sqrt{\frac{1+\alpha^2}{1-\alpha^2}},$$

$$C_0 = 1, \qquad C_1 = \frac{a^2 + ab + b}{\bar{\rho} - a^2}\bar{\rho}, \qquad C_2 = \frac{ab + (a+b)\bar{\rho}}{\bar{\rho} - a^2}, \qquad C_3 = \frac{ab + b + \bar{\rho}}{\bar{\rho} - a^2}.$$

Let x be a vertex of type i inserted or rejected by DelTetExudePLC$(\mathcal{P}, \bar{\rho}, \alpha)$. *Then $r_x \geq f(x)/C_i$.*

Proof. The proof proceeds by induction on the order by which DelTetExudePLC inserts or rejects vertices. Type 0 vertices are the inductive basis: for every vertex x in $\mathcal{P}$, $r_x \geq f(x)$ by Proposition 11.4(i).

Consider any vertex x inserted or rejected by DelTetExudePLC. For the inductive step, assume the proposition holds for x's parent p.

If x is of type 3, then $r_x \geq \bar{\rho} r_p$ by Proposition 11.4(iv). Observe that $C_1 > C_2 > C_3 > C_0 = 1$. By the inductive assumption, $f(p) \leq C_1 r_p$ irrespective of the type of p. By the Lipschitz property,

$$f(x) \leq f(p) + d(p,x) \leq C_1 r_p + r_x \leq \left(\frac{C_1}{\bar{\rho}} + 1\right) r_x = C_3 r_x.$$

Suppose that x is of type $i \in \{1,2\}$. If its parent p is of type $j \leq i$, then $r_x \geq f(x)/b > f(x)/C_i$ by Proposition 11.4(ii) and the result holds. If p is of type $j > i$, then $r_x \geq r_p/a$ by Proposition 11.4(iii). By the inductive assumption, $f(p) \leq C_j r_p \leq C_{i+1} r_p$. The proof of Proposition 11.4 shows that $d(p,x) \leq b r_x$. By the Lipschitz property,

$$f(x) \leq f(p) + d(p,x) \leq C_{i+1} r_p + b r_x \leq (a C_{i+1} + b)\, r_x = C_i r_x,$$

and the result holds. □

Recall Proposition 8.6, which states that for any two vertices p and q that DelTetPLC inserts, $d(p,q) \geq f(p)/(C_1 + 1)$. The proposition also holds for DelTetExudePLC, so long as we update the constant C_1 to be the larger value specified in Proposition 11.5. Thus these propositions show that the final mesh has provably good grading: there is a lower bound on its edge lengths proportional to the local feature size. For example, if $\bar{\rho} = 3$ and $\alpha = 1/\sqrt{8}$, then $C_1 \doteq 21.6$, so the spacing of vertices is at worst about 23 times smaller than the local feature size.

Any two vertices in S are separated by a distance of at least $f_{\min}/(C_1 + 1)$, where $f_{\min} = \inf_{x \in |\mathcal{P}|} f(x)$ is the infimum local feature size. Thus, the Packing Lemma shows that DelTetExudePLC must terminate.

Proposition 11.6. *Suppose that $\bar{\rho} > 2/(1-\alpha^2)$ and $\mathcal{P}$ satisfies the projection condition. Then* DelTetExudePLC$(\mathcal{P}, \bar{\rho}, \alpha)$ *terminates, and its first six steps construct a Steiner Delaunay triangulation of $\mathcal{P}$ whose tetrahedra have radius-edge ratios at most $\bar{\rho}$.*

Proof. Identical to that of Theorem 8.7, albeit for a different value of C_1. □

11.5 Finite triangulations and the Sliver Theorem

The Sliver Theorem and several results leading to it in Chapter 10 rely on two assumptions: the tetrahedra in Del S have bounded radius-edge ratios, and S is a countably infinite point set such that conv $S = \mathbb{R}^3$. Step 6 of DELTETEXUDEPLC enforces the first assumption, but the second is just a convenient mathematical conceit.

The assumption that Del S fills space is used in three places in Chapter 10: in the Orthoradius-Edge Lemma (Lemma 10.1), the Delaunay Length Lemma (Lemma 10.4), and the Edge Length Lemma (Lemma 10.5). Here, we give a version of the Delaunay Length Lemma that holds for finite Delaunay triangulations, at the cost of a weaker constant κ_{len}. We patch the other two lemmas by replacing the infinite triangulation with the assumption that the orthocenter of every tetrahedron in Del $S[\omega]$ lies in conv S. This assumption is indirectly enforced by TESTCONFORMITY, which ensures that no weighted vertex encroaches upon any subsegment or subpolygon, and it is guaranteed by the Orthocenter Containment Lemma (Lemma 7.7). With this assumption, the Orthoradius-Edge Lemma remains true with the same constant. The Edge Length Lemma holds with a weaker constant c_{len}, but only because it depends on the weakened constant κ_{len} in the Delaunay Length Lemma; the formula that specifies c_{len} in terms of κ_{len} in Lemma 10.5 still holds. The Vertex Degree Lemma and the Sliver Theorem remain true with no change to their proofs, with the constants c_{len}, c_{deg}, and $\psi_{\min}$ adjusted to reflect the weaker value of κ_{len}.

The Orthoradius-Edge and Edge Length Lemmas use the assumption that Del S fills space to show that for every tetrahedron in $K(S, \alpha)$ with orthoradius Z, there is a tetrahedron nearby in Del S with circumradius at least $Z/2$. Near the boundary of a finite triangulation, this claim is not always true: a tetrahedron in $K(S, \alpha)$ might have a large orthoball, most of it outside conv S, though all the tetrahedra in Del S have small circumballs. However, the claim remains true if every tetrahedron's orthocenter lies in conv S. The following proposition patches the Orthoradius-Edge Lemma; the Edge Length Lemma is easily patched with the same idea.

Proposition 11.7. *Let S be a point set in $\mathbb{R}^3$ such that no tetrahedron in* Del S *has a radius-edge ratio greater than a constant $\bar{\rho}$. Suppose that for every weight assignment $\omega \in \mathrm{W}(\alpha)$, the orthocenter of every tetrahedron in* Del $S[\omega]$ *lies in* conv S. *Then no tetrahedron in $K(S, \alpha)$ has an orthoradius-edge ratio greater than $\mathring{\rho} = 2\bar{\rho}$.*

PROOF. The proof is essentially the same as the proof of the Orthoradius-Edge Lemma (Lemma 10.1), except where we argue that for every tetrahedron $\tau \in K(S, \alpha)$ with shortest edge pq and orthocenter z, if the Voronoi polygon V_{pq} contains a point $c_{pq} \in pqz$, then V_{pq} has a Voronoi vertex w such that $d(w, p) \geq d(c_{pq}, p)$ (and an analogous claim for $V_{pp'}$ and $c_{pp'} \in pqz$). In Lemma 10.1, this claim follows because V_{pq} is bounded. Here, the claim follows because the orthocenter z lies in conv S, implying that $c_{pq} \in pqz \subseteq$ conv S.

To see that the polygon V_{pq} has a Voronoi vertex w for which $d(w, p) \geq d(c_{pq}, p)$, let $\vec{r}$ be any ray originating at c_{pq}, let $\Pi_{\vec{r}}$ be the plane through c_{pq} orthogonal to $\vec{r}$, and let $H_{\vec{r}}$ be the halfspace with boundary $\Pi_{\vec{r}}$ into which $\vec{r}$ points. As $c_{pq} \in$ conv S, some vertex $y \in S$ lies in $H_{\vec{r}}$. If $\vec{r}$ adjoins $\overrightarrow{c_{pq}p}$ at an angle greater than 90°, then $p \notin H_{\vec{r}}$, and $\vec{r} \not\subset V_{pq}$ because all points sufficiently far along $\vec{r}$ are closer to y than to p. Therefore, there is a finite upper bound on the maximum distance from $H_{\overrightarrow{c_{pq}p}}$ to any point on V_{pq}. One of the points maximizing this distance must be a vertex of V_{pq}, because no 1-face of V_{pq} is an entire line, as the vertices in S are not all coplanar. This vertex of V_{pq} is at least as far from p as c_{pq}. □

Lemma 11.8 (Delaunay Length Lemma II). *Let S be a point set in $\mathbb{R}^d$ for $d \geq 2$, and suppose that no d-simplex in* Del S *has a radius-edge ratio greater than a constant $\bar{\rho}$. For any two adjoining edges $pq, pr \in$* Del S, *$d(p,q) \leq e^{\sqrt{4\bar{\rho}^2-1}\angle qpr} d(p,r) \leq \kappa_{\text{len}} d(p,r)$, where $\kappa_{\text{len}} = e^{\sqrt{4\bar{\rho}^2-1}\pi}$ and the angle $\angle qpr$ is given in radians.*

Proof. Taking a coordinate system with p as its origin, consider the function $\iota(v) = v/\|v\|^2$, called an *inversion through the unit sphere*. Observe that $\iota(v)$ has the same polar angles relative to p as v; the map changes only the distance from p, so that $d(p, \iota(v)) = 1/d(p, v)$. Let T be the set of d-simplices adjoining p in Del S, and let $\iota(T)$ be the d-simplices produced by inverting p's neighbors while leaving p at the origin. Exercise 4 asks the reader to verify that $|\iota(T)|$ is convex. (Although we do not directly use this fact, it might be intuitively helpful to know that $|\iota(T)|$ is the polar dual of p's Voronoi cell V_p. Polar duality is a well-known concept in geometry that we do not discuss here.)

The proof is similar to that of Lemma 10.4, except that we replace V_p with $|\iota(T)|$. If p is on the boundary of conv S, then it is on the boundary of $|\iota(T)|$, so we ignore the faces of $|\iota(T)|$ that contain p. For every point x on a facet g of $|\iota(T)|$ that does not contain p, px cannot meet g at an angle less than arcsin $1/(2\bar{\rho})$; see Exercise 4.

Let Π be a plane that contains p, $\iota(q)$, and $\iota(r)$. Define a polar coordinate system for Π centered at p and oriented so that $\iota(q)$ lies at the angle $\theta = 0$ and $\iota(r)$ at an angle no greater than π. If the three points are collinear, choose Π so that it intersects the interior of conv S in the upper quadrants where $\theta \in [0, \pi]$.

The remainder of the proof is nearly identical to the proof of Lemma 10.4, except that we integrate r over the boundary of $\Pi \cap |\iota(T)|$, and $dr/d\theta$ may be as great as $\sqrt{4\bar{\rho}^2 - 1}r$ over the entire turn, so $d(p, \iota(r)) \leq e^{\sqrt{4\bar{\rho}^2-1}\angle qpr} d(p, \iota(q))$ and, hence, $d(p,q) \leq e^{\sqrt{4\bar{\rho}^2-1}\angle qpr} d(p,r)$. □

Like its predecessor, Lemma 11.8 holds for Delaunay triangulations of any dimension $d \geq 2$. We conjecture that the bounds in both lemmas become tight in the limit as d approaches infinity.

By patching the lemmas it depends on, we have shown that the Sliver Theorem generalizes to finite meshes produced by the first six steps of DelTetExudePLC, giving us the main result of this chapter: DelTetExudePLC can mesh PLCs that satisfy the projection condition with tetrahedra of bounded sliver measure.

Theorem 11.9. *Suppose that $\bar{\rho} > 2/(1 - \alpha^2)$ and $\mathcal{P}$ satisfies the projection condition. Then* DelTetExudePLC$(\mathcal{P}, \bar{\rho}, \alpha)$ *returns a Steiner triangulation of $\mathcal{P}$ in which every tetrahedron has a sliver measure greater than $\psi_{\min}$ and every triangle has a radius-edge ratio at most $2\bar{\rho}/\sqrt{1 - 4\alpha^2}$, for the value $\psi_{\min}$ specified in Theorem 10.10 with κ_{len} as specified in Lemma 11.8.* □

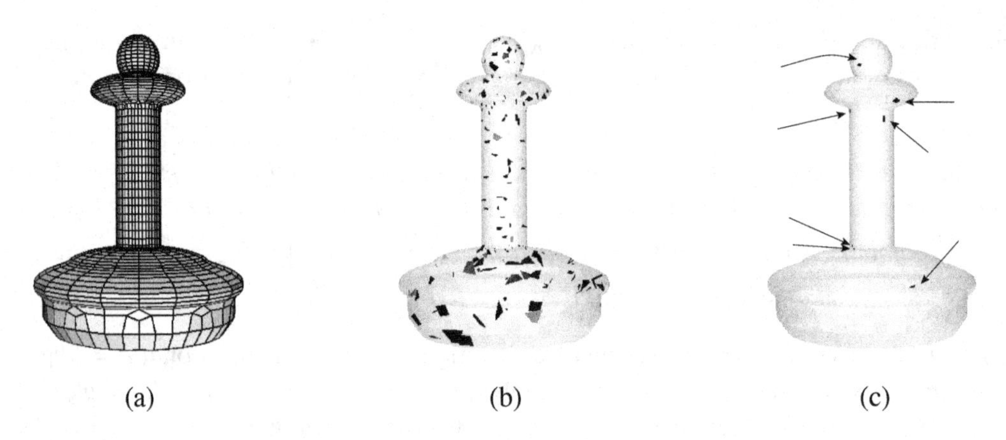

Figure 11.1: (a) A PLC for a chess pawn. (b) The mesh prior to sliver exudation; sliver tetrahedra are highlighted. (c) Few slivers survive exudation. All surviving slivers are near acute domain angles.

11.6 Notes and exercises

The algorithm and analysis in this chapter improves an early version by Cheng and Dey [48].

As we mentioned in the previous chapter, Chew [62] proposes a guaranteed-quality algorithm that removes slivers by inserting randomly placed vertices near their circumcenters. Li and Teng [137] extend this approach to work for polyhedral domains that satisfy the projection condition. Their encroachment rules insert vertices at randomly chosen locations near subsegment midpoints or subpolygon circumcenters, and reject vertices that would cause the formation of particularly bad slivers. These vertices are constrained to lie on the segments and polygons, but Li and Teng show that they are nevertheless random enough to prevent the skinniest of slivers from appearing.

Cheng, Dey, and Ray [55] extend the Delaunay refinement ideas in this chapter to polyhedral domains with small angles. They protect acute PLC segments by enclosing them in protecting balls in which the Delaunay refinement algorithm is forbidden to insert vertices. This strategy ensures domain conformity, but as in Chapter 9, some tetrahedra with large radius-edge ratios survive near acute segments because they cannot be split. When the algorithm assigns a weight to a vertex to eliminate slivers, it checks whether the weighted vertex would encroach upon a protecting ball if the radius of every protecting ball were doubled; if so, the weight of the vertex is reset to zero to ensure domain conformity. Slivers sufficiently far away from protecting balls are eliminated. Figure 11.1 shows the effect of sliver removal with this strategy.

Exercises

1. Prove the correctness of the last three sentences of Section 11.1. Exercise 1 in Chapter 6 is a good warm-up for this problem.

2. Show that Proposition 8.1 extends to weighted Delaunay triangulations where each vertex v has a weight in the range $[0, N(v)/2)$.

3. The algorithm DELTETACUTEPLC in Chapter 9 uses vertex weights to protect small domain angles. The algorithm DELTETEXUDEPLC in this chapter uses vertex weights for sliver exudation. Describe how to combine the two algorithms so that you can perform sliver exudation in meshes of domains with small angles. What numerical guarantees can you make on the orthoradius-edge ratios of the tetrahedra just prior to sliver exudation? What numerical guarantees can you make on the quality of the tetrahedra of the final mesh after sliver exudation? (These guarantees may be limited to a subset of the tetrahedra.)

4. Consider the inversion through the unit sphere $\iota(v) = v/\|v\|^2$ defined in Lemma 11.8.

   (a) Show that if $puvw$ is a tetrahedron whose vertex p lies at the origin, ι maps its circumsphere to the plane aff $\iota(u)\iota(v)\iota(w)$. Show also that if R is the circumradius of $puvw$, then $d(p, \text{aff } \iota(u)\iota(v)\iota(w)) = 1/(2R)$.

   (b) Let S be a finite point set in $\mathbb{R}^3$, and let $p \in S$ be a vertex that lies at the origin. Show that for every neighbor q of p in Del S, $\iota(q)$ lies on the boundary of the convex polyhedron $G = \text{conv}(\{p\} \cup \{\iota(q) : q \in S \setminus \{p\}\})$. (It follows that the polyhedron $|\iota(T)|$ in Lemma 11.8 is convex.) Show also that for every vertex $\iota(q)$ of G, q is a neighbor of p in *every* Delaunay triangulation of S.

   (c) If no tetrahedron in Del S has a radius-edge ratio greater than $\bar{\rho}$, show that for every point x on every facet g of G that does not contain p, px does not meet g at an angle less than $\arcsin 1/(2\bar{\rho})$.

5. Exercise 7 in Chapter 10 suggests a way to modify the Delaunay refinement algorithm DELTETPLC so that there is a bound on the ratio κ_{len} between the lengths of any two adjoining edges. That bound is the strongest bound that the algorithm can enforce without weakening the guarantees of Proposition 8.5.

   The Delaunay refinement algorithm DELTETEXUDEPLC offers weaker bounds on the radius-edge ratios than DELTETPLC, reflected by the differences between Proposition 8.5 and Proposition 11.5. Suggest a refinement rule like that in Exercise 7 that makes DELTETEXUDEPLC obtain the strongest possible bound on the edge length ratios without weakening the guarantees of Proposition 11.5.

6. In Section 6.5, we show that Ruppert's algorithm produces meshes that are size-optimal. In Section 8.3, we explain that DELTETPLC does not always produce size-optimal meshes because the algorithm does not guarantee that it will produce no sliver tetrahedra. By contrast, DELTETEXUDEPLC returns a size-optimal mesh in which the number of tetrahedra is at most a constant times greater than in any other mesh of the same PLC in which no tetrahedron has a sliver measure less than $\psi_{\min}$. Outline how you would prove that fact. In particular, which results in this book would you use?

Chapter 12

Smooth surfaces and point samples

Until now, we have studied how to mesh piecewise linear domains. For the remainder of the book we focus on curved domains, namely, smooth surfaces, piecewise smooth surfaces, and the volumes they bound. We mesh curved domains with linear simplices, so exact domain conformity is impossible: the underlying space of the output mesh cannot be that of the input complex. This begs for an explanation of what we mean by a triangulation of a curved domain, and how well a linear triangulation can approximate a curved domain.

We answer the first question with topological concepts such as homeomorphism and isotopy. We ask that the underlying space of the mesh be topologically equivalent, or homeomorphic, to the domain it triangulates. In Chapter 15, we treat piecewise smooth complexes composed of smooth cells, and we ask that each cell of a domain be homeomorphic to the union of the simplices that triangulate the cell.

A mesh should approximate a domain not only topologically, but geometrically. We define the notion of smoothness with basic concepts from differential geometry. We study conditions under which an edge or triangle whose vertices lie on a smooth surface is nearly parallel to the surface tangents at those vertices. Accurate approximations of tangent planes and normal vectors depend on the simplices being small relative to a function called the local feature size, which is fundamentally different from the function of the same name defined in Chapter 6 yet has many similarities. The local feature size of a smooth surface is defined in terms of an important geometric structure called the medial axis. Local distances from a manifold to its medial axis are good indicators of the local density of vertices required to construct a mesh that represents the manifold well.

12.1 Topological spaces

In Section 1.6, we state that the heart of topology is to ask what it means for a set of points to be connected, and we answer that question with the concept of limit points in metric spaces. Topological spaces provide a way to describe the topology of a point set without a metric or point coordinates, so they are more abstract but more general than metric spaces. In a topological space, points are abstract entities that might have no characteristics except that they can be distinguished from one other. However, topological spaces remain founded on the concept of limit points. In place of a metric, we encode the connectedness of a point set by supplying a list of all of the open sets. This list is called a *system* of subsets of the point set. The point set and its system together describe a topological space.

Definition 12.1 (topological space)**.** A *topological space* is a point set $\mathbb{T}$ endowed with a *system of subsets* T, which is a set of subsets of $\mathbb{T}$ that satisfies the following conditions.

- $\emptyset, \mathbb{T} \in T$.
- For every $U \subseteq T$, the union of the subsets in U is in T.
- For every finite $U \subseteq T$, the common intersection of the subsets in U is in T.

The system T is called a *topology* on $\mathbb{T}$. The sets in T are called the *open sets* in $\mathbb{T}$. A *neighborhood* of a point $p \in \mathbb{T}$ is an open set containing p.

The axioms of Definition 12.1 may seem puzzling; we will not use them explicitly. Mathematicians have found them to be a simple and general set of rules from which one can derive most of the topological concepts one expects from familiarity with metric space topology.

Topological spaces may seem baffling from a computational point of view, because a point set with an interesting topology has uncountably infinitely many open sets containing uncountably infinitely many points. But from a mathematical point of view, topological spaces are attractive because they exclude information that is not topologically essential. For instance, the act of stretching a rubber sheet changes the distances between points and thereby changes the metric, but it does not change the open sets or the topology of the rubber sheet. The charm of a pure topological space is that, for example, all 2-spheres are indistinguishable from each other, so we simply call them "the 2-sphere."

Of course, the easiest way to define a topological space is to inherit the open sets from a metric space. For example, we can construct a topology on the d-dimensional Euclidean space $\mathbb{R}^d$ by letting T be the set of all possible open sets in $\mathbb{R}^d$. We can make the idea of "all possible open sets in $\mathbb{R}^d$" more concrete. Every open set in $\mathbb{R}^d$ is a union of a set of open d-balls, and vice versa, although sometimes requiring uncountably many d-balls. Therefore, we can let T be the set of all possible unions of open balls. In this topology, every open ball is a neighborhood of the point at its center.

In Section 1.6, we build the concepts of topology around the idea of limit points. Topological spaces require a different definition of limit point, but with the new definition in place, concepts that are defined in terms of limit points such as connectedness and closure extend without change to topological spaces.

Definition 12.2 (limit point)**.** Let $Q \subseteq \mathbb{T}$ be a point set. A point $p \in \mathbb{T}$ is a *limit point* of Q if every open set that contains p also contains a point in $Q \setminus \{p\}$.

Recall from Definition 1.12 that the *closure* $\mathrm{Cl}\, Q$ of a point set $Q \subseteq \mathbb{T}$ is the set containing every point in Q and every limit point of Q, and a point set Q is *closed* if $Q = \mathrm{Cl}\, Q$. It is straightforward to prove (Exercise 2) that in a topological space, the *complement* $\mathbb{T} \setminus Q$ of every open set $Q \in \mathbb{T}$ is closed, and that $\mathrm{Cl}\, Q$ is the smallest closed set containing Q.

For every point set $\mathbb{U} \subseteq \mathbb{T}$, the topology T induces a *subspace topology* on $\mathbb{U}$, namely, the system of open subsets $U = \{P \cap \mathbb{U} : P \in T\}$. The point set $\mathbb{U}$ endowed with the system

U is said to be a *topological subspace* of $\mathbb{T}$. The topological spaces we consider in this book are subsets of $\mathbb{R}^d$ that inherit its topology as a subspace topology. Examples of topological subspaces are the Euclidean d-ball $\mathbb{B}^d$, Euclidean d-sphere $\mathbb{S}^d$, open Euclidean d-ball $\mathbb{B}^d_o$, and Euclidean halfball $\mathbb{H}^d$, where

$$\begin{aligned}
\mathbb{B}^d &= \{x \in \mathbb{R}^d : \|x\| \leq 1\}, \\
\mathbb{S}^d &= \{x \in \mathbb{R}^{d+1} : \|x\| = 1\}, \\
\mathbb{B}^d_o &= \{x \in \mathbb{R}^d : \|x\| < 1\}, \\
\mathbb{H}^d &= \{x \in \mathbb{R}^d : \|x\| < 1 \text{ and } x_d \geq 0\}.
\end{aligned}$$

12.2 Maps, homeomorphisms, and isotopies

Two topological spaces are considered to be the same if the points that constitute them are connected the same way. For example, the boundary of a cube can be deformed into a sphere without cutting or gluing it. They have the same topology. We formalize this notion of topological equality by defining a function that maps the points of one space to points of the other and preserves how they are connected. Specifically, the function preserves limit points.

A function from one space to another that preserves limit points is called a *continuous function* or a *map*.[1] Continuity is just a step on the way to topological equivalence, because a continuous function can map many points to a single point in the target space, or map no points to a given point in the target space. If the former does not happen—that is, if the function is injective—the function is called an *embedding* of the domain into the target space. True equivalence is marked by a *homeomorphism*, a bijective function from one space to another that possesses both continuity and a continuous inverse, so that limit points are preserved in both directions.

Definition 12.3 (continuous function; map). Let $\mathbb{T}$ and $\mathbb{U}$ be topological spaces. A function $g : \mathbb{T} \to \mathbb{U}$ is *continuous* if for every set $Q \subseteq \mathbb{T}$ and every limit point $p \in \mathbb{T}$ of Q, $g(p)$ is either a limit point of the set $g(Q)$ or in $g(Q)$. Continuous functions are also called *maps*.

Definition 12.4 (embedding). A map $g : \mathbb{T} \to \mathbb{U}$ is an *embedding* of $\mathbb{T}$ into $\mathbb{U}$ if g is injective.

A topological space can be *embedded* into a Euclidean space by assigning coordinates to its points such that the assignment is continuous. For example, a drawing of a square is an embedding of $\mathbb{S}^1$ into $\mathbb{R}^2$. Not every topological space has an embedding into a Euclidean space, or even into a metric space—there are spaces that cannot be represented by any metric—but we will have no need for such spaces.

A homeomorphism is an embedding whose inverse is also an embedding.

---

[1] There is a small caveat with this characterization: a function g that maps a neighborhood of x to a single point $g(x)$ may be continuous, but technically $g(x)$ is not a limit point of itself, so in this sense a continuous function might not preserve all limit points. This technicality does not apply to homeomorphisms because they are bijective; homeomorphisms preserve all limit points, in both directions.

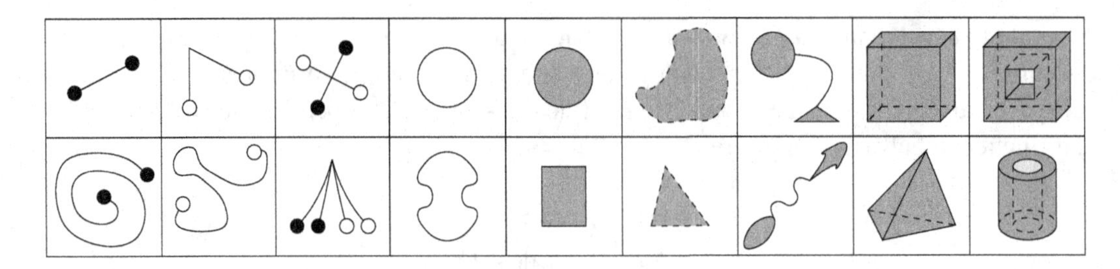

Figure 12.1: Each point set in this figure is homeomorphic to the point set above or below it, but not to any of the others. Open circles indicate points missing from the point set, as do the dashed edges in the point sets fourth from the right.

Definition 12.5 (homeomorphism). Let $\mathbb{T}$ and $\mathbb{U}$ be topological spaces. A *homeomorphism* is a bijective map $h : \mathbb{T} \to \mathbb{U}$ whose inverse is continuous too. Two topological spaces are *homeomorphic* if there exists a homeomorphism between them.

Homeomorphism induces an equivalence relation among topological spaces, which is why two homeomorphic topological spaces are called *topologically equivalent*. Figure 12.1 show pairs of topological spaces that are homeomorphic. A less obvious example is that the open d-ball $\mathbb{B}^d_o$ is homeomorphic to the Euclidean space $\mathbb{R}^d$, as demonstrated by the map $h(x) = \frac{1}{1-\|x\|}x$. The same map shows that the halfball $\mathbb{H}^d$ is homeomorphic to the Euclidean halfspace $\{x \in \mathbb{R}^d : x_d \geq 0\}$.

A subspace of a Euclidean space is said to be *compact* if it is bounded and closed with respect to the Euclidean space. Boundedness is a metric space concept; there is a purely topological definition of compactness, which we omit. If $\mathbb{T}$ and $\mathbb{U}$ are compact metric spaces, every bijective map from $\mathbb{T}$ to $\mathbb{U}$ has a continuous inverse. We will take advantage of this fact to prove that certain functions are homeomorphisms.

Homeomorphism gives us a purely topological definition of what it means to triangulate a domain.

Definition 12.6 (triangulation of a topological space). A simplicial complex $\mathcal{K}$ is a *triangulation* of a topological space $\mathbb{T}$ if its underlying space $|\mathcal{K}|$ is homeomorphic to $\mathbb{T}$.

When two topological spaces are subspaces of the same larger space, there is another notion of similarity that is stronger than homeomorphism, called *isotopy*. If two subspaces are isotopic, one can be continuously deformed into the other so that the deforming subspace remains always homeomorphic to its original form. For example, a cube can be continuously deformed into a ball.

Homeomorphic subspaces are not necessarily isotopic. Consider a torus embedded in $\mathbb{R}^3$, illustrated in Figure 12.2(a). One can embed the torus in $\mathbb{R}^3$ so that it is knotted, as shown in Figure 12.2(b). The knotted torus is homeomorphic to the standard, unknotted one. However, it is not possible to continuously deform one to the other while keeping it embedded in $\mathbb{R}^3$ and topologically unchanged. Any attempt to do so will cause the torus to pass through a state in which it is "self-intersecting" and not a manifold. The easiest way to recognize this fact is to look not at the topology of the tori, but at the topology of the space around them. Although the knotted and unknotted tori are homeomorphic, their

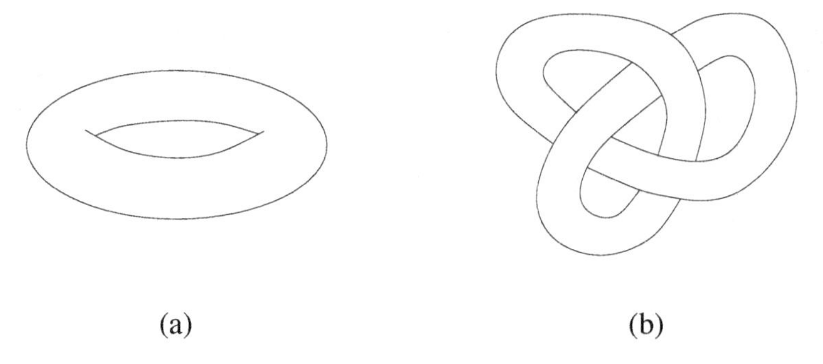

Figure 12.2: (a) Torus. (b) Knotted torus.

complements are not. Therefore, we consider both the notion of an *isotopy*, in which a torus deforms continuously, and the notion of an *ambient isotopy*, in which not only the torus deforms, but the entirety of $\mathbb{R}^3$ deforms with it.

Definition 12.7 (isotopy). An *isotopy* connecting two spaces $\mathbb{T} \subseteq \mathbb{R}^d$ and $\mathbb{U} \subseteq \mathbb{R}^d$ is a continuous map $\xi : \mathbb{T} \times [0, 1] \to \mathbb{R}^d$ where $\xi(\mathbb{T}, 0) = \mathbb{T}$, $\xi(\mathbb{T}, 1) = \mathbb{U}$, and for every $t \in [0, 1]$, $\xi(\cdot, t)$ is a homeomorphism between $\mathbb{T}$ and its image $\{\xi(x, t) : x \in \mathbb{T}\}$. An *ambient isotopy* connecting $\mathbb{T}$ and $\mathbb{U}$ is a map $\xi : \mathbb{R}^d \times [0, 1] \to \mathbb{R}^d$ such that $\xi(\cdot, 0)$ is the identity function on $\mathbb{R}^d$, $\xi(\mathbb{T}, 1) = \mathbb{U}$, and for each $t \in [0, 1]$, $\xi(\cdot, t)$ is a homeomorphism.

For example, the map

$$\xi(x, t) = \frac{1 - (1 - t)\|x\|}{1 - \|x\|} x$$

sends the open d-ball $\mathbb{B}^d_o$ to itself if $t = 0$, and to the Euclidean space $\mathbb{R}^d$ if $t = 1$. Think of the parameter t as the time, so $\xi(\mathbb{B}^d_o, t)$ deforms continuously from a ball at time zero to $\mathbb{R}^d$ at time one. Hence, the open d-ball and $\mathbb{R}^d$ are related by an isotopy.

Every ambient isotopy becomes an isotopy if its domain is restricted from $\mathbb{R}^d \times [0, 1]$ to $\mathbb{T} \times [0, 1]$. It is known that if two subspaces are related by an isotopy, there exists an ambient isotopy connecting them, so the two notions are equivalent. In Section 13.6, we will show that some algorithms generate triangular meshes whose underlying spaces are related by an isotopy to the surfaces they triangulate.

There is another notion of similarity among topological spaces that is weaker than homeomorphism, called *homotopy equivalence*. It relates spaces that can be continuously deformed to one another but may not be homeomorphic. For example, a ball can shrink to a point, but they are not homeomorphic; there is not even a bijective function from an infinite point set to a single point. However, homotopy preserves some aspects of connectedness, such as the number of connected components and the number of holes in a space. Thus, a coffee cup is homotopy equivalent to a circle, but not to a ball or a point.

To get to homotopy equivalence, we first need the concept of homotopies, which generalize isotopies so that homeomorphism is not required.

Definition 12.8 (homotopy). Let $g : \mathbb{X} \to \mathbb{U}$ and $h : \mathbb{X} \to \mathbb{U}$ be maps. A *homotopy* is a map $H : \mathbb{X} \times [0, 1] \to \mathbb{U}$ such that $H(\cdot, 0) = g$ and $H(\cdot, 1) = h$. Two maps are *homotopic* if there is a homotopy connecting them.

For example, if $h : \mathbb{B}^3 \to \mathbb{R}^3$ is the identity map on a unit ball and $g : \mathbb{B}^3 \to \mathbb{R}^3$ maps every point in the ball to the origin, the fact that g and h are homotopic is demonstrated by the homotopy $H(x, t) = t \cdot h(x)$; hence, $H(\mathbb{B}^3, t)$ deforms continuously from a point at time zero to a ball at time one. A key property of a homotopy is that, as H is continuous, at every time t the map $H(\cdot, t)$ is continuous.

It is more revealing to consider two maps that are not homotopic. Let $g : \mathbb{S}^1 \to \mathbb{S}^1$ be the identity map from the circle to itself, and let $h : \mathbb{S}^1 \to \mathbb{S}^1$ map every point on the circle to a single point $p \in \mathbb{S}^1$. Although it is easy to imagine contracting a circle to a point, that image is misleading: the map H is constrained by the requirement that every point on the circle at every time maps to a point on the circle. The circle can contract to a point only if it is cut somewhere, implying that H is not continuous.

Observe that whereas a homeomorphism is a topological relationship between two topological spaces $\mathbb{T}$ and $\mathbb{U}$, a homotopy or an isotopy (which is a special kind of homotopy) is a relationship between two maps, which indirectly establishes a relationship between two topological subspaces $g(\mathbb{X}) \subseteq \mathbb{U}$ and $h(\mathbb{X}) \subseteq \mathbb{U}$. That relationship is not necessarily an equivalence class, but the following relationship is.

Definition 12.9 (homotopy equivalent). Two topological spaces $\mathbb{T}$ and $\mathbb{U}$ are *homotopy equivalent* if there exist maps $g : \mathbb{T} \to \mathbb{U}$ and $h : \mathbb{U} \to \mathbb{T}$ such that $h \circ g$ is homotopic to the identity map $\iota_\mathbb{T} : \mathbb{T} \to \mathbb{T}$ and $g \circ h$ is homotopic to the identity map $\iota_\mathbb{U} : \mathbb{U} \to \mathbb{U}$.

Whereas homeomorphic spaces have the same dimension, homotopy equivalent spaces sometimes do not. To see that the 2-ball is homotopy equivalent to a single point p, construct a map $h : \mathbb{B}^2 \to \{p\}$ and a map $g : \{p\} \to \mathbb{B}^2$ where $g(p)$ is any point q in $\mathbb{B}^2$. Observe that $h \circ g$ is the identity map on $\{p\}$, which is trivially homotopic to itself. In the other direction, $g \circ h : \mathbb{B}^2 \to \mathbb{B}^2$ sends every point in $\mathbb{B}^2$ to q. There is a homotopy connecting $g \circ h$ to the identity map $\iota_{\mathbb{B}^2}$, namely, the map $H(x, t) = (1 - t)q + tx$.

The definition of homotopy equivalent is somewhat mysterious. A useful intuition for understanding it is the fact that two spaces $\mathbb{T}$ and $\mathbb{U}$ are homotopy equivalent if and only if there exists a third space $\mathbb{X}$ such that both $\mathbb{T}$ and $\mathbb{U}$ are *deformation retracts* of $\mathbb{X}$, illustrated in Figure 12.3.

Definition 12.10 (deformation retract). Let $\mathbb{T}$ be a topological space, and let $\mathbb{U} \subset \mathbb{T}$ be a subspace. A *retraction* r of $\mathbb{T}$ to $\mathbb{U}$ is a map from $\mathbb{T}$ to $\mathbb{U}$ such that $r(x) = x$ for every $x \in \mathbb{U}$. The space $\mathbb{U}$ is a *deformation retract* of $\mathbb{T}$ if the identity map on $\mathbb{T}$ can be continuously deformed to a retraction with no motion of the points already in $\mathbb{U}$: specifically, there is a homotopy $R : \mathbb{T} \times [0, 1] \to \mathbb{T}$ such that $R(\cdot, 0)$ is the identity map on $\mathbb{T}$, $R(\cdot, 1)$ is a retraction of $\mathbb{T}$ to $\mathbb{U}$, and $R(x, t) = x$ for every $x \in \mathbb{U}$ and every $t \in [0, 1]$.

If $\mathbb{U}$ is a deformation retract of $\mathbb{T}$, then $\mathbb{T}$ and $\mathbb{U}$ are homotopy equivalent. For example, any point on a line segment (open or closed) is a deformation retract of the line segment and is homotopy equivalent to it. The letter V is a deformation retract of the letter W, and also of a ball. Moreover, two spaces are homotopy equivalent if they are deformation retractions of a common space. The symbols ∅, ∞, and o–o (viewed as one-dimensional point sets) are deformation retracts of a double doughnut—a doughnut with two holes. Therefore, they are homotopy equivalent to each other, although none of them is a deformation retract of any of the others. They are not homotopy equivalent to X, O, ⊕, ⊙, ◎, a ball, or a coffee cup.

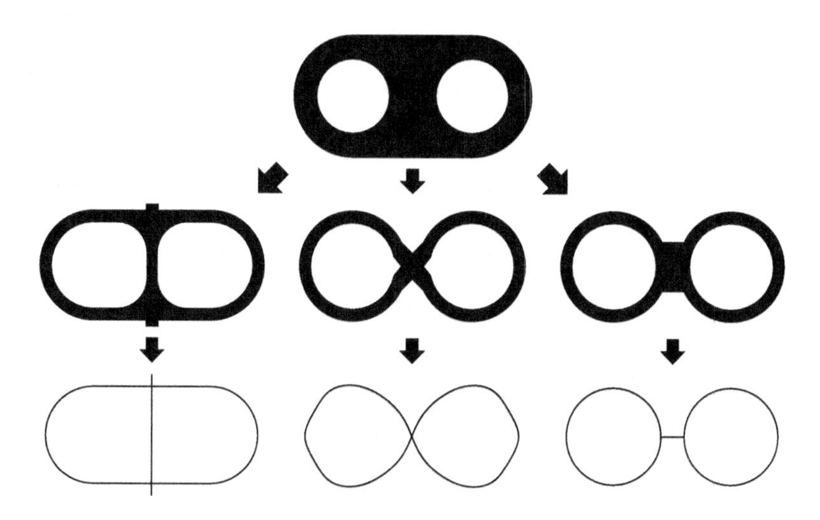

Figure 12.3: All seven of these point sets are homotopy equivalent, because they are all deformation retracts of the top point set.

12.3 Manifolds

A manifold is a set of points that is locally connected in a particular way. A 1-manifold has the structure of a piece of string, possibly with its ends tied in a loop, and a 2-manifold has the structure of a piece of paper or rubber sheet that has been cut and perhaps glued along its edges—a category that includes disks, spheres, tori, and Möbius bands.

Definition 12.11 (manifold)**.** A topological space Σ is a *k-manifold*, or simply *manifold*, if every point $x \in \Sigma$ has a neighborhood homeomorphic to $\mathbb{B}^k_o$ or $\mathbb{H}^k$. The *dimension* of Σ is k.

A manifold can be viewed as a purely abstract topological space, or it can be embedded into a metric space or a Euclidean space. Even without an embedding, every manifold can be partitioned into boundary and interior points. Observe that these words mean very different things for a manifold than they do for a metric space or topological space.

Definition 12.12 (boundary; interior)**.** The *interior* $\operatorname{Int}\Sigma$ of a manifold Σ is the set of points in Σ that have a neighborhood homeomorphic to $\mathbb{B}^k_o$. The *boundary* $\operatorname{Bd}\Sigma$ of Σ is the set of points $\Sigma \setminus \operatorname{Int}\Sigma$. The boundary $\operatorname{Bd}\Sigma$, if not empty, consists of the points that have a neighborhood homeomorphic to $\mathbb{H}^k$. If $\operatorname{Bd}\Sigma$ is the empty set, we say that Σ is *without boundary*.

A single point, a 0-ball, is a 0-manifold without boundary according to this definition. The closed disk $\mathbb{B}^2$ is a 2-manifold whose interior is the open disk $\mathbb{B}^2_o$ and whose boundary is the circle $\mathbb{S}^1$. The open disk $\mathbb{B}^2_o$ is a 2-manifold whose interior is $\mathbb{B}^2_o$ and whose boundary is the empty set. This highlights an important difference between Definitions 1.13 and 12.12 of "boundary": when $\mathbb{B}^2_o$ is viewed as a point set in the space $\mathbb{R}^2$, its boundary is $\mathbb{S}^1$ according to Definition 1.13; but viewed as a manifold, its boundary is empty according to Definition 12.12. The boundary of a manifold is *always* included in the manifold.

The open disk $\mathbb{B}^2_o$, the Euclidean space $\mathbb{R}^2$, the sphere $\mathbb{S}^2$, and the torus are all connected 2-manifolds without boundary. The first two are homeomorphic to each other, but

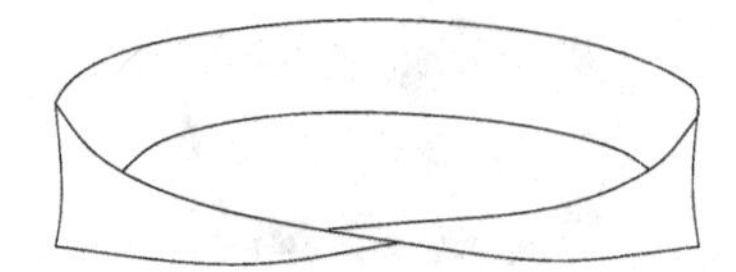

Figure 12.4: Möbius band.

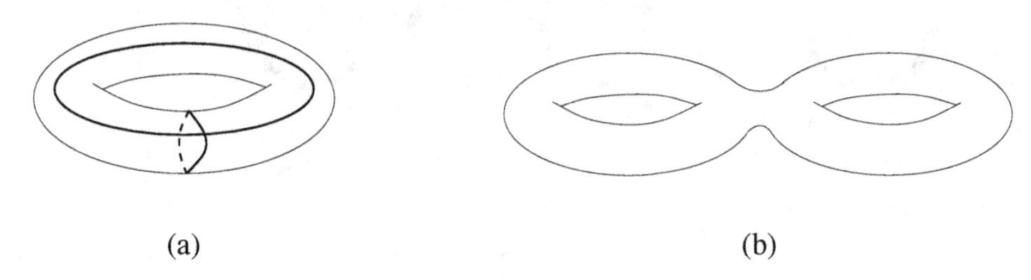

Figure 12.5: (a) Removal of the bold loops opens up the torus into a topological disk. (b) Every surface without boundary in $\mathbb{R}^3$ resembles a sphere or a conjunction of one or more tori.

the last two are topologically different from the others. The sphere and the torus are compact (bounded and closed with respect to $\mathbb{R}^3$) whereas $\mathbb{B}_o^2$ and $\mathbb{R}^2$ are not.

A 2-manifold Σ is *non-orientable* if, starting from a point p, one can walk on one side of Σ and end up on the opposite side of Σ upon returning to p. Otherwise, Σ is *orientable*. Spheres and balls are orientable, whereas the *Möbius band* in Figure 12.4 is a non-orientable 2-manifold.

A *surface* is a 2-manifold that is a subspace of $\mathbb{R}^d$. Any compact surface without boundary in $\mathbb{R}^3$ is an orientable 2-manifold. To be non-orientable, a compact surface must have a nonempty boundary (like the Möbius band) or be embedded in a 4- or higher-dimensional Euclidean space.

A surface can sometimes be disconnected by removing one or more *loops* (connected 1-manifolds without boundary) from it. The *genus* of a surface is g if $2g$ is the maximum number of loops that can be removed from the surface without disconnecting it; here the loops are permitted to intersect each other. For example, the sphere has genus zero as every loop cuts it into two balls. The torus has genus one: a circular cut around its neck and a second circular cut around its circumference, illustrated in Figure 12.5(a), allow it to unfold into a rectangle, which topologically is a disk. A third loop would cut it into two pieces. Figure 12.5(b) shows a 2-manifold without boundary of genus 2. Although a high-genus surface can have a very complex shape, all compact 2-manifolds in $\mathbb{R}^3$ that have the same genus and no boundary are homeomorphic to each other.

12.4 Smooth manifolds

A purely topological manifold has no geometry, but once embedded in a Euclidean space it may appear smooth or creased. Here we enrich the notion of a geometric manifold by imposing a differential structure on it. For the rest of this chapter, we are discussing only manifolds without boundary.

Consider a map $\phi : U \to W$ where U and W are open sets in $\mathbb{R}^k$ and $\mathbb{R}^d$, respectively. The map ϕ has d components, namely, $\phi(x) = (\phi_1(x), \phi_2(x), \ldots, \phi_d(x))$, where $x = (x_1, x_2, \ldots, x_k)$ denotes a point in $\mathbb{R}^k$. The *Jacobian* of ϕ at x is the $d \times k$ matrix of the first-order partial derivatives

$$\begin{bmatrix} \frac{\partial \phi_1(x)}{\partial x_1} & \cdots & \frac{\partial \phi_1(x)}{\partial x_k} \\ \vdots & \ddots & \vdots \\ \frac{\partial \phi_d(x)}{\partial x_1} & \cdots & \frac{\partial \phi_d(x)}{\partial x_k} \end{bmatrix}.$$

The map ϕ is *regular* if its Jacobian has rank k at every point in U. The map ϕ is C^i-continuous if the ith-order partial derivatives of ϕ are continuous.

The reader may be familiar with *parametric surfaces*, for which U is a k-dimensional *parameter space* and its image $\phi(U)$ in d-dimensional space is a parametric surface. Unfortunately, a single parametric surface cannot easily represent a manifold with a complicated topology. However, for a manifold to be smooth, it suffices that each point on the manifold has a neighborhood that looks like a smooth parametric surface.

Definition 12.13 (smooth manifold)**.** For any $i > 0$, a k-manifold Σ without boundary embedded in $\mathbb{R}^d$ is C^i*-smooth* if for every point $p \in \Sigma$, there exists an open set $U_p \subset \mathbb{R}^k$, a neighborhood $W_p \subset \mathbb{R}^d$ of p, and a map $\phi_p : U_p \to W_p \cap \Sigma$ such that (i) ϕ_p is C^i-continuous, (ii) ϕ_p is a homeomorphism, and (iii) ϕ_p is regular. If $k = 2$, we call Σ a C^i*-smooth surface*.

The first condition says that each map is continuously differentiable at least i times. The second condition requires each map to be bijective, ruling out "wrinkles" where multiple points in U map to a single point in W. The third condition prohibits any map from having a directional derivative of zero at any point in any direction. The first and third conditions together enforce smoothness, and imply that there is a well-defined tangent k-flat at each point in Σ. The three conditions together imply that the maps ϕ_p defined in the neighborhood of each point $p \in \Sigma$ overlap smoothly. There are two extremes of smoothness. We say that Σ is C^∞-smooth if for every point $p \in \Sigma$, the partial derivatives of ϕ_p of all orders are continuous. On the other hand, Σ is *nonsmooth* if Σ is a k-manifold (therefore C^0-smooth) but not C^1-smooth.

12.5 The medial axis and local feature size of a smooth manifold

To mesh a compact smooth surface Σ, we generate a discrete set of points in Σ called a *sample* of Σ. The sample points become the vertices of a triangulation whose underlying space is meant to be homeomorphic to Σ and a geometrically good approximation of Σ. Figure 12.6 shows two samples of different densities on a knotted torus in $\mathbb{R}^3$.

Recall that for a piecewise linear complex, the local feature size at a point x is the radius of the smallest ball centered at x that intersects two disjoint linear cells. A smooth manifold is not composed of discrete cells such as vertices, edges, and polygons. However, there are other features that dictate how small the elements of a mesh of a manifold must be: the local curvature of the manifold and the distances between portions of the manifold whose connections are not local. We define a local feature size function on a smooth manifold that

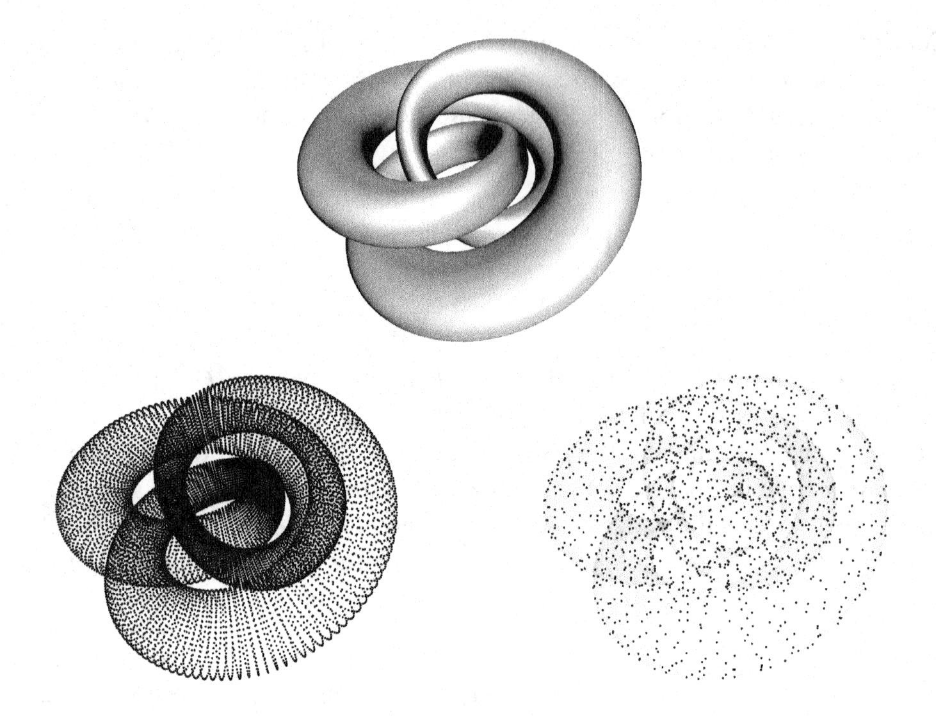

Figure 12.6: A knotted torus and two samples of it.

is smallest where the curvature is large or where a portion of the manifold comes close to another portion though the nearest connection between them is not close, like pleats in a curtain. Both these phenomena are captured by defining the local feature size at a point on a manifold to be its distance from a structure called the *medial axis* of the manifold. If points are sampled on a smooth surface with a spacing dictated by the local feature size, we can derive properties of the sample points and of the edges and triangles spanning the sample points, and show that the sample permits us to approximate the surface accurately.

Throughout this section, we use the shorthand term *smooth manifold* to refer to a compact C^2-smooth manifold without boundary, embedded in $\mathbb{R}^d$; and Σ is always such a smooth manifold. However, we urge the reader to observe that most definitions and results in this section also apply to manifolds with boundary.

12.5.1 The medial axis

The medial axis of a smooth manifold Σ without boundary is a sort of skeleton of the shape bounded by Σ. There are slightly different definitions of the medial axis in the literature; we adopt one suitable for defining the local feature size. For a point $x \in \mathbb{R}^d$, recall that $d(x, \Sigma)$ is the distance from x to the nearest point on the manifold, and let $\Lambda(x)$ denote the set of points in Σ at distance $d(x, \Sigma)$ from x. In other words,

$$\Lambda(x) = \{\mathrm{argmin}_{y \in \Sigma}\, d(x, y)\}.$$

Definition 12.14 (medial axis)**.** The *medial axis* M of a smooth manifold $\Sigma \subset \mathbb{R}^d$ is the closure of the set $\{x \in \mathbb{R}^d : \Lambda(x)$ contains more than one point$\}$.

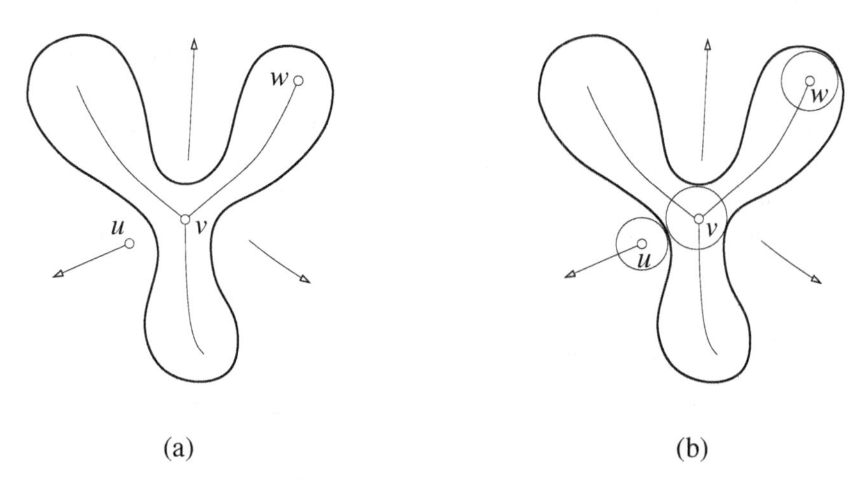

Figure 12.7: (a) A curve in two dimensions and its medial axis, leaving out portions that extend infinitely far away. (b) The medial ball centered at v meets the curve tangentially at three points, whereas the medial balls with centers u and w meet the curve at only one point. Infinitely many medial balls meet this curve at two points.

Figure 12.7 illustrates a medial axis of a 1-manifold. By definition, the medial axis contains every point that has at least two closest points on Σ, as well as the limit points of those points, such as u and w. For every point $x \in M$, the *medial ball* $B(x, d(x, \Sigma))$ centered at x touches Σ tangentially at one or more points, but its interior does not intersect Σ, so every contact is tangential, as Figure 12.7(b) shows. Most medial balls are tangent to Σ at two or more points; the medial balls of the points added by taking the closure are tangent at just one.

The medial axis is not necessarily a manifold, nor without boundary. It may have branching points such as v and boundary points such as u and w. A medial axis in $\mathbb{R}^3$ can have the appearance of a cluster of 2-manifolds and 1-manifolds glued together, each with or without boundary. Medial axes can also have isolated points at the centers of spheres.

The medial axis is not necessarily connected. A smooth $(d-1)$-manifold without boundary cuts $\mathbb{R}^d$ into a bounded inner face and an unbounded outer face. The component of the medial axis included in the inner face is the *inner medial axis*, the component included in the outer face is the *outer medial axis*, and the two are disjoint as illustrated. The outer medial axis is the empty set if Σ bounds a convex volume. A manifold of dimension $d-2$ or less does not cut $\mathbb{R}^d$, so its medial axis does not generally have inner and outer components.

A *medial ball at* $p \in \Sigma$ is a medial ball tangent to Σ at p. To find a medial ball at a point $p \in \Sigma$, imagine growing a ball from zero radius while keeping it tangent to Σ at p, stopping the growth when the ball's boundary touches a second point in Σ or cannot grow larger because of the curvature of Σ at p. There is a finite inner medial ball at every point of a bounded smooth $(d-1)$-manifold in $\mathbb{R}^d$, but not necessarily an outer one. A point on a manifold of dimension $d-2$ or less can have infinitely many medial balls.

An *osculating ball* at a point p on a 1-manifold Σ is the diametric ball of p and two other distinct points on Σ in the limit as they approach p. Intuitively, an osculating ball is not only tangent to Σ at p, but also has the same curvature. The *radius of curvature* of Σ at p is the radius of the osculating ball at p, and the *curvature* of Σ at p is the reciprocal of the radius. In general, an osculating ball is *not* a medial ball; typically its interior intersects Σ in a neighborhood of p. However, if a point p is at a local maximum of curvature on Σ, then the osculating ball at p can be a medial ball that intersects only a single point of Σ. The medial balls centered at u and w in Figure 12.7(b) are osculating balls whose centers are in the medial axis because we take the closure of the centers of the medial balls that have more than one contact point. We omit discussion of how curvature is defined on 2-manifolds, referring the reader instead to any introductory textbook on differential geometry.

12.5.2 Definition of the local feature size

A local feature size function for smooth manifolds should indicate the sample spacing necessary to resolve a manifold unambiguously. It should also have Lipschitz continuity, because a region that requires very dense sampling necessarily influences regions nearby. For example, if two disjoint regions requiring extremely dense sampling are separated by a distance d, the region between them necessarily has a sample spacing not much greater than d.

A smooth manifold Σ is the boundary of the union of all its open medial balls (Exercise 9). Hence, the medial axis M augmented with the distance of each point $m \in M$ from Σ captures the shape of Σ. One might think to capture the local feature size with the functions $\rho_i, \rho_o : \Sigma \to \mathbb{R}$ where $\rho_i(x)$ and $\rho_o(x)$ are the radii of the medial balls at x with centers on the inner and outer medial axes, respectively. These two functions are continuous for a large class of smooth manifolds. Either $\rho_i(x)$ or $\rho_o(x)$, or a combination of both, could be taken as the local feature size at x.

However, these functions are not k-Lipschitz for any fixed k. The following definition of local feature size (LFS) is 1-Lipschitz and has proven itself repeatedly as a powerful measure of the sample spacing required to accurately approximate a manifold.

Definition 12.15 (local feature size)**.** Let Σ be a smooth manifold in $\mathbb{R}^d$. Let $f : \Sigma \to \mathbb{R}$ be the function $f(x) = d(x, M)$, the distance from x to the medial axis M of Σ. The value $f(x)$ is the *local feature size* (LFS) of Σ at x, and f is the *LFS function* for Σ.

Figure 12.8 illustrates how this definition captures the details of the shape of a loop (a connected 1-manifold without boundary). Observe that the local feature size is smallest in regions of high curvature, and the local feature sizes on the legs are smaller than the local feature sizes near the middle of the torso—for instance, $f(d)$ is smaller than $f(a)$—according with our intuitive notion of features. A point's local feature size can be determined by either the inner or the outer medial axis; $f(c)$ is determined by the outer medial axis whereas $f(d)$ is determined by the inner one. For a manifold of dimension $d-2$ or less, the medial axis does not have inner and outer components, but the LFS function is defined the same way.

It is straightforward to see that $f(x) \leq \min\{\rho_i(x), \rho_o(x)\}$. For example, Figure 12.8 shows that $f(d)$ is smaller than the radii of the two medial balls at d. We will use this fact often. For example, Section 12.5.4 will show that the local feature size at a point cannot exceed its radius of curvature.

It follows easily from its definition that the LFS function is Lipschitz.

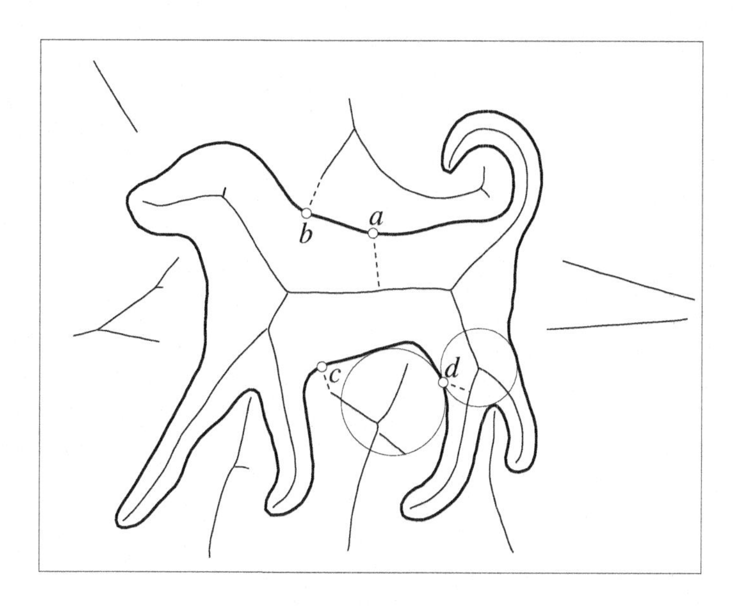

Figure 12.8: The local feature sizes at a, b, c, and d are the lengths of the dashed line segments adjoining them.

Proposition 12.1. *The LFS function $f : \Sigma \to \mathbb{R}$ is 1-Lipschitz—that is, for any two points x and y on Σ, $f(x) \leq f(y) + d(x, y)$.*

Proof. Let m be a point on the medial axis of Σ such that $f(y) = d(y, m)$. By the definition of f and the triangle inequality,

$$f(x) \leq d(x, m) \leq d(y, m) + d(x, y) = f(y) + d(x, y).$$

□

12.5.3 Point samples on manifolds

A *sample* S of Σ is a finite set of points $S \subset \Sigma$. One can accurately approximate a manifold by choosing a sample in which the local spacing of points does not exceed a fraction of the local feature size.

Definition 12.16 (ε-sample)**.** A sample S of Σ is an *ε-sample* of Σ if for every point $x \in \Sigma$, $d(x, S) \leq \varepsilon f(x)$.

A smaller value of ε implies a denser sample. However, ε sets only an upper bound on the point spacing. It indicates that every point $x \in \Sigma$ has at least one sample point at most $\varepsilon f(x)$ away, but there could be billions. An ε-sample is also an ε'-sample for any $\varepsilon' > \varepsilon$. Figure 12.9(a) illustrates a 0.2-sample of a circle, which is also a 0.3-sample. Figure 12.9(b) shows another dense sample of a curve, which respects the local feature size but is unnecessarily dense in places. In $\mathbb{R}^3$, a strong theory of surface sampling has been developed for roughly $\varepsilon = 0.09$, but the algorithms for mesh generation and surface reconstruction arising from this theory often work well for much larger values of ε. At the

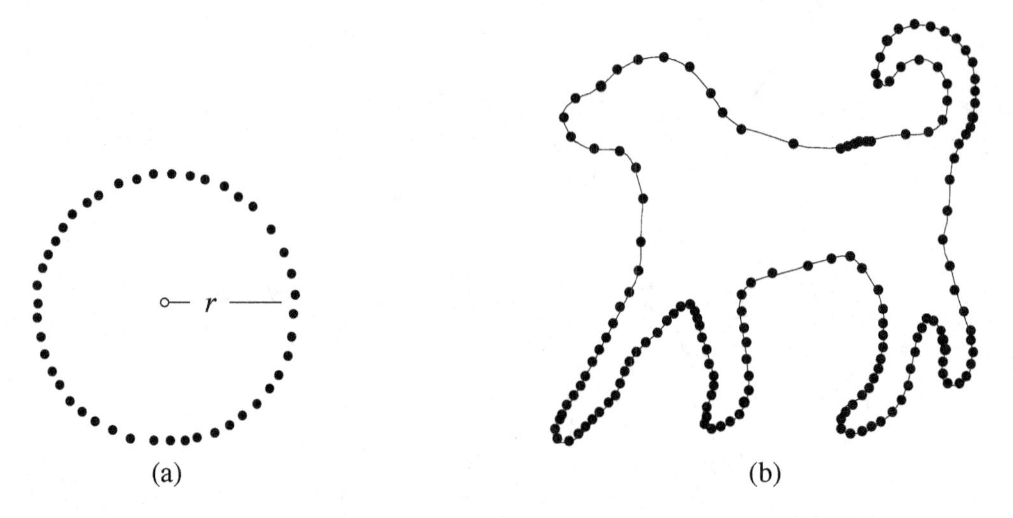

Figure 12.9: (a) The local feature size at every point on a circle is equal to its radius r. In the sample, every point on the circle has a sample point within a distance of $0.2r$. (b) A non-uniform sample on dog which is dense with respect to local feature size. Notice the oversampling on the back of the dog.

other extreme, it is known that 1-samples are not dense enough to capture the topology of a surface, even if the theory improves in the future.

Point samples are the reason we insist that Σ be C^2-smooth. If the local feature size $f(x)$ is zero at a point $x \in \Sigma$, the ε-sample condition cannot be satisfied by any finite sample. As meshes with infinitely many elements are rarely useful, we require that the LFS function does not vanish anywhere on Σ. If Σ is C^2-smooth, the LFS is strictly positive everywhere. If Σ is only C^1-smooth, the LFS can be zero at some points.

A useful application of the Lipschitz property of f is that the distance between two points, expressed in terms of the LFS of one, can also be expressed in terms of the LFS of the other.

Lemma 12.2 (Feature Translation Lemma). *Let x and y be any two points on Σ. If there is an $\varepsilon \in (0, 1)$ such that $d(x, y) \leq \varepsilon f(x)$, then*

(i) $f(x) \leq \frac{1}{1-\varepsilon} f(y)$ *and*

(ii) $d(x, y) \leq \frac{\varepsilon}{1-\varepsilon} f(y)$.

Proof. By the Lipschitz property,

$$\begin{aligned} f(x) &\leq f(y) + d(x, y) \\ &\leq f(y) + \varepsilon f(x). \end{aligned}$$

Rearranging terms gives $f(x) \leq f(y)/(1 - \varepsilon)$, proving (i); (ii) follows immediately. □

We use the Feature Translation Lemma to derive an upper bound for the distance between two points that are close to a third point with respect to the third point's LFS.

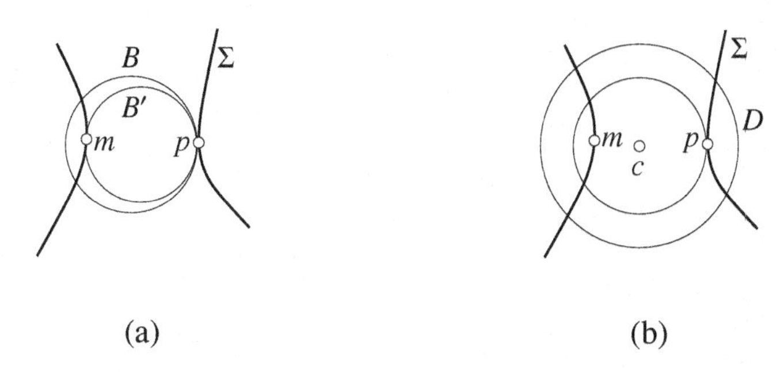

Figure 12.10: (a) By shrinking a ball B that is tangent to Σ at p, we find a medial ball B'. (b) The intersection $(\text{Bd}\, D) \cap \Sigma$ is not a topological $(k-1)$-sphere, so D contains at least two critical points.

Lemma 12.3 (Three Points Lemma)**.** *Let p, x, and y be any three points on Σ. If there is an $\varepsilon < 1$ such that $d(p,x) \leq \varepsilon f(p)$ and $d(p,y) \leq \varepsilon f(p)$, then $d(x,y) \leq 2\varepsilon f(p) \leq \frac{2\varepsilon}{1-\varepsilon} f(x)$.*

Proof. By the triangle inequality, $d(x,y) \leq d(p,x) + d(p,y) \leq 2\varepsilon f(p)$, which is at most $2\varepsilon f(x)/(1-\varepsilon)$ by the Feature Translation Lemma. □

12.5.4 Properties of a surface and its medial axis

In this section we prove two results about the medial axis, the Feature Ball Lemma and the Small Ball Lemma, that play crucial roles in the theory of surface sampling. We begin with two simple propositions that give some of the flavor of reasoning about the medial axis.

Proposition 12.4. *Let Σ be a smooth k-manifold in $\mathbb{R}^d$. If a Euclidean d-ball B intersects Σ at a tangential contact point p and at least one other point, then the medial axis of Σ intersects both B and the line segment pc, where c is the center of B.*

Proof. As B intersects two points of Σ, if its interior is empty, B is a medial ball and its center c lies on the medial axis. Otherwise, let $B' \subset B$ be the largest ball that is also tangent to Σ at p but whose interior is disjoint from Σ, as depicted in Figure 12.10(a). Because B' is the largest such ball, either its boundary intersects a second point of Σ and thus B' is a medial ball, or B' is the limit of a sequence of medial balls at points in a neighborhood of p, and the center of B' is in the closure of the centers of those medial balls, so again B' is a medial ball. (If $d = 2$, the latter case implies that p is a point of locally maximum curvature on Σ and B' has the same curvature.) The center of B' is in B, on pc, and on the medial axis. □

Proposition 12.5. *Suppose that at a point p on a 2-manifold embedded in $\mathbb{R}^3$, the maximum of the principal curvatures is κ. Then $f(p) \leq 1/\kappa$.*

Proof. Let B be a ball with radius $1/\kappa$ that is tangent to the manifold at p and determines the maximum principal curvature. Either B is a medial ball or the interior of B intersects Σ, in which case by Proposition 12.4, some point of the medial axis lies between p and the

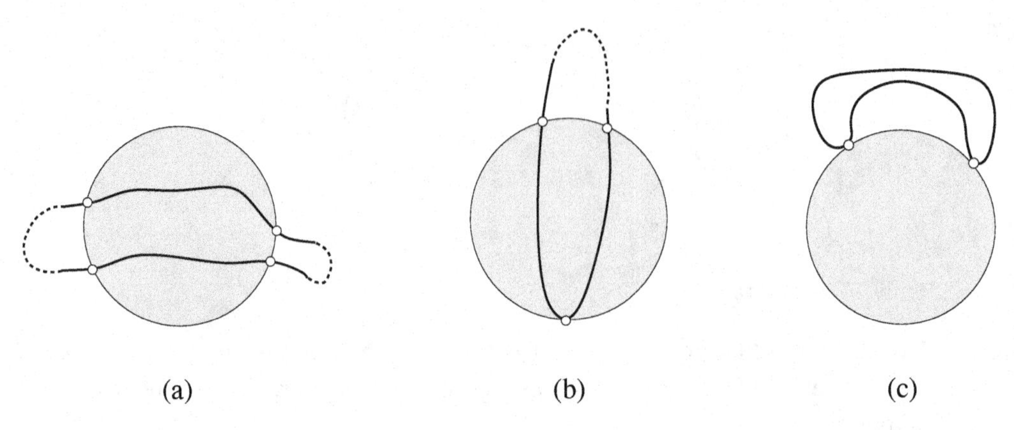

(a) (b) (c)

Figure 12.11: Three antecedents of the Feature Ball Lemma. (a) $D \cap \Sigma$ is not a topological 1-ball, and $(\text{Bd}\, D) \cap \Sigma$ is not a topological 0-sphere. (b) $D \cap \Sigma$ is a topological 1-ball, but $(\text{Bd}\, D) \cap \Sigma$ is not a topological 0-sphere. (c) $(\text{Bd}\, D) \cap \Sigma$ is a topological 0-sphere, but $D \cap \Sigma$ is not a topological 1-ball.

center of B. In either case, the distance from p to the medial axis is at most $1/\kappa$. □

Below, the Feature Ball Lemma generalizes Proposition 12.4 to a wider range of balls, such as balls for which $B \cap \Sigma$ is not connected. The proof of the lemma uses some concepts from Morse theory (critical point theory) in differential topology. We introduce the theory briefly here.

Let $h : \Sigma \to \mathbb{R}$ be a smooth scalar function defined over a smooth k-manifold Σ. For any point $x \in \Sigma$, let $T\Sigma_x$ denote its tangent space, that is, the k-flat spanned by all vectors tangent to Σ at x. The derivative of h at x, denoted Dh_x, is a linear map from $T\Sigma_x$ to $\mathbb{R}$. A point x is a *critical point* and the value $h(x)$ is a *critical value* if Dh_x is a zero map—that is, Dh_x maps all tangent vectors to zero. The *level set* of a value $a \in \mathbb{R}$ is $h^{-1}(a) = \{x \in \Sigma : h(x) = a\}$. It is a fact of Morse theory that if h is generic (its Hessian is nonsingular) and a is not critical, the level set $h^{-1}(a)$ is a smooth $(k-1)$-manifold. It is also known that two level sets $h^{-1}(a_1)$ and $h^{-1}(a_2)$ are homeomorphic if there is no critical value in the interval $[a_1, a_2]$.

The following result shows that a ball D centered on a k-manifold Σ must intersect the medial axis if $(\text{Bd}\, D) \cap \Sigma$ is not a topological $(k-1)$-sphere, as illustrated in Figures 12.11(a) and (b), or if $D \cap \Sigma$ is not a topological k-ball, as illustrated in Figures 12.11(a) and (c).

Lemma 12.6 (Feature Ball Lemma)**.** *Let Σ be a smooth k-manifold in $\mathbb{R}^d$, and let D be a Euclidean d-ball that intersects Σ at more than one point. If* $(\text{Bd}\, D) \cap \Sigma$ *is not a topological* $(k-1)$*-sphere, or $D \cap \Sigma$ is not a topological k-ball, then D intersects the medial axis of Σ.*

Proof. Let c be the center of the ball D. Let $h : \Sigma \to \mathbb{R}$ be the smooth function $h(x) = d(c, x)$. At every critical point x of h, the derivative of h vanishes and the ball $B(c, h(x))$ is tangent to Σ at x. Let m be a point in Σ that globally minimizes $h(m)$, which is necessarily a critical point of h. If there is more than one globally minimum point, then $B(c, h(m))$ is a medial ball, as its interior is empty and it meets Σ tangentially at each global minimum. In this case, c lies on the medial axis and the result follows. Assume for the rest of the proof that h has one global minimizer m only.

We first show that $D \cap \Sigma$ contains a critical point besides m, as Figure 12.10(b) illustrates. Suppose for the sake of contradiction that there is no critical point other than m. Because D intersects Σ at more than one point, $h(m) < \text{radius}(D)$ and $B(c, h(m)) \subset D$. As m is a sole global minimizer, for all sufficiently small $\delta > 0$, $B(c, h(m) + \delta) \cap \Sigma$ is a topological k-ball whose boundary, the level set $h^{-1}(h(m) + \delta)$, is a topological $(k-1)$-sphere. Because $D \cap \Sigma$ contains no critical point of h besides m, Morse theory implies that $h^{-1}(r)$ is a topological $(k-1)$-sphere for the entire range of radii r in the interval $(h(m), \text{radius}(D)]$. Clearly, $h^{-1}(r)$ is a single point m for the radius $r = h(m)$, and empty for $r < h(m)$.

It follows that $(\text{Bd}\, D) \cap \Sigma = h^{-1}(\text{radius}(D))$ is a topological $(k-1)$-sphere. Moreover, $D \cap \Sigma$ is homeomorphic to a space formed by taking the product of $\mathbb{S}^{k-1}$ with the closed unit interval $[0, 1] \subset \mathbb{R}$ and merging the points at the zero boundary into a single point—that is, the quotient space $(\mathbb{S}^{k-1} \times [0, 1])/(\mathbb{S}^{k-1} \times \{0\})$. This space is homeomorphic to a k-ball, so $D \cap \Sigma$ is a topological k-ball. These statements contradict the assumption that $(\text{Bd}\, D) \cap \Sigma$ is not a topological $(k-1)$-sphere or $D \cap \Sigma$ is not a topological k-ball. This completes the argument that h has a second critical point $p \neq m$ on $D \cap \Sigma$.

The ball $B(c, h(p))$ is tangent to Σ at p and contains $m \in \Sigma$. By Proposition 12.4, $B(c, h(p))$ intersects the medial axis, and therefore, so does $D \supseteq B(c, h(p))$. □

A useful observation is that a ball centered at a point on Σ with radius smaller than the local feature size satisfies the negation of the antecedent of the Feature Ball Lemma.

Lemma 12.7 (Small Ball Lemma). *For all $\varepsilon \in (0, 1)$ and every point $p \in \Sigma$, $B(p, \varepsilon f(p)) \cap \Sigma$ is a topological k-ball and $(\text{Bd}\, B(p, \varepsilon f(p))) \cap \Sigma$ is a topological $(k-1)$-sphere.*

Proof. If it were not true, by the Feature Ball Lemma, $B(p, \varepsilon f(p))$ would contain a point m on the medial axis. But the distance from p to the medial axis is $f(p) > \varepsilon f(p)$, a contradiction. □

12.6 The variation in normal vectors on smooth surfaces

For the rest of this chapter, we use the shorthand term *smooth surface* to refer to a compact C^2-smooth surface (2-manifold) without boundary, embedded in $\mathbb{R}^3$; and Σ is always such a smooth surface. Every point $p \in \Sigma$ has a well-defined tangent plane and a unit normal vector $\mathbf{n}_p$ that is orthogonal to the tangent plane. The surface Σ cuts $\mathbb{R}^3$ into at least two faces: an outer, unbounded face and one or more mutually disjoint, bounded faces. Our normals $\mathbf{n}_p$ are always oriented into an inner, bounded face.

An important theorem is that the normals at points on Σ cannot differ too much if they are close to each other relative to the local feature size.

Theorem 12.8 (Normal Variation Theorem). *Let q and q' be points on a smooth surface Σ. If there is an $\varepsilon < 1$ such that $d(q, q') \leq \varepsilon f(q)$, then measured in radians, $\angle(\mathbf{n}_q, \mathbf{n}_{q'}) \leq -\ln(1 - \varepsilon) < \varepsilon/(1 - \varepsilon)$.*

We introduce some notation and some differential geometry to assist the proof. Two points q and q' on a surface Σ are separated by both the usual Euclidean distance $d(q, q')$

and a *geodesic distance* $d_\Sigma(q, q')$, defined to be the length of the shortest curve on Σ joining q to q'. The curves of length $d_\Sigma(q, q')$ joining q to q' are called the *minimizing geodesics* between q and q'. One property of a minimizing geodesic curve on Σ is that the curvature of the geodesic at a point p cannot be larger than the largest principal curvature of Σ at p. The following two facts are well known in differential geometry.

Proposition 12.9. *For every point q on a smooth surface Σ,*

$$\lim_{q' \to q,\, q' \in \Sigma} \frac{d_\Sigma(q, q')}{d(q, q')} = 1.$$

Proposition 12.10. *Let q and q' be two points on a smooth surface Σ. Let κ be the maximum principal curvature of Σ among all points on a minimizing geodesic connecting q and q'. The angle between their normals, measured in radians, is $\angle(\mathbf{n}_q, \mathbf{n}_{q'}) \leq \kappa \cdot d_\Sigma(q, q')$.*

A weaker version of the Normal Variation Theorem is obtained from Proposition 12.10 by integrating the curvature κ over a minimizing geodesic, with the observation that at any point $p \in \Sigma$, the curvature is at most $1/f(p)$. However, we obtain a stronger bound by integrating over a straight line segment from q to q' in $\mathbb{R}^3$. To make this possible, we extend the field of normal vectors beyond Σ.

For every point $p \in \mathbb{R}^3$ not on the medial axis M of Σ, let $\tilde{p}$ denote the unique point nearest p on Σ. A point $\tilde{p} \in \Sigma$ has a well-defined normal $\mathbf{n}_{\tilde{p}}$ to Σ; we extend the normals to every point in $p \in \mathbb{R}^3 \setminus M$ by defining $\mathbf{n}_p = \mathbf{n}_{\tilde{p}}$.

Consider the distance function

$$h : \mathbb{R}^3 \to \mathbb{R}, \;\; h(p) = d(p, \Sigma).$$

The shortest path from any point $p \notin M$ to Σ is a straight line from p to $\tilde{p}$ that meets Σ orthogonally, so at every point $p \in \mathbb{R}^3 \setminus \Sigma \setminus M$, the gradient $\nabla h(p)$ is a vector parallel to $\mathbf{n}_p$ with magnitude $(h(p) - h(\tilde{p}))/d(p, \tilde{p}) = 1$. Thus, $\nabla h(p) = \pm \mathbf{n}_p$.

Let $\Sigma_\omega = h^{-1}(\omega) = \{p : d(p, \Sigma) = \omega\}$ be an *offset surface* of Σ—that is, the set containing every point at a distance ω from Σ. The following proposition shows that most points lie in smooth neighborhoods on an offset surface.

Proposition 12.11. *Let Σ be a smooth surface with medial axis M. Let p be a point in $\mathbb{R}^3 \setminus \Sigma \setminus M$. Let $\omega = d(p, \Sigma)$ and observe that $p \in \Sigma_\omega$. There is an open set $U \subset \mathbb{R}^3$ containing p such that $\sigma = \Sigma_\omega \cap U$ is a C^2-smooth 2-manifold.*

PROOF. Because Σ is C^2-smooth and p does not lie on Σ or M, the distance function h is C^2-smooth at p. According to the implicit function theorem from differential geometry, there exists a neighborhood $U \subset \mathbb{R}^3$ of p such that $\sigma = h^{-1}(\omega) \cap U$ is a C^2-smooth 2-manifold. □

PROOF OF THEOREM 12.8. Observe that because $d(q, q') \leq f(q)$, the line segment qq' does not intersect the medial axis, so for every point $p \in qq'$, $\tilde{p}$ and $\mathbf{n}_p$ are defined. Consider a linear parametrization $s : [0, d(q, q')] \to qq'$ of the line segment qq' by length, where $s(0) = q$ and $s(d(q, q')) = q'$. Let $p = s(t)$ and $p' = s(t + \Delta t)$ be two arbitrarily close points on qq' for an arbitrarily small $\Delta t > 0$, as illustrated in Figure 12.12. Let $\theta(u) = \angle(\mathbf{n}_q, \mathbf{n}_{s(u)})$ and

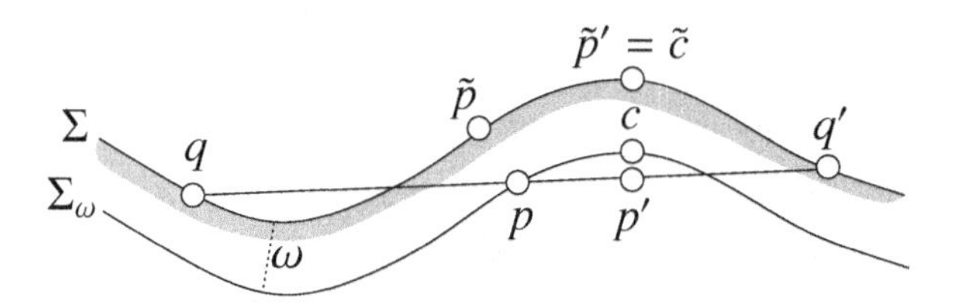

Figure 12.12: The Normal Variation Theorem.

$\Delta\alpha = \angle(\mathbf{n}_p, \mathbf{n}_{p'})$. By the triangle inequality, $\Delta\alpha \geq |\angle(\mathbf{n}_q, \mathbf{n}_{p'}) - \angle(\mathbf{n}_q, \mathbf{n}_p)| = |\theta(t + \Delta t) - \theta(t)|$ and

$$\left|\frac{d\theta(t)}{dt}\right| \leq \lim_{\Delta t \to 0} \frac{\Delta\alpha}{\Delta t}.$$

If we can show that $\lim_{\Delta t \to 0} \Delta\alpha/\Delta t$ is no more than $1/(f(q) - t)$, the result follows because

$$\begin{aligned}
\angle(\mathbf{n}_q, \mathbf{n}_{q'}) &\leq \int_{qq'} \left|\frac{d\theta(t)}{dt}\right| dt \\
&\leq \int_{qq'} \frac{1}{f(q) - t}\, dt \\
&= \ln \frac{f(q)}{f(q) - d(q, q')} \\
&\leq \ln \frac{1}{1 - \varepsilon} \\
&= -\ln(1 - \varepsilon).
\end{aligned}$$

The more easily understood bound of $\varepsilon/(1 - \varepsilon)$ holds by inspection of the Taylor series: $-\ln(1 - \varepsilon) = \varepsilon + \varepsilon^2/2 + \varepsilon^3/3 + \varepsilon^4/4 + \ldots < \varepsilon + \varepsilon^2 + \varepsilon^3 + \varepsilon^4 + \ldots = \varepsilon/(1 - \varepsilon)$.

Suppose without loss of generality that $d(p, \Sigma) \leq d(p', \Sigma)$; otherwise, swap p and p' in the following argument. Let $\omega = d(p, \Sigma)$ and observe that $p \in \Sigma_\omega$, as illustrated in Figure 12.12. By Proposition 12.11, there is a C^2-smooth neighborhood $\sigma \subset \Sigma_\omega$ of p.

Let c be the point nearest p' on Σ_ω, as illustrated, and suppose that Δt is small enough that $c \in \sigma$ and σ includes the minimizing geodesic connecting p to c in Σ_ω. Because $\omega \leq d(p', \Sigma)$, the unique shortest segment from p' to Σ begins with the shortest segment from p' to Σ_ω and, therefore, passes through c and includes the unique shortest segment from c' to Σ. It follows that $\tilde{p}' = \tilde{c}$, $\mathbf{n}_{p'} = \mathbf{n}_c$, and therefore, $\Delta\alpha = \angle(\mathbf{n}_p, \mathbf{n}_c)$.

For the sake of this proof, we extend the domain of the local feature size f to $\mathbb{R}^3$. For every $x \in \mathbb{R}^3$, let $f(x)$ be the distance from x to the medial axis M of Σ. Observe that f still has the 1-Lipschitz property; the proof of Proposition 12.1 extends to the larger domain without change.

Let m be the point on the minimizing geodesic between p and c in σ at which the largest principal curvature of σ is maximum, and let κ be the largest principal curvature of σ at m. By Proposition 12.5, $f(m) \leq 1/\kappa$. Recall that $d_\sigma(p, c)$ denotes the geodesic distance between p and c on σ. By Proposition 12.10,

$$\Delta\alpha \leq \kappa d_\sigma(p, c) \leq \frac{d_\sigma(p, c)}{f(m)}.$$

Claim 1. $\lim_{\Delta t \to 0} d(p, c)/\Delta t \leq 1$.

PROOF. Consider the triangle pcp'. The plane tangent to σ at c is orthogonal to cp', so if it separates p from p', the angle $\angle pcp'$ is obtuse, and $d(p, c) \leq d(p, p') = \Delta t$. Otherwise, let x be the orthogonal projection of p onto aff $p'c$, and observe that $d(p, c) \cos \angle cpx = d(p, x) \leq d(p, p') = \Delta t$. In either case, $d(p, c)/\Delta t \leq 1/\cos \angle cpx$. Because $\angle cpx$ approaches 0 as Δt approaches 0, we have $\lim_{\Delta t \to 0} d(p, c)/\Delta t \leq 1$. □

As Δt approaches zero, $d_\sigma(p, c)$ approaches $d(p, c)$ by Proposition 12.9, which approaches Δt or less by Claim 1. Therefore,

$$\lim_{\Delta t \to 0} \frac{\Delta \alpha}{\Delta t} \leq \lim_{\Delta t \to 0} \frac{d_\sigma(p, c)}{\Delta t\, f(m)} \leq \lim_{\Delta t \to 0} \frac{1}{f(m)}.$$

Because f is 1-Lipschitz, we have $f(m) \geq f(q) - d(q, m)$. As $d(q, m) \leq d(p, q) + d(p, m) \leq d(p, q) + d(p, c)$, Claim 1 implies that

$$\lim_{\Delta t \to 0} f(m) \geq \lim_{\Delta t \to 0} (f(q) - d(p, q) - d(p, c)) = f(q) - t - \lim_{\Delta t \to 0} \Delta t = f(q) - t.$$

It follows that

$$\lim_{\Delta t \to 0} \frac{\Delta \alpha}{\Delta t} \leq \lim_{\Delta t \to 0} \frac{1}{f(m)} \leq \frac{1}{f(q) - t},$$

which is what we needed to prove. □

12.7 Approximations of tangents by simplices

In this section, we prove that if an edge or a triangle whose vertices are on Σ has a small diametric ball, it lies nearly parallel to the surface. This implies that the vector normal to a triangle with a small diametric ball is a good approximation of the local normal vectors of the surface. Later in the book, we use these properties to prove topological properties of the underlying space of a mesh. The diametric ball of an edge is small if and only if the edge is short. But for a triangle, short edges do not guarantee a small diametric ball; recall that triangles with angles near 180° have large circumcircles. Such triangles approximate normals of surfaces poorly. We call edges and triangles *small* only if they have diametric balls that are small relative to the local feature size, regardless of their edge lengths.

Recall from Section 1.7 that for any two vectors, lines, or segments $\mathbf{u}$ and $\mathbf{v}$, $\angle_a(\mathbf{u}, \mathbf{v})$ denotes the acute angle between them, ignoring the directions of vectors.

12.7.1 Short edges are almost parallel to the surface

The Normal Variation Theorem (Theorem 12.8) shows that on a small enough scale, a smooth surface looks nearly flat. It implies that an edge connecting two nearby points on a surface is almost tangent to the surface and almost perpendicular to the normals at its vertices.

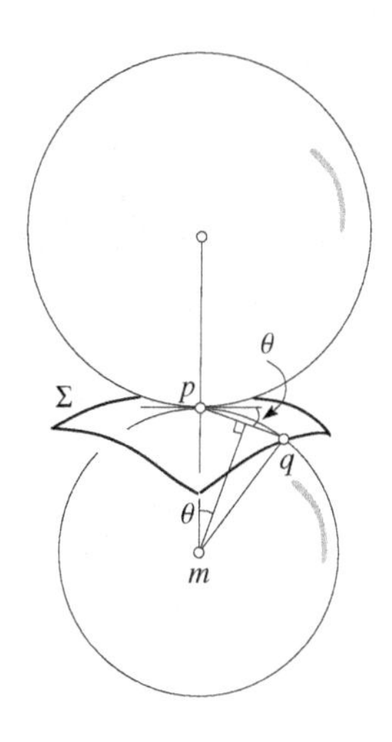

Figure 12.13: The Edge Normal Lemma.

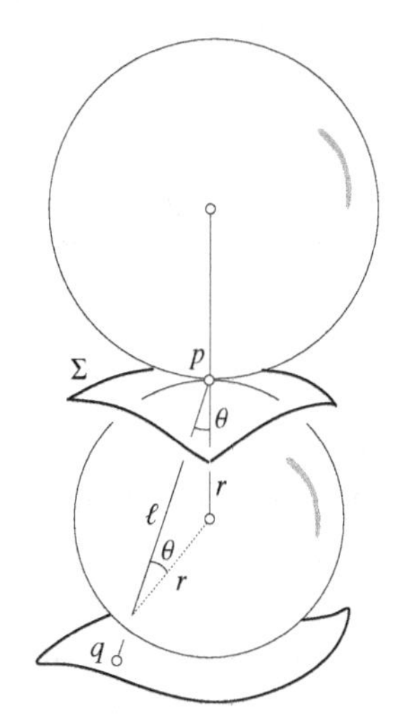

Figure 12.14: The Long Distance Lemma.

Lemma 12.12 (Edge Normal Lemma). *Let p and q be any two points on a smooth surface Σ. If $d(p,q) \leq 2f(p)$, then measured in radians, $\angle_a(pq, \mathbf{n}_p) \geq \pi/2 - \arcsin \frac{d(p,q)}{2f(p)}$.*

Proof. Consider the two medial balls at p sandwiching the surface Σ, and let m be the center of the smaller one. (One of the balls is a halfspace if p is on the boundary of conv Σ.) The point q is not inside either ball, as their interiors are disjoint from Σ. The diameters of both balls are at least $2f(p) \geq d(p,q)$. Let θ denote the angle between pq and the tangent plane at p. The acute angle between pq and $\mathbf{n}_p$ is smallest when q lies on the boundary of the smaller medial ball, as Figure 12.13 depicts, in which case

$$\sin\theta = \frac{d(p,q)}{2\,d(m,p)} \leq \frac{d(p,q)}{2f(p)}.$$

Therefore,

$$\angle_a(pq, \mathbf{n}_p) = \frac{\pi}{2} - \theta \geq \frac{\pi}{2} - \arcsin\frac{d(p,q)}{2f(p)}.$$

□

It is convenient to have a contrapositive version of the Edge Normal Lemma, stating that if a line segment connecting two points p and q in Σ makes a small angle with $\mathbf{n}_p$, then p and q are far apart with respect to the LFS.

Lemma 12.13 (Long Distance Lemma). *Let ℓ be a line intersecting Σ in two points p and q. Then $d(p,q) \geq 2f(p)\cos\angle_a(\ell, \mathbf{n}_p)$.*

Proof. Consider the two medial balls at p, shown in Figure 12.14. Let r be the radius of the smaller medial ball, and let $\theta = \cos\angle_a(\ell, \mathbf{n}_p)$. The intersection of ℓ and either ball is at least $2r\cos\theta$ long. As $r \geq f(p)$ and q is not inside a medial ball, we have $d(p,q) \geq 2r\cos\theta \geq 2f(p)\cos\theta$. □

12.7.2 Triangles with small circumradii are almost parallel

We have just seen that edges whose vertices are close together on a surface are almost parallel to the tangent planes at their vertices. The same is not necessarily true of triangles. Recall from Section 1.1.2 that triangles with angles arbitrarily close to 180° can cause arbitrarily large errors in the gradient of an interpolated piecewise linear function. For the same reasons, triangles with large angles can have wildly inaccurate normals and lie almost perpendicular to a surface. To guarantee that an interpolating triangle is nearly parallel to the surface and has accurate normals, it is not enough that it has short edges; but it suffices that it has a small circumradius relative to the local feature sizes at its vertices. The accuracy of an interpolating triangle depends on both its size and shape; the circumradius of the triangle indicates its accuracy, encapsulating both its size and shape and showing how they are balanced. The following lemma bounds the accuracy of a triangle's normal as an approximation of the surface normal at the vertex of the triangle opposite the triangle's longest edge.

Lemma 12.14 (Triangle Normal Lemma). *Let p, q, and r be three points on Σ, and suppose the largest angle of the flat triangle pqr is at p. Let $\mathbf{n}_{pqr}$ be the vector normal to the triangle pqr, and let R_{pqr} be the circumradius of pqr. If $R_{pqr} \leq f(p)/2$, then measured in radians,*

$$\angle_a(\mathbf{n}_{pqr}, \mathbf{n}_p) \leq \arcsin \frac{R_{pqr}}{f(p)} + \arcsin\left(\frac{2}{\sqrt{3}} \sin\left(2 \arcsin \frac{R_{pqr}}{f(p)}\right)\right).$$

Proof. Let $B = B(m, a)$ and $B' = B(m', a')$ be the medial balls at p. Let D be the diametric ball of pqr (its smallest circumscribing 3-ball), whose radius is R_{pqr}. Let C and C' be the circles in which the boundary of D intersects the boundaries of B and B', respectively, as illustrated in Figure 12.15(a). As Σ is sandwiched between B and B', the line through their centers m and m' is normal to Σ at p. Let α be the larger of the two angles this normal line makes with the normals to the planes $H = \operatorname{aff} C$ and $H' = \operatorname{aff} C'$. Because the radii of C and C' are at most R_{pqr},

$$\alpha \leq \arcsin \frac{R_{pqr}}{\min\{d(p, m), d(p, m')\}} \leq \arcsin \frac{R_{pqr}}{f(p)} \leq \arcsin \frac{1}{2} = \frac{\pi}{6}.$$

The planes H and H' meet at an acute dihedral angle no more than $2\alpha \leq \pi/3$. Let the wedge W be the intersection of the halfspace bounded by H that includes C' and the halfspace bounded by H' that includes C. Both q and r lie on the boundary of D, and $pqr \subset W$.

Let $H_{pqr} = \operatorname{aff} pqr$, and observe that H, H', and H_{pqr} all meet at p. To examine the dihedral angles at which these three planes intersect, we take their intersections with a unit sphere S centered at p. Each intersection of S with two of the three planes is two antipodal points, yielding six points total, but only four of them are in the wedge W: let u and u' be the two points in $S \cap H \cap H'$, let v be the point $S \cap H_{pqr} \cap H \cap W$, and let w be the point $S \cap H_{pqr} \cap H' \cap W$.

Figure 12.15(b) depicts two great arcs, which are $S \cap H \cap W$ and $S \cap H' \cap W$. Let θ_{uvw}, θ_{uwv}, and θ_{vuw} denote the spherical angles $\angle uvw$, $\angle uwv$, and $\angle vuw$, respectively, which are the dihedral angles at which H_{pqr} meets H, H_{pqr} meets H', and H meets H', respectively.

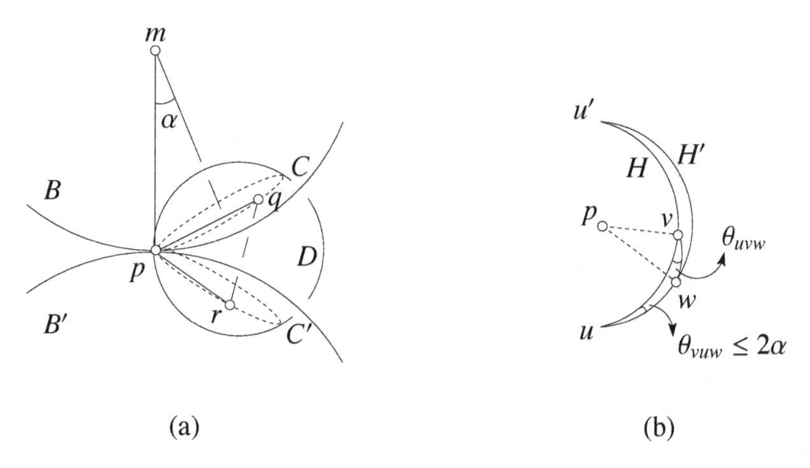

Figure 12.15: The Triangle Normal Lemma.

We have seen that $\theta_{vuw} \leq 2\alpha$. Without loss of generality, assume that $\theta_{uvw} \leq \theta_{uwv}$. Let $\mathbf{n}_H$ be the normal to H. Then

$$\angle_a(\mathbf{n}_p, \mathbf{n}_{pqr}) \leq \angle_a(\mathbf{n}_p, \mathbf{n}_H) + \angle_a(\mathbf{n}_H, \mathbf{n}_{pqr}) \leq \alpha + \theta_{uvw} \leq \arcsin \frac{R_{pqr}}{f(p)} + \theta_{uvw}.$$

It remains to show an upper bound on θ_{uvw}.

The 2-dimensional wedge $W \cap H_{pqr}$ is bounded by the rays $\vec{pv}$ and $\vec{pw}$ and includes pqr, so $\angle vpw \geq \angle qpr$, which is at least $\pi/3$ because it is the largest angle of pqr. Therefore, the length of the arc vw on the surface of S, denoted $|vw|$, is at least $\pi/3$. By the sine law of spherical geometry,

$$\begin{aligned} \sin \theta_{uvw} &= \sin |uw| \cdot \frac{\sin \theta_{vuw}}{\sin |wv|} \\ &\leq \sin |uw| \cdot \frac{\sin 2\alpha}{\sin |wv|}. \end{aligned} \tag{12.1}$$

If $\pi/3 \leq |wv| \leq \pi/2$, then $\sin |wv| \geq \sqrt{3}/2$ and

$$\sin \theta_{uvw} \leq \frac{2}{\sqrt{3}} \sin |uw| \cdot \sin 2\alpha \leq \frac{2}{\sqrt{3}} \sin 2\alpha$$

and hence

$$\theta_{uvw} \leq \arcsin \left(\frac{2}{\sqrt{3}} \sin \left(2 \arcsin \frac{R_{pqr}}{f(p)} \right) \right)$$

and the result follows.

Suppose instead that $\pi/2 < |wv| \leq \pi$. Observe that $\theta_{vu'w} = \theta_{vuw} \leq \pi/3$. The spherical angles of a spherical triangle always add to more than π, so $\theta_{vu'w}$ is not the largest angle of the triangle $vu'w$, and hence, vw is not its longest arc. If $|wv| \leq |wu'|$, then

$$|uw| + |wv| \leq |uw| + |wu'| = |uu'| = \pi.$$

Otherwise, $|wv| \leq |vu'|$. By assumption, $\theta_{uvw} \leq \theta_{uwv}$, so $|uw| \leq |uv|$ and again

$$|uw| + |wv| \leq |uv| + |vu'| = |uu'| = \pi.$$

In either case, because $|wv| > \pi/2$,

$$\frac{\sin |uw|}{\sin |wv|} \leq \frac{\sin(\pi - |wv|)}{\sin |wv|} = 1.$$

Substituting this ratio into (12.1) yields $\theta_{uvw} \leq 2\alpha$, and again the result follows. □

12.8 Notes and exercises

Hocking and Young [113] and Munkres [155] are good sources for point set topology with topological spaces. A recent book by Hatcher [110] provides intuitive explanations of maps and homotopies. Definitions of smooth surfaces appear in many introductory books on differential geometry, such as the one by do Carmo [82].

Medial axes were introduced by Blum [25] for image analysis. Important papers on medial axes include those by Matheron [143], Wolter [225], and Chazal and Lieutier [43]. The local feature size function in terms of the medial axis was introduced by Amenta, Bern, and Eppstein [5] to reconstruct smooth curves from point clouds, and was adapted for reconstruction of smooth surfaces in three dimensions by Amenta and Bern [3]. Definitions are rarely respected as much as theorems, but the most important observation in the theory of surface sampling is the fact that ε-samples of a surface, defined in terms of distance to the medial axis, determine the necessary sampling density. The Feature Ball Lemma (Lemma 12.6) proves to be a crucial ingredient for analyzing properties of smooth surfaces in the context of sampling. Amenta et al. [5] introduced a version of this lemma for curves in the plane, and Boissonnat and Cazals [26] give a version for surfaces. The proof given here, which treats smooth manifolds in spaces of arbitrary dimension, is taken from Dey's book [73].

The Normal Variation Theorem (Theorem 12.8) is an essential ingredient for proving approximation results for finite samples of a smooth surface. A version of this theorem was introduced by Amenta and Bern [3], but the proof was not completely correct. Amenta and Dey [8] corrected and improved it; we adopt their proof here.

Smooth curve discretization and smooth surface discretization differ in a fundamental way. A sequence of meshes of a smooth curve, each with shorter edges than the previous mesh, necessarily converges to the curve as the length of the longest edge approaches zero, and the edges converge to the tangents. However, Schwarz [189] shows that in a sequence of triangulations of a surface, each with shorter edges than the previous mesh, the tangent planes and normal vectors of the mesh may or may not converge to those of the surface. If the triangles have bounded aspect ratios, Schwarz shows that the sequence of triangulations does converge. Amenta, Choi, Dey, and Leekha [6] prove the Triangle Normal Lemma (Lemma 12.14), which shows that the requirement for bounded aspect ratios can be relaxed to the requirement that the circumradii of the triangles be small compared to the local feature sizes.

Exercises

1. The following definition of *continuous function* appears in many topology textbooks. Prove that it is equivalent to Definition 12.3—in other words, that any func-

tion deemed continuous by one definition is continuous by the other definition too. A function $g : \mathbb{T} \to \mathbb{U}$ is *continuous* if for every open set $Q \subseteq \mathbb{U}$, the set $g^{-1}(Q) = \{p : g(p) \in Q\}$ is open in $\mathbb{T}$.

2. Let $\mathbb{T}$ be a topological space. Prove that the complement $\mathbb{T} \setminus Q$ of every open set $Q \subseteq \mathbb{T}$ is closed, and that $\text{Cl}\, Q$ is the smallest closed set containing Q.

3. Show that the homotopy of maps is an equivalence relation.

4. Prove that any two surjective maps of the same space to $\mathbb{R}^k$ are homotopic.

5. Prove that every homeomorphism is a homotopy equivalence but the converse is not true.

6. Prove that homotopy equivalent spaces have the same number of connected components. Prove the same for homeomorphic spaces.

7. Show examples of the following.

   (a) A manifold without boundary in $\mathbb{R}^3$ that does not cut $\mathbb{R}^3$ into two pieces.

   (b) A manifold in $\mathbb{R}^3$ that is not compact but has a nonempty boundary.

   (c) A non-orientable manifold in $\mathbb{R}^3$ whose boundary is empty.

8. Show that any two homeomorphic curves in $\mathbb{R}^2$ are related by an isotopy, but this statement does not hold for homeomorphic curves in $\mathbb{R}^3$.

9. Show that a smooth manifold Σ is the boundary of the union of all its open medial balls, unless Σ is a flat.

10. Show that the functions ρ_i and ρ_o defined in Section 12.5.2 are continuous but not necessarily k-Lipschitz for any fixed k.

11. Show that the converse of the Feature Ball Lemma is not necessarily true. Specifically, construct an example where a Euclidean 3-ball B intersects a surface Σ in a topological disk, $\text{Bd}\, B$ intersects Σ in a loop (a connected 1-manifold without boundary), and B intersects the medial axis of Σ.

12. [42] For a smooth compact surface Σ, define the distance function $h : \mathbb{R}^3 \to \mathbb{R}$, $h(x) = d(x, \Sigma)$. We adopt the notion of a *critical value* that we use for the Feature Ball Lemma (Lemma 12.6), but we define it differently because h is not smooth on the medial axis of Σ. A value $a \in \mathbb{R}$ is a critical value if for every $\delta > 0$, there is an $\varepsilon \leq \delta$ such that the level set $h^{-1}(a - \varepsilon)$ has a different topology from the level set $h^{-1}(a + \varepsilon)$. Define the weak feature size $\text{wfs}(\Sigma)$ of Σ to be the smallest critical value of h. Prove that the consequence of the Small Ball Lemma (Lemma 12.7) holds for every Euclidean 3-ball with radius smaller than $\text{wfs}(\Sigma)$.

13. The Normal Variation Theorem (Theorem 12.8) can be strengthened if the local feature sizes at both points are given. Specifically, let q and q' be points on a smooth surface Σ. If $d(q, q') < f(q) + f(q')$, prove that $\angle(\mathbf{n}_q, \mathbf{n}_{q'}) \leq \ln(4\, f(q)\, f(q')/a^2)$, where

$a = f(q) + f(q') - d(q, q')$. Hint: split the line segment qq' into two pieces at the optimal point and integrate over each of them separately.

14. Suppose the function f in the Normal Variation Theorem is k-Lipschitz for some $k \geq 1$. Derive a normal variation bound under this assumption.

15. Explain how to construct a triangle and a smooth surface such that the triangle's vertices lie on the surface, the triangle's edges are arbitrarily short, and the triangle's normal is almost perpendicular to the surface normals at its vertices. If you make the triangle's edge lengths approach zero without changing the surface, how do you need to modify the triangle to maintain its awful normals as it shrinks?

16. The Normal Variation Theorem gives an upper bound on the angular difference in normals of $-\ln(1 - \varepsilon) = \varepsilon + \varepsilon^2/2 + \varepsilon^3/3 + \varepsilon^4/4 + \ldots$. We conjecture that the true bound is $\varepsilon + O(\varepsilon^3)$. Prove or disprove this conjecture.

17. Improve the bound stated in the Triangle Normal Lemma (Lemma 12.14). Any significant improvement in this bound improves various results derived in the next two chapters.

Chapter 13

Restricted Delaunay triangulations of surface samples

The restricted Delaunay triangulation is a subcomplex of the three-dimensional Delaunay triangulation that has proven itself as a mathematically powerful tool for surface meshing and surface reconstruction. Consider a smooth surface $\Sigma \subset \mathbb{R}^3$ and a finite sample S on Σ. If S is dense enough, as dictated by the local feature size over Σ, then the restricted Delaunay triangulation is a triangulation of Σ by Definition 12.6—it has an underlying space that is guaranteed to be topologically equivalent to Σ. Moreover, the triangulation lies close to Σ and approximates it geometrically. The theory of restricted Delaunay triangulations of surface samples lays the foundation for the design and analysis of Delaunay refinement mesh generation algorithms in the subsequent chapters, which use incremental vertex insertion not only to guarantee good element quality, but also to guarantee that the mesh is topologically correct and geometrically close to the surface, with accurate surface normals.

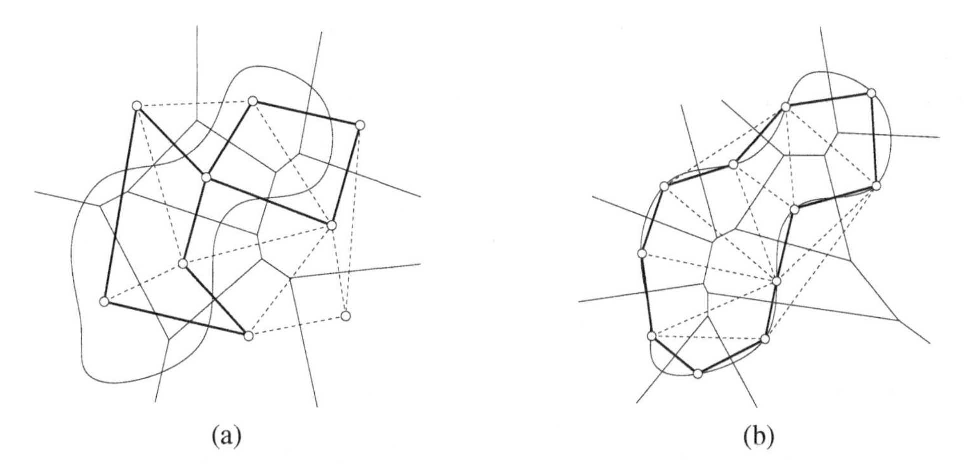

Figure 13.1: Restricted Delaunay triangulations, indicated by bold edges. These edges are included because the curve intersects their Voronoi dual edges. Dashed edges are Delaunay but not restricted Delaunay. The restricted Delaunay triangulation at right is a passable reconstruction of the curve.

13.1 Restricted Voronoi diagrams and Delaunay triangulations

Restricted Delaunay triangulations are defined in terms of their duals, the restricted Voronoi diagrams; we must understand the latter to understand the former. The *restriction* $Q|_{\mathbb{T}}$ of a point set $Q \subset \mathbb{R}^d$ to a topological subspace $\mathbb{T} \subset \mathbb{R}^d$ is simply $Q \cap \mathbb{T}$. The *restriction* $\mathcal{C}|_{\mathbb{T}}$ of a complex $\mathcal{C}$ to a topological subspace $\mathbb{T}$ is $\{g \cap \mathbb{T} : g \in \mathcal{C}\}$, the complex found by taking the restriction of each cell in $\mathcal{C}$. The restricted Voronoi diagram is defined accordingly.

Definition 13.1 (restricted Voronoi diagram). Let $S \subseteq \mathbb{R}^d$ be a finite set of sites, and let $\mathbb{T} \subseteq \mathbb{R}^d$ be a topological subspace of a Euclidean space. The *restricted Voronoi cell* $V_u|_{\mathbb{T}}$ of a site $u \in S$ is $V_u \cap \mathbb{T} = \{p \in \mathbb{T} : \forall w \in S, d(u, p) \leq d(w, p)\}$, the restriction of u's Euclidean Voronoi cell to $\mathbb{T}$. A *restricted Voronoi face* $V_{u_1 \ldots u_j}|_{\mathbb{T}}$ is a nonempty restriction of a Voronoi face $V_{u_1 \ldots u_j} \in \operatorname{Vor} S$ to $\mathbb{T}$—that is, $V_{u_1 \ldots u_j}|_{\mathbb{T}} = V_{u_1 \ldots u_j} \cap \mathbb{T}$. The *restricted Voronoi diagram* of S with respect to $\mathbb{T}$, denoted $\operatorname{Vor}|_{\mathbb{T}} S$, is the cell complex containing every restricted Voronoi face.

Restricted Delaunay triangulations are defined *not* by restricting Delaunay simplices to a topological subspace, but by dualizing the restricted Voronoi diagram.

Definition 13.2 (restricted Delaunay). A simplex is *restricted Delaunay* if its vertices are in S and together they generate a nonempty restricted Voronoi face. In other words, $\operatorname{conv}\{u_1, \ldots, u_j\}$ is restricted Delaunay if $V_{u_1 \ldots u_j}|_{\mathbb{T}}$ is nonempty. If S is generic, the *restricted Delaunay triangulation* of S with respect to $\mathbb{T}$, denoted $\operatorname{Del}|_{\mathbb{T}} S$, is the simplicial complex containing every restricted Delaunay simplex. If S is not generic, a restricted Delaunay triangulation of S is a simplicial complex containing every restricted Delaunay simplex in some particular $\operatorname{Del} S$.

Figure 13.1(a) shows a restricted Delaunay triangulation with respect to a loop in the plane. Typically, no Voronoi vertex lies on the curve, so there are no restricted Delaunay triangles. Figure 13.1(b) shows a different restricted Delaunay triangulation with respect to the same loop. The vertices in the latter example are a fairly good sample of the curve, and the restricted triangulation is also a loop. In general, our goal is to sample a space so that $|\operatorname{Del}|_{\mathbb{T}} S|$ is homeomorphic to $\mathbb{T}$, as it is here. More powerfully, there is a homeomorphism that maps every sample point to itself.

It is not obvious that the set of all restricted Delaunay simplices is really a complex. Observe that if $\operatorname{Del}|_{\mathbb{T}} S$ contains a simplex σ, it contains every face of σ because every subset of σ's vertices generates a Voronoi face that includes the face dual to σ and, therefore, intersects Σ. If S is generic, then $\operatorname{Del}|_{\mathbb{T}} S \subseteq \operatorname{Del} S$, because every face of $\operatorname{Del}|_{\mathbb{T}} S$ dualizes to a face of $\operatorname{Vor}|_{\mathbb{T}} S$, which is induced by restricting a face of $\operatorname{Vor} S$, which dualizes to a face of $\operatorname{Del} S$. Therefore, $\operatorname{Del}|_{\mathbb{T}} S$ is a simplicial complex, and a standard Delaunay triangulation algorithm is a useful first step in constructing it.

If S has multiple Delaunay triangulations, the set of all restricted Delaunay simplices might not form a complex, so we compute a particular $\operatorname{Del} S$ and choose its restricted Delaunay simplices to form $\operatorname{Del}|_{\mathbb{T}} S$. Alternatively, we could use a subcomplex of the Delaunay subdivision.

In this chapter, the topological space $\mathbb{T}$ is usually a smooth surface Σ embedded in three dimensions. (In later chapters $\mathbb{T}$ will sometimes be a volume or a piecewise smooth complex.) In most applications of restricted Delaunay triangulations, the site set S is a sample of Σ, but our definitions do not require it; a set of sites $S \not\subset \Sigma$ is sometimes useful to model noisy point samples or to remesh a polyhedral surface that approximates a smooth surface. In this book's applications, however, we will usually generate S on Σ.

We say that a Voronoi k-face intersects Σ *generically* or *transversally* if at each point of the intersection, the plane tangent to Σ does not include the affine hull of the Voronoi face. Such an intersection is a $(k-1)$-manifold with or without boundary. For example, a Voronoi edge may intersect Σ at one or more distinct points, and a Voronoi polygon may intersect Σ in one or more curves or loops. By contrast, intersections can be *degenerate* or *non-transverse*, such as a Voronoi polygon whose intersection with Σ is a single point or a 2-ball or a figure-8 curve, or a Voronoi vertex that intersects Σ. If a k-flat Π has a non-transverse intersection with a C^2-smooth surface Σ, there is a point in $\Pi \cap \Sigma$ at which Π is tangent to Σ. Non-transverse intersections can be eliminated by perturbing the surface; for example, if a Voronoi vertex lies on Σ, simply pretend it lies infinitesimally inside the surface.

Even though $\text{Del}|_\Sigma S$ is always a simplicial complex, it is not guaranteed to be a triangulation of Σ, or coherent in any way, unless we impose strong constraints on S. Ideally, each restricted Voronoi cell $V_p|_\Sigma$ would be a simple region homeomorphic to a disk. But if the sample S is not dense enough, a Voronoi cell V_p can reach through space to touch other portions of the surface, so $V_p|_\Sigma$ can have multiple connected components. Odder problems can occur; for example, imagine a surface in the shape of a sausage with just three sample points, one in the middle and one at each end. The restricted Voronoi cell of the sample point in the middle is topologically equivalent to an annulus instead of a disk.

This chapter studies how the topologies of $\text{Vor}|_\Sigma S$ and $\text{Del}|_\Sigma S$ are determined by the ways in which the faces of $\text{Vor}\, S$ intersect Σ, and how a dense sample S tames those intersections. Specifically, we require S to be a 0.08-sample of Σ (recall Definition 12.16)—though we hope the constant can be improved, and we observe that the algorithms are more forgiving in practice.

We begin with a well-known theorem in computational topology, the Topological Ball Theorem, which states that if every face of $\text{Vor}\, S$ intersects Σ nicely enough, then the underlying space of $\text{Del}|_\Sigma S$ is homeomorphic to Σ. Thus by Definition 12.6, $\text{Del}|_\Sigma S$ is a triangulation of Σ. This establishes the goal of most of the rest of the chapter: to establish how finely we must sample Σ to guarantee that the antecedents of the theorem hold.

Next, we prove a local result about restricted Voronoi vertices, which are the points where Voronoi edges pass through the surface. If every vertex of a restricted Voronoi cell $V_p|_\Sigma$ is close to its generating site $p \in S$, and every connected component of $V_p|_\Sigma$ has a vertex, then $V_p|_\Sigma$ and its restricted Voronoi faces are topological closed balls of appropriate dimensions. Specifically, if g is a k-face of V_p, the restricted Voronoi face $g|_\Sigma$ is a topological $(k-1)$-ball. A Voronoi edge intersects Σ at a single point; a Voronoi polygon intersects Σ in a single curve that is not a loop; and a Voronoi 2-cell intersects Σ in a topological disk.

The local result says nothing about a restricted Voronoi face that does not adjoin a restricted Voronoi vertex—for example, a circular face where a Voronoi polygon cuts a fingertip off Σ. We extend the local result to a global result. If S is dense enough—for example, if it is a 0.08-sample of Σ—and at least one restricted Voronoi vertex exists, then every Voronoi k-face either intersects Σ in a topological $(k-1)$-ball or does not intersect Σ

at all.

Having established that every face of Vor S intersects Σ nicely, we apply the Topological Ball Theorem to establish the homeomorphism of $|\text{Del}|_\Sigma S|$ and Σ. We strengthen this result with another correspondence that is both topological and geometric. We construct an explicit isotopy relating $|\text{Del}|_\Sigma S|$ and Σ, and show that each point in $|\text{Del}|_\Sigma S|$ is displaced only slightly by this isotopy. Hence, $\text{Del}|_\Sigma S$ is a geometrically good triangulation of Σ.

Finally, we prove additional results about the geometric quality of $\text{Del}|_\Sigma S$: a bound on the circumradii of its triangles, a small deviation between the normal of a triangle and the normals to the surface at its vertices, and a large lower bound (near π) on the dihedral angles between adjoining triangles. These properties of $\text{Del}|_\Sigma S$ are summarized in the Surface Discretization Theorem (Theorem 13.22) at the end of the chapter.

For the rest of this chapter, Σ is a *smooth surface*, our shorthand for a compact C^2-smooth 2-manifold without boundary, embedded in $\mathbb{R}^3$. Moreover, we assume that Σ is connected. A unit normal vector, written $\mathbf{n}_p$, is normal to Σ at a point p. All sets of restricted Delaunay vertices are samples $S \subset \Sigma$. Because we are exclusively concerned with three-dimensional space, *facet* will always mean 2-face, and a Voronoi facet is a Voronoi polygon. We use the following notation for many results and their proofs.

$$\begin{aligned} \alpha(\varepsilon) &= \frac{\varepsilon}{1-\varepsilon}, \\ \beta(\varepsilon) &= \alpha(2\varepsilon) + \arcsin\frac{\varepsilon}{1-2\varepsilon} + \arcsin\left(\frac{2}{\sqrt{3}}\sin\left(2\arcsin\frac{\varepsilon}{1-2\varepsilon}\right)\right). \end{aligned}$$

The first expression arises from both the Feature Translation Lemma (Lemma 12.2) and the Normal Variation Theorem (Theorem 12.8), and the second in part from the Triangle Normal Lemma (Lemma 12.14). We use them partly for brevity and partly in the hope they can be improved. Several proofs in this chapter are obtained by contradicting the following two conditions.

$$\begin{aligned} \text{Condition A.1}: \quad \alpha(\varepsilon) &< \cos(\alpha(\varepsilon) + \beta(\varepsilon)) \\ \text{Condition A.2}: \quad \alpha(\varepsilon) &< \cos(\alpha(\varepsilon) + \arcsin\varepsilon + 2\beta(\varepsilon)) \end{aligned}$$

Condition A.1 is satisfied for $\varepsilon \leq 0.15$, and Condition A.2 is satisfied for $\varepsilon \leq 0.09$.

13.2 The Topological Ball Theorem

The goal of the next several sections is to establish that if a sample of a smooth surface Σ has the right properties, its restricted Delaunay triangulation is a triangulation of the surface. Our key tool for achieving this is an important result from computational topology called the Topological Ball Theorem, which states that the underlying space of $\text{Del}|_\Sigma S$ is homeomorphic to Σ if every Voronoi k-face that intersects Σ does so transversally in a topological ball of dimension $k-1$. To state the theorem formally, we introduce the topological ball property.

Definition 13.3 (topological ball property). Let $g \in \text{Vor}\, S$ be a Voronoi face of dimension k, $0 \leq k \leq 3$. Let $g|_\Sigma$ be the restricted Voronoi face $g \cap \Sigma$. The face g satisfies the *topological ball property* (TBP) if $g|_\Sigma$ is empty or

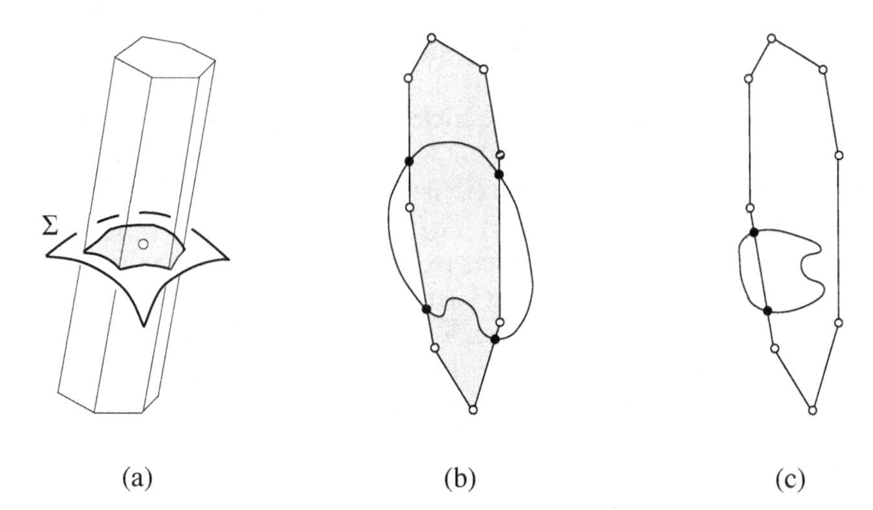

Figure 13.2: (a) Good: the intersection between Σ and each Voronoi cell or face is a topological ball of appropriate dimension. (b) Bad: the intersection between Σ and a Voronoi facet is not a single topological 1-ball. (c) Bad: the intersection between Σ and a Voronoi edge is not a single point.

(i) $g|_\Sigma$ is a topological $(k-1)$-ball (homeomorphic to $\mathbb{B}^{k-1}$), and

(ii) $(\operatorname{Int} g) \cap \Sigma = \operatorname{Int}(g|_\Sigma)$. Here we use the definition of "interior" for manifolds, Definition 12.12, not Definition 1.13.

The pair (S, Σ) satisfies the *TBP* if every Voronoi face $g \in \operatorname{Vor} S$ satisfies the *TBP*.

Figure 13.2 illustrates condition (i), which means that Σ intersects a Voronoi cell in a single topological disk, a Voronoi facet in a single topological 1-ball (a curve with two endpoints), a Voronoi edge in a single point, and a Voronoi vertex not at all. Condition (ii) rules out Σ intersecting the boundary of a Voronoi face without intersecting its interior infinitesimally close by. If all the intersections between Σ and Voronoi faces are transverse, condition (ii) is automatically satisfied. To prove that the TBP holds when S is sufficiently dense, we will explicitly prove that the Voronoi edges and Voronoi facets intersect Σ transversally. The fact that the Voronoi vertices do (i.e. no Voronoi vertex lies on Σ) follows because if a Voronoi vertex lay on Σ, then not all the Voronoi facets adjoining it could intersect Σ transversally in an interval. Trivially, all Voronoi 3-cells intersect Σ transversally.

Theorem 13.1 (Topological Ball Theorem). *If the pair (S, Σ) satisfies the topological ball property, the underlying space of* $\operatorname{Del}|_\Sigma S$ *is homeomorphic to* Σ.

At first, this guarantee might not seem very impressive; after all, an elephant-shaped mesh of an airplane impresses nobody, even if the boundaries of both are topological spheres. The point, however, is that a subcomplex of $\operatorname{Del} S$ need not be a manifold without boundary, or even a manifold at all. $\operatorname{Del}|_\Sigma S$ can be a jumble of simplices. The Topological Ball Theorem will help us to guarantee that for a dense enough sample, $\operatorname{Del}|_\Sigma S$ is surprisingly well behaved. Sections 13.3–13.5 establish sampling conditions under which the TBP holds.

13.3 Distances and angles in ε-samples

We begin by bounding the sizes of the restricted Delaunay edges and triangles of an ε-sample. If pq is a restricted Delaunay edge, it is dual to a Voronoi facet g that intersects Σ and lies on the bisector of pq. For any intersection point $x \in g \cap \Sigma$, the length of pq cannot be more than twice $d(p, x)$. Thus, if $d(p, x) \leq \varepsilon f(p)$, then $d(p, q) \leq 2\varepsilon f(p)$.

We can extend this argument to restricted Delaunay triangles too. A restricted Delaunay triangle τ is dual to a Voronoi edge τ^* that intersects Σ. An intersection point $x \in \tau^* \cap \Sigma$ is in every Voronoi cell having edge τ^*. If p is a vertex of τ, V_p is such a cell. The point x is the center of a circumball of the triangle τ. The circumradius of τ, being the radius of its smallest circumball, is at most $d(p, x)$. The following proposition is thus immediate.

Proposition 13.2. *For any $\varepsilon < 1$, the following properties hold.*

(i) *Let e be a restricted Delaunay edge with a vertex p. If the dual Voronoi facet of e intersects Σ in a point x such that $d(p, x) \leq \varepsilon f(p)$, the length of e is at most $2\varepsilon f(p)$.*

(ii) *Let τ be a restricted Delaunay triangle with a vertex p. If the dual Voronoi edge of τ intersects Σ in a point x such that $d(p, x) \leq \varepsilon f(p)$, the circumradius of τ is at most $\varepsilon f(p)$.*

In the previous chapter, we show that if edges and triangles connecting points on a smooth surface have small circumradii with respect to the local feature size, they lie nearly parallel to the surface. Hence, their dual Voronoi facets and edges intersect Σ almost orthogonally. The circumradius of a restricted Delaunay simplex is the distance from its vertices to the affine hull of its dual Voronoi face, so if the circumradius is small, the affine hull of the dual face intersects the surface near a sample point. The next two propositions quantify these statements.

Proposition 13.3. *Let g be a facet of a Voronoi cell V_p. Let $\Pi = \text{aff } g$. Let $\mathbf{n}_g$ be the normal to g. If there is an $\varepsilon < 1$ and a point $x \in \Pi \cap \Sigma$ such that $d(p, x) \leq \varepsilon f(p)$, then*

(i) $\angle_a(\mathbf{n}_g, \mathbf{n}_p) \geq \pi/2 - \arcsin \varepsilon$ *and*

(ii) $\angle_a(\mathbf{n}_g, \mathbf{n}_x) \geq \pi/2 - \arcsin \varepsilon - \alpha(\varepsilon)$, *which is greater than $\pi/6$ for $\varepsilon < 1/3$.*

Proof. Refer to Figure 13.3. Let pq be the Delaunay edge dual to g. By Proposition 13.2(i), $d(p, q) \leq 2\varepsilon f(p)$. Because pq is orthogonal to Π, $\angle_a(\mathbf{n}_g, \mathbf{n}_p) = \angle_a(pq, \mathbf{n}_p)$, and by the Edge Normal Lemma (Lemma 12.12), $\angle_a(pq, \mathbf{n}_p) \geq \pi/2 - \arcsin \varepsilon$, yielding (i).

By the Normal Variation Theorem (Theorem 12.8), $\angle(\mathbf{n}_p, \mathbf{n}_x) \leq \alpha(\varepsilon)$. By the triangle inequality,

$$\begin{aligned} \angle_a(\mathbf{n}_g, \mathbf{n}_x) &\geq \angle_a(\mathbf{n}_g, \mathbf{n}_p) - \angle_a(\mathbf{n}_p, \mathbf{n}_x) \\ &\geq \frac{\pi}{2} - \arcsin \varepsilon - \alpha(\varepsilon). \end{aligned}$$

□

The next proposition shows that if a Voronoi edge e intersects the surface Σ at a point close to the vertices of e's Delaunay dual triangle, then e is nearly orthogonal to Σ at that

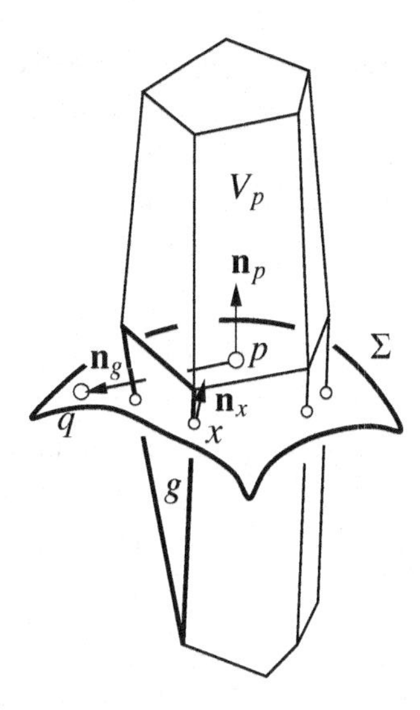

Figure 13.3: The normals $\mathbf{n}_p$ and $\mathbf{n}_x$ are almost orthogonal to $\mathbf{n}_g$.

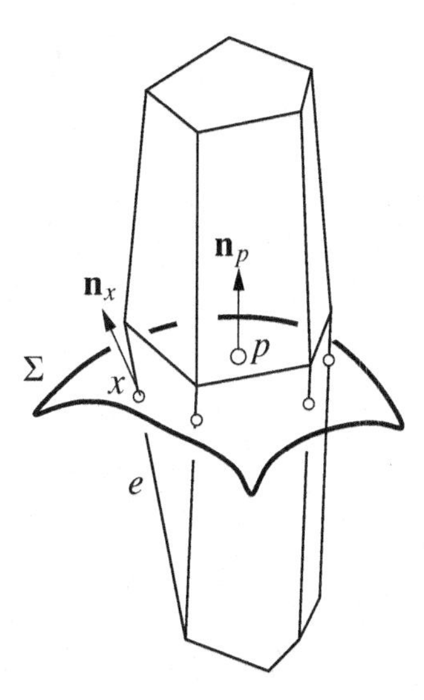

Figure 13.4: By Proposition 13.4, $\mathbf{n}_x$, e, and $\mathbf{n}_p$ are almost parallel.

point, as illustrated in Figure 13.4. It is also nearly parallel to the normals at the dual triangle's vertices.

Proposition 13.4. *Let e be an edge of a Voronoi cell V_p. If there is a point $x \in e \cap \Sigma$ such that $d(p, x) \leq \varepsilon f(p)$ for some $\varepsilon < 1/2$, then*

(i) $\angle_a(e, \mathbf{n}_p) \leq \beta(\varepsilon)$ *and*

(ii) $\angle_a(e, \mathbf{n}_x) \leq \alpha(\varepsilon) + \beta(\varepsilon)$.

Proof. Let τ be the Delaunay triangle dual to e, and observe that e is parallel to the normal $\mathbf{n}_\tau$ to τ. Let q be the vertex of τ where the angle is largest. The circumradius of τ cannot exceed $d(p,x) \le \varepsilon f(p)$. Hence $d(p,q) \le 2\varepsilon f(p)$. By the Feature Translation Lemma (Lemma 12.2), $f(p) \le f(q)/(1-2\varepsilon)$. Therefore, the circumradius of τ is at most $\frac{\varepsilon}{1-2\varepsilon} f(q)$. We apply the Triangle Normal Lemma (Lemma 12.14) to τ to obtain

$$\angle_a(\mathbf{n}_\tau, \mathbf{n}_q) \le \arcsin \frac{\varepsilon}{1-2\varepsilon} + \arcsin\left(\frac{2}{\sqrt{3}} \sin\left(2\arcsin\frac{\varepsilon}{1-2\varepsilon}\right)\right).$$

Because $2\varepsilon < 1$ by assumption, the Normal Variation Theorem (Theorem 12.8) applies, giving

$$\angle(\mathbf{n}_p, \mathbf{n}_q) \le \alpha(2\varepsilon) \quad \text{and} \quad \angle(\mathbf{n}_p, \mathbf{n}_x) \le \alpha(\varepsilon).$$

Therefore,

$$\begin{aligned} \angle_a(e, \mathbf{n}_p) &\le \angle_a(\mathbf{n}_p, \mathbf{n}_q) + \angle_a(e, \mathbf{n}_q) \\ &\le \alpha(2\varepsilon) + \angle_a(\mathbf{n}_\tau, \mathbf{n}_q) \\ &= \beta(\varepsilon), \end{aligned}$$

proving (i). The correctness of (ii) follows by the triangle inequality: $\angle_a(e, \mathbf{n}_x) \le \angle_a(\mathbf{n}_p, \mathbf{n}_x) + \angle_a(e, \mathbf{n}_p) \le \alpha(\varepsilon) + \beta(\varepsilon)$. □

The next proposition proves a fact about distances given an angle constraint. It is similar to the Long Distance Lemma (Lemma 12.13), which is used in the proof.

Proposition 13.5. *Let p, x, and y be three points on Σ. If there is an $\varepsilon \le 0.15$ such that $\angle_a(xy, \mathbf{n}_x) \le \alpha(\varepsilon) + \beta(\varepsilon)$, then at least one of $d(p,x)$ or $d(p,y)$ is larger than $\varepsilon f(p)$.*

Proof. Assume to the contrary that both $d(p,x)$ and $d(p,y)$ are at most $\varepsilon f(p)$. By the Feature Translation Lemma (Lemma 12.2), $f(x) \le f(p)/(1-\varepsilon)$. By the Three Points Lemma (Lemma 12.3),

$$d(x,y) \le 2\varepsilon f(p) \le \frac{2\varepsilon}{1-\varepsilon} f(x) = 2\alpha(\varepsilon) f(x).$$

On the other hand, by the Long Distance Lemma (Lemma 12.13),

$$d(x,y) \ge 2f(x)\cos\angle(xy, \mathbf{n}_x) \ge 2f(x)\cos(\alpha(\varepsilon) + \beta(\varepsilon)).$$

The lower and upper bounds on $d(x,y)$ together yield

$$\alpha(\varepsilon) \ge \cos(\alpha(\varepsilon) + \beta(\varepsilon)).$$

This contradicts condition A.1, which is satisfied for $\varepsilon \le 0.15$. □

13.4 Local properties of restricted Voronoi faces

The goal of this section is to prove that a restricted Voronoi face has a simple topology if it has a vertex (i.e. a restricted Voronoi vertex) and all its vertices are close to a sample point.

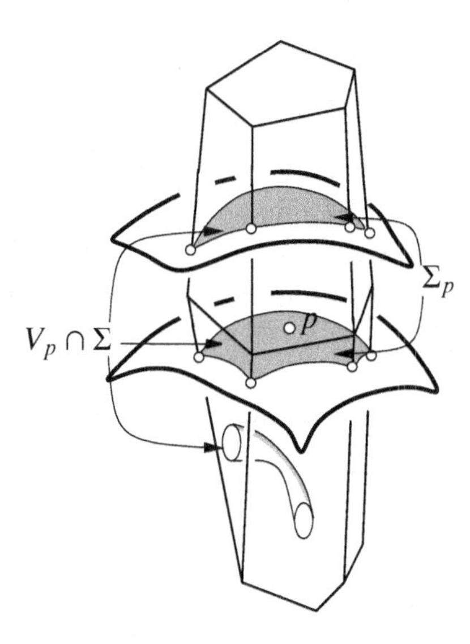

Figure 13.5: The large polyhedron is V_p, and the three curved components inside it are $V_p \cap \Sigma$. Σ_p consists of the two shaded components, which intersect Voronoi edges. The cylindrical component at the bottom is not part of Σ_p, because it does not intersect any Voronoi edge.

By "simple topology," we mean that Σ intersects a Voronoi edge in a single point, a Voronoi facet in a single curve that is not a loop, and a Voronoi cell in a single topological disk. We prove these three cases separately in results called the Voronoi Edge Lemma (Lemma 13.7), the Voronoi Facet Lemma (Lemma 13.9), and the Voronoi Cell Lemma (Lemma 13.12). We collect these results together as the Small Intersection Theorem (Theorem 13.6).

The main difficulty to overcome is that a Voronoi cell V_p can intersect the surface Σ in multiple connected components. The characterization of the results above is not quite correct, because they apply only to the connected components that adjoin a restricted Voronoi vertex. We introduce a notation for these components because they play an important role in the forthcoming analysis.

Definition 13.4 (ε-small)**.** Let Σ_p be the union of the connected components of $V_p \cap \Sigma$ that adjoin at least one restricted Voronoi vertex. We say that Σ_p is *ε-small* if Σ_p is empty or the distances from p to every restricted Voronoi vertex in Σ_p are less than $\varepsilon f(p)$.

Figure 13.5 shows Σ_p for a sample point p. Observe that Σ_p is a subset of p's restricted Voronoi cell $V_p|_\Sigma$. The Small Intersection Theorem formalizes the properties of the intersections between Σ_p and the faces of V_p when Σ_p is 0.09-small. The connected components of a 1-manifold are either *loops*, that is, connected curves with no boundary, or *topological intervals*, also known as 1-balls: curves having two boundary points each.

Theorem 13.6 (Small Intersection Theorem)**.** *If there is an $\varepsilon \leq 0.09$ such that Σ_p is nonempty and ε-small, then the following properties hold.*

(i) *Σ_p is a topological disk and every point in Σ_p is a distance less than $\varepsilon f(p)$ from p.*

(ii) *Every edge of V_p that intersects Σ does so transversally in a single point.*

(iii) *Every facet g of V_p that intersects Σ_p does so in a topological interval, and the intersection is transverse at every point of $g \cap \Sigma_p$.*

Proof. (i) Follows directly from the Voronoi Cell Lemma (Lemma 13.12). (ii) Follows directly from the Voronoi Edge Lemma (Lemma 13.7). (iii) Because of (i), Σ_p has a single boundary and it intersects a Voronoi edge. Thus, the boundary of Σ_p cannot lie completely in the interior of any facet of V_p without intersecting one of its edges. Then, the Voronoi Facet Lemma (Lemma 13.9) applies. □

The three lemmas mentioned above occupy the remainder of this section. The first of them is the easiest to prove.

Lemma 13.7 (Voronoi Edge Lemma)**.** *Suppose that Σ intersects an edge e of a Voronoi cell V_p. If there is an $\varepsilon \leq 0.15$ such that $d(p, x) \leq \varepsilon f(p)$ for every point $x \in e \cap \Sigma$, then e intersects Σ transversally in a single point.*

Proof. Let x be a point in $e \cap \Sigma$. By Proposition 13.4(ii), we have

$$\angle_a(e, \mathbf{n}_x) \leq \alpha(\varepsilon) + \beta(\varepsilon).$$

Assume for the sake of contradiction that e does not intersect Σ transversally in a single point. Either e intersects Σ tangentially at x or there is a point $y \in e \cap \Sigma$ other than x; Figure 13.6 shows both cases. In the first case, $\angle_a(e, \mathbf{n}_x) = \pi/2$, which implies that $\alpha(\varepsilon) + \beta(\varepsilon) \geq \pi/2$, which is impossible for any $\varepsilon \leq 0.15$. In the second case, $\angle_a(xy, \mathbf{n}_x) = \angle_a(e, \mathbf{n}_x) \leq \alpha(\varepsilon) + \beta(\varepsilon)$. But then Proposition 13.5 states that $d(p, x)$ or $d(p, y)$ is larger than $\varepsilon f(p)$, contradicting the assumption. □

We use the next proposition to prove the Voronoi Facet Lemma (Lemma 13.9) and the Voronoi Cell Lemma (Lemma 13.12). It says that if the intersection of a Voronoi facet with Σ includes a curve with nonempty boundary whose boundary points are close to a sample point, then the entire curve is close to the sample point.

Proposition 13.8. *Let g be a facet of a Voronoi cell V_p. Suppose that $g \cap \Sigma$ contains a topological interval I. If there is an $\varepsilon \leq 0.15$ such that the distances from p to the endpoints of I are less than $\varepsilon f(p)$, then the distance from p to any point in I is less than $\varepsilon f(p)$.*

Proof. Let B be the Euclidean 3-ball centered at p with radius $\varepsilon f(p)$. Let $\Pi = \text{aff } g$. As the distances from p to the endpoints of I are less than $\varepsilon f(p)$, the endpoints of I lie in $\text{Int } D$, where D is the Euclidean disk $B \cap \Pi$. To prove the proposition, it suffices to show that $I \subseteq \text{Int } D$. Assume to the contrary that $I \not\subseteq \text{Int } D$. Refer to Figure 13.7.

Let C be the connected component of $\Pi \cap \Sigma$ containing I. Observe that $C \not\subseteq \text{Int } D$ because $I \subset C$ and $I \not\subseteq \text{Int } D$ by assumption. For any point $x \in \Pi \cap \Sigma \cap D$, because $d(p, x) \leq \varepsilon f(p)$, the point x cannot be a tangential contact point between Π and Σ as that would contradict Proposition 13.3(ii). Thus, $C \cap \text{Int } D$ is a collection of disjoint topological intervals.

We claim that $C \cap \text{Int } D$ consists of at least two topological intervals. Assume to the contrary that $C \cap \text{Int } D$ is one topological interval. Recall that $I \subset C$, $I \not\subseteq \text{Int } D$, and the

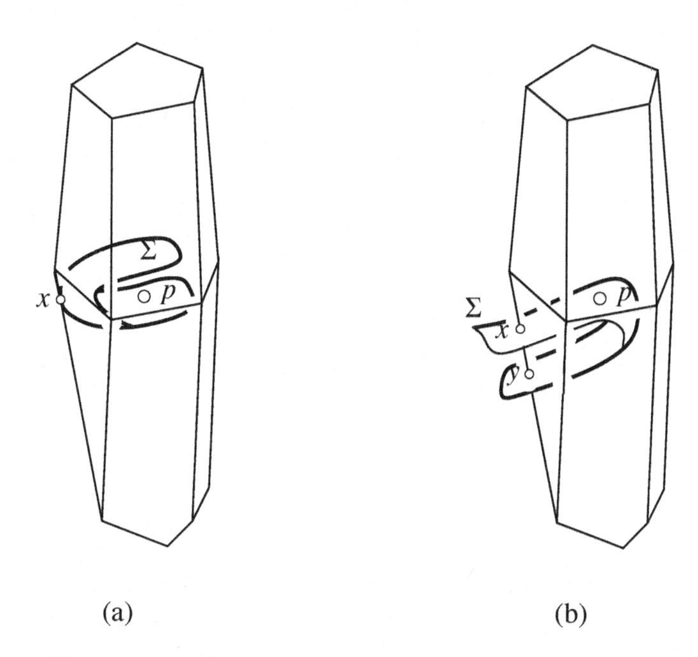

Figure 13.6: (a) A Voronoi edge intersects Σ tangentially at a single point. (b) A Voronoi edge intersects Σ in two points.

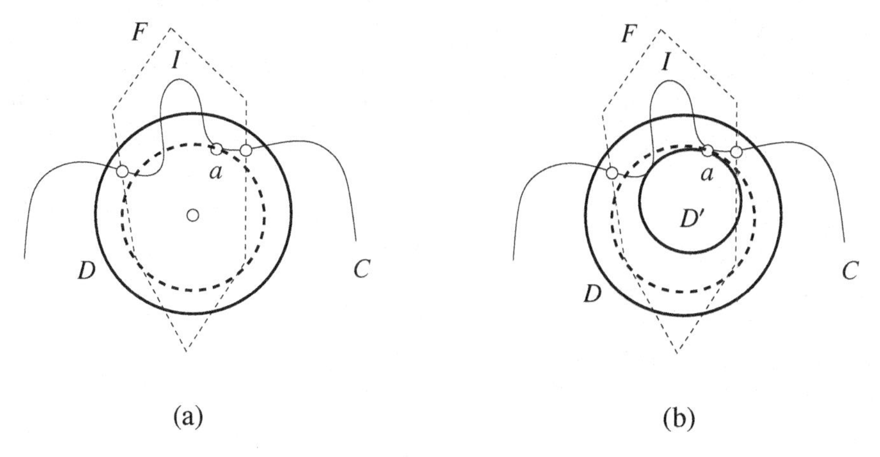

Figure 13.7: (a) D is shrunk radially until it is tangent to C at some point a. (b) The shrinking of D is continued by moving the center toward a.

endpoints of I lie in $C \cap \operatorname{Int} D$. It follows that $(C \cap \operatorname{Int} D) \cup I$ is a loop. Take an edge e of g that contains an endpoint x of I. Since Σ meets e transversally, the affine hull ℓ of e crosses $C \cap \operatorname{Int} D$ at x. Since g is convex, it lies on one side of ℓ. After $C \cap \operatorname{Int} D$ leaves the side of ℓ containing g, it must return to g in order to form a loop with I, which lies in g. It means that $C \cap \operatorname{Int} D$ has to cross ℓ at least twice. Hence, ℓ intersects $C \cap \operatorname{Int} D$ at x and another point y. Since $x, y \in D$, both $d(p, x)$ and $d(p, y)$ are at most $\varepsilon f(p)$. By Proposition 13.4, $\angle_a(xy, \mathbf{n}_x) = \angle_a(e, \mathbf{n}_x) \leq \alpha(\varepsilon) + \beta(\varepsilon)$. But then Proposition 13.5 implies that $d(p, x)$ or $d(p, y)$ is larger than $\varepsilon f(p)$, a contradiction.

So we can assume that $C \cap \operatorname{Int} D$ consists of at least two disjoint topological intervals.

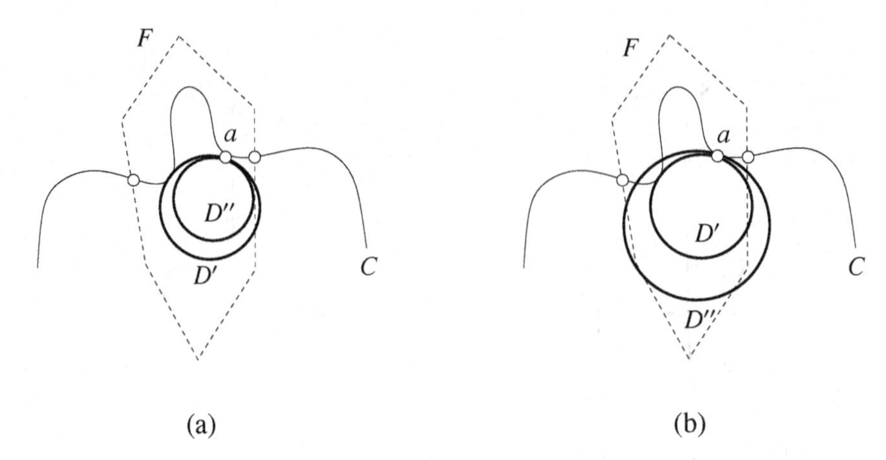

Figure 13.8: D' and D''.

Then, D can be shrunk to a smaller disk D' as follows so that D' meets C tangentially at two points and $C \cap \mathrm{Int}\, D' = \emptyset$. First, shrink D radially until it meets C tangentially at some point a. Refer to Figure 13.7(a). It follows that $d(p,a) < \varepsilon f(p)$. If this shrunk D does not meet the requirement of D' yet, shrink it further by moving its center toward a until a disk D' is obtained as required. Refer to Figure 13.7(b). Observe that a is one of the contact points between D' and C.

The affine hull Π of g intersects the two medial balls of Σ at a in two disks. Among these two disks, let D'' be the one that intersects D'. Let B'' be the medial ball such that $D'' = B'' \cap \Pi$. The boundaries of D' and D'' meet tangentially at a. So either $D'' \subseteq D'$ (Figure 13.8(a)) or $D' \subset D''$ (Figure 13.8(b)).

We claim that $D'' \subseteq D'$ and the radius of D'' is greater than $\varepsilon f(p)$. Suppose that $D' \subset D''$. By construction, D' meets Σ tangentially at two points. So one of these contact points must lie in $\mathrm{Int}\, D''$. This is a contradiction because $D'' = B'' \cap \Pi$ and $\mathrm{Int}\, B'' \cap \Sigma = \emptyset$ as B'' is a medial ball. This shows that $D'' \subseteq D'$. By Proposition 13.3(ii), the acute angle between Π and $\mathbf{n}_a$ is at most $\alpha(\varepsilon) + \arcsin \varepsilon$. The angle between the diametric segments of B'' and D'' adjoining a is equal to the angle between $\mathbf{n}_a$ and Π. Therefore,

$$
\begin{aligned}
\mathrm{radius}(D'') &\geq \mathrm{radius}(B'') \cdot \cos(\alpha(\varepsilon) + \arcsin \varepsilon) \\
&\geq f(a) \cdot \cos(\alpha(\varepsilon) + \arcsin \varepsilon).
\end{aligned}
$$

Observe that $\beta(\varepsilon) > \arcsin \varepsilon$. Also, $\cos(\alpha(\varepsilon) + \beta(\varepsilon)) > \alpha(\varepsilon)$ by condition A.1 as $\varepsilon \leq 0.15$. Thus,

$$
\begin{aligned}
\mathrm{radius}(D'') &\geq f(a) \cdot \cos(\alpha(\varepsilon) + \arcsin \varepsilon) \\
&> f(a) \cdot \cos(\alpha(\varepsilon) + \beta(\varepsilon)) \\
&> \alpha(\varepsilon) \cdot f(a) \\
&= \frac{\varepsilon}{1-\varepsilon} f(a).
\end{aligned}
$$

It follows from the Feature Translation Lemma (Lemma 12.2) that $\mathrm{radius}(D'') > \varepsilon f(p)$. This completes the proof of our claim that $D'' \subseteq D'$ and $\mathrm{radius}(D'') > \varepsilon f(p)$; therefore, radius$(D') > \varepsilon f(p)$. But D' is obtained by shrinking $D = B \cap \Pi$ where radius$(B) = \varepsilon f(p)$. We reach a contradiction. In all, the contrapositive assumption that $I \not\subseteq \text{Int}\, D$ cannot hold. It follows that the distance from p to any point in I is less than $\varepsilon f(p)$. □

With this preparation, we prove the Voronoi Facet Lemma: the intersection between a Voronoi facet and Σ is a single topological interval if each component of the intersection contains at least one restricted Voronoi vertex and every such restricted Voronoi vertex is close to a sample point.

Lemma 13.9 (Voronoi Facet Lemma). *Let g be a facet of a Voronoi cell V_p. Let C_g be the union of the curves in $g \cap \Sigma$ that each contain a restricted Voronoi vertex. If there is an $\varepsilon \le 0.09$ such that every restricted Voronoi vertex in* Bd g *is at a distance less than $\varepsilon f(p)$ from p, then C_g consists of exactly one topological interval at which g intersects Σ transversally.*

Proof. First we claim that C_g includes no loop. If there is one, it must intersect a Voronoi edge by the assumption that it contains a restricted Voronoi vertex. Since the distance from p to every restricted Voronoi vertex in C_g is less than $\varepsilon f(p)$, the affine hull of g intersects Σ transversally at each such vertex by Proposition 13.3(ii). Thus, a loop in C_g can only meet a Voronoi edge tangentially. But this would mean that the edge meets Σ tangentially, which is forbidden by the Voronoi Edge Lemma (Lemma 13.7) for $\varepsilon \le 0.09$. So C_g includes only topological intervals.

Assume for the sake of contradiction that there are two or more intervals and let I and I' be any two of them. Let u and v be the endpoints of I. Let x and y be the endpoints of I'. By the Voronoi Edge Lemma, no edge of g intersects Σ in two or more points, so the four edges of g containing u, v, x, and y are distinct. Let Q be the convex quadrilateral on aff g bounded by the affine hulls of these four edges. We call Q's edges e_u, e_v, e_x, and e_y according to the interval endpoints that they contain. Refer to Figure 13.9(a).

The distances $d(p,u)$, $d(p,v)$, $d(p,x)$, and $d(p,y)$ are less than $\varepsilon f(p)$ by assumption. Consider the Delaunay triangles dual to the edges of g containing u, v, x, and y. Their circumradii are at most $\varepsilon f(p)$ by Proposition 13.2(ii). By Proposition 13.4(i), the angles $\angle_a(e_u, \mathbf{n}_p)$, $\angle_a(e_v, \mathbf{n}_p)$, $\angle_a(e_x, \mathbf{n}_p)$, and $\angle_a(e_y, \mathbf{n}_p)$ are at most $\beta(\varepsilon)$. By Proposition 13.3(i), the acute angle between aff g and $\mathbf{n}_p$ is at most $\arcsin \varepsilon$. Let $\tilde{\mathbf{n}}_p$ be the projection of $\mathbf{n}_p$ onto aff g. So $\angle_a(\mathbf{n}_p, \tilde{\mathbf{n}}_p) \le \arcsin \varepsilon$. It follows that

$$\angle_a(e_u, \tilde{\mathbf{n}}_p) \le \angle_a(e_u, \mathbf{n}_p) + \angle_a(\mathbf{n}_p, \tilde{\mathbf{n}}_p) \le \beta(\varepsilon) + \arcsin \varepsilon.$$

Similarly, the angles $\angle_a(e_v, \tilde{\mathbf{n}}_p)$, $\angle_a(e_x, \tilde{\mathbf{n}}_p)$, and $\angle_a(e_y, \tilde{\mathbf{n}}_p)$ are at most $\beta(\varepsilon) + \arcsin \varepsilon$. The convexity of Q implies that one of its interior angles must be at least $\pi - 2(\beta(\varepsilon) + \arcsin \varepsilon)$, say, the interior angle between e_v and e_x. In this case, a line parallel to $\tilde{\mathbf{n}}_p$ may either cut through or be tangent to the corner of Q between e_v and e_x. Figures 13.9(b) and 13.9(c) illustrate these two possibilities. In both configurations, $\angle_a(vx, e_v) \le \beta(\varepsilon) + \arcsin \varepsilon$ or $\angle_a(vx, e_x) \le \beta(\varepsilon) + \arcsin \varepsilon$. Without loss of generality, assume that $\angle_a(vx, e_x) \le \beta(\varepsilon) + \arcsin \varepsilon$.

By Proposition 13.4(ii), $\angle_a(e_x, \mathbf{n}_x) \le \alpha(\varepsilon) + \beta(\varepsilon)$. Then

$$\begin{aligned} \angle_a(vx, \mathbf{n}_x) &\le \angle_a(vx, e_x) + \angle_a(e_x, \mathbf{n}_x) \\ &\le \alpha(\varepsilon) + \arcsin \varepsilon + 2\beta(\varepsilon). \end{aligned}$$

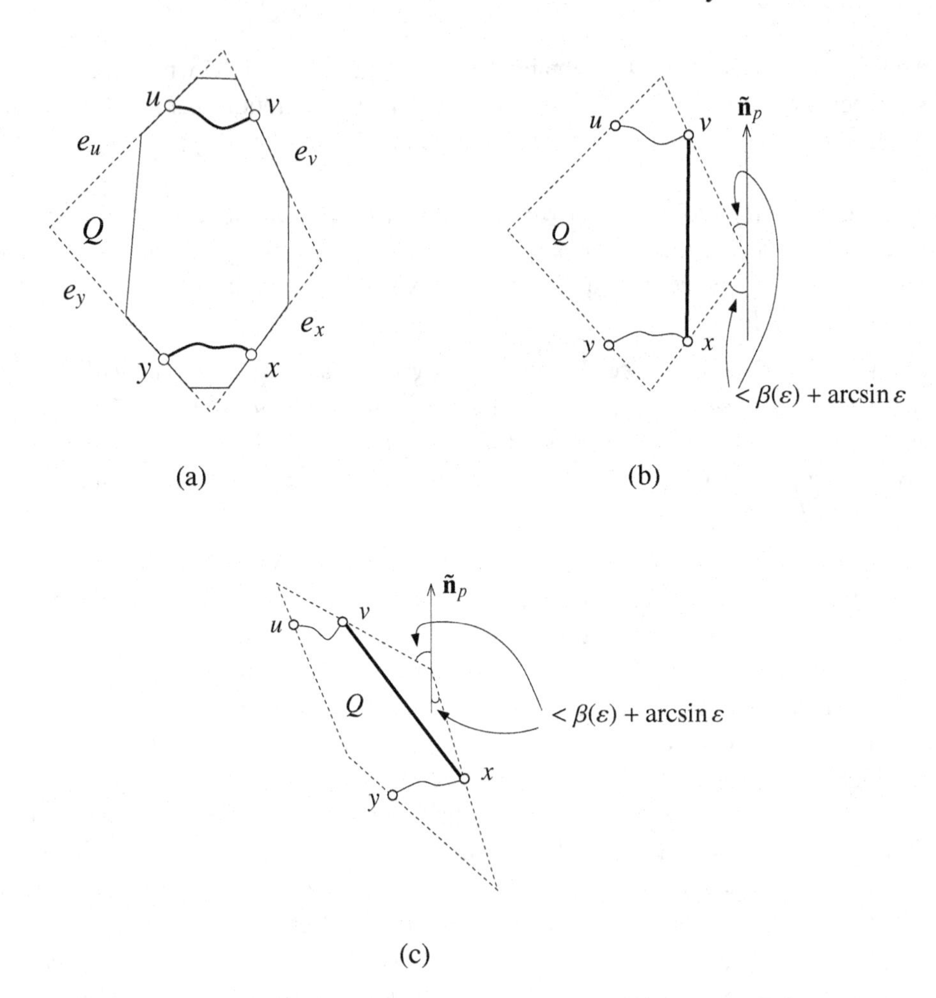

Figure 13.9: A Voronoi facet (bounded by solid line segments) intersects Σ in two topological intervals (shown as curves). The convex quadrilateral Q is bounded by dashed line segments.

On the other hand, $d(v, x) \leq 2\varepsilon f(x)/(1 - \varepsilon)$ by the Three Points Lemma (Lemma 12.3). Then, the Edge Normal Lemma (Lemma 12.12) implies that

$$\begin{aligned} \angle_a(vx, \mathbf{n}_x) &\geq \frac{\pi}{2} - \arcsin \frac{\varepsilon}{1 - \varepsilon} \\ &= \frac{\pi}{2} - \arcsin \alpha(\varepsilon) \\ &= \arccos \alpha(\varepsilon). \end{aligned}$$

The upper and lower bounds on $\angle_a(vx, \mathbf{n}_x)$ together give

$$\begin{aligned} \arccos \alpha(\varepsilon) &\leq \alpha(\varepsilon) + \arcsin \varepsilon + 2\beta(\varepsilon) \\ \Rightarrow \qquad \alpha(\varepsilon) &\geq \cos(\alpha(\varepsilon) + \arcsin \varepsilon + 2\beta(\varepsilon)). \end{aligned}$$

This contradicts condition A.2, which is satisfied for $\varepsilon \leq 0.09$. Therefore, it contradicts the assumption that C_g includes more than one topological interval.

It remains to prove that g intersects Σ transversally at C_g. Suppose to the contrary Σ intersects g tangentially at some point s in C_g. It follows from Proposition 13.8 that $d(p, s) < \varepsilon f(p)$. Therefore, $\angle_a(\mathbf{n}_p, \mathbf{n}_s) \leq \alpha(\varepsilon)$ by the Normal Variation Theorem (Theorem 12.8). The triangle inequality and Proposition 13.3(i) give

$$\begin{aligned}\angle_a(\mathbf{n}_g, \mathbf{n}_s) &\geq \angle_a(\mathbf{n}_g, \mathbf{n}_p) - \angle_a(\mathbf{n}_p, \mathbf{n}_s) \\ &\geq \frac{\pi}{2} - \arcsin \varepsilon - \alpha(\varepsilon).\end{aligned}$$

The quantity $\pi/2 - \arcsin \varepsilon - \alpha(\varepsilon)$ is positive for $\varepsilon \leq 0.09$. We reach a contradiction because $\angle_a(\mathbf{n}_g, \mathbf{n}_s)$ should be zero as Σ is tangent to g at s. □

Recall the surface Σ_p from Definition 13.4. If Σ_p is ε-small, the Voronoi Edge and Voronoi Facet Lemmas state that the edges and facets of V_p intersect Σ transversally, so Σ_p is a 2-manifold whose boundary is a 1-manifold without boundary, i.e. a collection of loops. Classify these boundary loops into two types.

- Type 1: A loop that intersects one or more Voronoi edges. The intersection points are restricted Voronoi vertices.
- Type 2: A loop that does not intersect any Voronoi edge.

Although every connected component of Σ_p adjoins a restricted Voronoi vertex by the definition of Σ_p, a connected component can have more than one boundary loop (e.g. if it is homeomorphic to an annulus), only one of which must be of Type 1.

The forthcoming Voronoi Cell Lemma says that for a sufficiently small ε, Σ_p is a topological disk. This requires us to argue that Σ_p has only one boundary loop.

We show some properties of Type 1 loops in Proposition 13.10 below. The result following it, Proposition 13.11, shows that Σ_p has exactly one Type 1 loop.

Proposition 13.10. *Let Σ_p be ε-small for some $\varepsilon \leq 0.09$. Let C be a loop of Type 1 in* $\text{Bd}\,\Sigma_p$. *Then, C bounds a topological disk $D \subset \Sigma$ such that $d(p, x) < \varepsilon f(p)$ for every point $x \in D$. Furthermore, if D does not include any other Type 1 loop, then*

(i) *D does not include any other loop of* $\text{Bd}\,\Sigma_p$,

(ii) *D is included in V_p, and*

(iii) *D is a connected component of Σ_p.*

Proof. Proposition 13.8 and the definition of Type 1 loop imply that there is a Voronoi facet of V_p that contains a single topological interval of each Type 1 loop. Continuing the same argument for other facets on which the two end points of the topological interval lie, we conclude that each Type 1 loop consists of a set of topological intervals residing in Voronoi facets of V_p all of whose points are a distance less than $\varepsilon f(p)$ away from p. It follows that all Type 1 loops lie strictly inside a Euclidean 3-ball $B = B(p, \varepsilon f(p))$. The Voronoi Edge Lemma (Lemma 13.7) and the Voronoi Facet Lemma (Lemma 13.9) show that Σ intersects $\text{Bd}\, V_p$ transversally at these loops, so the intersection is a 1-manifold. By the Small Ball Lemma (Lemma 12.7), $B \cap \Sigma$ is a topological disk. It follows that each Type 1

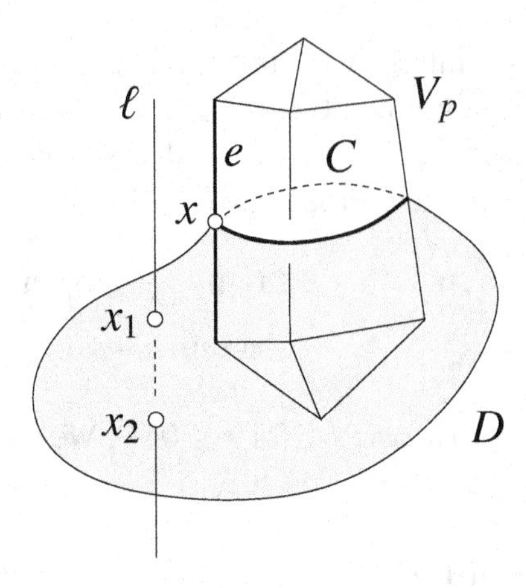

Figure 13.10: Proof of Proposition 13.10: disk D with the opening C is like a "sack" that encloses a part of V_p.

loop bounds a topological disk in $B \cap \Sigma$ that lies strictly inside B. This proves the first part of the proposition: C bounds a topological disk $D \subset \Sigma$ and $d(p, x) < \varepsilon f(p)$ for every point $x \in D$.

Consider (i). By assumption, the topological disk D does not include other loops of Type 1. Assume to the contrary that D includes a loop C' of Type 2. So C' is contained in a facet, say g, of V_p. Since D lies strictly inside B, so does C'. Take any point $x \in C'$. Consider the line ℓ_x through x parallel to the projection of $\mathbf{n}_x$ onto g. By Proposition 13.3(ii) $\angle_a(\ell_x, \mathbf{n}_x) \leq \alpha(\varepsilon) + \arcsin \varepsilon \leq \pi/3$. It means that ℓ_x intersects C' transversally and thus intersects it at another point x'. Both $d(p, x)$ and $d(p, x')$ are less than $\varepsilon f(p)$. Also, $\angle_a(\ell_x, \mathbf{n}_x) \leq \alpha(\varepsilon) + \arcsin \varepsilon \leq \alpha(\varepsilon) + \beta(\varepsilon)$. Therefore, by Proposition 13.5, $d(p, x)$ or $d(p, x')$ is greater than $\varepsilon f(p)$, a contradiction.

Consider (ii). Because V_p is a closed set, it is sufficient to show that $\operatorname{Int} D$ is included in V_p. Suppose not. Then, $\operatorname{Int} D$ must lie completely outside V_p; otherwise, $\operatorname{Int} D$ would include a boundary loop of Σ_p, which is prohibited by (i). Refer to Figure 13.10. Let e be an edge of V_p that intersects C. Let x be an intersection point of e and C. By Proposition 13.4(ii), $\angle_a(e, \mathbf{n}_x) \leq \alpha(\varepsilon) + \beta(\varepsilon)$, which is less than $\pi/3$ for $\varepsilon \leq 0.09$. Thus, $\operatorname{aff} e$ intersects Σ transversally at x. Let ℓ be a line outside V_p that is parallel to and arbitrarily close to $\operatorname{aff} e$. Then ℓ must intersect $\operatorname{Int} D$ transversally at a point x_1 arbitrarily close to x.

As $\operatorname{Bd} V_p$ is a topological sphere, C cuts it into two topological disks. Let T be one of them. The union $T \cup D$ is a topological sphere. Because ℓ intersects $\operatorname{Int} D$ at x_1, ℓ must intersect $T \cup D$ at another point $x_2 \neq x_1$.

The point x_2 must lie in D because $T \subsetneq \operatorname{Bd} V_p$ and ℓ lies outside V_p. By the Long Distance Lemma (Lemma 12.13), $d(x_1, x_2) \geq 2f(x_1)\cos(\angle_a(\ell, \mathbf{n}_{x_1}))$. Observe that x_1 is arbitrarily close to x and $\angle_a(\ell, \mathbf{n}_x) = \angle_a(e, \mathbf{n}_x) < \pi/3$. Thus, $\angle_a(\ell, \mathbf{n}_{x_1}) < \pi/3$ and so $d(x_1, x_2) > 2f(x_1)\cos \pi/3 = f(x_1)$. As x_1 is arbitrarily close to x and f is continuous, $d(x_1, x_2) \geq f(x)$; thus, $d(x_1, x_2) \geq (1 - \varepsilon)f(p)$ by the Feature Translation Lemma (Lemma 12.2). But this is a contradiction because D lies inside B and B has diameter $2\varepsilon f(p) < (1 - \varepsilon)f(p)$ for $\varepsilon \leq 1/3$.

The correctness of (iii) follows immediately from (i) and (ii). □

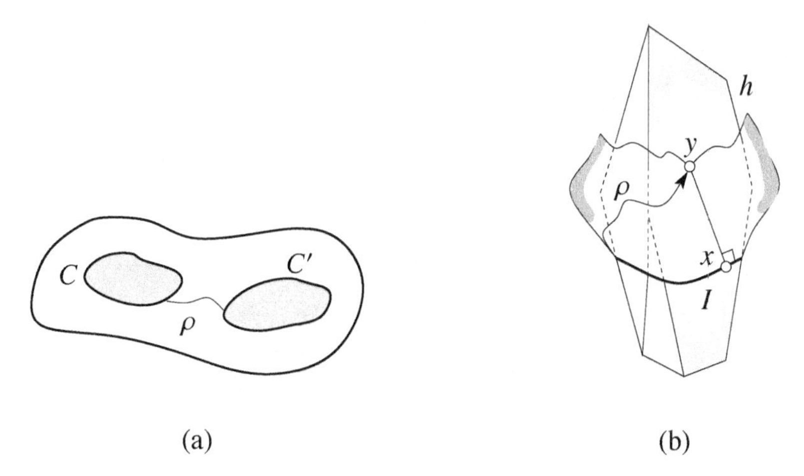

Figure 13.11: (a) Two loops C and C' drawn schematically on the patch $B \cap \Sigma$. The path ρ starting from C goes outside V_p and then has to reach V_p again to reach C'. (b) A different view with the polytope P. The lower bold curve denotes C, whose intersection with the shaded facet h is a topological interval I. The curved patch shown is part of $(B \cap \Sigma) \setminus \text{Int}\, V_p$. The curved path on it is ρ.

We use Proposition 13.10 to show that Σ_p has only one Type 1 boundary loop if Σ_p is ε-small for $\varepsilon \leq 0.09$, which will prepare us to prove the Voronoi Cell Lemma.

Proposition 13.11. *If Σ_p is ε-small for some $\varepsilon \leq 0.09$, then* Bd Σ_p *has exactly one Type 1 loop.*

Proof. The definition of Σ_p implies that its boundary has at least one Type 1 loop. Each Type 1 loop bounds a topological disk in Σ by Proposition 13.10. Because the loops are disjoint, the topological disks bounded by them are either disjoint or nested. So there is a loop C of Type 1 bounding a topological disk D in Σ such that D contains no Type 1 loop. By Proposition 13.10(iii), D is a connected component of Σ_p.

If Bd Σ_p does not have any Type 1 loop other than C, there is nothing to prove, so assume to the contrary that there is another loop C' of Type 1 in Bd Σ_p. Consider the set of facets of V_p that intersect C. Each facet in this set bounds a halfspace containing p. The intersection of these halfspaces is a convex polytope P that includes V_p. As D lies in V_p, it lies in P too.

Let $B = B(p, \varepsilon f(p))$. By Proposition 13.8, both loops C and C' lie inside $B \cap \Sigma$. Let ρ be a curve in $(B \cap \Sigma) \setminus \text{Int}\, V_p$ that connects C to C'. Since $D \subset P$ and the contact between D and Bd P is not tangential, ρ leaves P where ρ leaves D. Since $C' \subset V_p \subseteq P$, the path ρ must return to some facet of P to meet C'. Let h be a facet of P that ρ intersects after leaving D. Let y be a point in $\rho \cap h$. Let g be the facet of V_p included in h. By the definition of P, C must intersect g.

The Voronoi Facet Lemma (Lemma 13.9) implies that $C \cap g$ is one topological interval. Call this topological interval I. Refer to Figure 13.11.

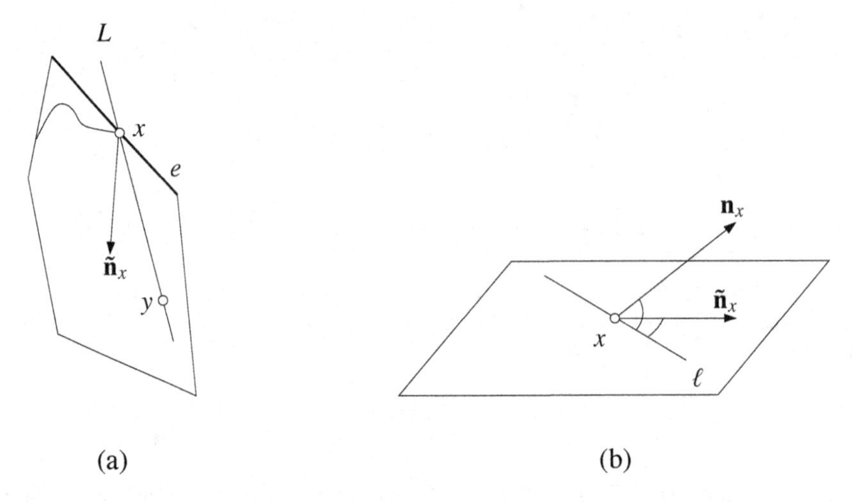

Figure 13.12: (a) The curve is $C \cap g$ and $\angle_a(L, \tilde{\mathbf{n}}_x) \le \angle_a(e, \tilde{\mathbf{n}}_x)$. (b) The angle $\angle_a(\ell, \mathbf{n}_x)$ is an increasing function of $\angle_a(\ell, \tilde{\mathbf{n}}_x)$.

Claim 1. *Every edge of g that contains an endpoint of I is included in some edge of h.*

Proof. Consider an endpoint z of I. The point z lies on the boundary of g, which means that the other facet(s) of V_p that share z with g are intersected by C. So the affine hulls of these facets bound P. It follows that the edges of g containing z are included in some edges of h. □

By Claim 1, the endpoints of I lie on the boundary of h, and hence, $C \cap h = C \cap g = I$. Let x be the point closest to y on I. Then $x \in I \subset C \subset B \cap \Sigma$ and $y \in \rho \subset B \cap \Sigma$. So the distances from p to x and y are at most radius$(B) = \varepsilon f(p)$. Let L be the line through x and y. There are two cases to consider.

- Case 1: x lies in the interior of h. Then L intersects I at x at a right angle. This means that L is the projection of $\mathbf{n}_x$ onto aff h. By Proposition 13.3(ii), $\angle_a(L, \mathbf{n}_x) \le \alpha(\varepsilon) + \arcsin \varepsilon \le \alpha(\varepsilon) + \beta(\varepsilon)$. But then $d(p, x)$ or $d(p, y)$ is greater than $\varepsilon f(p)$ by Proposition 13.5, a contradiction.

- Case 2: x lies on the boundary of h. Let e be an edge of h containing x. By Claim 1, e includes an edge of g that contains x. Let $\tilde{\mathbf{n}}_x$ be the projection of $\mathbf{n}_x$ onto aff h. Refer to Figure 13.12(a). Because x is the point closest to y on I, $\angle_a(L, \tilde{\mathbf{n}}_x) \le \angle_a(e, \tilde{\mathbf{n}}_x)$. For any line ℓ in aff h through x, the angle $\angle_a(\ell, \mathbf{n}_x)$ increases as the angle $\angle(\ell, \tilde{\mathbf{n}}_x)$ increases. Refer to Figure 13.12(b). We conclude that $\angle_a(L, \mathbf{n}_x) \le \angle(e, \mathbf{n}_x)$, which is at most $\alpha(\varepsilon) + \beta(\varepsilon)$ by Proposition 13.4(ii). But then $d(p, x)$ or $d(p, y)$ is greater than $\varepsilon f(p)$ by Proposition 13.5, a contradiction.

It follows from the contradiction that Bd Σ_p includes only one Type 1 loop. □

We now have all the ingredients to prove the Voronoi Cell Lemma.

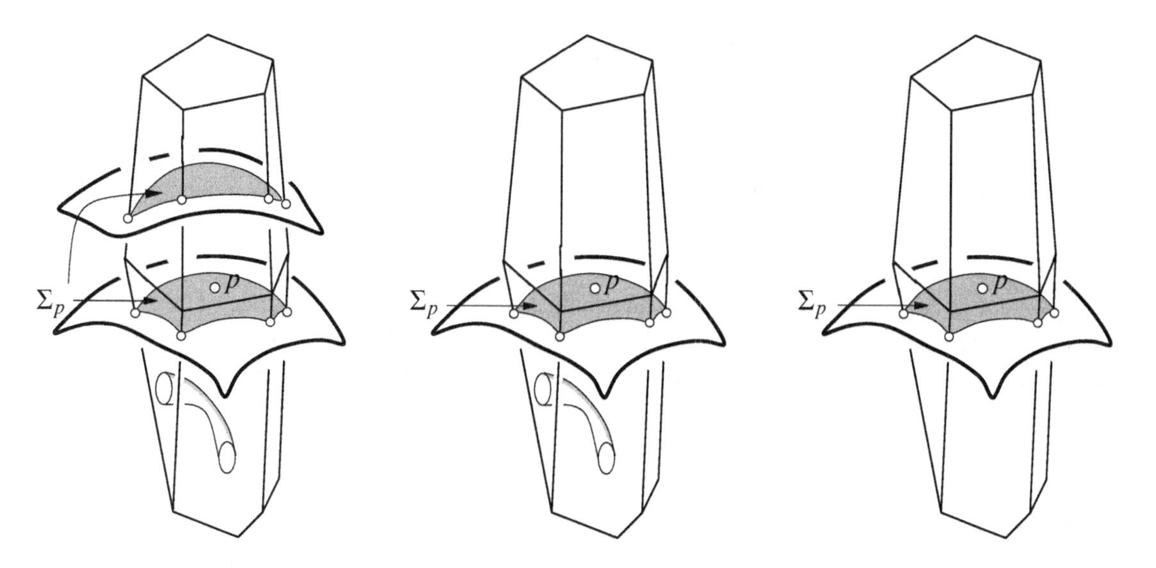

Figure 13.13: Let $\varepsilon \leq 0.09$. If Σ_p is ε-small, the configuration at left is impossible by the Small Intersection Theorem, but the configuration at center is possible because the small cylindrical component is not part of Σ_p. If Σ_p is ε-small for *every* $p \in S$ and at least one Voronoi edge intersects Σ, the configuration at center is impossible by the Voronoi Intersection Theorem (Theorem 13.14), and only the configuration at right is possible.

Lemma 13.12 (Voronoi Cell Lemma). *If Σ_p is nonempty and ε-small for some $\varepsilon \leq 0.09$, then Σ_p is a topological disk and $d(p, x) < \varepsilon f(p)$ for every point $x \in \Sigma_p$.*

Proof. By Propositions 13.10 and 13.11, Σ_p has exactly one boundary loop C of Type 1 and C bounds a topological disk D, which is a connected component of Σ_p. There is no other connected component in Σ_p because, by the definition of Σ_p, such a component would have another Type 1 boundary loop, contradicting Proposition 13.11. Hence, $D = \Sigma_p$. By Proposition 13.10, the distance from every point in D to p is less than $\varepsilon f(p)$. □

13.5 Global properties of restricted Voronoi faces

The Small Intersection Theorem does not rule out the possibility that a restricted Voronoi face may have misbehaved components that do not adjoin a restricted Voronoi vertex, as illustrated at center in Figure 13.13. It only guarantees that, if certain local constraints are imposed, a restricted Voronoi face has at most one connected component that adjoins a restricted Voronoi vertex, ruling out the configuration at left in Figure 13.13; and that component is a topological ball. The forthcoming Voronoi Intersection Theorem shows that if Σ_p is ε-small for *every* sample point p and at least one edge in Vor S intersects the surface Σ, then every restricted Voronoi face is a topological ball, as illustrated at right in Figure 13.13. Moreover, it guarantees that the topological ball property holds.

Proposition 13.13. *Let S be a finite sample of a connected, smooth surface Σ. If some edge in* Vor S *intersects Σ and there is an $\varepsilon \leq 0.09$ such that Σ_p is ε-small for every $p \in S$, then $V_p \cap \Sigma = \Sigma_p$ for every $p \in S$.*

Proof. Assume for the sake of contradiction that there is a sample point $q \in S$ such that $V_q \cap \Sigma \neq \Sigma_q$. It follows from the Small Intersection Theorem that some connected component σ_q of $V_q \cap \Sigma$ does not intersect any Voronoi edge.

By assumption, there exists a sample point s such that some edge of V_s intersects Σ. By the Voronoi Cell Lemma (Lemma 13.12), Σ_s is a topological disk. Consider a path in Σ connecting a point in σ_q to a point in Σ_s. Let $q = p_0, p_1, ..., p_k = s$ be the sequence of sample points in S whose Voronoi cells are visited along this path. Let σ_{p_i} be the connected component of $V_{p_i} \cap \Sigma$ visited by the path when it visits V_{p_i} in the sequence. There must be two consecutive points p_i and p_{i+1} in this sequence such that σ_{p_i} does not intersect any Voronoi edge and $\sigma_{p_{i+1}}$ intersects a Voronoi edge, because σ_{p_0} does not intersect any Voronoi edge and $\sigma_{p_k} = \Sigma_s$ does.

The boundaries of σ_{p_i} and $\sigma_{p_{i+1}}$ intersect. Because the former boundary intersects no Voronoi edge, the intersection must be one or more complete boundary loops. By the Voronoi Cell Lemma (Lemma 13.12), $\sigma_{p_{i+1}}$ is the topological disk $\Sigma_{p_{i+1}}$; hence, $\sigma_{p_{i+1}}$ has only one boundary loop. But this loop intersects a Voronoi edge, a contradiction. □

Theorem 13.14 (Voronoi Intersection Theorem). *Let S be a finite sample of a connected, smooth surface Σ. If some edge in* Vor S *intersects Σ and there is an $\varepsilon \leq 0.09$ such that Σ_p is ε-small for every $p \in S$, then the following properties hold for every $p \in S$.*

(i) $V_p|_\Sigma = \Sigma_p$.

(ii) *Σ_p is a topological disk and the distance from p to every point in Σ_p is less than $\varepsilon f(p)$.*

(iii) *Every edge of* Vor S *that intersects Σ does so transversally at a single point.*

(iv) *Every facet of* Vor S *that intersects Σ does so transversally in a topological interval.*

(v) *The point set S is an $\varepsilon/(1-\varepsilon)$-sample of Σ.*

Proof. Assertions (i)–(iv) follow from the Small Intersection Theorem (Theorem 13.6) and Proposition 13.13. For assertion (v), let x be a point in Σ. It follows from (i) that $x \in \Sigma_p$ for some $p \in S$. Assertion (ii) and the Feature Translation Lemma (Lemma 12.2) give $d(p, x) < \varepsilon f(p) \leq \frac{\varepsilon}{1-\varepsilon} f(x)$. Hence, S is an $\varepsilon/(1-\varepsilon)$-sample of Σ. □

The Voronoi Intersection Theorem offers stronger guarantees than the Small Intersection Theorem (Theorem 13.6). First, every restricted Voronoi cell $V_p|_\Sigma$ is a topological disk. Second, a facet of Vor S cannot intersect Σ in a loop or degenerate curve. Third, the topological ball property holds. Fourth, S is guaranteed to be a dense sample of Σ.

The preconditions of the Voronoi Intersection Theorem imply that S is dense, being an $\varepsilon/(1-\varepsilon)$-sample. The implication can be reversed, thereby showing that a sufficiently dense sample guarantees the same good consequences.

Theorem 13.15 (Dense Sample Theorem). *Let S be an ε-sample of Σ for some $\varepsilon \leq 0.08$. Then, for every $p \in S$, Σ_p is nonempty and* 0.09*-small. Moreover, the consequences of the Voronoi Intersection Theorem (Theorem 13.14) hold.*

Proof. Suppose for the sake of contradiction that some Σ_p is empty. Then the loops in $(\text{Bd } V_p) \cap \Sigma$ reside in the relative interiors of the facets of V_p. Let C be such a loop and let g be the facet of V_p that includes C. Consider a line $\ell_x \subset \text{aff } g$ that is perpendicular to C at a point $x \in C$. The line ℓ_x intersects C at another point y because C is a loop in a plane. As S is an ε-sample and p is a nearest sample point to x, $d(p, x) \leq \varepsilon f(x) \leq \frac{\varepsilon}{1-\varepsilon} f(p)$ by the Feature Translation Lemma (Lemma 12.2). Similarly, $d(p, y) \leq \frac{\varepsilon}{1-\varepsilon} f(p)$. As $\varepsilon \leq 0.08$, $d(p, x) < 0.09 f(p)$ and $d(p, y) < 0.09 f(p)$. The line ℓ_x contains the projection of $\mathbf{n}_x$ onto g because it is perpendicular to C at x. By Proposition 13.3(ii), we have $\angle_a(xy, \mathbf{n}_x) = \angle_a(\ell_x, \mathbf{n}_x) \leq \alpha(0.09) + \arcsin 0.09$, which is less than $\alpha(0.09) + \beta(0.09)$. But then Proposition 13.5 implies that one of $d(p, x)$ or $d(p, y)$ is greater than $0.09 f(p)$, a contradiction.

Because Σ_p is nonempty for every $p \in S$, it follows that Σ intersects a Voronoi edge. For any $p \in S$, let x be a restricted Voronoi vertex of Σ_p. As S is an ε-sample and p is a sample point nearest x, $d(p, x) \leq \frac{\varepsilon}{1-\varepsilon} f(p)$. It follows that Σ_p is $\varepsilon/(1 - \varepsilon)$-small and thus 0.09-small for $\varepsilon \leq 0.08$.

The assumptions of the Voronoi Intersection Theorem are satisfied and its consequences follow. □

13.6 The fidelity of the restricted Delaunay triangulation

The meshing algorithms described in subsequent chapters generate the restricted Delaunay triangulation $\text{Del}|_\Sigma S$ or a tetrahedral mesh bounded by $\text{Del}|_\Sigma S$. Here we use the Voronoi Intersection Theorem (Theorem 13.14) to show that the underlying space of $\text{Del}|_\Sigma S$ is topologically equivalent to Σ and that the two manifolds are geometrically similar. For a precise statement of the nature of the geometric similarity, see the Surface Discretization Theorem (Theorem 13.22) at the end of this chapter.

Observe that the topological ball property is one consequence of the Voronoi Intersection Theorem (Theorem 13.14). We put it together with the Topological Ball Theorem.

Theorem 13.16 (Homeomorphism Theorem). *Let S be a sample of a connected, smooth surface $\Sigma \subset \mathbb{R}^3$. If some edge in* Vor S *intersects Σ and there is an $\varepsilon \leq 0.09$ such that for every $p \in S$, Σ_p is ε-small, then the underlying space of* $\text{Del}|_\Sigma S$ *is homeomorphic to Σ.*

Thus $\text{Del}|_\Sigma S$ is a triangulation of Σ by Definition 12.6.

13.6.1 The nearest point map is a homeomorphism

The Homeomorphism Theorem is purely topological; it establishes that, under the right conditions, $\text{Del}|_\Sigma S$ is a triangulation of Σ, but it does not say that they are geometrically similar. Most of the remainder of this chapter is devoted to showing that $|\text{Del}|_\Sigma S|$ and Σ can be continuously deformed to one another by small perturbations that maintain a homeomorphism between them; thus, they are related by an isotopy. Let M be the medial axis of Σ. To construct the homeomorphism, we recall from Section 12.6 the function that maps every point $z \in \mathbb{R}^3 \setminus M$ to the unique point $\tilde{z}$ nearest z on Σ. We call this function the *nearest point map*.

Definition 13.5 (nearest point map). Let $\mathcal{T}$ be a triangulation of a 2-manifold $\Sigma \subset \mathbb{R}^3$. The *nearest point map* on $\mathcal{T}$ maps every point $z \in |\mathcal{T}|$ to the nearest point $\tilde{z}$ on Σ.

For the nearest point map to be defined, $|\mathcal{T}|$ must be disjoint from the medial axis M. If that restriction is satisfied, the nearest point map is continuous. For the nearest point map to be bijective, and therefore a homeomorphism, the triangulation must satisfy additional conditions, which we encapsulate in the notion of *ε-dense triangulations*.

Definition 13.6 (ε-dense). A triangulation $\mathcal{T}$ of a 2-manifold $\Sigma \subset \mathbb{R}^3$ is *ε-dense* if

(i) the vertices in $\mathcal{T}$ lie on Σ,

(ii) for every triangle τ in $\mathcal{T}$ and every vertex p of τ, the circumradius of τ is at most $\varepsilon f(p)$, and

(iii) $|\mathcal{T}|$ can be oriented such that for every triangle τ in $\mathcal{T}$ and every vertex p of τ, the angle between $\mathbf{n}_p$ and the vector $\mathbf{n}_\tau$ normal to τ is at most $\pi/2$.

One implication of this definition is that for sufficiently small ε, an ε-dense triangulation of Σ does not intersect the medial axis.

Proposition 13.17. *For $\varepsilon < 0.5$, an ε-dense triangulation $\mathcal{T}$ of Σ does not intersect the medial axis of Σ.*

Proof. Let τ be an arbitrary triangle in $\mathcal{T}$. Let p be a vertex of τ. Because $\mathcal{T}$ is ε-dense, τ lies in $B(p, 2\varepsilon f(p))$, which is included in the interior of $B(p, f(p))$ as $\varepsilon < 0.5$. By the definition of LFS, the medial axis of Σ intersects the boundary of $B(p, f(p))$ but not its interior. It follows that τ is disjoint from the medial axis of Σ. □

Another implication is that for $\varepsilon \leq 0.09$, the nearest point map is bijective and so it is a homeomorphism. We will also use it to define an isotopy relating $|\mathrm{Del}|_\Sigma S|$ to Σ.

Proposition 13.18. *If there is an $\varepsilon \leq 0.09$ such that $\mathcal{T}$ is an ε-dense triangulation of Σ, then the nearest point map on $\mathcal{T}$ is bijective.*

Proof. For every point $x \in \Sigma$, let ℓ_x be the line segment connecting the centers of the two medial balls at x. Observe that ℓ_x is orthogonal to Σ at x. If there is only one medial ball at x, let ℓ_x be the ray originating at the center of the medial ball and passing through x.

Let $\nu : |\mathcal{T}| \to \Sigma$ be the nearest point map of $\mathcal{T}$. Let z be a point in $\ell_x \cap |\mathcal{T}|$. Observe that the ball $B(z, d(x, z))$ is included in the medial ball at x containing z. Recall that Σ does not intersect the interiors of the medial balls, so x is the point nearest z on Σ. Hence, $\tilde{z} = x$ for every point $z \in \ell_x \cap |\mathcal{T}|$.

Now we show that ν is injective: for every point $x \in \Sigma$, ℓ_x intersects $|\mathcal{T}|$ in at most one point. Assume to the contrary that for some $x \in \Sigma$, ℓ_x intersects $|\mathcal{T}|$ at two points y_1 and y_2.

We find a triangle τ intersecting ℓ_x such that $\mathbf{n}_x$ makes an angle at least $\pi/2$ with the normal $\mathbf{n}_\tau$ to τ. Such a triangle always exists. This claim is trivial if ℓ_x is parallel to a triangle that it intersects. Suppose that it is not. Then y_1 and y_2 lie in distinct triangles not parallel to ℓ_x. We can thus choose y_1 and y_2 to be consecutive intersection points in $\ell_x \cap |\mathcal{T}|$. Without loss of generality, we assume that ℓ_x enters the volume bounded by $|\mathcal{T}|$ at y_1 and leaves this volume at y_2. Then, $\mathbf{n}_x$ makes an angle more than $\pi/2$ with the normal to one of the two triangles containing y_1 and y_2.

Let y be an intersection point in $\ell_x \cap \tau$. Let v be the vertex of τ having the largest angle in τ. Because $\mathcal{T}$ is ε-dense, the circumradius of τ is at most $\varepsilon f(v)$. By the Triangle Normal Lemma (Lemma 12.14), $\angle(\mathbf{n}_v, \mathbf{n}_\tau) < \arcsin \varepsilon + \arcsin\left(\frac{2}{\sqrt{3}} \sin(2 \arcsin \varepsilon)\right)$. Therefore,

$$\begin{aligned} \angle(\mathbf{n}_v, \mathbf{n}_x) &\geq \angle(\mathbf{n}_x, \mathbf{n}_\tau) - \angle(\mathbf{n}_v, \mathbf{n}_\tau) \\ &\geq \frac{\pi}{2} - \arcsin \varepsilon - \arcsin\left(\frac{2}{\sqrt{3}} \sin(2 \arcsin \varepsilon)\right). \end{aligned} \tag{13.1}$$

On the other hand, since the circumradius of τ is at most $\varepsilon f(v)$, $d(v, y) \leq 2\varepsilon f(v)$. We have argued that $\tilde{y} = x$. So $d(x, y) \leq d(v, y)$. This gives $d(v, x) \leq d(x, y) + d(v, y) \leq 4\varepsilon f(v)$ and so $\angle(\mathbf{n}_v, \mathbf{n}_x) \leq \alpha(4\varepsilon)$ by the Normal Variation Theorem (Theorem 12.8). We reach a contradiction because $\alpha(4\varepsilon)$ is less than the lower bound (13.1). We conclude that, for any point $x \in \Sigma$, there is at most one point in $\ell_x \cap |\mathcal{T}|$.

It remains to show that ν is surjective: for every point $x \in \Sigma$, ℓ_x intersects $|\mathcal{T}|$. Assume to the contrary that for some point $x_0 \in \Sigma$, ℓ_{x_0} does not intersect $|\mathcal{T}|$. As $|\mathcal{T}|$ is disjoint from the medial axis of Σ, we can pick any point $z' \in |\mathcal{T}|$, and there is a unique point $x_1 \in \Sigma$ closest to z', implying that ℓ_{x_1} intersects $|\mathcal{T}|$ at z'. Move a point x continuously from x_0 to x_1 on Σ, and stop when ℓ_x intersects $|\mathcal{T}|$ for the first time. When we stop, the segment ℓ_x is tangent to $|\mathcal{T}|$ at a point y. Let τ be a triangle in $\mathcal{T}$ containing y. The point y cannot lie in the interior of τ because ℓ_x would then intersect τ in more than one point, violating the injectivity of ν. So y lies in a boundary edge e of τ. We can move to a point $x' \in \Sigma$ arbitrarily close to x such that $\ell_{x'}$ intersects $|\mathcal{T}|$ in more than one point because each edge in $\mathcal{T}$ lies in two triangles. We reach a contradiction to the injectivity of ν. □

13.6.2 Proximity and isotopy

The geometric similarity between two point sets is often measured in terms of Hausdorff distances. The *Hausdorff distance* between two point sets $A, B \subseteq \mathbb{R}^d$ is

$$d_H(A, B) = \max\{\max_{y \in B} d(y, A), \max_{x \in A} d(x, B)\};$$

that is, we find the point that is farthest from the nearest point in the other object. Observe that if the maxima are replaced by minima, we have $d(A, B)$ instead.

We will show that the nearest point map sends a point $z \in |\mathcal{T}|$ over a distance that is tiny compared to the local feature size at $\tilde{z}$, the point nearest to z on Σ. It follows that the Hausdorff distance between $|\mathcal{T}|$ and Σ is small in terms of the maximum local feature size.

Proposition 13.19. *Let $\mathcal{T}$ be an ε-dense triangulation of Σ for some $\varepsilon < 0.09$. Then, $d(z, \tilde{z}) < 15\varepsilon^2 f(\tilde{z})$ for every point $z \in |\mathcal{T}|$.*

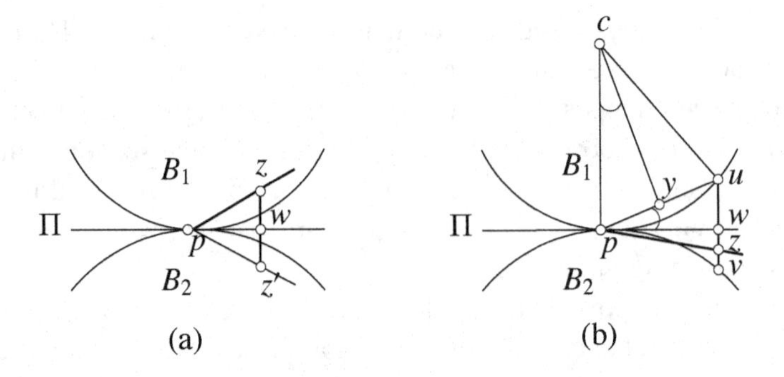

Figure 13.14: Proposition 13.19. (a) Case 1: z is in B_1. (b) Case 2: z is outside B_1 and B_2.

Proof. Let z be a point in $|\mathcal{T}|$. Let τ be a triangle in $\mathcal{T}$ that contains z. Let p be the vertex of τ nearest to z. Therefore, $d(p,z) \leq \varepsilon f(p)$. By the Triangle Normal Lemma (Lemma 12.14) and the Normal Variation Theorem (Theorem 12.8),

$$\begin{aligned} \angle_a(\mathbf{n}_\tau, \mathbf{n}_p) &\leq \frac{2\varepsilon}{1-2\varepsilon} + \arcsin\varepsilon + \arcsin\left(\frac{2}{\sqrt{3}}\sin(2\arcsin\varepsilon)\right) \\ &< 6\varepsilon \text{ for } \varepsilon < 0.09. \end{aligned}$$

This implies that for any point $r \in \tau$, the segment pr makes an angle less than 6ε with the plane tangent to Σ at p.

Let B_1 and B_2 be two balls with radii $f(p)$, tangent to Σ at p, and lying on opposite sides of Σ. The medial balls at p are also tangent to Σ at p and their radii are $f(p)$ or larger. So B_1 and B_2 are included in these two medial balls, and Σ does not intersect the interiors of B_1 and B_2.

There are two cases in bounding the distance $d(z,\tilde{z})$, depending on whether $z \in B_1 \cup B_2$.

- Case 1: z lies in $B_1 \cup B_2$. Assume without loss of generality that z lies in B_1. Let Π be the tangent plane at p. Let z' be the reflection of z with respect to Π. By symmetry, the point z' lies in B_2. Refer to Figure 13.14(a). As B_1 and B_2 lie on opposite sides of Σ, the segment zz' intersects Σ. Let x be an intersection point in $zz' \cap \Sigma$. Clearly, $d(z,\tilde{z}) \leq d(z,x) \leq d(z,z')$. By construction, zz' also intersects the tangent plane Π orthogonally at a point w. Consider the right-angled triangle zpw. We have

$$\begin{aligned} d(z,w) &= d(p,z)\sin\angle zpw \\ &\leq \varepsilon f(p)\sin(6\varepsilon) \\ &< 5.72\varepsilon^2 f(p). \end{aligned} \tag{13.2}$$

  It follows that

$$d(z,\tilde{z}) \leq d(z,z') = 2d(z,w) < 11.5\varepsilon^2 f(p).$$

- Case 2: z lies outside $B_1 \cup B_2$. Extend a line segment through z perpendicular to Π until the extension stops at points u and v on the boundaries of B_1 and B_2, respectively. Again, uv intersects Σ, which is sandwiched between B_1 and B_2. Refer to

Figure 13.14(b). Either the segment zu or zv intersects Σ. It follows that $d(z, \tilde{z})$ is at most $\max\{d(z, u), d(z, v)\}$. Assume without loss of generality that $d(z, u) \geq d(z, v)$, so u and z lie on opposite sides of the plane Π.

Let $w = zu \cap \Pi$. The same analysis that obtains (13.2) applies here, so

$$d(z, w) < 5.72\varepsilon^2 f(p).$$

If we can bound $d(u, w)$, then a bound on $d(z, u)$ follows since $d(z, u) = d(z, w) + d(u, w)$. Consider the triangle puw in Figure 13.14(b). Then

$$d(p, u) \cos \angle upw = d(p, w) \leq d(p, z) \leq \varepsilon f(p).$$

At the same time,

$$d(p, u) = 2f(p) \sin \angle pcy = 2f(p) \sin \angle upw.$$

Thus,

$$\begin{aligned} 2f(p) \sin \angle upw \cos \angle upw &\leq \varepsilon f(p) \\ \Rightarrow \qquad \angle upw &\leq (\arcsin \varepsilon)/2. \end{aligned}$$

It follows that

$$d(u, w) \leq d(p, w) \tan \angle upw \leq \varepsilon f(p) \tan \left(\frac{1}{2} \arcsin \varepsilon \right) \leq \varepsilon^2 f(p).$$

Hence, $d(z, u) \leq d(z, w) + d(u, w) < 6.72\varepsilon^2 f(p)$.

Cases 1 and 2 together yield $d(z, \tilde{z}) \leq 11.5\varepsilon^2 f(p)$. By the triangle inequality and the fact that $\mathcal{T}$ is ε-dense, $d(p, \tilde{z}) \leq d(p, z) + d(z, \tilde{z}) \leq 2.1\varepsilon f(p)$. The Feature Translation Lemma (Lemma 12.2) implies that $11.5\varepsilon^2 f(p) \leq 11.5\varepsilon^2 f(\tilde{z})/(1 - 2.1\varepsilon) < 15\varepsilon^2 f(\tilde{z})$. □

Next, we use the nearest point map of an ε-dense triangulation $\mathcal{T}$ to construct an isotopy relating $|\mathcal{T}|$ and Σ. Each point in $|\mathcal{T}|$ is very close to its image in Σ under the isotopy.

Proposition 13.20. *If there is an* $\varepsilon < 0.09$ *such that* $\mathcal{T}$ *is an* ε*-dense triangulation of* Σ*, then the nearest point map of* $\mathcal{T}$ *induces an ambient isotopy that moves each point* $x \in \Sigma$ *by a distance less than* $15\varepsilon^2 f(x)$*, and each point* $z \in |\mathcal{T}|$ *by a distance less than* $15\varepsilon^2 f(\tilde{z})$.

Proof. We define a map $\xi : \mathbb{R}^3 \times [0, 1] \to \mathbb{R}^3$ such that $\xi(|\mathcal{T}|, 0) = |\mathcal{T}|$ and $\xi(|\mathcal{T}|, 1) = \Sigma$ and $\xi(\cdot, t)$ is a continuous and bijective map for all $t \in [0, 1]$. Consider the following tubular neighborhood of Σ.

$$N_\Sigma = \{z \in \mathbb{R}^3 : d(z, \Sigma) \leq 15\varepsilon^2 f(\tilde{z})\}.$$

Proposition 13.19 implies that $|\mathcal{T}| \subset N_\Sigma$. We define an ambient isotopy ξ as follows. For $z \in \mathbb{R}^3 \setminus N_\Sigma$, ξ is the identity on z for all $t \in [0, 1]$; that is, $\xi(z, t) = z$. For $z \in N_\Sigma$, we use a map that moves points on $|\mathcal{T}|$ toward their closest points on Σ. Consider the line segment s that is normal to Σ at $\tilde{z}$, thus passing through z, and has endpoints s_i and s_o on the two boundary surfaces of N_Σ. Because $15\varepsilon^2 < 1$, the tubular neighborhood N_Σ is disjoint from the medial axis of Σ; thus, s intersects Σ only at $\tilde{z}$. Likewise, by Proposition 13.18, s intersects $|\mathcal{T}|$ only

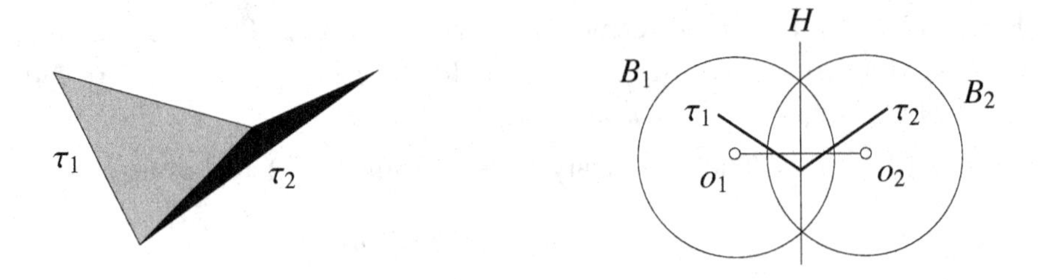

Figure 13.15: Two triangles that share an edge.

at a single point $\dot{z}$. Let $z_t = (1-t)\dot{z} + t\tilde{z}$ be a point that moves linearly from $\dot{z} \in |\mathcal{T}|$ at time zero to $\tilde{z} \in \Sigma$ at time 1. Let $\xi(\cdot, t)$ linearly map the segments $s_i\dot{z}$ to $s_i z_t$ and $s_o\dot{z}$ to $s_o z_t$. That is,

$$\xi(z,t) = \begin{cases} s_i + \dfrac{d(z_t, s_i)}{d(\dot{z}, s_i)}(z - s_i) & \text{if } z \in s_i\dot{z}, \\ \\ s_o + \dfrac{d(z_t, s_o)}{d(\dot{z}, s_o)}(z - s_o) & \text{if } z \in s_o\dot{z}. \end{cases}$$

The map ξ is continuous and bijective for all $z \in \mathbb{R}^3$ and $t \in [0,1]$, so it is an ambient isotopy, and $\xi(|\mathcal{T}|, 0) = |\mathcal{T}|$ and $\xi(|\mathcal{T}|, 1) = \Sigma$. □

13.6.3 Fidelity and dihedral angles of the discretized surface

We wish to prove that the restricted Delaunay triangulation is ε-dense and apply Proposition 13.20 to conclude that $\mathrm{Del}|_\Sigma S$ approximates Σ. It requires some work to show that $\mathrm{Del}|_\Sigma S$ satisfies the orientation property (iii) in Definition 13.6. The first step is to prove that the dihedral angles between adjoining triangles are obtuse.

Proposition 13.21. *Let τ_1 and τ_2 be two triangles in $\mathrm{Del}|_\Sigma S$ that share an edge. Suppose that there is an $\varepsilon \le 0.09$ such that for every vertex p of each $\tau \in \{\tau_1, \tau_2\}$, there is an empty circumball of τ with radius at most $\varepsilon f(p)$ centered at a point on Σ. Then τ_1 and τ_2 meet at a dihedral angle greater than $\pi/2$.*

Proof. For $i \in \{1,2\}$, let B_i be the smallest empty circumball of τ_i whose center o_i lies on Σ. The boundaries of B_1 and B_2 intersect in a circle C. Let $H = \mathrm{aff}\, C$. The plane H contains the common edge of τ_1 and τ_2, as illustrated in Figure 13.15. The triangles τ_1 and τ_2 must lie on opposite sides of H; otherwise, either the interior of B_1 would contain a vertex of τ_2 or the interior of B_2 would contain a vertex of τ_1.

We have $d(o_1, o_2) \le d(p, o_1) + d(p, o_2) \le 2\varepsilon f(p)$, which implies that $d(o_1, o_2) \le \frac{2\varepsilon}{1-\varepsilon} f(o_1)$ by the Three Points Lemma (Lemma 12.3). Hence the segment $o_1 o_2$ meets $\mathbf{n}_{o_1}$ at an angle of at least $\pi/2 - \arcsin \frac{\varepsilon}{1-\varepsilon}$ by the Edge Normal Lemma (Lemma 12.12). In turn, $\angle_a(\mathbf{n}_{o_1}, \mathbf{n}_p) \le \varepsilon/(1-\varepsilon)$ by the Normal Variation Theorem (Theorem 12.8). It follows that the angle between H and $\mathbf{n}_p$ is at most $\varepsilon/(1-\varepsilon) + \arcsin \varepsilon/(1-\varepsilon)$, which is less than 0.2 radians for $\varepsilon \le 0.09$.

As $\varepsilon \leq 0.09$, the Triangle Normal Lemma (Lemma 12.14) and the Normal Variation Theorem (Theorem 12.8) imply that the normal of τ_1 differs from $\mathbf{n}_p$ by at most

$$\arcsin 0.09 + \arcsin\left(\frac{2}{\sqrt{3}} \sin(2 \arcsin 0.09)\right) + \frac{0.18}{1 - 0.18} < 0.52.$$

The angle between the normal of τ_2 and $\mathbf{n}_p$ is also less than 0.52 radians.

Therefore, τ_1 and τ_2 lie on opposite sides of H and the angle between H and either triangle is greater than $\pi/2 - (0.52 + 0.2) > 0.85$. So, the dihedral angle between τ_1 and τ_2 is greater than $1.7 > \pi/2$ radians. □

We come to the climax of the chapter, the Surface Discretization Theorem, which states that the underlying space of $\text{Del}|_\Sigma S$ is related by isotopy and geometrically close to Σ. It also states several other geometric properties of $\text{Del}|_\Sigma S$, including an upper bound on the circumradii of the triangles, an upper bound on the deviation between the surface normal at a sample point and the normal of any triangle having it as a vertex, and a lower bound on the dihedral angle between two adjoining triangles.

Theorem 13.22 (Surface Discretization Theorem). *Let S be a sample of a connected, smooth surface $\Sigma \subset \mathbb{R}^3$. If some edge in* Vor S *intersects Σ and there is an $\varepsilon \leq 0.09$ such that Σ_p is ε-small for every $p \in S$, then $\text{Del}|_\Sigma S$ is a triangulation of Σ with the following properties.*

(i) *For every triangle τ in $\text{Del}|_\Sigma S$ and every vertex p of τ, the circumradius of τ is at most $\varepsilon f(p)$.*

(ii) *$|\text{Del}|_\Sigma S|$ can be oriented so that for every triangle $\tau \in \text{Del}|_\Sigma S$ and every vertex p of τ, the angle between $\mathbf{n}_p$ and the oriented normal $\mathbf{n}_\tau$ of τ is less than 7ε.*

(iii) *Any two triangles in $\text{Del}|_\Sigma S$ that share an edge meet at a dihedral angle greater than $\pi - 14\varepsilon$.*

(iv) *The nearest point map $\nu : |\text{Del}|_\Sigma S| \to \Sigma, z \mapsto \tilde{z}$ is a homeomorphism that induces an isotopy relating $|\text{Del}|_\Sigma S|$ to Σ.*

(v) *For every point z in $|\text{Del}|_\Sigma S|$, $d(z, \tilde{z}) < 15\varepsilon^2 f(\tilde{z})$.*

Proof. We argue that $\text{Del}|_\Sigma S$ is a ε-dense triangulation of Σ. The correctness of (i)–(v) is established along the way.

By the Homeomorphism Theorem (Theorem 13.16), the underlying space of $\text{Del}|_\Sigma S$ is homeomorphic to Σ and $\text{Del}|_\Sigma S$ is a triangulation of Σ. Let τ be a triangle in $\text{Del}|_\Sigma S$ and let p be a vertex of τ. Because Σ_p is ε-small for every $p \in S$, the dual edge of τ in $\text{Del}|_\Sigma S$ intersects Σ at a point within a distance of $\varepsilon f(p)$ from p. Therefore, τ has an empty circumball with radius at most $\varepsilon f(p)$ whose center lies on Σ. As this is true for every vertex of every triangle in $\text{Del}|_\Sigma S$, Proposition 13.21 applies.

By Proposition 13.2, the circumradius of τ is at most $\varepsilon f(p)$, proving (i). By Proposition 13.4, $\angle_a(\mathbf{n}_\tau, \mathbf{n}_p) \leq \beta(\varepsilon)$, which is less than 7ε for $\varepsilon \leq 0.09$. Furthermore, as no dihedral angle is less than $\pi/2$ by Proposition 13.21, one can orient the triangles in $\mathrm{Del}|_\Sigma S$ consistently so that for any $p \in S$ and for any triangle τ adjoining p, the angle between $\mathbf{n}_p$ and the oriented normal of τ is less than 7ε, proving (ii). The correctness of (iii) follows by comparing the oriented normals of two adjoining triangles with the surface normal at a common vertex.

We conclude that $\mathrm{Del}|_\Sigma S$ is an ε-dense triangulation of Σ. Thus Proposition 13.20 gives (iv) and (v). □

13.7 Notes and exercises

The earliest appearance of the restricted Delaunay triangulation of a surface is in a paper by Chew [61] on Delaunay surface meshing. Given a sufficiently fine triangulation of a surface, Chew presents methods to make the triangulation "Delaunay" through edge flips and to refine the triangulation so it satisfies quality guarantees. He did not recognize that his triangulations are usually a subcomplex of the Delaunay tetrahedralization. Restricted Delaunay triangulations were formally introduced by Edelsbrunner and Shah [92], who also introduced the topological ball property and proved the Topological Ball Theorem (Theorem 13.1).

The earliest applications of restricted Delaunay triangulations were in algorithms for reconstructing curves and surfaces from point clouds. These algorithms try to recover the shape and topology of an object from a dense sample of points collected from the object's surface by a laser range scanner or stereo photography. Amenta and Bern [3] show that for a sufficiently dense sample, the Voronoi cells are elongated along the directions normal to the surface, and that the Voronoi diagram satisfies the topological ball property. Hence, the restricted Delaunay triangulation is homeomorphic to the surface.

Cheng, Dey, Edelsbrunner, and Sullivan [47] adapt these ideas to develop a theory of surface sampling for generating an ε-sample whose restricted Delaunay triangulation is homeomorphic to a surface, which they use to triangulate specialized surfaces for molecular modeling. This paper includes early versions of the Voronoi Edge, Voronoi Facet, and Voronoi Cell Lemmas that require an ε-sample, which is a global property. The stronger, local versions of the Voronoi Edge and Voronoi Facet Lemmas presented here are adapted from Cheng, Dey, Ramos, and Ray [53], who use them to develop a surface meshing algorithm described in the next chapter. The versions of the Voronoi Cell Lemma and the Small Intersection Theorem (Theorem 13.6) in this chapter appear for the first time here.

Boissonnat and Oudot [29, 30] independently developed a similar theory of surface sampling and applied it to the more general problem of guaranteed-quality Delaunay refinement meshing of smooth surfaces. The notion of an ε-small Σ_p, used heavily here, is related to Boissonnat and Oudot's notion of a loose ε-sample [30]. They prove the Surface Discretization Theorem (Theorem 13.22) not by using the topological ball property, but by showing that restricted Delaunay triangulations of loose ε-samples satisfy the preconditions for homeomorphic meshing outlined by Amenta, Choi, Dey, and Leekha [6]: a simplicial complex $\mathcal{T}$ has $|\mathcal{T}|$ homeomorphic to Σ if it satisfies the following four conditions. (i) All the vertices in $\mathcal{T}$ lie on Σ. (ii) $|\mathcal{T}|$ is a 2-manifold. (iii) All the triangles in $\mathcal{T}$ have small circumradii compared to their local feature sizes. (iv) The triangle normals approximate closely the surface normals at their vertices. Boissonnat and Oudot [29, 30] extend the homeomorphism to an isotopy with a nearest point map similar to that of Definition 13.5. Here we prove the Discretization Theorem more directly by showing that the topological ball property holds for loose ε-samples, then extending the homeomorphism to isotopy.

Exercises

1. Let Σ be a 2-manifold without boundary in $\mathbb{R}^3$. Let $S \subset \Sigma$ be a finite point sample. Let τ_1 and τ_2 be two tetrahedra in Del S that share a triangular face $\sigma \subset \Sigma$. In other words, the surface Σ just happens to have a flat spot that includes the shared face.

   Give a specific example that shows why σ might nonetheless not appear in the restricted Delaunay triangulation $\mathrm{Del}|_\Sigma S$.

2. [3] Let S be an ε-sample of a smooth surface Σ without boundary. For a sample point $p \in S$, the Voronoi vertex $v \in V_p$ farthest from p is called the *pole* of p and the vector $\mathbf{v}_p = v - p$ is called the *pole vector* of p. Prove that if $\varepsilon < 1$, then $\angle_a(\mathbf{n}_p, \mathbf{v}_p) \leq 2 \arcsin \frac{\varepsilon}{1-\varepsilon}$.

3. Let S be an ε-sample of a smooth surface Σ without boundary for some $\varepsilon \leq 0.1$. Prove that the intersection of Σ and a Voronoi facet in Vor S is either empty or a topological interval.

4. Let S be an ε-sample of a smooth surface Σ without boundary for some $\varepsilon \leq 0.1$. Prove that for any $p \in S$, the restricted Voronoi cell $V_p \cap \Sigma$ is a topological disk.

5. Construct an example where Σ_p is 0.09-small for exactly one sample point p in S.

6. Show that the TBP is not necessary for the underlying space of $\mathrm{Del}|_\Sigma S$ to be homeomorphic to Σ.

7. Let S be a sample of a smooth, compact manifold Σ without boundary in $\mathbb{R}^3$.

   (a) Show an example where Σ is a 1-manifold and the TBP holds, but there exists no isotopy that continuously deforms $|\mathrm{Del}|_\Sigma S|$ to Σ.

   (b) If Σ is a surface, prove or disprove that if the TBP holds, there is an isotopy relating $|\mathrm{Del}|_\Sigma S|$ to Σ.

8. [6] Let S be a sample of a smooth surface Σ without boundary. Let $\mathcal{T}$ be a subcomplex of Del S such that $|\mathcal{T}|$ is a 2-manifold and the Voronoi edge dual to each triangle $\tau \in \mathcal{T}$ intersects Σ, and all the intersection points are within a distance of $0.09f(p)$ from a vertex p of τ. Prove that $|\mathcal{T}|$ is homeomorphic to Σ.

9. [93] Let Σ be a smooth surface without boundary. Let $\lambda : \Sigma \to \mathbb{R}$ be a 1-Lipschitz function that specifies an upper bound on the spacing of points in a sample $S \subset \Sigma$: for every point $x \in \Sigma$, $d(x, S) \leq \lambda(x)$. Prove that if $\lambda(x) \leq f(x)/5$ for every $x \in \Sigma$, then S contains $\Omega\left(\int_\Sigma dx/\lambda(x)^2\right)$ sample points. For example, every ε-sample of Σ for any $\varepsilon \leq 1/5$ has $\Omega\left(\varepsilon^{-2} \int_\Sigma dx/f(x)^2\right)$ sample points.

Chapter 14

Meshing smooth surfaces and volumes

The theory of surface sampling and restricted Delaunay triangulations developed in the last two chapters seems to mark a clear path to designing a Delaunay refinement algorithm for triangular mesh generation on a smooth surface: maintain a restricted Delaunay triangulation by maintaining a Delaunay tetrahedralization, refine it by inserting new vertices at the centers of circumballs of restricted Delaunay triangles, and continue refining until the sample is dense enough to guarantee topological correctness, geometric accuracy, and high triangle quality. Upon termination, the algorithm returns a mesh that is related to the input surface by an isotopy and enjoys all the geometric guarantees offered by the Surface Discretization Theorem (Theorem 13.22).

The fly in the ointment is that it is very difficult to know when the algorithm has succeeded and can stop refining. In theory, we can achieve these guarantees by generating a 0.08-sample or making sure that every restricted Voronoi cell is 0.09-small. In practice, there are two problems. First, it is both difficult and expensive to compute the local feature size function. The Delaunay refinement algorithms we have studied for polygonal and polyhedral domains use the local feature size in their analysis, but do not need to compute it. Curved domains are different: without computing the local feature size, it is difficult to be certain that a mesh generator has not overlooked some high-curvature feature of the domain. Second, in practice it is usually possible to recover a surface with a vertex set much less dense than a 0.08-sample, and we would prefer to have a mechanism for knowing when we can stop early.

The first surface meshing algorithm we present assumes that the user can somehow compute the LFS function, has domain knowledge that makes it possible to specify a lower bound on the LFS function, or simply wants triangles small enough that resolution is not an issue. If a mesh of uniformly-sized elements is acceptable, a constant lower bound on the feature size will do.

For many applications, however, no such approximation is known, or the concomitant overrefinement is unacceptable. Fortunately, we can appeal directly to the Topological Ball Theorem (Theorem 13.1), which says that if the faces in $\mathrm{Vor}\, S$ intersect Σ in topological closed balls of appropriate dimensions, then the underlying space of $\mathrm{Del}|_\Sigma S$ is homeomor-

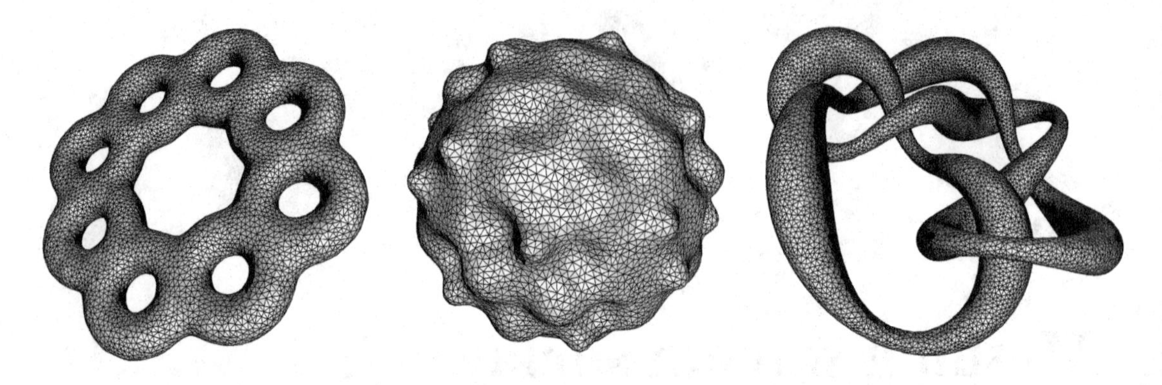

Figure 14.1: Surface meshes generated by DelSurf.

phic to Σ. A Delaunay refinement algorithm can be driven directly by diagnosing violations of the conditions of the theorem. These violations can be detected by critical point computations and combinatorial checks, which underlie our second surface meshing algorithm.

Neither algorithm is very practical, because both algorithms require expensive computations of either local feature sizes or critical points. Our third surface meshing algorithm combines ideas from both algorithms in a more pragmatic way. In this algorithm, a user-specified size field determines the fineness of the mesh and combinatorial tests guarantee that the final mesh is a 2-manifold, but a correct topology is guaranteed only if the size field is sufficiently small. Figure 14.1 depicts three Delaunay surface meshes produced by the algorithm. All three algorithms can refine the mesh to guarantee that no angle is less than 30° or greater than 120°.

The surface meshing algorithms generate a restricted Delaunay triangulation, which is a subcomplex of a Delaunay tetrahedralization, so the volume enclosed by the surface is already meshed by Delaunay tetrahedra. Of course, these tetrahedra have poor shapes unless we refine them by inserting new vertices at their circumcenters, as we do for polyhedral domains. These vertex insertions may delete some surface triangles, so an encroachment rule and several other refinement rules for surface triangles act to maintain domain conformity. We finish the chapter with a guaranteed-quality tetrahedral mesh generation algorithm for volumes bounded by smooth surfaces, which enforces an upper bound on the radius-edge ratios just a little worse than 2.

Throughout this chapter, Σ is a *smooth surface*, our shorthand for a compact (bounded), C^2-smooth 2-manifold without boundary, embedded in $\mathbb{R}^3$. We also assume that Σ is connected. If a surface has multiple connected components, each component can be meshed separately.

14.1 Delaunay surface meshing with a known local feature size

For our first Delaunay refinement surface meshing algorithm, the user supplies a 1-Lipschitz function $\lambda : \Sigma \to \mathbb{R}$, called a *size field*, that reflects the locally desired spacing of vertices on the surface. The algorithm promises to produce no triangle having an empty circumball centered at $c \in \Sigma$ with radius greater than $\lambda(c)$, which implies that no triangle

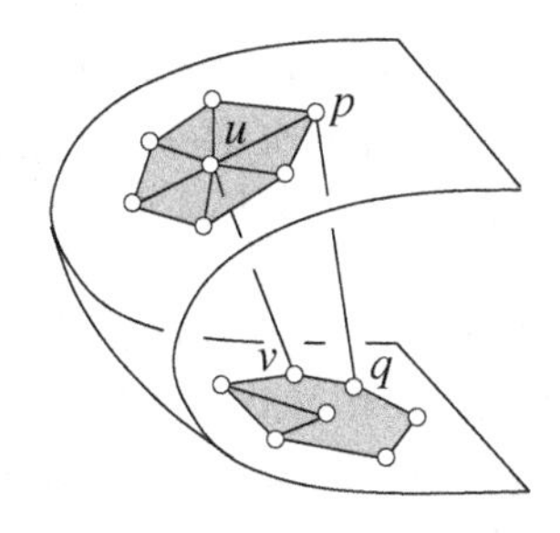

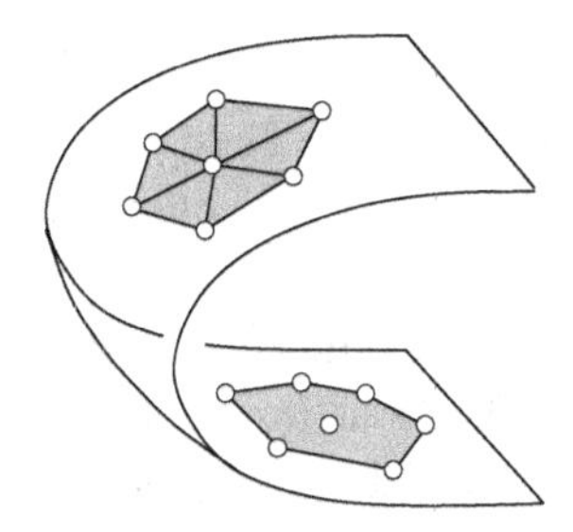

Figure 14.2: The restricted Delaunay edges pq and uv, which are not faces of any restricted Delaunay triangle, are present in $\mathrm{Del}|_\Sigma S$ (left) but absent by definition from $\mathrm{Del}|^2_\Sigma S$ (right).

has a circumradius greater than $\lambda(c)$. We require that $\inf_{x\in\Sigma} \lambda(x) > 0$. To guarantee that the algorithm will produce a topologically correct mesh, we also require that $\lambda(x) \leq \varepsilon f(x)$ for some $\varepsilon \in (0, 0.08]$. As we have mentioned, it can be quite difficult to estimate the value of f, but in many practical applications, the user desires elements smaller than those necessary for correctness.

The algorithm refines a restricted Delaunay triangulation by sampling new points on Σ at the centers of circumballs of restricted Delaunay triangles, and adding them to the vertex set. Early in the algorithm's progress, when the sample S is still sparse, $\mathrm{Del}|_\Sigma S$ can contain edges and vertices that are not faces of any triangle. There is no need for the algorithm to keep track of these dangling simplices, so we focus on the subcomplex of $\mathrm{Del}|_\Sigma S$ consisting of restricted Delaunay triangles and their faces. We call this subcomplex the *restricted Delaunay* 2*-subcomplex*, denoted

$$\mathrm{Del}|^2_\Sigma S = \{\tau : \tau \text{ is a face of a triangle in } \mathrm{Del}|_\Sigma S\}.$$

Recall that a simplex is a face of itself; thus, $\mathrm{Del}|^2_\Sigma S$ contains every triangle in $\mathrm{Del}|_\Sigma S$ with its edges and vertices, but omits dangling edges and vertices, as Figure 14.2 illustrates. Later, we will introduce a *restricted Delaunay* 3*-subcomplex* for tetrahedral mesh generation.

Definition 14.1 (surface Delaunay ball). A *surface Delaunay ball* of a restricted Delaunay triangle $\sigma \in \mathrm{Del}|_\Sigma S$ is a circumball of σ whose center lies both on Σ and on the Voronoi edge σ^* dual to σ.

Every restricted Voronoi vertex is the center of a surface Delaunay ball. Ideally, for every triangle $\sigma \in \mathrm{Del}|_\Sigma S$ there would be exactly one point in $\Sigma \cap \sigma^*$, but for a sparse sample there can be many, so a triangle can have many surface Delaunay balls. A surface Delaunay ball's interior contains no vertex of S because it is centered on σ^*, the Voronoi edge whose generating sites are the vertices of σ.

The Delaunay refinement algorithm splits triangles whose surface Delaunay balls are too large. Specifically, if a surface Delaunay ball $B(c, r)$ has $r > \lambda(c)$, the algorithm inserts its center c into S and updates the Delaunay tetrahedralization $\mathrm{Del}\, S$ and the surface triangulation $\mathrm{Del}|^2_\Sigma S$. Note that there is no need to store two separate triangulations; it is only necessary to mark which triangles in $\mathrm{Del}\, S$ are restricted Delaunay. When no large ball survives, every restricted Voronoi cell $V_p|_\Sigma$ has vertices close to its generating site p;

specifically, Σ_p is ε-small for some appropriate ε (recall Definition 13.4). By the Surface Discretization Theorem (Theorem 13.22), the underlying space of $\mathrm{Del}|_\Sigma S$ is homeomorphic to Σ and related by an isotopy, and $\mathrm{Del}|_\Sigma S$ approximates Σ well geometrically. At this time, $\mathrm{Del}|_\Sigma S = \mathrm{Del}|^2_\Sigma S$.

To bootstrap the algorithm, the sample S is initially a set containing the three vertices of a triangle chosen to be small enough that it remains restricted Delaunay throughout execution. We call it the *persistent triangle*. Its presence guarantees that at least one Voronoi edge intersects the surface as required by the Surface Discretization Theorem, namely, the persistent triangle's dual edge. Initially, the dual edge is a line that intersects Σ in at least two points, one inducing a small surface Delaunay ball and one inducing a large one, at whose center the fourth vertex is inserted. We summarize the algorithm in pseudocode.

DELSURF1(Σ, λ)

1. Compute a persistent triangle with vertices on Σ. Let S be the set of vertices of the persistent triangle. Compute $\mathrm{Del}\, S$ and $\mathrm{Del}|^2_\Sigma S$.
2. While some triangle in $\mathrm{Del}|^2_\Sigma S$ has a surface Delaunay ball $B(c, r)$ with $r > \lambda(c)$, insert c into S, update $\mathrm{Del}\, S$ and $\mathrm{Del}|^2_\Sigma S$, and repeat Step 2.
3. Return $\mathrm{Del}|^2_\Sigma S$.

DELSURF1 is not concerned with the quality of the triangles it generates, but a small modification can change that by refining skinny triangles as well as oversized ones; see Section 14.4.

A sore point is how to compute a persistent triangle. It is unclear how to do so deterministically; the following randomized heuristic performs well in practice. Pick a point $x \in \Sigma$. Randomly pick three points in $\Sigma \cap B(x, \lambda(x)/7)$ to form a triangle σ. Accept σ as the persistent triangle if $\Sigma \cap B(x, \lambda(x)/7)$ intersects σ's dual line, the line perpendicular to σ through its circumcenter. Otherwise, throw away the vertices of σ and try again. We justify this heuristic in the next section (Proposition 14.3). The only reason we require λ to be 1-Lipschitz is to ensure that σ is persistent.

For DELSURF1 to run quickly, it should maintain a queue containing all the surface Delaunay balls whose radii exceed the threshold. When a new vertex is inserted into $\mathrm{Del}\, S$, each new triangular face σ is tested to see if its dual Voronoi edge intersects Σ, and if so, at what point or points. This computation determines both whether σ is restricted Delaunay and whether its surface Delaunay balls are too large; oversized balls are enqueued. Step 2 of the algorithm is a loop that begins by dequeueing a ball and checking whether it is still a surface Delaunay ball. Be forewarned that it is not enough to check whether σ is still in $\mathrm{Del}|^2_\Sigma S$; refinement may have shortened σ's dual edge without eliminating it, in which case σ may still be restricted Delaunay but have fewer surface Delaunay balls than it had when it was created.

14.1.1 Proof of termination and guarantees

We first show that DELSURF1 terminates.

Proposition 14.1. *If Σ is compact and* $\inf_{x \in \Sigma} \lambda(x) > 0$, DELSURF1 *terminates.*

Proof. DelSurf1 inserts vertices only at the centers of empty open balls whose radii exceed $\inf_{x \in \Sigma} \lambda(x)$, so no two vertices in S are ever closer to each other than that—except that the vertices of the persistent triangle may be closer to each other. As Σ is bounded, the Packing Lemma (Lemma 6.1) states that there is an upper bound on the number of vertices S can contain. □

The following theorem lists properties of a mesh generated by DelSurf1.

Theorem 14.2. *Let $\Sigma \subset \mathbb{R}^3$ be a C^2-smooth, compact, connected surface without boundary. Let $\lambda : \Sigma \to \mathbb{R}$ be a 1-Lipschitz function and $\varepsilon \leq 0.08$ a value such that for every point $x \in \Sigma$, $\lambda(x) \leq \varepsilon f(x)$. Suppose $\inf_{x \in \Sigma} \lambda(x) > 0$. Then DelSurf1$(\Sigma, \lambda)$ returns a mesh $\mathcal{T}$ with the following properties.*

(i) *$|\mathcal{T}|$ can be oriented so that for every triangle $\sigma \in \mathcal{T}$ and every vertex p of σ, the angle between $\mathbf{n}_p$ and the oriented normal $\mathbf{n}_\sigma$ of σ is less than $7\varepsilon/(1 - \varepsilon) < 0.61$ radians.*

(ii) *Every two distinct triangles in $\mathcal{T}$ that share an edge meet at a dihedral angle greater than $\pi - 14\varepsilon/(1 - \varepsilon)$.*

(iii) *The nearest point map $\nu : |\mathcal{T}| \to \Sigma, z \mapsto \tilde{z}$ is a homeomorphism between $|\mathcal{T}|$ and Σ that induces an isotopy.*

(iv) *For every point $z \in |\mathcal{T}|$, $d(z, \tilde{z}) < \frac{15\varepsilon^2}{(1-\varepsilon)^2} f(\tilde{z}) < 0.12 f(\tilde{z})$.*

(v) *The set of vertices in $\mathcal{T}$ is an $\varepsilon/(1 - 2\varepsilon)$-sample of Σ.*

Proof. Let S be the set of vertices in $\mathcal{T}$, and recall that $\mathcal{T} = \mathrm{Del}|_\Sigma^2 S$. Let p be a vertex in S whose restricted Voronoi cell $V_p|_\Sigma$ has a restricted Voronoi vertex x, which is therefore a vertex of $\Sigma_p \subseteq V_p|_\Sigma$. Upon termination, $d(p, x) \leq \lambda(x)$; otherwise, DelSurf1 would have inserted a vertex at x. Because $\lambda(x) \leq \varepsilon f(x)$,

$$d(p, x) \leq \varepsilon f(x) \leq \frac{\varepsilon}{1 - \varepsilon} f(p),$$

the last step following from the Feature Translation Lemma (Lemma 12.2). Therefore, Σ_p is $\varepsilon/(1 - \varepsilon)$-small for every $p \in S$. Because $\varepsilon \leq 0.08$, $\varepsilon/(1 - \varepsilon) \leq 0.09$.

By Proposition 14.3 below, the presence of the persistent triangle guarantees that some edge in Vor S intersects Σ. Therefore, the premises of the Surface Discretization Theorem (Theorem 13.22) are satisfied, and $\mathrm{Del}|_\Sigma S$ has properties (i)–(iv). Property (iii) implies that $\mathrm{Del}|_\Sigma S$ has no dangling simplex that is not a face of a triangle, so $\mathrm{Del}|_\Sigma S = \mathrm{Del}|_\Sigma^2 S = \mathcal{T}$. Because Σ_p is $\varepsilon/(1 - \varepsilon)$-small for every $p \in S$, the Voronoi Intersection Theorem (Theorem 13.14) states that S is an $\varepsilon/(1 - 2\varepsilon)$-sample of Σ. □

Proposition 14.3. *The triangle σ computed as described at the end of Section 14.1 is persistent—it is present in the output mesh.*

PROOF. By construction, $\Sigma \cap B(x, \lambda(x)/7)$ contains the vertices of σ and intersects the line perpendicular to σ through the circumcenter of σ. Therefore, every surface Delaunay ball of σ centered at a point in $\Sigma \cap B(x, \lambda(x)/7)$ has radius at most $2\lambda(x)/7$, and is included in $B(x, 3\lambda(x)/7)$. We will see that DELSURF1 never inserts a vertex in $B(x, 3\lambda(x)/7)$. Therefore, there is a surface Delaunay ball of σ that remains empty throughout the algorithm, and σ is a persistent triangle.

Let p be a vertex inserted by DELSURF1 other than σ's vertices. Assume for the sake of contradiction that $d(p, x) \leq 3\lambda(x)/7$. Let v be a vertex of σ. The 1-Lipschitz property of λ implies that $\lambda(p) \geq \lambda(x) - d(p, x) \geq 4\lambda(x)/7 \geq d(p, x) + d(v, x) \geq d(p, v)$.

As DELSURF1 inserts p at the center of a surface Delaunay ball with radius greater than $\lambda(p)$, the vertex v lies inside this ball, contradicting the fact that every surface Delaunay ball is empty. □

14.1.2 Deleting two vertices of the persistent triangle

The vertices of the persistent triangle are much closer to each other than the user requested. We can fix this by deleting two of them from Del S. Unfortunately, their deletion may yield a surface mesh with the wrong topology, or create restricted Delaunay triangles that have surface Delaunay balls a bit larger than requested and a bit too large for the guarantees of Theorem 14.2 to apply. We can mitigate this problem by choosing the vertices of the persistent triangle to be very close together, but not so close as to cause numerical problems. We can fix the problem entirely by refining the mesh a little bit more after removing the two vertices. Unfortunately, it is possible that this refinement can create a sample set whose Voronoi diagram has no edge that intersects Σ—that is, $\text{Del}|_\Sigma S$ contains no triangle and the algorithm is stuck. At any rate, we do not know how to prove that this never happens, though it seems utterly unlikely in practice. For the sake of theoretical certainty, we discuss here a simple way to delete vertices of the persistent triangle while maintaining the guarantees of Theorem 14.2.

Let v and w be the two vertices of the persistent triangle we wish to delete. When DELSURF1 terminates, it returns a triangulation of a 2-manifold, so $\text{Del}|_\Sigma S$ contains at least one triangle σ that does not have v for a vertex. The Voronoi edge σ^* dual to σ intersects Σ. Deleting v from S may have the effect of lengthening σ^*, but not of shortening or deleting it, so σ^* still intersects Σ. If the deletion of v causes the mesh to no longer satisfy the antecedents of Theorem 14.2, or if it creates a triangle that is too large, then continue to refine the mesh.

It is possible, as we have said, for the algorithm to get stuck if $\text{Del}|_\Sigma S$ contains no triangles after a vertex insertion. In that unlikely event, fix it by inserting v again, thereby reintroducing the persistent triangle. The key observation is that after v is deleted, it is always possible to insert at least one vertex besides v before getting stuck. Perhaps v will be deleted and reinserted many times, but the algorithm will make progress, and by the Packing Lemma it will terminate.

After producing a satisfactory mesh lacking v, delete w and continue refinement until the mesh is again satisfactory or the algorithm gets stuck. In the latter case, reinsert *both* v and w and continue refining until v can be removed again. Again, we argue that after w

is deleted, at least one new vertex can be inserted before v and w must be reintroduced, so even if both vertices are deleted and reinserted many times, the algorithm will make progress and eventually succeed.

14.1.3 Computing edge-surface intersections

The crucial numerical computation in most surface meshing algorithms, including DELSURF1, is finding the points on a surface that intersect a line or edge. There is no universal algorithm to perform this query, because there are many different ways to represent surfaces, each requiring a different algorithm. A surface meshing program should abstract this computation as a black box subroutine whose replacement allows the mesher to work with different surface representations. Most solid modeling programs include an interface by which an application can ask the solid modeler to perform this query.

For most surface representations, the intersection points are solutions of a system of equations. For a triangulated or polygonal surface, the equations are linear and easily solved. For a parametrized spline, the equations are polynomial, and are usually best solved by iterative methods.

Many applications work with an important class of surfaces called *isosurfaces*, also known as *implicit surfaces*. An isosurface is induced by a smooth function $h : \mathbb{R}^3 \to \mathbb{R}$. For any real number η, the point set $\Sigma = h^{-1}(\eta) = \{p : h(p) = \eta\}$ is an isosurface of h having isovalue η. If η is not a critical value of h, Σ is a smooth surface.

Depending on the nature of h, evaluating an isosurface can be arbitrarily difficult. But if $h(p)$ is polynomial in p, the intersection of a line with an isosurface of h is computed by finding the roots of a univariate polynomial. More commonly, $h(p)$ is piecewise polynomial, and a root finding computation must be done for each piece that intersects the line or edge. Isosurfaces of piecewise polynomials are rarely C^2-smooth, but Delaunay surface meshing algorithms tend to work well in practice anyway.

Isosurfaces are often created from *voxel data* such as medical images. Voxel data specifies the value of h at each vertex of a cubical grid, but h is not known anywhere else. There are many triangulation algorithms for voxel data, some very fast, but most of them produce some skinny triangles. Algorithms such as DELSURF1 can extract a high-quality mesh of a kidney from a medical image by first extending the domain of h from the vertices to the cubes by interpolation, then meshing an isosurface of the interpolated function. The interpolation basis is usually piecewise polynomial, most commonly piecewise trilinear. Of course, not all discrete point data sets use a cubical grid. For instance, isosurfaces can be generated from irregularly placed data points by constructing their Delaunay triangulation and interpolating the data with piecewise linear functions or more sophisticated methods such as natural neighbor interpolation.

Call a Voronoi edge *bipolar* if the values of h at its two vertices have opposite signs, indicating that one vertex is enclosed by Σ and one is not. If all intersections are transverse, a bipolar edge intersects Σ an odd number of times. If a more sophisticated root finding procedure is not available, a point where a bipolar edge intersects Σ can be approximated by repeated bisection, the secant method, or if h is sufficiently smooth, the Newton–Raphson method.

A good strategy for speed is to delay computing edge-surface intersections until they are needed and to avoid identifying non-bipolar edges that intersect Σ until it becomes necessary. Maintain a queue of Delaunay triangles that dualize to bipolar edges and have not yet been checked. Repeatedly remove a triangle σ from the queue and check whether its dual Voronoi edge σ^* is still bipolar. If so, compute the intersection points in $\sigma^* \cap \Sigma$ and decide if any of the surface Delaunay balls centered at those points are too large. DELSURF1 often succeeds in practice without considering non-bipolar edges (except when inserting the fourth vertex; there are no bipolar edges when S contains only the vertices of the persistent triangle). Only if the mesh is inadequate when the queue runs empty must the algorithm resort to testing non-bipolar edges for surface intersections.

Figure 14.3 shows meshes of voxel data obtained by using trilinear interpolation to define isosurfaces and a variant of DELSURF1 described in Section 14.3 to create surface meshes. Instead of computing a persistent triangle, we took intersections of the isosurface with the lines of the voxel grid as seed vertices.

14.2 Topology-driven surface meshing

The algorithm DELSURF1 can fail if λ does not satisfy the condition $\lambda(x) \leq 0.08 f(x)$ for all $x \in \Sigma$. Unfortunately, it is difficult to estimate $f(x)$, so it is difficult to know whether a user's size field is acceptable. Moreover, the constant 0.08 arising from the theory is conservative; much sparser samples often suffice in practice. It is important to know when the algorithm can stop early.

Our second surface meshing algorithm does not need to know anything about the local feature sizes. For simplicity, we will dispense with the size field λ and suppose that the user desires a topologically correct mesh with no more vertices than the algorithm reasonably requires. Refinement is driven solely by the wish to satisfy the topological ball property (TBP). Of course, the algorithm can optionally incorporate a size field too.

Unfortunately, the algorithm requires that Σ be C^4-smooth, whereas DELSURF1 requires only C^2-smoothness. A collection of critical point computations is essential to the algorithm, so Σ must have a representation that makes them possible.

Recall from Section 13.2 that the pair (S, Σ) satisfies the TBP if every Voronoi k-face in $\text{Vor}\, S$ that intersects Σ does so transversally in a topological $(k-1)$-ball. In that case, the Topological Ball Theorem (Theorem 13.1) states that the underlying space of $\text{Del}|_\Sigma S$ is homeomorphic to Σ. The topology-driven surface meshing algorithm searches for violations of the TBP and attempts to repair them by sampling points from the surface. It terminates only when (S, Σ) satisfies the TBP. The algorithm maintains a minimum distance between sample points, thereby guaranteeing termination. Before we describe the algorithm, we study the subroutines that diagnose violations of the TBP.

14.2.1 Diagnosing violations of the topological ball property

To find violations of the topological ball property, we define four subroutines VOREDGE, TOPODISK, VORFACET, and SILHOUETTE. When one of them identifies a violation, it samples a new point from Σ for use as a mesh vertex. For simplicity we assume that no Voronoi vertex lies on Σ.

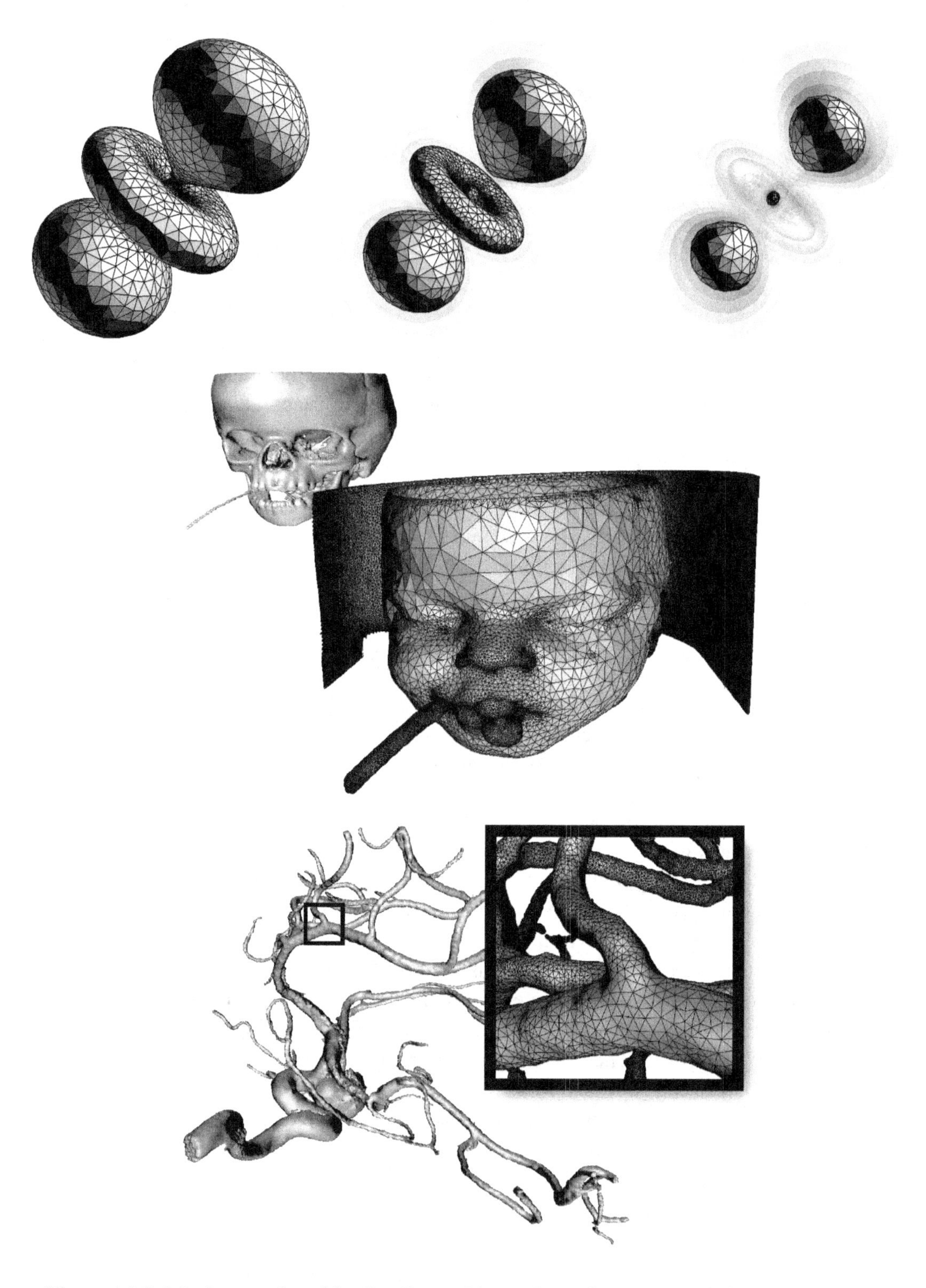

Figure 14.3: Meshes produced by DelSurf of isosurfaces from the Atom data set. In the top row, three different isovalues provide three different isosurfaces. The bottom two meshes are generated from medical images.

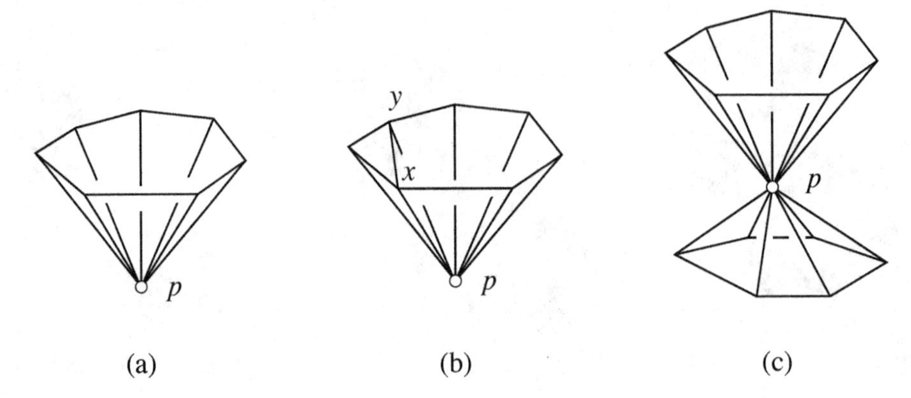

Figure 14.4: A test for whether the triangles adjoining p form a topological disk. (a) $|T_p|$ is a topological disk. (b) The edges px and py are not edges of exactly two triangles, so $|T_p|$ is not a topological disk. (c) There is more than one cycle of triangles, so $|T_p|$ is a union of topological disks joined at p.

VorEdge checks whether a Voronoi edge e satisfies the TBP. A Voronoi edge should either be disjoint from Σ or intersect Σ transversally at a single point, a 0-ball. Let σ be a restricted Delaunay triangle dual to e, and let $c_{\max}(\sigma) \in e \cap \Sigma$ and $r_{\max}(\sigma)$ be the center and the radius of σ's largest surface Delaunay ball.

VorEdge(e)

> If e intersects Σ tangentially or at multiple points, let σ be the triangle dual to e and return $c_{\max}(\sigma)$. Otherwise, return null.

TopoDisk checks whether the set T_p of restricted Delaunay triangles adjoining a vertex p forms a topological disk. Figure 14.4 illustrates the test: $|T_p|$ is a topological disk if and only if every edge adjoining p in a triangle in T_p is an edge of exactly two triangles in T_p, and there is only one cycle of triangles around p, rather than two or more.

If every Voronoi edge intersects Σ in at most one point (enforced by VorEdge), then TopoDisk detects two types of TBP violations. First, it diagnoses a Voronoi facet g that intersects Σ in two or more topological intervals. When this violation happens, the endpoints of the topological intervals each lie on a different edge of g (thanks to VorEdge), so there are at least four such edges. Their dual Delaunay triangles all share the edge dual to g, so the dual edge is not an edge of exactly two restricted Delaunay triangles. Second, TopoDisk diagnoses a Voronoi cell V_p whose boundary intersects Σ in two or more loops that meet Voronoi edges. This violation implies that T_p is a union of two or more topological disks joined at p, as in Figure 14.4(c).

TopoDisk(p)

1. Let T_p be the set of triangles in $\mathrm{Del}|^2_\Sigma S$ that adjoin p.
2. If $|T_p|$ is empty or a topological disk, return null.
3. Let σ be the triangle that maximizes $r_{\max}(\sigma)$ among T_p. Return $c_{\max}(\sigma)$.

If a Voronoi facet intersects Σ, the intersection should be transverse and it should be a single topological interval. This property is violated if (i) the Voronoi facet intersects Σ tangentially, (ii) the Voronoi facet intersects Σ in two or more topological intervals, or (iii) the Voronoi facet intersects Σ in a loop. All three violations (i), (ii), and (iii) can happen simultaneously.

As we have discussed, violation (ii) is detected by TopoDisk. The next subroutine, VorFacet, detects violations (i) and (iii). VorFacet uses critical point computations. Let C be a smooth loop on a plane. Given a direction d parallel to the plane, the critical points of C in direction d are the points where d is normal to C.

VorFacet(g)

1. If there exists a point in $g \cap \Sigma$ where g and Σ have the same tangent plane, return that point.
2. Let $\Pi = \text{aff}\, g$. Choose a random vector d parallel to Π. Compute the set X of critical points of the curves in $\Pi \cap \Sigma$ in the direction d.
3. If no point in X lies on g, return null.
4. As g intersects Σ transversally, $g \cap \Sigma$ is a collection of disjoint simple curves (intervals or loops) and $X \cap g$ is the set of critical points of these curves in direction d. Let V_p be a Voronoi cell with face g. For each $x \in X \cap g$,
   (a) Compute the line $\ell_x \subset \Pi$ through x parallel to d. The line ℓ_x is normal to $g \cap \Sigma$ at x.
   (b) Compute $Y \leftarrow \ell_x \cap g \cap \Sigma$. If Y contains two or more points, return the point in Y farthest from the vertex p dual to V_p.
5. Return null.

Step 2 of VorFacet assumes that $\Pi \cap \Sigma$ has finitely many critical points in a direction. It is known that the critical points in any direction are isolated if Σ is C^3-smooth, and the number of critical points is finite for algebraic surfaces of fixed degree.

The three subroutines VorEdge, TopoDisk, and VorFacet ensure that the Voronoi edges and Voronoi facets in Vor S satisfy the TBP. The TBP has one other requirement: that each Voronoi cell intersect Σ in a topological disk or not at all. The subroutines TopoDisk and VorFacet ensure that for each Voronoi cell V_p, $(\text{Bd}\, V_p) \cap \Sigma$ is either empty or a single loop that crosses more than one facet of V_p. Hence, the only ways that V_p can violate the TBP are if $V_p \cap \Sigma = \Sigma$ (we assume Σ is connected) or $V_p \cap \Sigma$ has a handle. We diagnose these cases by checking the *silhouettes* of Σ.

Definition 14.2 (silhouette)**.** For a smooth surface Σ and a specified direction d, the silhouette J_d is the set of points $\{x \in \Sigma : \mathbf{n}_x \cdot d = 0\}$. That is, the normal to Σ at each point in J_d is orthogonal to the direction d.

The following result motivates the use of silhouettes.

Proposition 14.4. *Let δ be a connected component of $V_p \cap \Sigma$ for some Voronoi site p. If* Bd δ *is empty, then for every direction d, $\delta \cap J_d \neq \emptyset$. If* Bd δ *is a single loop and $\delta \cap J_d = \emptyset$ for some direction d, then δ is a topological disk.*

PROOF. If Bd δ is empty, then δ is the surface Σ and so $\delta \cap J_d = J_d$. It is clear in this case that J_d is nonempty for every direction d. Suppose that Bd δ is a single loop. Let H be a plane perpendicular to d. Consider the map $\varphi : \delta \to H$ that projects each point of δ orthogonally to H. Since δ is connected and compact, it suffices to prove that φ is injective.

Assume to the contrary that φ is not injective. Then there is a line ℓ parallel to d that intersects δ in two or more points. Let x and y be two consecutive intersection points along ℓ.

As $\delta \cap J_d = \emptyset$, neither x nor y belongs to J_d, which means that neither $\mathbf{n}_x$ nor $\mathbf{n}_y$ is orthogonal to d. Because x and y are consecutive in $\ell \cap \delta$, $\mathbf{n}_x$ and $\mathbf{n}_y$ are oppositely oriented in the sense that the inner products $\mathbf{n}_x \cdot d$ and $\mathbf{n}_y \cdot d$ have opposite signs. Because δ is connected, there is a smooth curve $\rho \subset \delta$ connecting x and y. The normal to Σ changes smoothly from $\mathbf{n}_x$ to $\mathbf{n}_y$ along ρ. By the mean value theorem, there is a point $z \in \rho$ such that $\mathbf{n}_z$ is orthogonal to d. But then $z \in \delta \cap J_d$, contradicting the emptiness of $\delta \cap J_d$. □

The subroutine SILHOUETTE takes advantage of Proposition 14.4. It checks if a Voronoi cell V_p intersects the silhouette J_d, where $d = \mathbf{n}_p$ is normal to Σ at p. If $V_p \cap J_d \neq \emptyset$, either J_d intersects some facets of V_p, or V_p contains a component of J_d. The first possibility is checked by a subroutine SILHOUETTEPLANE($\Sigma, \Pi, \mathbf{n}_p$) that returns the points where J_d intersects the affine hull Π of a facet of V_p. The second possibility can be detected by checking if V_p contains a critical point of J_d in a direction d' orthogonal to d. To perform this check, we require that J_d be a set of smooth loops, and that the number of critical points on J_d along a direction be finite. The first requirement is likely true if d is obtained by applying a tiny, random perturbation to $\mathbf{n}_p$; for simplicity, assume it holds. The second requirement holds if Σ is C^4-smooth. Let SILHOUETTECRITICAL(Σ, d, d') be the subroutine that returns the critical points of J_d in the direction d' (see Exercise 5).

SILHOUETTE(p)

1. Let d be $\mathbf{n}_p$ after a tiny random perturbation.
2. Choose a random direction d' orthogonal to d.
3. Compute $X \leftarrow$ SILHOUETTECRITICAL(Σ, d, d').
4. If X contains a point inside V_p, return it.
5. Otherwise, for each facet g of V_p:
   (a) Compute $Y \leftarrow$ SILHOUETTEPLANE(Σ, Π, d) where $\Pi = \text{aff } g$.
   (b) If Y contains a point on g, return it.
6. Return null.

The following proposition gives a lower bound on the distance between a site p and every silhouette point in V_p.

Proposition 14.5. *Let p be a Voronoi site. For every point $x \in V_p \cap J_{\mathbf{n}_p}$, $d(p, x) \geq f(p)/3$.*

PROOF. If $d(p, x) < f(p)/3$, the Normal Variation Theorem (Theorem 12.8) implies that $\angle(\mathbf{n}_p, \mathbf{n}_x) \leq \alpha(1/3)$, contradicting the fact that $\mathbf{n}_x$ is orthogonal to $\mathbf{n}_p$ by definition. □

14.2.2 A topology-driven Delaunay refinement algorithm

The topology-driven Delaunay refinement algorithm initializes the sample S with a set of points called *seeds*, which are critical points of Σ in a chosen direction. Then it repeatedly calls the subroutines of Section 14.2.1 to search for violations of the topological ball property. When a violation is found, a new vertex is added to S. When no violation survives, the underlying space of $\mathrm{Del}|_\Sigma S$ is homeomorphic to Σ, and the algorithm returns $\mathrm{Del}|_\Sigma^2 S$, which must be equal to $\mathrm{Del}|_\Sigma S$. It is possible that some seeds are too close together, so that the surface triangulation may have unnecessarily short edges adjoining the seeds. This problem can be fixed by deleting the seeds, then refining again to restore the correct topology.

DELSURF2(Σ)

1. Let S be a set of seed points on Σ. Compute $\mathrm{Vor}\, S$ and $\mathrm{Del}|_\Sigma^2 S$.
2. Perform Steps (a)–(d) below in order. As soon as a non-null point x is returned, terminate the current loop, skip the remaining steps, and go to Step 3.
   (a) For every Voronoi edge e in $\mathrm{Vor}\, S$, compute $x \leftarrow$ VOREDGE(e).
   (b) For every $p \in S$, compute $x \leftarrow$ TOPODISK(p).
   (c) For every Voronoi facet g in $\mathrm{Vor}\, S$, compute $x \leftarrow$ VORFACET(g).
   (d) For every $p \in S$, compute $x \leftarrow$ SILHOUETTE(p).
3. If x is non-null, insert x into S, update $\mathrm{Vor}\, S$ and $\mathrm{Del}|_\Sigma^2 S$, and go to Step 2.
4. Otherwise, return $\mathrm{Del}|_\Sigma^2 S$.

Instead of computing the seed set, DELSURF2 could begin with just a single sample point in S; SILHOUETTE would generate more sites on each component of Σ and the algorithm would succeed. However, the computation of critical points of a silhouette in a specified direction is more expensive than computing critical points of Σ.

14.2.3 Proof of termination and homeomorphism

To show that DELSURF2 terminates, we use the Small Intersection Theorem (Theorem 13.6) to guarantee a lower bound on the distances among the sites inserted into S, excepting distances between pairs of seed points.

Proposition 14.6. DELSURF2 *terminates.*

PROOF. Let $f_{\min} = \inf_{x \in \Sigma} f(x)$. We claim that every point returned by a subroutine called by DELSURF2 is at a distance of at least $0.09 f_{\min}$ from every site in S, so the Packing Lemma (Lemma 6.1) implies that DELSURF2 terminates.

If VOREDGE(e) returns a point, the Voronoi edge e intersects Σ either tangentially or at more than one point. Let p be a site whose Voronoi cell V_p has the edge e. By the Voronoi Edge Lemma (Lemma 13.7), the distance from p to the farthest point in $e \cap \Sigma$ is greater than $0.15 f(p)$. VOREDGE returns that furthest point; call it x. The claim follows because no site is closer to x than p.

If TopoDisk(p) returns a point, T_p is nonempty and not a topological disk. So Σ_p is nonempty. Σ_p is not 0.09-small; if it were, the Small Intersection Theorem (Theorem 13.6) would imply that $|T_p|$ is a topological disk. It follows that some triangle in T_p has a surface Delaunay ball with radius greater than $0.09f(p)$. The claim follows because TopoDisk returns the center of the largest surface Delaunay ball.

If VorFacet(g) returns a point, let $p \in S$ be a site whose Voronoi cell V_p has the face g. If g intersects Σ tangentially, VorFacet returns a tangential contact point x such that $\mathbf{n}_x = \pm\mathbf{n}_g$, and the contrapositive of Proposition 13.3(ii) implies that $d(x, p) \geq f(p)/3$, thus $d(x, S) \geq f(p)/3$. If g intersects Σ transversally, then VorFacet returns a point x in the mutual intersection of g, Σ, and a line ℓ. There must be at least two such intersection points for VorFacet to return a point. Let y be the intersection point where ℓ is perpendicular to the silhouette $\Sigma \cap g$, and let z be another intersection point. Proposition 13.3(ii) implies that $\angle(yz, \mathbf{n}_y) = \pi/2 - \angle_a(\mathbf{n}_g, \mathbf{n}_y) \leq \alpha(0.09) + \arcsin 0.09 < \alpha(0.09) + \beta(0.09)$. Proposition 13.5 implies that $\max\{d(p, x), d(p, z)\} > 0.09f(p)$. As VorFacet returns the intersection point furthest from p, the claim follows.

By Proposition 14.5, Silhouette(p) returns a point at a distance of $f(p)/3$ or more from p. □

DelSurf2 inserts sites as long as the TBP is violated, so (S, Σ) satisfies the TBP when DelSurf2 terminates. We appeal to the Homeomorphism Theorem (Theorem 13.16).

Theorem 14.7. *Let $\Sigma \subset \mathbb{R}^3$ be a C^4-smooth, compact, connected surface without boundary.* DelSurf2(Σ) *returns a restricted Delaunay triangulation whose underlying space is homeomorphic to Σ.*

This theorem does not guarantee that the Hausdorff distance between Σ and the output mesh is small—a guarantee that cannot be made unless feature sizes are computed. However, upper bounds on the triangle sizes and lower bounds on the dihedral angles at which triangles meet can be enforced by simply refining triangles that do not meet the bounds; see Section 14.4.

14.3 A practical surface meshing algorithm

Both DelSurf1 and DelSurf2 employ computationally expensive predicates. For DelSurf1, it is rarely practical to implement an oracle that computes local feature sizes, or even a decent lower bound on them. Computing the exact medial axis of a specified implicit surface is known to be hard. The medial axis can be approximated from a dense sample of the surface, but that requires a dense sample to be available in the first place.

For DelSurf2, the predicates that compute critical points are computationally expensive and difficult to perform stably. On the bright side, the topological disk test TopoDisk performs easy combinatorial checks, but it alone does not suffice to guarantee the homeomorphism of the input surface and the output mesh. Here we discuss a practical compromise between the two algorithms.

Our third surface meshing algorithm uses two tests to drive refinement, one geometric and one topological: the user-supplied size field $\lambda : \Sigma \to \mathbb{R}$ from DelSurf1 and the test

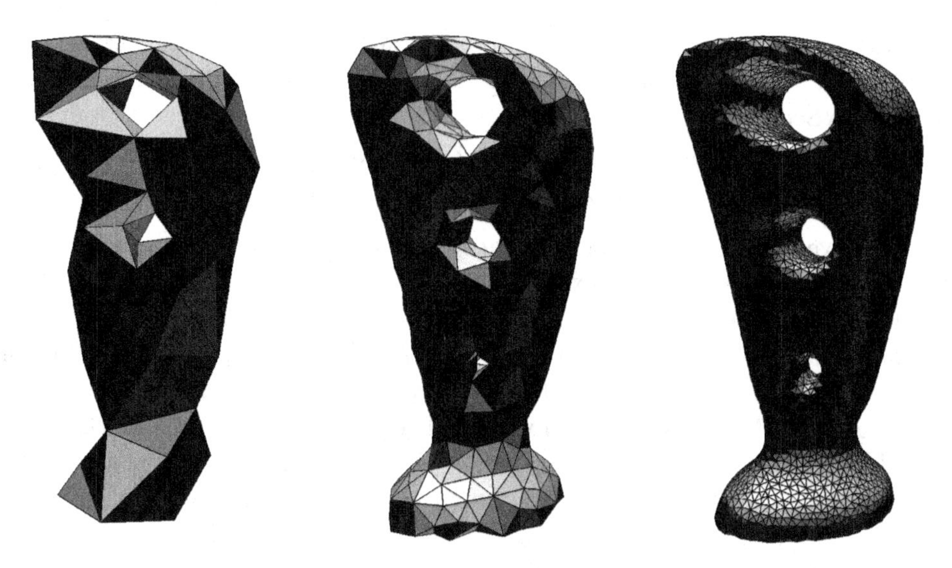

Figure 14.5: Both geometric and topological fidelity improve as the size field decreases.

TopoDisk from DelSurf2. A triangle is refined if its surface Delaunay ball is excessively large, or if the triangle is the largest among a group of triangles that adjoin a vertex in a neighborhood that is not a topological disk. If $\lambda(x) \leq \varepsilon f(x)$ for some $\varepsilon \in (0, 0.08]$, the algorithm offers the same guarantees as DelSurf1, enumerated in Theorem 14.2. Otherwise, the output mesh might not have the same topology as the input surface, but it is guaranteed to be a 2-manifold. In practice, this usually suffices to achieve homeomorphism as well.

DelSurf(Σ, λ)

1. Compute a persistent triangle with vertices on Σ. Let S be the set of vertices of the persistent triangle. Compute Del S and $\mathrm{Del}|_{\Sigma}^{2} S$.
2. While some triangle in $\mathrm{Del}|_{\Sigma}^{2} S$ has a surface Delaunay ball $B(c, r)$ with $r > \lambda(c)$, insert c into S, update Del S and $\mathrm{Del}|_{\Sigma}^{2} S$, and repeat Step 2.
3. For every $p \in S$, compute $c \leftarrow$ TopoDisk(p); if some c is non-null, stop looping, insert c into S, update Del S and $\mathrm{Del}|_{\Sigma}^{2} S$, and go to Step 2.
4. If S contains more than one vertex of the persistent triangle, delete one, update Del S and $\mathrm{Del}|_{\Sigma}^{2} S$, and go to Step 2.
5. Return $\mathrm{Del}|_{\Sigma}^{2} S$.

Figure 14.5 depicts meshes generated by DelSurf for three different constant values of λ. As λ decreases, the mesh better captures both the geometry and the topology of the surface.

The next proposition shows that, like Ruppert's algorithm, DelSurf generates a mesh whose size is proportional to the integral over the domain of the inverse squared local feature size—or the inverse squared size field, if the latter is more demanding. There is an asymptotically matching lower bound on the number of vertices needed to respect the size field, so in this sense DelSurf and DelSurf1 produce size-optimal meshes. Observe that if the size field is constant, this integral is proportional to the area of the surface. In the following bound, we assume that two vertices of the persistent triangle have been deleted, as discussed in Section 14.1.2.

Proposition 14.8. *If λ is 1-Lipschitz and for every point $x \in \Sigma$, $\lambda(x) \leq 0.08 f(x)$, then the mesh generated by* DelSurf(Σ, λ) *has fewer than $8 \cdot \int_\Sigma dx/\lambda(x)^2$ vertices.*

Proof. Every vertex c that Step 2 of DelSurf inserts into S is at a distance of at least $\lambda(c)$ from every previously inserted vertex. In Step 3, if TopoDisk(p) returns a point c, then Σ_p is not 0.09-small; if it were, the Small Intersection Theorem (Theorem 13.6) implies that $|T_p|$ would be a topological disk. Therefore, $d(c, p) \geq 0.09 f(p)$, and as f is 1-Lipschitz, $d(c, p) \geq 0.08 f(c)$. Because c is inserted in p's Voronoi cell, c is a distance of at least $0.08 f(c)$ from every previously inserted vertex. Therefore, every vertex c that DelSurf inserts is a distance of at least $\min\{\lambda(c), 0.08 f(c)\} = \lambda(c)$ from every previous vertex. Because λ is 1-Lipschitz, every vertex $p \in S$ is a distance of at least $\lambda(p)/2$ from every other vertex in S.

If we center balls of radii $\lambda(p)/4$ at every vertex $p \in S$, their interiors are pairwise disjoint. Let $\Gamma_p = \Sigma \cap B(p, \lambda(p)/4)$, and let a_p be the area of Γ_p. Because λ is 1-Lipschitz, $\lambda(x) \leq 5\lambda(p)/4$ for every $x \in \Gamma_p$. Thus,

$$\begin{aligned} \int_\Sigma \frac{1}{\lambda(x)^2}\, dx &> \sum_{p \in S} \int_{\Gamma_p} \frac{1}{\lambda(x)^2}\, dx \\ &> \sum_{p \in S} \frac{16}{25\lambda(p)^2} a_p \\ \Rightarrow \qquad |S| &< \frac{25}{16} \left(\min_{p \in S} \frac{a_p}{\lambda(p)^2} \right)^{-1} \int_\Sigma \frac{1}{\lambda(x)^2}. \end{aligned}$$

The remainder of the proof shows that $a_p/\lambda(p)^2 \geq 0.196$, from which the proposition follows.

Let B_1 and B_2 be the two medial balls sandwiching the surface Σ at p, with B_1 not larger than B_2. Let D be the geometric disk bounded by the circle at the intersection of the boundaries of $B(p, \lambda(p)/4)$ and B_1, as illustrated in Figure 14.6. Consider the orthogonal projections $\widetilde{D}$ and $\widetilde{\Gamma}_p$ of D and Γ_p, respectively, onto the plane tangent to Σ at p. Because Σ has no boundary, $\widetilde{D}$ is included in $\widetilde{\Gamma}_p$. Therefore,

$$a_p = \text{area}(\Gamma_p) \geq \text{area}(\widetilde{\Gamma}_p) \geq \text{area}(\widetilde{D}) = \text{area}(D). \tag{14.1}$$

We derive a lower bound on the radius of D to bound a_p from below. From the similar triangles pmz and pxy,

$$\begin{aligned} \frac{d(p, y)}{d(p, x)} &= \frac{d(p, z)}{d(p, m)} = \frac{d(p, x)}{2\, d(p, m)} \\ \Rightarrow \quad d(p, y) &= \frac{d(p, x)^2}{2\, d(p, m)} = \frac{\lambda(p)^2}{32\, d(p, m)}. \end{aligned}$$

Therefore,

$$\begin{aligned} \text{radius}(D)^2 &= d(p, x)^2 - d(p, y)^2 \\ &= \frac{\lambda(p)^2}{16} - \frac{\lambda(p)^4}{1{,}024\, \text{radius}(B_1)^2}. \end{aligned}$$

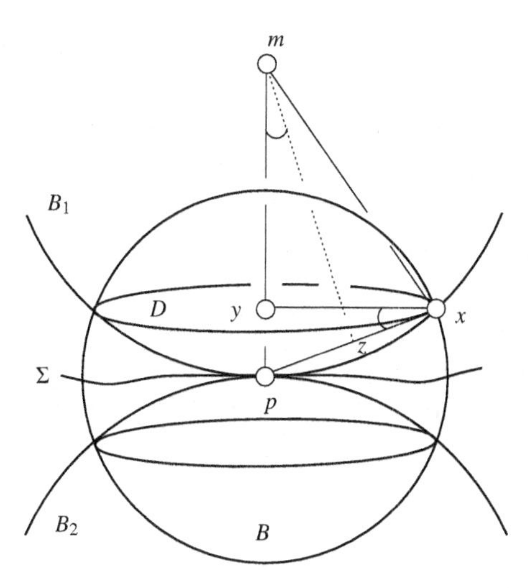

Figure 14.6: Two medial balls B_1 and B_2 sandwich the surface Σ. The projection of D onto the plane tangent to Σ at p is included in the projection of $\Sigma \cap B$ onto the same plane.

As radius$(B_1) \geq f(p)$, which is greater than $12\,\lambda(p)$ by our assumption that $\lambda(p) \leq 0.08 f(p)$,

$$\text{radius}(D)^2 \geq \frac{\lambda(p)^2}{16} - \frac{\lambda(p)^2}{147{,}456} > 0.0624\,\lambda(p)^2.$$

Hence, area(D) is greater than $0.0624\pi\,\lambda(p)^2$, implying that $a_p > 0.196\,\lambda(p)^2$ by (14.1). □

We summarize the properties of meshes generated by DELSURF.

Theorem 14.9. *Let $\Sigma \subset \mathbb{R}^3$ be a C^2-smooth, compact, connected surface without boundary. Suppose $\inf_{x\in\Sigma} \lambda(x) > 0$. Then* DELSURF$(\Sigma, \lambda)$ *returns a mesh $\mathcal{T}$ with the following properties.*

(i) *The underlying space of $\mathcal{T}$ is a 2-manifold.*

(ii) *If there is a value $\varepsilon \leq 0.08$ such that $\lambda(x) \leq \varepsilon f(x)$ for every point $x \in \Sigma$, then the nearest point map $\nu : |\mathcal{T}| \to \Sigma, z \mapsto \tilde{z}$ is a homeomorphism between $|\mathcal{T}|$ and Σ that induces an isotopy that moves each point $z \in |\mathcal{T}|$ by a distance $d(z, \tilde{z}) < \frac{15\varepsilon^2}{(1-\varepsilon)^2} f(\tilde{z}) < 0.12 f(\tilde{z})$.*

(iii) *If $\lambda(x) \leq 0.08 f(x)$ for every point $x \in \Sigma$ and λ is 1-Lipschitz, then there are fewer than $8 \cdot \int_\Sigma dx/\lambda(x)^2$ vertices in $\mathcal{T}$. This number is asymptotically optimal: every sample $S \subset \Sigma$ such that $d(x, S) \leq \lambda(x)$ for every $x \in \Sigma$ has $\Omega\left(\int_\Sigma dx/\lambda(x)^2\right)$ vertices. In this sense, $\mathcal{T}$ is size-optimal.*

PROOF. When DELSURF terminates, every vertex in $\mathcal{T}$ adjoins a set of triangles whose underlying space is a topological disk. By a well-known result of piecewise linear topology, $|\mathcal{T}|$ is a 2-manifold, proving (i). The proof of (ii) is the same as that of Theorem 14.2(iii, iv). The first part of (iii) is Proposition 14.8. The second part of (iii) is Exercise 9 in Chapter 13. □

14.4 Extensions: quality, smoothness, and polyhedral surfaces

It is easy to incorporate refinement steps into the three surface meshing algorithms we have studied to ensure that the triangles of the output mesh have good quality or adjoin each other at nearly flat dihedral angles, thereby better capturing the geometry of Σ.

Let $\rho(\sigma)$ be the radius-edge ratio of a triangle σ, and let $\bar{\rho}$ be an upper bound on the acceptable radius-edge ratio. Equivalently, $\arcsin 1/(2\bar{\rho})$ is a lower bound on the smallest acceptable angle. The radius-edge ratio threshold can be as small as 1, in which case no triangle in the final mesh has an angle less than $30° = \pi/6$ radians nor an angle greater than $120° = 2\pi/3$ radians. If there is a skinny triangle σ in $\text{Del}|_\Sigma S$, the subroutine QUALITY inserts $c_{\max}(\sigma)$, the center of σ's largest surface Delaunay ball.

QUALITY($S, \bar{\rho}$)

> If there is a restricted Delaunay triangle σ in $\text{Del}|^2_\Sigma S$ for which $\rho(\sigma) > \bar{\rho}$, then insert $c_{\max}(\sigma)$ into S, update Del S, and update $\text{Del}|^2_\Sigma S$.

As Σ is a smooth surface, the triangles of the mesh adjoin each other at dihedral angles that approach π as the edge lengths approach zero, as Theorem 14.2(ii) shows. The *roughness* of an edge e in $\text{Del}|_\Sigma S$, denoted rough(e), is π radians minus the dihedral angle at e. If the roughness of an edge exceeds a threshold $\bar{r}$, the subroutine SMOOTH refines one of the two adjoining triangles. SMOOTH thereby enforces the restriction that the dihedral angles are at least $\pi - \bar{r}$ radians.

SMOOTH($S, \bar{\rho}$)

> If there is an edge e in $\text{Del}|^2_\Sigma S$ such that rough(e) $> \bar{r}$, let σ be the restricted Delaunay triangle with the largest surface Delaunay ball among the two triangles having e for an edge, insert $c_{\max}(\sigma)$, update Del S, and update $\text{Del}|^2_\Sigma S$.

In the algorithm DELSURF, the calls to QUALITY and SMOOTH can be added just after Step 4. Whenever either call inserts a new vertex, control should return to Step 2 to ensure that the mesh respects the size field and is topologically valid. The persistent triangle must be removed before QUALITY is called, or its short edges may cause the region around it to be overrefined.

Figure 14.7 illustrates the effects of SMOOTH and QUALITY. The mesh at center shows that the quality of the triangles can suffer if we use SMOOTH alone. The quality can be corrected by using QUALITY as well.

When QUALITY($S, \bar{\rho}$) splits a skinny triangle with $\bar{\rho} \geq 1$, the newly inserted vertex is at least as far from every other vertex in S as the vertices of the skinny triangle's shortest edge are from each other. Therefore, QUALITY never creates a shorter edge in Del S than the shortest existing edge, so it cannot cause DELSURF to run forever. Neither can SMOOTH, because for any $\bar{r} > 0$, Theorem 14.2(ii) guarantees that every edge will satisfy the roughness bound once sufficient refinement has taken place.

An alternative way to achieve high quality is to choose a size field λ that does not change too quickly. For example, if λ is 0.5-Lipschitz, no triangle of the final mesh has

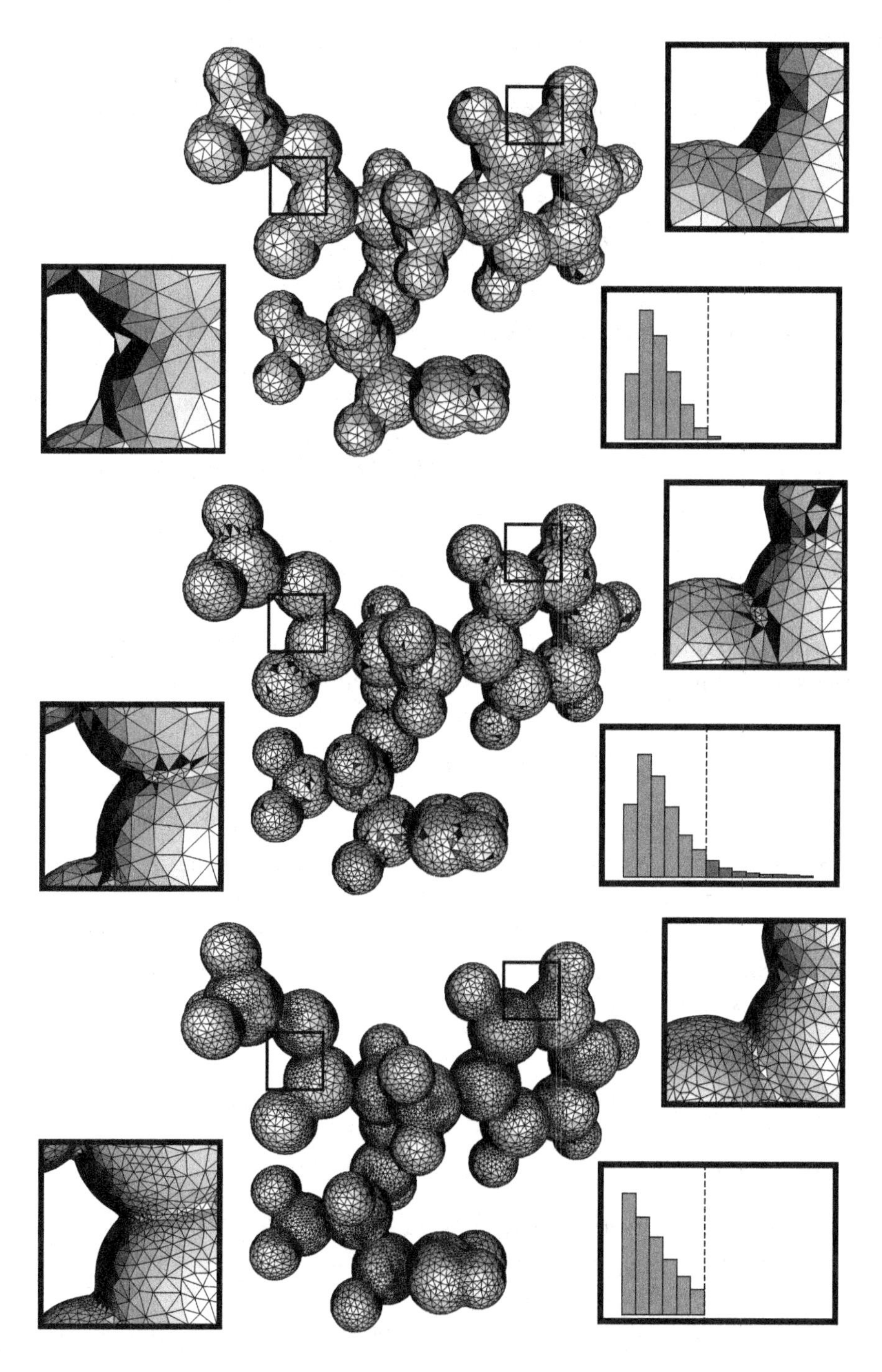

Figure 14.7: The topmost mesh is generated by DELSURF with no attention to triangle quality or the dihedral angles at which adjoining triangles meet. The histogram tabulates the triangles' radius-edge ratios, with only a few triangles (at right) exceeding the threshold. The center mesh shows that refinement intended to improve the dihedral angles succeeds at the expense of triangle quality. The bottom mesh shows the mesh after further refinement to eliminate all the poor-quality triangles.

a radius-edge ratio exceeding 2 nor, therefore, an angle less than 14.47°. A constant size field guarantees no angle less than 30°, at the cost of losing the flexibility to create a graded mesh. We will use this observation to analyze the grading of the QUALITY subroutine when the size field is not so well-behaved.

Proposition 14.10. *If the size field λ is δ-Lipschitz for $\delta < 1$, and $\lambda(x) \leq 0.08f(x)$ for every point $x \in \Sigma$,* DELSURF(Σ, λ) *returns a mesh whose triangles have radius-edge ratios at most* $1/(1-\delta)$.

PROOF. Because $\lambda(x) \leq 0.08f(x)$, DELSURF refines no triangle solely for topological reasons; every vertex $p \in S$ has no vertex within a distance of $\lambda(p)$ at the time p is inserted.

Let σ be an output triangle whose shortest edge has length ℓ and whose surface Delaunay ball has center c and radius r. The circumradius of σ is less than or equal to r. Let p be the most recently inserted vertex of the shortest edge of σ; because it was inserted after the other vertex, $\ell > \lambda(p)$. Because no vertex was inserted at c and λ is δ-Lipschitz, $r \leq \lambda(c) \leq \lambda(p) + \delta r$; rearranging terms gives $r \leq \lambda(p)/(1-\delta)$. Thus, $\rho(\sigma) \leq r/\ell \leq 1/(1-\delta)$. □

We use this proposition to analyze QUALITY$(S, \bar{\rho})$, particularly how the grading of the mesh depends on $\bar{\rho}$, when the size field is arbitrary. Define the $(\bar{\rho}-1)/\bar{\rho}$-Lipschitz *regularized size field*

$$\mu(x) = \inf_{y \in \Sigma} \left(\min\{\lambda(y), 0.08f(y)\} + \frac{\bar{\rho}-1}{\bar{\rho}} d(x,y) \right),$$

and observe that $\mu(x) \leq \lambda(x)$ and $\mu(x) \leq 0.08f(x)$. We will see that the Lipschitz property captures how refining for quality can affect the edge lengths in the mesh. The more $\bar{\rho}$ exceeds 1, the more strongly the mesh can be graded.

The idea is that if we call DELSURF(Σ, μ) with the regularized size field μ but no QUALITY option, it splits every triangle that would be split by DELSURF(Σ, λ) with the original size field λ and the QUALITY$(S, \bar{\rho})$ option, and possibly other triangles as well. Therefore, the latter does not refine more than the former.

Theorem 14.11. *If $\bar{\rho} \geq 1$,* DELSURF(Σ, λ) *with the* QUALITY$(S, \bar{\rho})$ *option terminates and returns a mesh in which every vertex p is at a distance greater than $\mu(p) \cdot \bar{\rho}/(2\bar{\rho}-1)$ from the nearest distinct vertex.*

PROOF. First, we claim that every vertex c is at a distance greater than $\mu(c)$ from the nearest vertex inserted before it. Suppose for the sake of contradiction that a vertex c is inserted such that $d(c,p) \leq \mu(c)$, where p was inserted prior to c. Suppose that c is the first such vertex. As $\mu(c) \leq \min\{\lambda(c), 0.08f(c)\}$, c must have been inserted by the subroutine QUALITY at the center of a surface Delaunay ball of a skinny triangle.

If DELSURF had been called with the regularized size field $\mu(p)$ but without the QUALITY option, it could have created the same skinny triangle, because c is the first vertex whose insertion is not justified by μ, but DELSURF would have declined to split the skinny triangle. But this contradicts Proposition 14.10. Therefore, $d(c,p) > \mu(c)$.

Figure 14.8: Polygonal surfaces (top row) remeshed by DelSurf (bottom row).

Consider the same distance from the point of view of a point p inserted before c. By the Lipschitz property of μ, $d(c, p) > \mu(c) \geq \mu(p) - d(c, p) \cdot (\bar{\rho} - 1)/\bar{\rho}$. Rearranging terms gives $d(c, p) > \mu(p) \cdot \bar{\rho}/(2\bar{\rho} - 1)$. □

Although Quality($S, \bar{\rho}$) and a $(\bar{\rho} - 1)/\bar{\rho}$-Lipschitz size field appear to be two theoretically equivalent paths to quality, the former is lazier—it refines a triangle only when really necessary to improve its quality—so it tends to produce meshes with fewer triangles than the theory suggests. It also has the advantages of working well in conjunction with TopoDisk and Smooth and not requiring the user to implement a Lipschitz size field.

A *polygonal surface* is a PLC whose underlying space is a 2-manifold without boundary. Although DelSurf is designed to mesh smooth surfaces, it often succeeds in meshing or remeshing polygonal surfaces that closely approximate smooth surfaces, in the sense that their faces meet at dihedral angles close to π. *Remeshing* is the act of replacing an existing mesh with a new one to improve its quality, to make it finer or coarser, or to make its faces be Delaunay, as illustrated in Figure 14.8. The initial Delaunay triangulation can be built from the vertices of the input surface, but if the goal is to produce a coarser mesh, or if the input surface has closely spaced vertices that are harmful to mesh quality, a well-spaced subset of the input vertices is a better start. Note that in practice, we have never needed to use a persistent triangle for remeshing.

One of the authors of this book has software available that implements the algorithm DelSurf. There are two programs: SurfRemesh[1] takes polygonal surfaces as input and

[1]http://www.cse.ohio-state.edu/~tamaldey/surfremesh

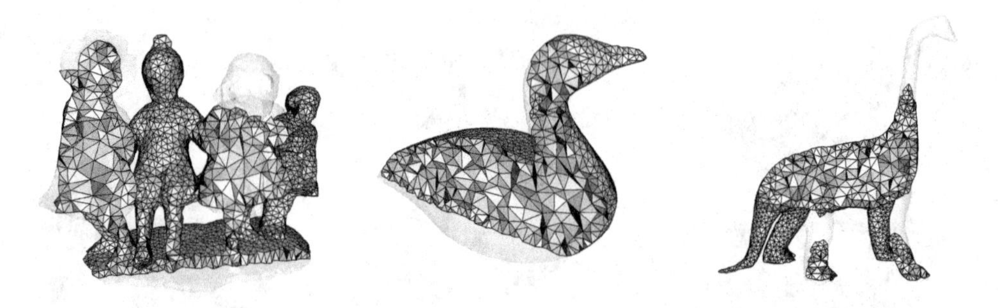

Figure 14.9: Tetrahedral meshes generated by DELTETSURF.

remeshes them, whereas DELISO[2] takes voxel data as input, interpolates an isosurface over the grid, and constructs a surface mesh.

14.5 Tetrahedral meshing of volumes with smooth boundaries

The beauty of using a restricted Delaunay triangulation to generate a surface mesh is that it comes with a byproduct: a Delaunay tetrahedralization of the region it bounds. The tetrahedra may have poor quality, but we can refine them, as the meshes in Figure 14.9 illustrate; observe that their boundary triangulations are the same polygonal meshes shown in Figure 14.8. This section discusses an algorithm guaranteed to produce meshes in which no tetrahedron has a radius-edge ratio greater than 2.

Tetrahedron refinement introduces vertices in the interior of the domain that can damage the surface triangulation. We solve this problem by adopting an encroachment rule similar to those used for polyhedral domains in Chapters 6 and 8. Unfortunately, curved surfaces bring a new hazard: internal vertices can be inserted too close to the boundary before the boundary is fully resolved. Imagine a concave dent in a surface. As surface refinement places new vertices on the dent, an internal vertex may become exposed because it is too close to the dent, perhaps even outside the domain. A consequence is that restricted Delaunay triangles can appear whose vertices do not all lie on the surface. We fix these triangles by deleting their internal vertices.

Let $\mathcal{O}$ be a volume enclosed by a smooth surface Σ. The definitions of restricted Voronoi diagram and restricted Delaunay triangulation work with $\mathcal{O}$ as they do for Σ, though $\mathcal{O}$ is not a surface. Thus $\text{Vor}|_{\mathcal{O}}\, S$ is the Voronoi diagram restricted to the domain $\mathcal{O}$, and $\text{Del}|_{\mathcal{O}}\, S$ is a subcomplex of $\text{Del}\, S$ containing the simplices dual to faces of $\text{Vor}|_{\mathcal{O}}\, S$. These include all the simplices in $\text{Del}|_{\Sigma}\, S$ and more. A tetrahedron is in $\text{Del}|_{\mathcal{O}}\, S$ if its circumcenter is in $\mathcal{O}$. The algorithm does not need to keep track of dangling simplices, so we define the *restricted Delaunay* 3*-subcomplex*, denoted

$$\text{Del}|^3_{\mathcal{O}}\, S = \{\tau : \tau \text{ is a face of a tetrahedron in } \text{Del}|_{\mathcal{O}}\, S\}.$$

When refinement is done, the tetrahedral meshing algorithm returns the mesh $\text{Del}|^3_{\mathcal{O}}\, S$.

The *boundary complex* of $\text{Del}|^3_{\mathcal{O}}\, S$, written $\partial\text{Del}|^3_{\mathcal{O}}\, S$, is the subcomplex containing all the triangles that are faces of exactly one tetrahedron in $\text{Del}|^3_{\mathcal{O}}\, S$, and the faces of those

[2]http://www.cse.ohio-state.edu/~tamaldey/deliso.html

triangles. It is easy to see that $\partial \text{Del}|^3_{\mathcal{O}} S \subseteq \text{Del}|^2_{\Sigma} S$, because every triangle in $\partial \text{Del}|^3_{\mathcal{O}} S$ dualizes to a Voronoi edge that has one vertex in $\mathcal{O}$ and one outside and, therefore, intersects Σ. However, $\text{Del}|^2_{\Sigma} S$ can contain additional triangles whose dual Voronoi edges intersect Σ tangentially or at an even number of points, so we refine these triangles by inserting vertices at those intersection points. This refinement is essentially a special case of using vertices returned by the subroutine VorEdge to refine. Ideally, we would like to have $\partial \text{Del}|^3_{\mathcal{O}} S = \text{Del}|^2_{\Sigma} S$ and for both to be a triangulation of the surface Σ.

The algorithm begins by computing a persistent triangle. Just like DelSurf, it refines triangles in $\text{Del}|^2_{\Sigma} S$ by inserting the centers of their largest surface Delaunay balls. The presence of the persistent triangle ensures that surface refinement never gets stuck. The persistent triangle is deleted as described in Section 14.1.2 once a surface mesh is built. It is important to delete the persistent triangle before refining tetrahedra; otherwise, unnecessarily small tetrahedra will be generated around its vertices.

Tetrahedra whose radius-edge ratios exceed $\bar{\rho}$ are split by new vertices inserted at their circumcenters, but a circumcenter that lies inside a surface Delaunay ball is rejected, and an encroached surface triangle is split instead. The quality threshold $\bar{\rho}$ must be at least 2 to guarantee that the algorithm terminates. Pseudocode for the main algorithm follows.

DelTetSurf($\Sigma, \lambda, \bar{\rho}$)

1. Compute a persistent triangle with vertices on Σ. Let S be the set of vertices of the persistent triangle. Compute Del S and $\text{Del}|^2_{\Sigma} S$.
2. While $\text{Del}|^2_{\Sigma} S$ has a vertex $p \notin \Sigma$, delete p from S; update Del S, $\text{Del}|^2_{\Sigma} S$, and $\text{Del}|^3_{\mathcal{O}} S$; and repeat Step 2.
3. If some triangle in $\text{Del}|^2_{\Sigma} S$ has a surface Delaunay ball $B(c, r)$ with $r > \lambda(c)$, insert c into S; update Del S, $\text{Del}|^2_{\Sigma} S$, and $\text{Del}|^3_{\mathcal{O}} S$; and go to Step 2.
4. If some triangle σ in $\text{Del}|^2_{\Sigma} S$ is a face of zero or two tetrahedra in $\text{Del}|^3_{\mathcal{O}} S$, insert $c_{\max}(\sigma)$ into S; update Del S, $\text{Del}|^2_{\Sigma} S$, and $\text{Del}|^3_{\mathcal{O}} S$; and go to Step 2.
5. For every $p \in S$, compute $c \leftarrow$ TopoDisk(p); if some c is non-null, stop looping, insert c into S; update Del S, $\text{Del}|^2_{\Sigma} S$, and $\text{Del}|^3_{\mathcal{O}} S$; and go to Step 2.
6. If S contains more than one vertex of the persistent triangle, delete one; update Del S, $\text{Del}|^2_{\Sigma} S$, and $\text{Del}|^3_{\mathcal{O}} S$; and go to Step 2.
7. If there is a tetrahedron $\tau \in \text{Del}|^3_{\mathcal{O}} S$ for which $\rho(\tau) > \bar{\rho}$, then
   (a) If the circumcenter z of τ lies inside a surface Delaunay ball $B(c, r)$, insert c into S; update Del S, $\text{Del}|^2_{\Sigma} S$, and $\text{Del}|^3_{\mathcal{O}} S$; and go to Step 2.
   (b) Insert z into S and update Del S, $\text{Del}|^2_{\Sigma} S$, and $\text{Del}|^3_{\mathcal{O}} S$.
   (c) If the insertion of z creates a new triangle in $\text{Del}|^2_{\Sigma} S$, let $B(c, r)$ be the largest new surface Delaunay ball, delete z from S, insert c into S, and go to Step 2.
   (d) Go to Step 7.
8. Return $\text{Del}|^3_{\mathcal{O}} S$.

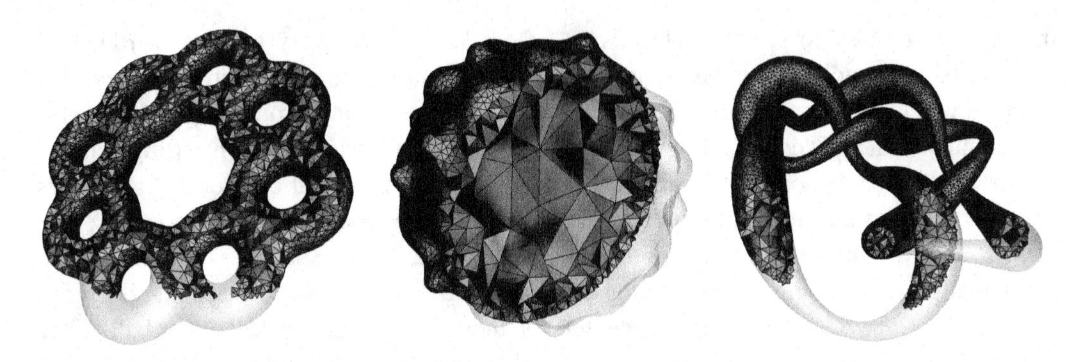

Figure 14.10: More tetrahedral meshes generated by DelTetSurf.

Occasionally, the refinement of a boundary triangle can expose an internal vertex, making it become a vertex of $\mathrm{Del}|^2_\Sigma S$; when this happens, Step 2 purges all such internal vertices before any further refinement can occur. Step 3 refines oversized surface triangles as DelSurf does. Step 4 refines any surface triangle in $\mathrm{Del}|^2_\Sigma S$ that is missing from $\partial\mathrm{Del}|^3_\mathcal{O} S$. Step 5 ensures that $\mathrm{Del}|^2_\Sigma S$ is a 2-manifold. Step 6 deletes vertices of the persistent triangle, which could be unnecessarily close to each other. Step 7 refines skinny tetrahedra, but if a tetrahedron circumcenter threatens to delete or create a triangle in $\mathrm{Del}|^2_\Sigma S$, the circumcenter is rejected and a restricted Delaunay triangle is split instead. Step 7(a) splits restricted Delaunay triangles that are encroached upon by rejected circumcenters. Step 7(c) identifies circumcenters whose insertion creates new restricted Delaunay triangles without deleting any old ones; this occurs only when the surface mesh is coarse and the shape of Σ is still being discovered.

Figure 14.10 shows more volume meshes generated by DelTetSurf.

DelTetSurf raises the question of how to identify which tetrahedra in $\mathrm{Del}\, S$ are in $\mathrm{Del}|^3_\mathcal{O} S$, which is necessary for Step 7. These tetrahedra are the ones whose dual Voronoi vertices are in $\mathcal{O}$. Recall from Section 14.1.3 that there are at least two different ways to represent a surface: as a collection of patches or polygons, or as an isosurface.

Suppose the input is a set of patches whose underlying space is a 2-manifold Σ without boundary, enclosing the volume $\mathcal{O}$ we wish to mesh. The edges of $\mathrm{Vor}\, S$ form a graph whose vertices we wish to label as being either inside the volume—these vertices dualize to tetrahedra in $\mathrm{Del}|^3_\mathcal{O} S$—or outside the volume. Initially, S contains only the three vertices of the persistent triangle, and $\mathrm{Vor}\, S$ contains only one vertex—the vertex at infinity—which is labeled "outside." When a new vertex is inserted into S, we compute labels for the new Voronoi vertices. If one vertex of a Voronoi edge is labeled and the other is not, we can normally compute the label for the second vertex by counting the number of points where the Voronoi edge intersects the patches. However, tangential intersections are tricky because the edge may or may not cross from outside to inside the volume. One way to simplify this problem is to use the subroutine VorEdge to refine these intersections away, modifying Step 4.

Isosurfaces, by contrast, make the identification of $\mathrm{Del}|^3_\mathcal{O} S$ easy—even easier than identifying $\mathrm{Del}|^2_\Sigma S$. Recall that an isosurface of a function $h : \mathbb{R}^3 \to \mathbb{R}$ is a level set $\Sigma = h^{-1}(\eta)$ for a selected isovalue η. Suppose that h is greater than η inside the volume to be meshed;

hence, $\mathcal{O} = \{p : h(p) \geq \eta\}$. Then determining whether the dual of a Voronoi vertex v is a restricted Delaunay tetrahedron is a simple matter of evaluating $h(v)$.

For an isosurface, computing $\mathrm{Del}|_\Sigma^2 S$ is more difficult than computing $\mathrm{Del}|_\mathcal{O}^3 S$, as it necessitates extra computation to identify Voronoi edges that cross the isosurface an even number of times. It is natural to ask whether these computations can be eliminated by replacing $\mathrm{Del}|_\Sigma^2 S$ with $\partial\mathrm{Del}|_\mathcal{O}^3 S$ in DELTETSURF (and deleting Step 4). This question remains open (see Exercise 9), although it is clear that the persistent triangle must be replaced with a stronger strategy; $\partial\mathrm{Del}|_\mathcal{O}^3 S$ is empty when S contains only the vertices of the persistent triangle. It is always possible to fall back on edge-surface intersection computations when $\partial\mathrm{Del}|_\mathcal{O}^3 S$ is empty.

14.5.1 Proof of termination and guarantees

As usual, our goal is to prove that there is a positive lower bound on the distances among the vertices; hence, DELTETSURF terminates, and it returns a mesh in which every tetrahedron has a radius-edge ratio at most $\bar{\rho}$.

The size field λ and local feature size function f are defined only on Σ. We define a function that extends λ's domain over the entirety of $\mathcal{O}$ and forces it to be $(\bar{\rho} - 2)/(3\bar{\rho})$-Lipschitz everywhere in $\mathcal{O}$. Let $\mu : \mathcal{O} \to \mathbb{R}$ be the regularized size field

$$\mu(x) = \inf_{y \in \Sigma} \left(\min\{\lambda(y), 0.08 f(y)\} + \frac{\bar{\rho} - 2}{3\bar{\rho}} d(x, y) \right),$$

and observe that $\mu(x) \leq \lambda(x)$ and $\mu(x) \leq 0.08 f(x)$ for every point $x \in \Sigma$. Note that we use μ solely to analyze the algorithm, not as an input. As in Section 14.4, the Lipschitz property captures how refining for quality affects the edge lengths. The more $\bar{\rho}$ exceeds 2, the more strongly the mesh can be graded.

Recall from Section 8.3 that the *insertion radius* r_x of a vertex x is the distance from x to the nearest distinct vertex in the sample S when x is inserted into S. Immediately after its insertion, r_x is the length of the shortest edge adjoining x in Del S.

Proposition 14.12. *Let S be the set of vertices in a mesh returned by* DELTETSURF$(\Sigma, \lambda, \bar{\rho})$. *If $\bar{\rho} \geq 2$, then every vertex $p \in S$ has insertion radius $r_p > \mu(p)$, and every internal vertex $p \in S \setminus \Sigma$ has insertion radius $r_p > \mu(p) \cdot 3\bar{\rho}/(\bar{\rho} + 1)$.*

PROOF. When Step 3 of DELTETSURF inserts a vertex c, it is at the center of a surface Delaunay ball whose interior contains no vertex and whose radius exceeds $\lambda(c)$. Therefore, $r_c > \lambda(c) \geq \mu(c)$.

When Step 4 of DELTETSURF inserts a vertex c, it is at the center of a surface Delaunay ball of a triangle $\sigma \in \mathrm{Del}|_\Sigma^2 S$ that is a face of either zero or two tetrahedra in $\mathrm{Del}|_\mathcal{O}^3 S$. Therefore, the vertices of its dual Voronoi edge σ^* are either both in $\mathcal{O}$ or both outside $\mathcal{O}$, so σ^* intersects Σ either tangentially or at more than one point. The new vertex c is the same vertex that would be returned by VOREDGE(σ^*). Let p be a vertex of σ; thus, σ^* is an edge of the Voronoi cell V_p. By the same reasoning given in the proof of Proposition 14.6, $d(c, p) > 0.15 f(p)$. Therefore, $r_c > 0.15 f(p) > 0.13 f(c) \geq \mu(c)$ by the Feature Translation Lemma.

When Step 5 of DELTETSURF inserts a vertex c, some vertex p adjoins a nonempty set of triangles in $\text{Del}|^2_\Sigma S$ that do not form a topological disk. So Σ_p is nonempty. Σ_p is not 0.09-small; otherwise, the Small Intersection Theorem (Theorem 13.6) would imply that the triangles adjoining p form a topological disk. It follows that some triangle adjoining p has a surface Delaunay ball with radius $d(c, p) > 0.09f(p) > 0.08f(c)$ by the Feature Translation Lemma. Therefore, $r_c > 0.08f(c) \geq \mu(c)$.

We use induction to show that a vertex inserted by Step 7 also satisfies the claims. Assume that $r_p > \mu(p)$ for every previously inserted vertex p. This assumption holds before the first tetrahedron is refined because Steps 3, 4, and 5 ensure it for every vertex except the vertices of the persistent triangle, two of which are deleted by Step 6. Step 7 inserts a vertex either at the circumcenter z of a tetrahedron τ whose radius-edge ratio exceeds $\bar{\rho}$ or at the center c of a surface Delaunay ball that contains z. The circumradius of τ is $R_\tau > \bar{\rho}\ell$, where ℓ is the length of the shortest edge of τ. Let p be the most recently inserted vertex of that shortest edge. Then $r_p \leq \ell$ and

$$R_\tau \; > \; \bar{\rho} r_p \; \geq \; \bar{\rho} \cdot \mu(p).$$

As μ is $(\bar{\rho} - 2)/(3\bar{\rho})$-Lipschitz,

$$\begin{aligned} R_\tau \; &> \; \bar{\rho}\left(\mu(z) - \frac{\bar{\rho}-2}{3\bar{\rho}} d(p, z)\right) \\ &= \; \bar{\rho}\left(\mu(z) - \frac{\bar{\rho}-2}{3\bar{\rho}} R_\tau\right) \\ \Rightarrow R_\tau \; &> \; \frac{3\bar{\rho}}{\bar{\rho}+1} \mu(z). \end{aligned} \tag{14.2}$$

If z does not encroach upon a surface Delaunay ball, then Step 7(b) inserts z and $r_z = R_\tau > \mu(z) \cdot 3\bar{\rho}/(\bar{\rho}+1)$, as claimed.

Suppose the circumcenter z of τ encroaches upon a surface Delaunay ball B, and Step 7(a) inserts a vertex c at its center; or the insertion of z creates a new surface Delaunay ball B, and Step 7(c) inserts a vertex c at its center. In either case, there is no vertex in B's interior, so r_c is the radius of B. Because $z \in B$, $d(c, z) \leq r_c$, so

$$\begin{aligned} \mu(z) \; &\geq \; \mu(c) - \frac{\bar{\rho}-2}{3\bar{\rho}} d(c, z) \\ &\geq \; \mu(c) - \frac{\bar{\rho}-2}{3\bar{\rho}} r_c \end{aligned} \tag{14.3}$$

The ball B circumscribes some triangle in $\text{Del}|^2_S \Sigma$. Let p be a vertex of this triangle other than z. Observe that p lies on the boundary of B, but not in the open circumball of the Delaunay tetrahedron τ. As $z \in B$,

$$\begin{aligned} 2r_c \; &\geq \; d(p, z) \\ &\geq \; R_\tau \\ &\overset{(14.2)}{>} \; \frac{3\bar{\rho}}{\bar{\rho}+1} \mu(z) \\ 2(\bar{\rho}+1) r_c \; &\overset{(14.3)}{\geq} \; 3\bar{\rho} \cdot \mu(c) - (\bar{\rho}-2) r_c. \end{aligned}$$

Rearranging terms gives $r_c > \mu(c)$. □

Proposition 14.13. *If* $\bar{\rho} \geq 2$ *and* $\inf_{x \in \Sigma} \lambda(x) > 0$, DelTetSurf$(\Sigma, \lambda, \bar{\rho})$ *terminates and returns a mesh in which every vertex p is at a distance greater than* $\mu(p) \cdot 3\bar{\rho}/(4\bar{\rho} - 2)$ *from the nearest distinct vertex.*

Proof. Observe that $\mu_{\min} = \inf_{x \in \mathcal{O}} \mu(x)$ is strictly positive because $\inf_{x \in \Sigma} \lambda(x)$ and $\inf_{x \in \Sigma} f(x)$ are. By Proposition 14.12, every vertex p is at a distance greater than $\mu(p)$ from every vertex inserted before it. For any vertex q inserted after p, $d(p, q) > \mu(q) \geq \mu(p) - d(p, q) \cdot (\bar{\rho} - 2)/(3\bar{\rho})$. Rearranging terms gives $d(p, q) > \mu(p) \cdot 3\bar{\rho}/(4\bar{\rho} - 2)$.

By the Packing Lemma (Lemma 6.1), there is an upper bound on the number of vertices in the mesh. The Packing Lemma alone does not guarantee termination, because vertices are sometimes deleted. Step 2 can delete internal vertices, but only vertices of $\mathrm{Del}|_{\Sigma}^2 S$. Step 7 never inserts an internal vertex that changes $\mathrm{Del}|_{\Sigma}^2 S$; an internal vertex can become part of $\mathrm{Del}|_{\Sigma}^2 S$ only during the insertion of a new surface vertex. Therefore, after Step 2 finishes looping, it cannot run again until at least one new surface vertex is inserted. Surface vertices, except two vertices of the persistent triangle, are never deleted. Eventually there will be no room for new surface vertices, then no room for new internal vertices, whereupon DelTetSurf terminates. □

DelTetSurf returns the tetrahedral subcomplex $\mathrm{Del}|_{\mathcal{O}}^3 S$ and enforces the requirement that $\partial \mathrm{Del}|_{\mathcal{O}}^3 S = \mathrm{Del}|_{\Sigma}^2 S$, regardless of the size field λ. If λ is sufficiently small, the two complexes coincide with the subcomplex $\mathrm{Del}|_{\Sigma}^2 (S \cap \Sigma)$ obtained by discarding the vertices not on the surface.

Proposition 14.14. *After* DelTetSurf *terminates, the following statements hold.*

(i) $\partial \mathrm{Del}|_{\mathcal{O}}^3 S = \mathrm{Del}|_{\Sigma}^2 S$.

(ii) *If* $\lambda(x) \leq 0.08 f(x)$ *for every* $x \in \Sigma$, *then* $\partial \mathrm{Del}|_{\mathcal{O}}^3 S = \mathrm{Del}|_{\Sigma}^2 S = \mathrm{Del}|_{\Sigma}^2 (S \cap \Sigma) = \mathrm{Del}|_{\Sigma} (S \cap \Sigma)$, *and all four have an underlying space homeomorphic to* Σ.

Proof. (i) DelTetSurf terminates only if every triangle in $\mathrm{Del}|_{\Sigma}^2 S$ is in $\partial \mathrm{Del}|_{\mathcal{O}}^3 S$ too. Every triangle σ in $\partial \mathrm{Del}|_{\mathcal{O}}^3 S$ is a face of exactly one tetrahedron in $\mathrm{Del}|_{\mathcal{O}}^3 S$, so the Voronoi edge dual to σ has one vertex in $\mathcal{O}$ and one outside. Hence, the Voronoi edge intersects Σ, and σ is in $\mathrm{Del}|_{\Sigma}^2 S$.

(ii) Any time execution reaches Step 7—that is, just before tetrahedron refinement begins or resumes—the restricted Delaunay 2-subcomplex is a triangulation of Σ by Theorem 14.2. As $\lambda(x) \leq 0.08 f(x)$, Σ_p is 0.09-small for every vertex p in the mesh. By the Dense Sample Theorem (Theorem 13.15), every edge in $\mathrm{Vor}\,(S \cap \Sigma)$ that intersects Σ does so transversally at a single point, and by the Homeomorphism Theorem (Theorem 13.16), $|\mathrm{Del}|_{\Sigma} (S \cap \Sigma)|$ is homeomorphic to Σ, which implies that $\mathrm{Del}|_{\Sigma}^2 (S \cap \Sigma) = \mathrm{Del}|_{\Sigma} (S \cap \Sigma)$.

Consider the Voronoi diagram with the interior vertices as well. Steps 2 and 7 of the algorithm ensure that when DelTetSurf terminates, every triangle in $\mathrm{Del}|_{\Sigma}^2 S$ has all three of its vertices on Σ. Therefore, every edge in $\mathrm{Vor}\, S$ that intersects Σ is generated by vertices on Σ and is, thus, included in some edge of $\mathrm{Vor}\,(S \cap \Sigma)$ and intersects Σ transversally at a single point. It follows that $\mathrm{Del}|_{\Sigma}^2 S \subseteq \mathrm{Del}|_{\Sigma}^2 (S \cap \Sigma)$.

If the inclusion $\mathrm{Del}|_\Sigma^2 S \subset \mathrm{Del}|_\Sigma^2 (S \cap \Sigma)$ is strict, then some edge in $\mathrm{Del}|_\Sigma^2 S$ is an edge of a single triangle, because $|\mathrm{Del}|_\Sigma^2 (S \cap \Sigma)|$ is a connected 2-manifold without boundary. But $\partial \mathrm{Del}|_\mathcal{O}^3 S$ cannot contain an edge of exactly one triangle, so $\mathrm{Del}|_\Sigma^2 S = \mathrm{Del}|_\Sigma^2 (S \cap \Sigma)$. □

We summarize the properties of meshes generated by DelTetSurf.

Theorem 14.15. *Let $\mathcal{O}$ be a volume bounded by a C^2-smooth, compact, connected surface $\Sigma \subset \mathbb{R}^3$ without boundary. Suppose $\inf_{x \in \Sigma} \lambda(x) > 0$ and $\bar{\rho} \geq 2$. For properties (iv)–(vi) below, suppose also that $\lambda(x) \leq \varepsilon f(x)$ for some $\varepsilon \leq 0.08$ and every $x \in \Sigma$. Then* DelTetSurf$(\Sigma, \lambda, \bar{\rho})$ *terminates and returns a mesh $\mathcal{T} = \mathrm{Del}|_\mathcal{O}^3 S$ with the following properties.*

(i) *$|\mathcal{T}|$ is a 3-manifold whose boundary is a 2-manifold without boundary.*

(ii) *No tetrahedron in $\mathcal{T}$ has a radius-edge ratio exceeding $\bar{\rho}$.*

(iii) *Every edge adjoining each vertex p is longer than $\mu(p) \cdot 3\bar{\rho}/(4\bar{\rho} - 2)$.*

(iv) *The underlying space of $\mathcal{T}$ is homeomorphic to $\mathcal{O}$, and the two are related by an isotopy.*

(v) *The boundary complex of $\mathcal{T}$ is $\partial \mathcal{T} = \mathrm{Del}|_\Sigma^2 S = \mathrm{Del}|_\Sigma (S \cap \Sigma)$.*

(vi) *The Hausdorff distance between $|\mathcal{T}|$ and $\mathcal{O}$ is less than $15 \frac{\varepsilon^2}{(1-\varepsilon)^2} f_{\max}$, where $f_{\max} = \sup_{p \in \Sigma} f(p)$.*

Proof. (i) By Proposition 14.13, DelTetSurf terminates. At that time, $|\mathrm{Del}|_\Sigma^2 S|$ is a 2-manifold without boundary because TopoDisk does not permit termination otherwise. By Proposition 14.14(i), the boundary complex is $\partial \mathcal{T} = \partial \mathrm{Del}|_\mathcal{O}^3 S = \mathrm{Del}|_\Sigma^2 S$. The space bounded by a 2-manifold embedded in $\mathbb{R}^3$ is a 3-manifold.

(ii, iii) DelTetSurf terminates only when no tetrahedron has a radius-edge ratio exceeding $\bar{\rho}$. The bound on edge lengths is taken from Proposition 14.13.

(iv, v) By Proposition 14.14, $\partial \mathcal{T} = \partial \mathrm{Del}|_\mathcal{O}^3 S = \mathrm{Del}|_\Sigma^2 S = \mathrm{Del}|_\Sigma (S \cap \Sigma)$ has an underlying space homeomorphic to Σ. As discussed in the proof of that proposition, Σ_p is 0.09-small for every $p \in S \cap \Sigma$, so the Surface Discretization Theorem (Theorem 13.22) implies that $|\mathrm{Del}|_\Sigma (S \cap \Sigma)|$ and Σ are related by an ambient isotopy. Specifically, Proposition 13.20 constructs an ambient isotopy $\xi : \mathbb{R}^3 \times [0, 1] \to \mathbb{R}^3$ such that $\xi(|\partial \mathcal{T}|, 0) = |\partial \mathcal{T}|$ and $\xi(|\partial \mathcal{T}|, 1) = \Sigma$. Because $|\partial \mathcal{T}|$ and Σ are both connected 2-manifolds without boundary that partition $\mathbb{R}^3$ into two pieces, the map $\xi(\cdot, 1)$ is a homeomorphism between the regions they enclose, $|\mathcal{T}|$ and $\mathcal{O}$. Moreover, ξ is an ambient isotopy that takes the identity map $\xi(\cdot, 0)|_{|\mathcal{T}|} : |\mathcal{T}| \to |\mathcal{T}|$ to the homeomorphism $\xi(\cdot, 1)|_{|\mathcal{T}|} : |\mathcal{T}| \to \mathcal{O}$.

(vi) The Hausdorff distance between $|\mathcal{T}|$ and $\mathcal{O}$ is realized by points on their boundaries and, therefore, equals the Hausdorff distance between $|\mathrm{Del}|_\Sigma (S \cap \Sigma)|$ and Σ. With $\lambda(x) \leq \varepsilon f(x)$ for some $\varepsilon \leq 0.08$, Σ_p is $\frac{\varepsilon}{1-\varepsilon}$-small for every $p \in \Sigma$. By the Surface Discretization Theorem, the Hausdorff distance between $|\mathrm{Del}|_\Sigma (S \cap \Sigma)|$ and Σ is less than $15 \frac{\varepsilon^2}{(1-\varepsilon)^2} f_{\max}$. □

Recall from Section 14.5 that if Σ is an isosurface, DelTetSurf becomes simpler and faster if we replace $\text{Del}|_\Sigma^2 S$ with $\partial\text{Del}|_O^3 S$ and do not compute $\text{Del}|_\Sigma^2 S$ at all. Unfortunately, we do not know whether the guarantees (iv)–(vi) of Theorem 14.15 hold for $\partial\text{Del}|_O^3 S$ when λ is sufficiently small, although it seems likely to be true. We leave it as an open question in Exercise 9.

14.6 Notes and exercises

Early surface meshing algorithms developed for computer graphics and medical imaging were not concerned with triangle quality. By far the most famous and heavily used of these is the 1987 *marching cubes* algorithm of Lorensen and Cline [141], from the most-cited paper in the history of the conference SIGGRAPH. The marching cubes algorithm triangulates an isosurface of a function h by computing the value of h at each vertex of a cubical grid, then the intersections of the isosurface with the edges of the grid, then a few triangles in each cube that span the intersection points. It can also create isosurfaces from voxel data such as medical images, in which case the intersections of an unknown surface with the grid edges can be estimated by linear interpolation. The algorithm is very fast and performs interpolation only over edges, not over whole cubes. Unfortunately, it can create arbitrarily skinny triangles. Bloomenthal [24] proposes a similar method that uses a tetrahedral background grid.

After the development of Delaunay refinement meshing in the late 1980s, Chew [61] in 1993 presented an algorithm for triangular surface meshing that flips the edges of an existing surface triangulation so that if its vertices are sufficiently dense, the final triangulation is, in some sense, Delaunay. Then the algorithm refines the mesh by inserting vertices at the centers of surface Delaunay balls of poor-quality triangles, until no triangle has an angle less than 30°. In retrospect, the meshes this algorithm generates are usually subsets of a restricted Delaunay triangulation.

The introduction of restricted Delaunay triangulations by Edelsbrunner and Shah [92], their Topological Ball Theorem, and the sampling theory of Amenta and Bern [3] opened a path to developing algorithms with topological guarantees. By connecting Chew's refinement strategy with the ε-sampling theory, Cheng, Dey, Edelsbrunner, and Sullivan [47] give a Delaunay refinement algorithm for generating a restricted Delaunay triangulation that is homeomorphic to a specialized smooth surface for molecular modeling, called a *skin surface* [86].

Boissonnat and Oudot [29, 30] present a Delaunay refinement algorithm for homeomorphic meshing of a more general class of smooth surfaces. Our DelSurf1 pseudocode is essentially this algorithm without support for surface boundaries, which we address in the next chapter. Dey and Levine [76] show that once the sample is sufficiently dense, one may discard the tetrahedralization and continue to refine the surface triangulation; see Exercise 3.

Cheng, Dey, Ramos, and Ray [53] propose a topology-driven surface meshing algorithm that uses critical point computations and the TopoDisk subroutine. Violations of the topological ball property are detected and repaired. The algorithm does not need to estimate local feature sizes. This algorithm is embodied in our DelSurf2 pseudocode.

Our DELTETSURF pseudocode is a variation of the tetrahedral mesh generation algorithm of Oudot, Rineau, and Yvinec [162] for volumes bounded by smooth surfaces. We have modified it by using vertex deletions, thereby improving the radius-edge ratio bounds, permitting the algorithm to work reasonably well with sparse samples, and removing the requirement for a Lipschitz size field. The value of vertex deletion for improving both the theoretical guarantees and the practical performance of Delaunay refinement algorithms was introduced by Chew's pioneering paper [61].

A related tetrahedral meshing algorithm called *variational tetrahedral meshing* combines Delaunay triangulations and vertex smoothing in an interesting way. Recall that the Delaunay triangulation of a fixed vertex set minimizes the volume bounded between the triangulation lifted to the parabolic lifting map and the paraboloid itself, compared with all other triangulations of the same vertex set. What if the domain's interior vertices are not fixed? Chen and Xu [46] suggest globally smoothing the interior vertices by minimizing the same volume. Alliez, Cohen-Steiner, Yvinec, and Desbrun [2] implemented an iterative tetrahedral mesh improvement method that alternates between this global smoothing step and recomputing the Delaunay triangulation—in other words, it alternates between numerical and combinatorial optimization of the triangulation with respect to the same objective function. During smoothing steps, vertices on the boundary are constrained to maintain domain conformity.

A shortcoming of Delaunay refinement is that it does not scale well to meshes that are too large to fit in main memory, because its random patterns of access to a mesh can cause virtual memory to thrash. Dey, Levine, and Slatton [78, 80] study how to make algorithms in this chapter more local in their data access patterns so that they can generate huge meshes without heavy thrashing.

Although most guaranteed-quality algorithms for meshing surfaces and the volumes they enclose are founded on Delaunay triangulations, an exception is the *isosurface stuffing* algorithm of Labelle and Shewchuk [128], which uses an octree to fill a smooth isosurface with tetrahedra whose dihedral angles are guaranteed to be bounded between 10.7° and 164.8°. Unfortunately, no known Delaunay algorithm offers similar guarantees in theory, although some often achieve better angles in practice.

There are many non-Delaunay surface meshing methods that focus on obtaining correct topology, but not element quality, especially for isosurfaces and implicit surfaces. A few examples are the small normal variation method of Plantinga and Vegter [171], the sweep approach of Mourrain and Técourt [153], the adaptive approach of Snyder [207], and the critical point method of Boissonnat, Cohen-Steiner and Vegter [28].

Isosurfaces of smooth functions can be nonmanifold and nonsmooth for isovalues that are critical values. Such isosurfaces cannot be meshed by the methods described in this chapter, but recent research [138, 153] has begun to address this problem.

Edge-surface intersection computations are expensive. Boissonnat and Oudot [30] observe an idea for reducing the number of these computations when the surface Σ is an isosurface of a function h. Boissonnat and Oudot show that, instead of maintaining $\mathrm{Del}|_{\Sigma}^{2} S$, DELSURF1 can maintain a complex containing only the Delaunay triangles dual to the bipolar Voronoi edges, and compute only a single intersection point (and a single surface Delaunay ball) for each such edge, thereby reducing the computation time considerably.

Implementations of algorithms involving curved surfaces are plagued by floating-point roundoff error, which often causes algorithms to fail. Researchers are making geometric algorithms more numerically robust with sophisticated methods from computational real algebraic geometry. The literature on geometric robustness is too large to survey here. The software produced by the CGAL project[3] is making continuing progress in robust computing with curved geometries.

Exercises

1. [30] Let $\mathcal{B}$ be the set of surface Delaunay balls for a sample on a smooth surface Σ without boundary. Prove that Σ is included in the union of the balls in $\mathcal{B}$ if each ball with circumcenter c has a circumradius of at most $0.09f(c)$.

2. Suppose we initialize DelSurf1 with a set of n random points on Σ instead of the vertices of a persistent triangle. Show that the algorithm might be unable to insert additional vertices by showing that for every n, there is a surface and a sample of size n such that no Voronoi edge intersects Σ.

3. [76] Suppose we have computed $\text{Del}|_\Sigma S$ for a 0.09-sample S of a smooth surface Σ without boundary. Design an algorithm that can insert the center of a surface Delaunay ball into S and update $\text{Del}|_\Sigma S$ without accessing the three-dimensional Delaunay triangulation $\text{Del}\, S$. Then design a Delaunay refinement procedure for making a 0.09-sample even denser without computing $\text{Del}\, S$.

4. Suppose that Σ is given as an isosurface $h(x, y, z) = 0$. Explain in detail the numerical computations needed to compute the critical set $X \cap g$ in Step 4 of the subroutine VorFacet.

5. [53] Explain in detail the numerical computations needed for SilhouettePlane and SilhouetteCritical.

6. Some of the meshing algorithms in this chapter use vertex deletions. Suppose that the Voronoi diagram $\text{Vor}\, S$ satisfies the TBP for a sample S of a smooth surface. Prove that for every site $s \in S$ and for every site $p \in S \setminus \{s\}$, each restricted Voronoi cell $V_p \cap \Sigma$ in $\text{Vor}\,(S \setminus \{s\})$ remains a topological disk. (Note that this does not imply that the TBP still holds.)

7. Theorem 14.15 proves that the Hausdorff distance between $|\text{Del}|^3_{\mathcal{O}}\, S|$ and $\mathcal{O}$ is small. Prove that under the same conditions, there is an isotopy relating $|\text{Del}|^3_{\mathcal{O}}\, S|$ to $\mathcal{O}$ that moves each point $x \in |\text{Del}|^3_{\mathcal{O}}\, S|$ by a distance less than $15\varepsilon^2 f(x)$.ğ

8. Suppose we modify DelTetSurf as follows. Explain what happens in each case.

   (a) Exclude Step 6. Theorem 14.15 needs to be changed in this case.

   (b) Exclude Step 7(c). The algorithm might not terminate; why?

[3] http://www.cgal.org

9. Recall that if Σ is an isosurface, it is desirable to replace $\mathrm{Del}|_\Sigma^2 S$ with $\partial\mathrm{Del}|_\mathcal{O}^3 S$ in DelTetSurf. Also recall that $\partial\mathrm{Del}|_\mathcal{O}^3 S \subseteq \mathrm{Del}|_\Sigma^2 S$. Suppose the modified algorithm is started with a point sample $S \subset \Sigma$ such that while it runs, $\partial\mathrm{Del}|_\mathcal{O}^3 S$ never becomes empty. Suppose that $\lambda(x) \leq 0.08 f(x)$ for each $x \in \Sigma$. It is an open problem whether the properties (iv)–(vi) of Theorem 14.15 hold for the modified algorithm. Prove or disprove them.

Chapter 15

Meshing piecewise smooth complexes

Piecewise smooth complexes, PSCs in short, are a natural way to express the well-known idea of a piecewise smooth surface as a complex whose constituent cells are smooth patches and ridges. Where two surface patches meet at a ridge, the ridge in isolation is smooth, but the union of the patches is not necessarily smooth across the ridge. Two patches may meet at an arbitrarily small angle at a ridge. In mesh generation, PSCs present all the difficulties of curved surfaces and all the difficulties of polyhedral domains.

This chapter presents an algorithm for generating triangular surface meshes conforming to PSCs, with a very brief discussion of its extension to tetrahedral volume meshes. We adopt from the algorithm DELTETACUTEPLC in Chapter 9 the idea of using weighted vertices to protect ridges, so that they are respected by the weighted Delaunay triangulation, and the idea of dividing the algorithm into separate protection and refinement stages. Each weighted vertex represents a protecting ball in which unweighted vertices are never inserted. We adopt from the algorithm DELSURF in Chapter 14 the idea of using a restricted Delaunay triangulation—albeit a weighted one—and the idea of refining surface patches by identifying vertices around which the restricted Delaunay triangulation does not form a topological disk. It is possible for many patches to meet at a single vertex or along a single ridge, so each vertex should participate in one topological disk for each patch it adjoins. Figure 15.1 shows the protecting balls and the final mesh returned by our algorithm for a cast model.

As usual, we prove that the algorithm is guaranteed to terminate by proving that there

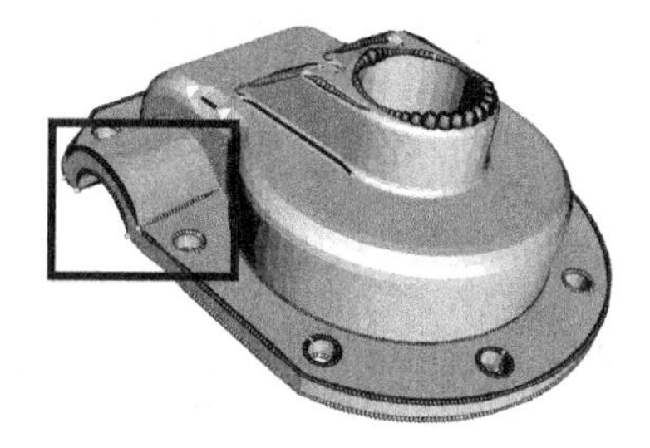

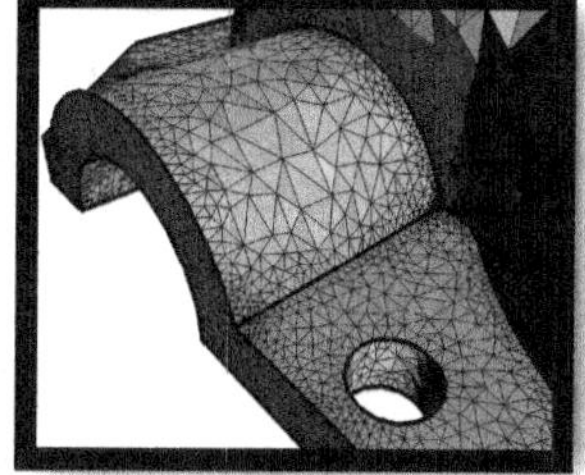

Figure 15.1: Protecting balls (left and center) and final mesh for a cast.

is a positive lower bound on the distances between the mesh vertices. The algorithm terminates only if the topological disk condition holds, so the restriction of the final mesh to any single surface patch is a manifold with boundary. The mesh maintains the incidence relations among the cells of the PSC. As with DELSURF, the mesh might not be homeomorphic to the domain because the algorithm may fail to detect a small topological feature such as a small handle. However, the mesh is guaranteed to be homeomorphic to the domain if the triangle size parameter is sufficiently small. In practice, the disk condition often suffices to obtain a homeomorphic output, just as with DELSURF.

15.1 Piecewise smooth complexes and their triangulations

This chapter is concerned mainly with surface meshing, so we limit our attention to two-dimensional piecewise smooth complexes embedded in three-dimensional space. A PSC is a set of *cells*, each being a smooth connected manifold, possibly with boundary. The 0-cells, 1-cells, and 2-cells are called vertices, ridges, and patches, respectively. A PSC could also contain 3-cells that designate regions to be meshed with tetrahedra, but we will not need them here.

It might seem that the natural way to define smoothly curved cells is to have them meet the smoothness criteria described in Section 12.4. However, such a definition would not rule out certain pathologies that make manifolds with boundaries difficult to mesh. To avoid those problems, we require that each cell of a PSC can be extended so that it has no boundary but is still smooth. In other words, each PSC cell is a subset of some smooth manifold *without boundary*. Recall the definitions of boundary and interior in Definition 12.12.

Definition 15.1 (ridge; patch). A *ridge* is a closed, connected subset of a smooth 1-manifold without boundary in $\mathbb{R}^3$. A *patch* is a 2-manifold that is a closed, connected subset of a smooth 2-manifold without boundary in $\mathbb{R}^3$.

Definition 15.2 (piecewise smooth complex; face of a cell). A *piecewise smooth complex* (PSC) $\mathcal{S}$ is a finite set of vertices, ridges, and patches (*cells* of dimensions zero, one, and two, respectively) that satisfy the following conditions.

- A ridge in $\mathcal{S}$ can intersect a vertex in $\mathcal{S}$ only at its boundary.
- The boundary of each cell in $\mathcal{S}$ is a union of cells in $\mathcal{S}$.
- If two distinct cells in $\mathcal{S}$ intersect, their intersection is a union of cells in $\mathcal{S}$, all having lower dimension than at least one of the two original cells.

The *faces* of a cell $c \in \mathcal{S}$ are the cells $\{g \in \mathcal{S} : g \subseteq c\}$. The *proper faces* of c are the faces of c except c, i.e. $\{g \in \mathcal{S} : g \subset c\}$; they have lower dimension than c.

As a ridge can intersect vertices only at its two endpoints, the third condition implies that two ridges can intersect each other only at their endpoints, in one or two vertices. The boundary of a patch consists of ridges and vertices. If a ridge intersects a patch, either it is included in the patch, or it intersects the patch in one or two vertices. If two patches meet, their intersection is a union of ridges and vertices.

Our main goal in this chapter is to develop an algorithm that generates a triangulation of a PSC. Element quality is a secondary consideration here, although as usual, good radius-edge ratios can be attained by refinement except near small domain angles.

Definition 15.3 (triangulation of a piecewise smooth complex). A simplicial complex $\mathcal{T}$ is a *triangulation* of a PSC $\mathcal{S}$ if there is a homeomorphism h from $|\mathcal{S}|$ to $|\mathcal{T}|$ such that $h(v) = v$ for each vertex $v \in \mathcal{S}$, and for each cell $\xi \in \mathcal{S}$, there is a subcomplex $\mathcal{T}_\xi \subseteq \mathcal{T}$ such that h is a homeomorphism from ξ to $|\mathcal{T}_\xi|$.

A triangulation of a PSC, like a triangulation of a topological space (Definition 12.6), is defined in terms of a homeomorphism; but the homeomorphism also induces a triangulation $\mathcal{T}_\xi$ of each patch, ridge, and vertex ξ of the PSC, and those triangulations preserve the incidence relationships in the PSC. Definition 15.3 requires that the vertices of the PSC are not moved by the homeomorphism, but the algorithm in this chapter enforces the stronger constraint that $h(v) = v$ for each mesh vertex v in $\mathcal{T}$.

Definition 15.2 represents a general ideal of the kind of complex we aspire to mesh—an ideal that will not be fully realized with the tools we currently have. In particular, the definition permits ridges and lone vertices to float in patch interiors. Unfortunately, our main tool for proving that a mesh is a triangulation of a PSC, an extension of the Topological Ball Theorem of Section 13.2, does not extend quite that far, though this extension is likely possible; see Exercise 8. Hence, we define a class of PSCs that are restricted so that cells intersect each other only at their boundaries; vertices and ridges cannot intersect patch interiors.

Definition 15.4 (limited piecewise smooth complex). A *limited piecewise smooth complex* $\mathcal{S}$ is a finite set of vertices, ridges, and patches that satisfy the following conditions.

- The boundary of each cell in $\mathcal{S}$ is a union of cells in $\mathcal{S}$.
- If two distinct cells c_1 and c_2 in $\mathcal{S}$ intersect, their intersection is a union of cells in $\mathcal{S}$ included in the boundary of c_1 or c_2.

To simplify our presentation, we impose several other restrictions on PSCs: every ridge is a topological interval with exactly two vertices in its boundary (a loop must be subdivided into ridges by two or more added vertices); every patch has a nonempty boundary; and every vertex and every ridge is a face of some patch. These definitions and restrictions are tailored for the algorithm in this chapter, but we do not intend to rule out more general conceptions of piecewise smooth complexes elsewhere, having higher-dimensional cells and fewer restrictions. For the rest of this chapter, every reference to a PSC is shorthand for a limited PSC.

15.2 An algorithm for meshing PSCs

Let $\mathcal{S}_i \subseteq \mathcal{S}$ denote the subset containing the i-cells in a PSC $\mathcal{S}$. $\mathcal{S}_0$ contains the vertices, $\mathcal{S}_1$ contains the ridges, and $\mathcal{S}_2$ contains the patches. Let $\mathcal{S}_{\leq i} = \mathcal{S}_0 \cup \cdots \cup \mathcal{S}_i$.

Our PSC meshing algorithm, like the algorithm DelTetAcutePLC in Chapter 9, is divided into two stages, a protection stage and a refinement stage. Both stages maintain a weighted Delaunay triangulation of a weighted point set $S[\omega]$ that includes the vertices in $\mathcal{S}$. The protection stage places a finite set $S[\omega]$ of weighted vertices on the ridges of the input PSC $\mathcal{S}$, satisfying $|\mathcal{S}_0| \subseteq S \subset |\mathcal{S}_{\leq 1}|$. These weighted vertices represent protecting balls in which unweighted vertices are not placed.

The refinement stage generates unweighted vertices on the patches of $\mathcal{S}$—specifically, on $|\mathcal{S}| \setminus |\mathcal{S}_{\leq 1}|$—and inserts them into S, with the invariant that $S \subset |\mathcal{S}|$. Like DELSURF, the PSC meshing algorithm places these vertices at the centers of surface Delaunay balls, but just as DELTETACUTEPLC places vertices at orthocenters rather than circumcenters, the present algorithm uses surface Delaunay balls that are weighted: their centers lie at intersection points of patches with weighted Voronoi edges. Unlike with DELTETACUTEPLC, the refinement stage of the PSC meshing algorithm sometimes replaces some of the protecting balls on the ridges with smaller ones. This adaptive refinement of protecting balls is necessary because we do not use any numerical procedure to directly estimate curvatures or feature sizes, so the algorithm may discover that some protecting balls are too big as the execution progresses.

Recall from Section 7.2 that $W_{p_1 \cdots p_j}$ denotes the weighted Voronoi face in $\text{Vor}\, S[\omega]$ generated by the vertices $p_1, p_2, \ldots,$ and p_j, and $W_{p_1 \cdots p_j}$ is the dual of the weighted Delaunay simplex $\tau = \text{conv}\{p_1, \ldots, p_j\}$. Given a cell $\xi \in \mathcal{S}$, let $\text{Del}|_\xi\, S[\omega]$ denote the restricted weighted Delaunay triangulation of S with respect to ξ, as defined in Section 13.1; i.e. the set of simplices $\tau \in \text{Del}\, S[\omega]$ such that $W_\tau \cap \xi \neq \emptyset$. Define

$$\text{Del}|_{\mathcal{S}_i}\, S[\omega] = \bigcup_{\xi \in \mathcal{S}_i} \text{Del}|_\xi\, S[\omega] \qquad \text{and} \qquad \text{Del}|_{\mathcal{S}}\, S[\omega] = \bigcup_{\xi \in \mathcal{S}} \text{Del}|_\xi\, S[\omega].$$

An i-dimensional cell $\xi \in \mathcal{S}_i$ is usually meshed with i-simplices. However, some simplices in $\text{Del}|_\xi\, S[\omega]$ of dimension less than i may not be faces of any restricted i-simplex, and there is no reason for the algorithm to keep track of those simplices. Therefore, as we define the restricted Delaunay subcomplexes $\text{Del}|^2_\Sigma\, S$ and $\text{Del}|^3_\mathcal{O}\, S$ in Section 14.1, we define the *restricted weighted Delaunay i-subcomplexes*

$$\begin{aligned} \text{Del}|^i_\xi\, S[\omega] &= \{\tau : \tau \text{ is an } i\text{-simplex or a face of an } i\text{-simplex in } \text{Del}|_\xi\, S[\omega]\} \quad \text{and} \\ \text{Del}|^i_{\mathcal{S}_i}\, S[\omega] &= \bigcup_{\xi \in \mathcal{S}_i} \text{Del}|^i_\xi\, S[\omega]. \end{aligned}$$

Our algorithm operates on $\text{Del}|^2_{\mathcal{S}_2}\, S[\omega]$, which equals $\text{Del}|^2_\mathcal{S}\, S[\omega]$, instead of $\text{Del}|_\mathcal{S}\, S[\omega]$.

15.2.1 Protection and the relaxed ball properties

The ridges and vertices in $\mathcal{S}_{\leq 1}$, and their neighborhoods, are regions of particular difficulty for Delaunay refinement. If two ridges with a common vertex meet at a small angle or two patches sharing a common ridge meet at a small angle, ordinary Delaunay refinement may fail to produce a mesh that conforms to the PSC. To overcome these problems, we protect the vertices and ridges with balls, represented by weighted vertices in a weighted Delaunay triangulation.

Throughout its execution, our PSC meshing algorithm maintains a set $S[\omega]$ of weighted mesh vertices with the following invariants.

I1. $|\mathcal{S}_0| \subseteq S \subset |\mathcal{S}|$.

I2. Every vertex $p[\omega_p] \in S[\omega]$ has nonnegative weight ω_p, and its weight is positive if and only if $p \in |\mathcal{S}_{\leq 1}|$.

I3. Every unweighted vertex in $S[\omega]$ has a positive power distance to every weighted vertex in $S[\omega]$.

Invariant I3 implies that no protecting ball contains an unweighted vertex. Consequently, every unweighted point $p \in S$ has a nonempty weighted Voronoi cell W_p in $\text{Vor}\, S[\omega]$.

The protection stage ensures that the protecting balls satisfy what we call the *relaxed ball properties*. The subsequent refinement stage augments the sample and refines the mesh according to a triangle size parameter and a *disk condition*. Whenever the disk condition is violated, some protecting balls may be replaced by smaller ones, or an unweighted vertex may be inserted to refine the triangulation. The relaxed ball properties are maintained throughout the refinement stage. If the protecting balls are refined to a sufficiently small size, the relaxed ball properties imply a stronger set of conditions called the *ball properties*, described in Section 15.3.1. If the ball properties are satisfied and the triangles are refined to a sufficiently small size, then the weighted Voronoi diagram of the mesh vertices has topological properties analogous to those specified in the Voronoi Intersection Theorem (Theorem 13.14). We present the analogous result, called the PSC Lemma, in Section 15.3.2. The ball properties and the PSC Lemma imply that the refinement stage will eventually terminate. The disk condition ensures that the output mesh is topologically valid; if the size parameter is small enough, the output mesh is guaranteed to be topologically equivalent to $\mathcal{S}$.

The relaxed ball properties use the following definitions.

Definition 15.5 (subridge)**.** For a ridge $\gamma \in \mathcal{S}$ and two distinct points $x, y \in \gamma$, let $\gamma(x, y)$ denote the interval of γ whose boundary is $\{x, y\}$, called a *subridge* of γ. The points x and y are called the *endpoints* of $\gamma(x, y)$. Let $d_\gamma(x, y)$ denote the length of the subridge $\gamma(x, y)$. For every subridge $\gamma' \subseteq \gamma$, let $d_\gamma(x, \gamma')$ denote $\inf_{y \in \gamma'} d_\gamma(x, y)$.

Definition 15.6 (interval of a ball in a ridge)**.** For a ridge $\gamma \in \mathcal{S}$, a point $p \in \gamma$, and a ball B_p with center p, the intersection $B_p \cap \gamma$ is a union of one or more topological intervals. Let $\text{seg}_\gamma(B_p)$ denote the interval containing p, illustrated in Figure 15.2. We call $\text{seg}_\gamma(B_p)$ the *interval of B_p in γ*.

Definition 15.7 (relaxed ball properties)**.** Let $\mathcal{S}$ be a PSC. Let $S[\omega]$ be a weighted point set that respects invariants I1 through I3. Each point $p[\omega_p] \in S[\omega]$ with positive weight represents a protecting ball $B_p = B(p, \sqrt{\omega_p})$. Let $r_{\min}$ be the minimum radius among all the protecting balls.

The protecting balls centered at the vertices in $\mathcal{S}_0$ are called *corner-balls*. Two protecting balls are said to be *adjacent* if their centers are consecutive along a ridge, with no other ball center between them. For example, in Figure 15.2, B_u and B_p are adjacent. The following properties are *the relaxed ball properties*.

R1. For every ridge $\gamma \in \mathcal{S}$ and every pair of weighted vertices $p, q \in \gamma \cap S$ that are adjacent along γ, $\text{seg}_\gamma(B_p) \cap \text{seg}_\gamma(B_q) \neq \emptyset$.

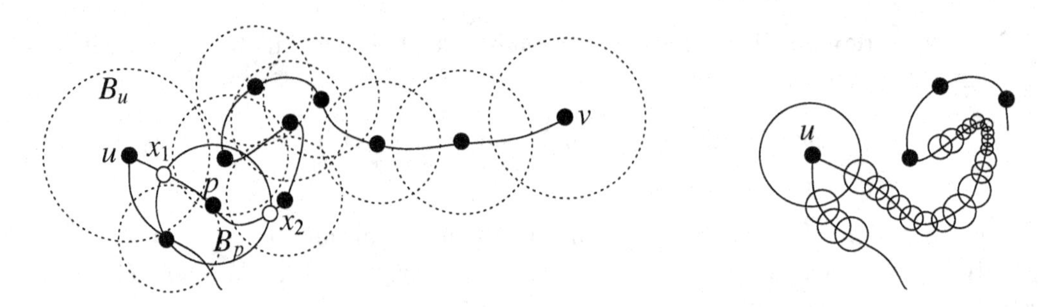

Figure 15.2: Protecting the ridges. At left, a ridge γ with vertices u and v is covered by protecting balls. The ball B_p (solid) intersects γ in two topological intervals, but $\text{seg}_\gamma(B_p)$ is a single interval extending from x_1 to x_2. At right, the balls after further refinement.

R2. Every ridge $\gamma \in \mathcal{S}$ is covered by the intervals of protecting balls centered on γ; i.e. $\gamma = \bigcup_{p \in S \cap \gamma} \text{seg}_\gamma(B_p)$. Both vertices u and v of γ are in S, and are thus protected by corner-balls B_u and B_v. A protecting ball B_p, where $p \in S \cap \gamma$ and p is adjacent to a vertex v of γ, satisfies $\text{radius}(B_p) < \text{radius}(B_v)/3$.

R3. For every ridge $\gamma \in \mathcal{S}$ and every pair of weighted vertices $p, q \in \gamma \cap S$ that are adjacent along γ with $\text{radius}(B_p) \geq \text{radius}(B_q)$,

(a) $d(p,q) \leq \max\{\text{radius}(B_p), \text{radius}(B_q)\} + \frac{14}{15}\min\{\text{radius}(B_p), \text{radius}(B_q)\}$.

(b) One of the following three statements holds.

(i) $\text{radius}(B_p) > 3\,\text{radius}(B_q)$ and $d_\gamma(q, \text{seg}_\gamma(B_p)) \geq r_{\min}/12$;

(ii) $\text{radius}(B_p) = \text{radius}(B_q)$ and $\max\{d_\gamma(q, \text{seg}_\gamma(B_p)), d_\gamma(p, \text{seg}_\gamma(B_q))\} \geq r_{\min}/12$; or

(iii) $\text{radius}(B_p) = \frac{5}{4}\text{radius}(B_q)$ and $d_\gamma(p, \text{seg}_\gamma(B_q)) \geq \text{radius}(B_q)/3 = \frac{4}{15}\text{radius}(B_p)$.

15.2.2 The protection stage

Our algorithm for meshing a PSC $\mathcal{S}$ begins with a protection stage that covers the vertices and ridges in $\mathcal{S}_1$ with protecting balls, as illustrated in Figure 15.3. The center of each protecting ball is a vertex of the mesh.

The protection stage begins by constructing corner-balls, one centered at each vertex of the PSC, and ensuring that these balls are small enough to leave some room between them; then it covers each ridge with additional protecting balls centered on the ridge. The radius of each corner-ball is either a user-specified parameter λ_{prot} or one third the distance from the vertex at its center to the nearest distinct vertex in the PSC, whichever is smaller.

Each ridge is covered by a call to a procedure named COVER, which is also used during the refinement stage to replace protecting balls that are too large. Let γ be a ridge. Let B_p and B_q be two protecting balls centered on γ, and suppose that $\text{seg}_\gamma(B_p)$ and $\text{seg}_\gamma(B_q)$ are disjoint. Let α be a target radius for new balls to be created between B_p and B_q, satisfying the constraints $\alpha \leq \text{radius}(B_p)/4$, $\alpha \leq \text{radius}(B_q)/4$, and $\alpha \leq d(B_p, B_q)/2$. (Note that

Figure 15.3: Protecting balls.

$d(B_p, B_q)$ is the distance between the balls, not their centers.) COVER creates new protecting balls whose intervals cover the subridge $\gamma(p, q) \setminus \mathrm{seg}_\gamma(B_p) \setminus \mathrm{seg}_\gamma(B_q)$.

Let $\gamma(x, z)$ denote the subridge $\gamma(p, q) \setminus \mathrm{seg}_\gamma(B_p) \setminus \mathrm{seg}_\gamma(B_q)$, where x is an endpoint of $\mathrm{seg}_\gamma(B_p)$ and z is an endpoint of $\mathrm{seg}_\gamma(B_q)$. The procedure COVER(x, z, α) covers $\gamma(x, z)$ by generating a sequence of balls $b_1, \ldots, b_{k-1}$ between $b_0 = B_p$ and $b_k = B_q$, as illustrated in Figure 15.4. The balls $b_1, \ldots, b_{k-2}$ have radii α, and b_{k-1} has radius α or $5\alpha/4$. Let c_i be the center of b_i.

The protecting ball b_{i+1} is computed from the previous ball $b_i = B(c_i, r_i)$ with the help of a sequence of up to five *aiding balls* of radius $\alpha/12$. The first aiding ball is centered on the boundary of b_i at the endpoint $y_{i,0}$ of $\mathrm{seg}_\gamma(b_i)$. (If $i = 0$, then $y_{0,0}$ equals x; if $i > 0$, then $y_{i,0}$ is further from x along γ.) Each subsequent aiding ball is centered on the boundary of the previous one, as illustrated; for $j \in [1, 5]$, let $y_{i,j}$ be the endpoint of $\mathrm{seg}_\gamma(B(y_{i,j-1}, \alpha/12))$ that is further from x along γ. Some of these points may be beyond z along γ. Our construction of b_{i+1} depends on where z lies among them. Of the following four rules, the first three are terminal rules that complete the covering, and the last is the most commonly executed.

- If $z \in \gamma(y_{i,0}, y_{i,2})$, then our constraint that $2\alpha \leq d(B_p, B_q)$ implies that b_i is not $b_0 = B_p$, so the radius r_i of b_i is α. We enlarge b_i by replacing it with $B(c_i, 5\alpha/4)$, which overlaps B_q, and COVER terminates. (Figure 15.4, bottom.)

- If $z \in \gamma(y_{i,2}, y_{i,5})$, we place one last protecting ball $B(y_{i,1}, \alpha)$, which overlaps B_q, and COVER terminates. (Figure 15.4, middle-right.)

- If $z \notin \gamma(y_{i,0}, y_{i,5})$ but $z \in \mathrm{seg}_\gamma(B(y_{i,4}, \alpha))$, we place one last protecting ball $B(y_{i,4}, 5\alpha/4)$, and COVER terminates. (We construct a ball of radius $5\alpha/4$ because

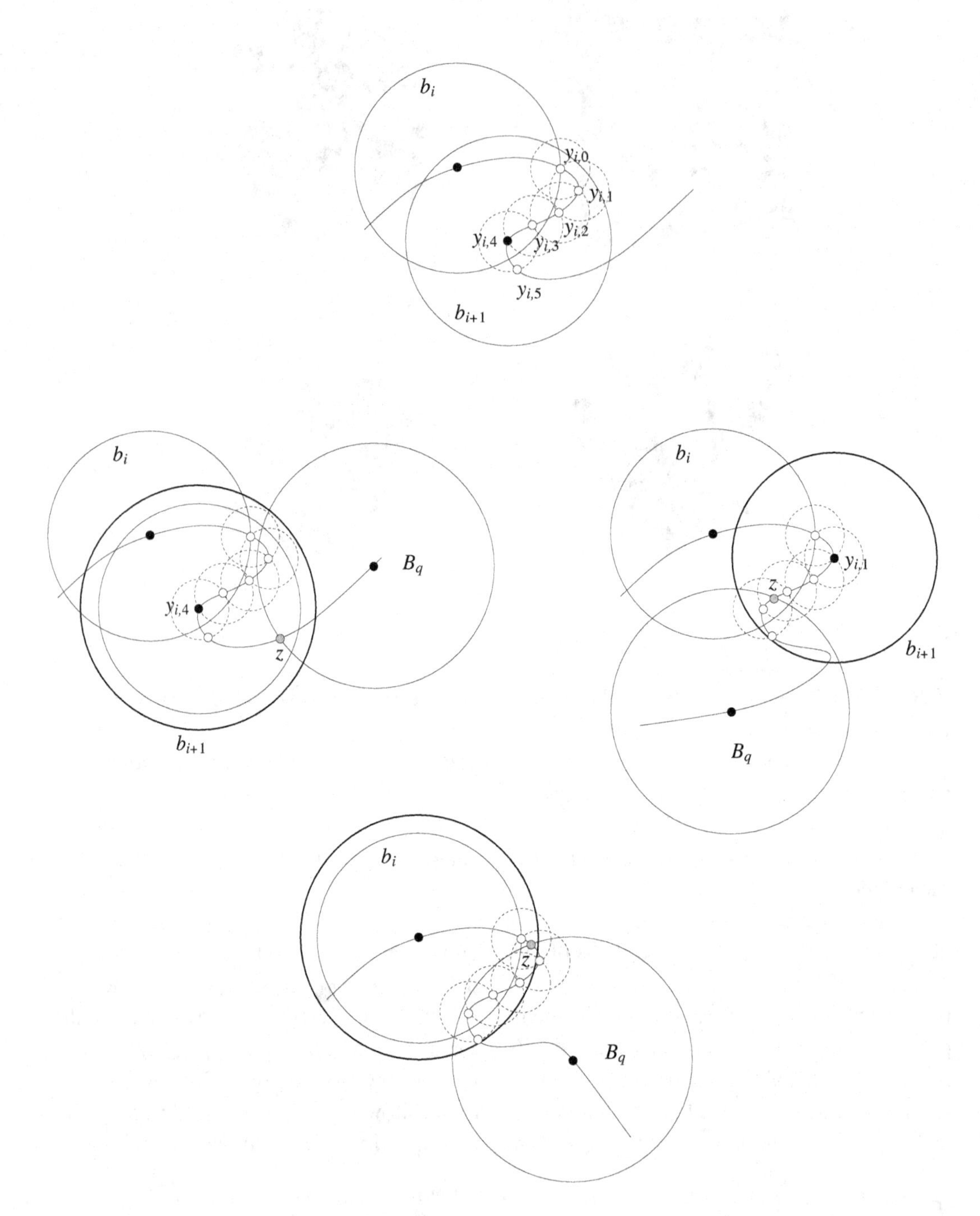

Figure 15.4: Covering the subridge with endpoints $x = y_{i,0}$ and z. Aiding balls are depicted as small dashed circles. The top figure shows the most common case, in which the new ball b_{i+1} is centered at $y_{i,4}$ with radius α. The other three figures show the three terminal cases. The last ball placed has a bold boundary. Note that the actual disparity between the radii of the balls is greater than depicted here.

a ball of radius α might not overlap B_q enough to maintain the relaxed ball property R3(a).) (Figure 15.4, middle-left.)

- If $z \notin \gamma(y_{i,0}, y_{i,5})$ and $z \notin \text{seg}_\gamma(B(y_{i,4}, \alpha))$, then construct $b_{i+1} = B(y_{i,4}, \alpha)$ and continue. (Figure 15.4, top.)

The protection stage is performed by the subroutine ProtectPSC, which takes as input $\mathcal{S}$ and a size parameter λ_{prot}, and returns a set of protecting balls whose radii do not exceed λ_{prot}.

ProtectPSC($\mathcal{S}$, λ_{prot})

1. For each vertex v in $\mathcal{S}$, construct a corner-ball $B_v = B(v, \min\{r_v, \lambda_{\text{prot}}\})$, where r_v is one third the distance from v to the nearest distinct vertex in $\mathcal{S}$. Let $\mathcal{B}$ be the set of corner-balls.
2. For each ridge γ,
   (a) Let u and v be the vertices of γ. Let x and z be the endpoints of the subridge $\gamma \setminus \text{seg}_\gamma(B_u) \setminus \text{seg}_\gamma(B_v)$.
   (b) Let $\alpha = \min\left\{4\lambda_{\text{prot}}/5, \text{radius}(B_u)/4, \text{radius}(B_v)/4\right\}$.
   (c) Set $\mathcal{B} \leftarrow \mathcal{B} \cup \text{Cover}(x, z, \alpha)$.
3. Return $\mathcal{B}$.

Step 2(b) of ProtectPSC does not explicitly enforce $\alpha \leq d(B_u, B_v)/2$ as we require. This condition is already satisfied by α, which is chosen to be at most one fourth the radii of B_u and B_v, which in turn are at most one third of $d(u, v)$ by construction. The following two propositions verify the correctness of Cover and ProtectPSC.

Proposition 15.1. *Let B_p and B_q be protecting balls whose centers p and q lie on a ridge γ, and suppose that $\text{seg}_\gamma(B_p)$ and $\text{seg}_\gamma(B_q)$ are disjoint. Let x and z be the endpoints of the subridge $\gamma(p, q) \setminus \text{seg}_\gamma(B_p) \setminus \text{seg}_\gamma(B_q)$. If $\alpha \leq \min\{\text{radius}(B_p)/4, \text{radius}(B_q)/4, d(B_p, B_q)/2\}$, then* Cover$(x, z, \alpha)$ *terminates and covers $\gamma(x, z)$ with balls of radii α or $5\alpha/4$. In conjunction with B_p and B_q, the balls placed by* Cover(x, z, α) *satisfy the relaxed ball properties R1 and R3.*

Proof. Let $b_1, \ldots, b_{k-1}$ be the protecting balls placed by Cover between $b_0 = B_p$ and $b_k = B_q$. Each has radius α except possibly the last one, which might instead have radius $5\alpha/4$. For $i \in [1, k-1]$, the center of b_i is $y_{i-1,1}$ or $y_{i-1,4}$. Therefore, the centers of b_{i-1} and b_i are separated by a distance of at least $13\alpha/12$ as measured along γ, and Cover(x, z, α) must terminate, as $13(k-1)\alpha/12$ is at most the length of $\gamma(x, z)$.

Conversely, because the distance between B_p and B_q is at least 2α, and the center of each new ball is at most a distance of $\alpha/3$ from the boundary of the previous ball, Cover always places at least two new balls between B_p and B_q.

For $i \in [0, k]$, let c_i and r_i be the center and radius of b_i, respectively. By construction, adjacent balls among $b_0, b_1, \ldots, b_k$ have overlapping intervals in γ, satisfying the relaxed ball property R1. Observe that for all $i \in [0, k-2]$, b_{i+1} includes every aiding ball from

$B(y_{i,0}, \alpha/12)$ to $B(c_{i+1}, \alpha/12)$; therefore, $\gamma(y_{i,0}, c_{i+1}) \subset b_{i+1}$ and $\gamma(c_i, c_{i+1}) \subset b_i \cup b_{i+1}$. By construction, $\text{seg}_\gamma(b_{k-1})$ contains z. We conclude that the intervals of the balls placed by COVER(x, z, α) cover $\gamma(x, z)$.

For $i \in [0, k-1]$, we prove that properties R3(a) and R3(b) hold for b_i and b_{i+1} by checking the following four cases.

- $i = 0$: COVER always places at least two new balls, so b_1 is not the last ball, $c_1 = y_{0,4}$, and $r_1 = \alpha$. The distance between the centers of $b_0 = B_p$ and b_1 is $d(c_0, c_1) \leq r_0 + \alpha/3 = r_0 + r_1/3$, so property R3(a) holds. The two balls also satisfy property R3(b)(i) because $r_0 = \text{radius}(B_p) \geq 4\alpha = 4r_1$ and $d_\gamma(y_{0,0}, c_1) \geq \sum_{j=0}^{3} d(y_{0,j}, y_{0,j+1}) = \alpha/3$.

- $1 \leq i \leq k-3$: As b_{i+1} is not the last ball, $r_i = r_{i+1} = \alpha$ and $c_{i+1} = y_{i,4}$. Property R3(a) holds because $d(c_i, c_{i+1}) \leq r_i + \alpha/3 = 4\alpha/3$, and Property R3(b)(ii) holds because $d_\gamma(y_{i,0}, c_{i+1}) \geq \sum_{j=0}^{3} d(y_{i,j}, y_{i,j+1}) = \alpha/3$.

- $i = k-2$: COVER always places at least two new balls, so $i \geq 1$ and $r_i = \alpha$. The ball b_{i+1} is the last protecting ball and may have radius α or $5\alpha/4$. In either case, Property R3(a) holds because $d(c_i, c_{i+1}) \leq r_i + \alpha/3 = 4\alpha/3$. If $r_{i+1} = 5\alpha/4$, then $c_{i+1} = y_{i,4}$, which implies that $d_\gamma(y_{i,0}, c_{i+1}) \geq \sum_{j=0}^{3} d(y_{i,j}, y_{i,j+1}) = \alpha/3 = r_i/3$, and property R3(b)(iii) holds. If $r_{i+1} = \alpha$, then $c_{i+1} = y_{i,1}$, which implies that $d_\gamma(y_{i,0}, c_{i+1}) \geq \alpha/12$, and property R3(b)(ii) holds.

- $i = k-1$: The ball b_{k-1} is the last protecting ball, so it is created by one of the three terminal rules. In all three cases, $d(c_{k-1}, c_k) \leq r_k + \frac{14}{15} r_{k-1}$, and property R3(a) holds: the three cases have $d(c_{k-1}, z) \leq 7\alpha/6$ and $r_{k-1} = 5\alpha/4$ as shown in Figure 15.4 bottom, or $d(c_{k-1}, z) \leq \alpha/3$ and $r_{k-1} = \alpha$ as shown in Figure 15.4 middle-right, or $d(c_{k-1}, z) \leq \alpha$ and $r_{k-1} = 5\alpha/4$ as shown in Figure 15.4 middle-left. Moreover, $r_k = \text{radius}(B_q) \geq 4\alpha > 3r_{k-1}$ and $d_\gamma(c_{k-1}, z) \geq \alpha/12$, so property R3(b)(i) holds for the balls b_{k-1} and b_k.

□

Proposition 15.2. PROTECTPSC *returns balls that have radii at most λ_{prot} and satisfy the relaxed ball properties.*

PROOF. By Proposition 15.1, COVER(x, y, α) returns balls that satisfy R1 and R3. Every ridge is covered and R2 is satisfied at the end of Step 2. The radius of each corner-ball is at most λ_{prot} by construction. As $\alpha \leq 4\lambda_{\text{prot}}/5$ in Step 2(b), by Proposition 15.1, the radii of the balls placed by COVER are at most $5\alpha/4 = \lambda_{\text{prot}}$. □

15.2.3 Refining the protecting balls

Let $\mathcal{B}$ be the set of protecting balls maintained by the meshing algorithm. Suppose that the refinement stage decides that some protecting ball $b \in \mathcal{B}$ is too large. It calls the procedure REFINEBALL$(\mathcal{B}, b)$ to replace b and possibly some contiguous protecting balls with smaller balls.

RefineBall($\mathcal{B}$, b)

1. If b is not a corner-ball, perform the following steps.
   (a) Let γ be the ridge containing the center of b. Let $\alpha = \text{radius}(b)/4$.
   (b) Let b' and b'' be the two balls in $\mathcal{B}$ adjacent to b along γ. Set $\alpha \leftarrow \min\{\alpha, \text{radius}(b')/4, \text{radius}(b'')/4\}$.
   (c) Delete b from $\mathcal{B}$.
   (d) If $d(b', b'') < 2\alpha$, then set b to be b' or b'', whichever is not a corner-ball, and go to Step 1(b).
   (e) Otherwise, let c' and c'' be the centers of b' and b'', let x and z be the endpoints of the subridge $\gamma(c', c'') \setminus \text{seg}_\gamma(b') \setminus \text{seg}_\gamma(b'')$, call Cover($x, z, \alpha$), insert into $\mathcal{B}$ the balls constructed by the call, and remove the unweighted vertices in S that lie in any ball constructed by the call.
2. Otherwise, b is a corner-ball; perform the following steps.
   (a) Reduce the radius of b by half.
   (b) For each ridge γ adjoining the center of b, let b' be the ball adjacent to b along γ, set $\mathcal{B} \leftarrow$ RefineBall($\mathcal{B}$, b'), and remove all unweighted vertices in S that lie in any ball constructed by the call.
3. Return $\mathcal{B}$.

Steps 1(e) and 2(b) of RefineBall are the only two parts of our algorithm that can remove unweighted vertices from S. They maintain the invariant that every unweighted vertex in S has positive power distances from all the protecting balls, which is necessary to prove that the algorithm terminates. The following proposition verifies the correctness of RefineBall—in particular, that it preserves the relaxed ball properties.

Proposition 15.3. *Let $\mathcal{B}$ be a finite set of protecting balls that satisfy the relaxed ball properties R1, R2, and R3. For any protecting ball b, the call* RefineBall($\mathcal{B}$, b) *returns a modified set of protecting balls that preserve the relaxed ball properties R1, R2, and R3. The radii of the new balls constructed by* RefineBall *are at most* $\text{radius}(b)/2$.

Proof. If b is not a corner-ball, then Steps 1(a), 1(b), and 1(d) ensure that α is no greater than any of $\text{radius}(b)/4$, $\text{radius}(b')/4$, $\text{radius}(b'')/4$, and $d(b', b'')/2$ when Step 1(e) calls Cover. Proposition 15.1 states that the balls created by Cover satisfy properties R1 and R3 and have radii no greater than $\frac{5}{4}\alpha \leq \frac{5}{16}\text{radius}(b)$. Because property R2 was true before the call to Cover, Proposition 15.1 implies that it remains true afterward.

If b is a corner-ball, it is replaced by a new corner-ball with half the radius. The invocations of RefineBall on the balls adjacent to b have the same effects as explained in the previous paragraph. □

15.2.4 The refinement stage

After the protection stage covers $\mathcal{S}_{\leq 1}$ with protecting balls, the refinement stage constructs the weighted Delaunay 2-subcomplex $\mathrm{Del}|^2_{\mathcal{S}}\, S\,[\omega]$ and then refines it by generating vertices on the patches. Like the meshing algorithms for smooth surfaces discussed in Chapter 14, our PSC meshing algorithm recovers and refines patches by inserting new vertices at the centers of surface Delaunay balls. However, the use of weighted vertices necessitates the introduction of a weighted version of surface Delaunay balls.

Consider a patch σ and a triangle $\tau \in \mathrm{Del}|^2_{\sigma}\, S\,[\omega]$. The triangle τ is restricted weighted Delaunay because it is the dual of a weighted Voronoi edge W_τ that intersects σ. The intersection $W_\tau \cap \sigma$ may contain multiple points, each of which is the center of a surface Delaunay ball that is orthogonal to the vertices of τ. The PSC meshing algorithm, like the algorithms for smooth surfaces, can refine τ by inserting a new vertex at the point in $W_\tau \cap \sigma$ that is at the maximum power distance from each vertex of τ. The radius of the surface Delaunay ball centered at that point is

$$\mathrm{size}(\tau, \sigma) = \sqrt{\max\left\{\pi(x, p[\omega_p]) : x \in W_\tau \cap \sigma \;\wedge\; p \text{ is a vertex of } \tau\right\}}.$$

Refinement is driven primarily by checking that every vertex satisfies a disk condition, specified below, which ensures that the restricted Delaunay 2-subcomplex with respect to each patch is a 2-manifold. When a vertex fails the disk condition, the algorithm refines either a protecting ball or a surface Delaunay ball. In the latter case, it inserts a vertex at the center of the largest surface Delaunay ball, for which $\mathrm{size}(\tau, \sigma)$ is a global maximum. Refinement is also driven by surface Delaunay balls whose radii exceed the user-specified size parameter λ_{ref}, which should be less than or equal to λ_{prot}.

Recall that the procedure TopoDisk in Section 14.2.1 tests whether the union of the restricted Delaunay triangles adjoining a vertex is a topological disk with the vertex in its interior. Here, the disk condition requires a change because patches can have boundaries. In particular, if a vertex lies on the boundary of a patch, the restricted Delaunay triangles adjoining it should form a fan that goes partway around the vertex, rather than all the way around. A vertex may lie on the boundaries of several patches, necessitating a fan of triangles for each patch.

Definition 15.8 (disk condition). For a vertex $p \in S$ and a patch $\sigma \in \mathcal{S}$, let $\mathrm{Umb}_\sigma(p)$ denote the set of triangles in $\mathrm{Del}|^2_{\sigma}\, S\,[\omega]$ that adjoin p. A vertex $p \in S$ satisfies the *disk condition* if all four of the following statements are true.

D1. For every patch $\sigma \in \mathcal{S}$, if two weighted points are connected in $\mathrm{Umb}_\sigma(p)$, then they are adjacent along some ridge in $\mathcal{S}$.

D2. Every patch $\sigma \in \mathcal{S}$ contains every vertex of $\mathrm{Umb}_\sigma(p)$.

D3. For every patch $\sigma \in \mathcal{S}$ that contains p, $|\mathrm{Umb}_\sigma(p)|$ is a topological disk.

D4. For every patch $\sigma \in \mathcal{S}$, $p \in \mathrm{Int}\,|\mathrm{Umb}_\sigma(p)|$ if and only if $p \in \mathrm{Int}\,\sigma$.

Figure 15.5 illustrates the disk condition. Observe that the disk condition can be diagnosed with purely combinatorial tests on $\mathrm{Del}|^2_{\sigma}\, S\,[\omega]$.

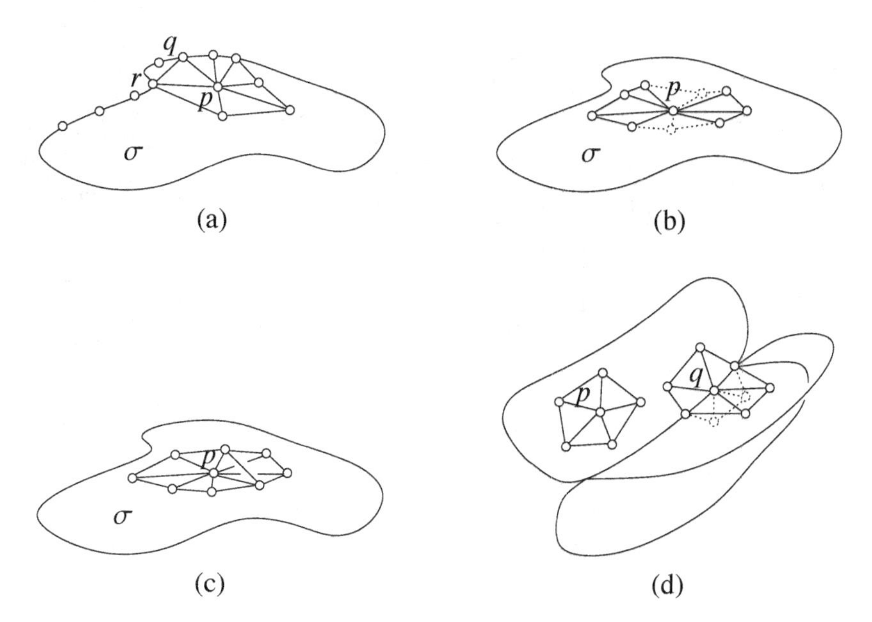

Figure 15.5: Three vertices that do not satisfy the disk condition, and two vertices that do. (a) The triangles in $\mathrm{Del}|^2_\sigma S[\omega]$ adjoining the vertex $p \in \sigma$ form a topological disk. However, they connect two vertices q and r that are not adjacent along a ridge, so p violates disk condition D1. (b) The triangles adjoining $p \in \sigma$ form a topological disk. However, some of their vertices (dashed) do not lie on σ, so p violates disk condition D2. (c) The triangles adjoining $p \in \sigma$ do not form a topological disk, violating disk condition D3. (d) The vertices p and q satisfy the disk condition. The vertex p lies in the interior of σ and also in the interior of the topological disk formed by the triangles adjoining p, as disk condition D4 requires. The vertex q lies on a ridge that is a face of three patches. For each of those patches, the triangles adjoining q form a topological disk with q on its boundary.

In the meshing algorithm, the maximum radius among all surface Delaunay balls, the minimum radius among all protecting balls, and the minimum distance among unweighted vertices play important roles. These values are

$$
\begin{aligned}
s_{\max} &= \max\left\{\mathrm{size}(\tau,\sigma) : \sigma \in \mathcal{S}_2, \tau \in \mathrm{Del}|^2_\sigma S[\omega]\right\}, \\
r_{\min} &= \min\left\{\sqrt{\omega_p} : p[\omega_p] \in S[\omega] \text{ and } \omega_p > 0\right\}, \\
d_{\min} &= \min\left\{d(p,q) : p, q \in S \text{ and } \omega_p = \omega_q = 0\right\}.
\end{aligned}
$$

The pseudocode for our PSC meshing algorithm follows.

DELPSC($\mathcal{S}, \lambda_{\mathrm{prot}}, \lambda_{\mathrm{ref}}$)

1. Let $\mathcal{B}$ = PROTECTPSC($\mathcal{S}, \lambda_{\mathrm{prot}}$).
2. Let $S[\omega] = \{p[\mathrm{radius}(B_p)^2] : B_p \in \mathcal{B}\}$ be a set of weighted vertices representing the balls in $\mathcal{B}$. Construct or update $\mathrm{Del}\, S[\omega]$. Let $r_{\min}$ be the radius of the smallest ball in $\mathcal{B}$. Compute the center x and radius $s_{\max}$ of the largest surface Delaunay ball; thus, x is a point in $W_\tau \cap \sigma$ for some

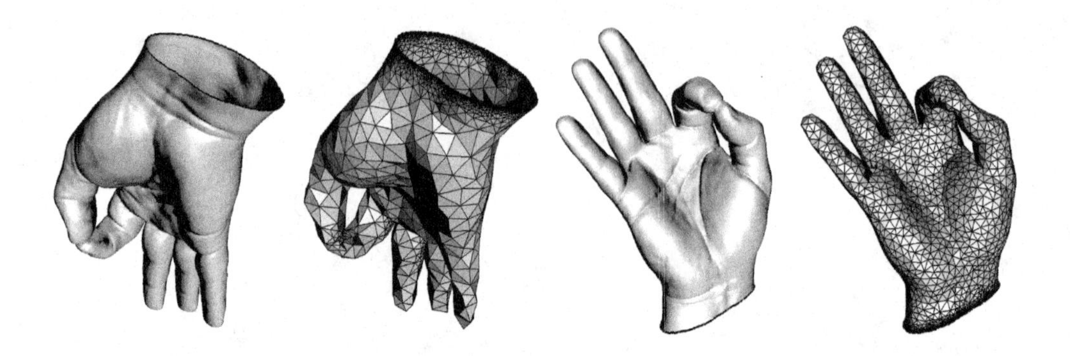

Figure 15.6: A manifold with boundary, meshed at two different resolutions.

patch $\sigma \in \mathcal{S}_2$ and some triangle $\tau \in \mathrm{Del}|^2_\sigma S[\omega]$ with $\mathrm{size}(\tau, \sigma) = s_{\max}$. Compute the minimum distance $d_{\min}$ among the unweighted vertices in S.

(a) If some corner-ball contains the center of a non-corner-ball, then let B be the larger of the two balls, set $\mathcal{B} \leftarrow$ REFINEBALL($\mathcal{B}, B$), and repeat Step 2.

(b) If two balls not centered on a common ridge intersect, then let B be the larger of the two balls, set $\mathcal{B} \leftarrow$ REFINEBALL($\mathcal{B}, B$), and repeat Step 2.

(c) If some vertex in S violates the disk condition D1, then

  (i) If $s_{\max} \geq 0.035\, r_{\min}$, then insert x into S and repeat Step 2.

  (ii) Otherwise, pick a pair of nonadjacent weighted vertices that are connected in $\mathrm{Umb}_\sigma(p)$ for some patch $\sigma \in \mathcal{S}_2$ and vertex $p \in S$, let q be the vertex with greater weight in the pair, set $\mathcal{B} \leftarrow$ REFINEBALL($\mathcal{B}, B_q$), and repeat Step 2.

(d) If some vertex $p \in S$ violates the disk condition D2, then there is a patch σ such that some vertex q of $\mathrm{Umb}_\sigma(p)$ does not lie on σ.

  (i) If $s_{\max} \geq \sqrt{\omega_q}$, then insert x into S and repeat Step 2.

  (ii) Otherwise, set $\mathcal{B} \leftarrow$ REFINEBALL($\mathcal{B}, B_q$) and repeat Step 2.

(e) If a point in S violates the disk condition D3 or D4, then:

  (i) If $s_{\max} \geq \min\{r_{\min}, d_{\min}\}$, then insert x into S and repeat Step 2.

  (ii) Otherwise, let B_m be the largest ball in $\mathcal{B}$, set $\mathcal{B} \leftarrow$ REFINEBALL($\mathcal{B}, B_m$), and repeat Step 2.

(f) If $s_{\max} > \lambda_{\mathrm{ref}}$, then insert x into S and repeat Step 2.

3. Return $\mathrm{Del}|^2_{\mathcal{S}} S[\omega]$.

Figure 15.6 shows meshes returned by DELPSC for two different resolutions specified by the parameter λ_{ref}. Although DELPSC ensures that the final mesh has no triangle whose circumradius exceeds λ_{ref}, some triangles may be much smaller because of refinement near vertices that violate the disk condition.

15.3 The ball properties and the PSC Lemma

The Voronoi Intersection Theorem (Theorem 13.14) and its consequences provide the main theoretical foundation for showing that the surface meshing algorithms in Chapter 14 terminate and return a mesh whose underlying space is a 2-manifold that, under the right conditions, is homeomorphic to the input surface. Here we derive a similar result for PSCs. The *PSC Lemma* states that if the protecting balls satisfy a set of *ball properties* and the triangles are refined to a sufficiently small size, then the PSC intersects Vor $S[\omega]$ in topologically desirable ways. The following sections state the ball properties, show that they are satisfied if the disk condition and the parameter λ_{ref} drive DELPSC to perform sufficient refinement, state the PSC Lemma, and prove its correctness. The algorithm DELPSC can be understood as an effort to achieve the preconditions of the PSC Lemma. A subsequent section uses the PSC Lemma to show that the refinement stage terminates and returns a topologically valid mesh.

As is usual for analyses of Delaunay refinement algorithms, the theory in these sections overestimates the amount of refinement necessary to obtain a conforming mesh in practice. Refinement stops when the disk condition is satisfied, which typically occurs long before the protecting balls are as small as the following theory suggests.

15.3.1 The ball properties

Definition 15.9 (ball properties). Let $\mathcal{S}$ be a PSC. Let $S[\omega]$ be a set of weighted points that represent protecting balls. Let $\delta = 1.17^\circ$. Consider the following *ball properties*.

B1. For every pair of protecting balls B_p and B_q that are adjacent along some ridge,

- (a) $B_p \cap B_q$ is nonempty, and $B_p \cup B_q$ includes the subridge $\gamma(p, q)$;
- (b) $p \notin B_q$ and $q \notin B_p$; and
- (c) for every point $x \in (\text{Bd}\, B_p) \cap (\text{Bd}\, B_q)$, the angle $\angle pxq$ is less than 166°.

B2. Every ridge $\gamma \in \mathcal{S}$ is covered by the intervals of the protecting balls centered on γ; i.e. $\gamma = \bigcup_{p \in S \cap \gamma} \text{seg}_\gamma(B_p)$. Both vertices u and v of γ are in S and are thus protected by corner-balls B_u and B_v. The radius of every corner-ball is larger than or equal to the radius of any adjacent protecting ball.

B3. Let $p \in S$ be a positively weighted point on a ridge $\gamma \in \mathcal{S}$. By Definition 15.1, there exists a 1-manifold without boundary Γ that includes γ. For every $r \le \text{radius}(B_p)$,

- (a) $B(p, 3r) \cap \gamma$ and $B(p, 3r) \cap \Gamma$ are topological intervals;
- (b) for all points $y, z \in B(p, 3r) \cap \Gamma$, $\angle_a(\mathbf{h}_p, yz) > 90^\circ - \delta$, where $\mathbf{h}_p$ is the plane orthogonal to Γ at p.

B4. Let $p \in S$ be a positively weighted point on a ridge γ of a patch $\sigma \in \mathcal{S}$. By Definition 15.1, there exists a 2-manifold without boundary Σ that includes σ. For every $r \le \text{radius}(B_p)$,

- (a) $B(p, 4r) \cap \sigma$ and $B(p, 4r) \cap \Sigma$ are topological disks;

(b) for every point $z \in B(p, 4r) \cap \Sigma$, $\angle(\mathbf{n}_p, \mathbf{n}_z) < \delta$, where $\mathbf{n}_p$ and $\mathbf{n}_z$ are the normals to Σ at p and z; and

(c) for all points $y, z \in B(p, 3r) \cap \Sigma$, $\angle_a(\mathbf{n}_p, yz) > 90° - \delta$.

B5. Any two protecting balls that are not centered on a common ridge are disjoint.

The relaxed ball properties imply the ball properties B1–B4 (albeit not necessarily B5) when the protecting balls are sufficiently small; therefore, DELPSC will eventually attain these ball properties unless it has the good fortune to terminate early.

Proposition 15.4. *There exists a constant $\lambda_b > 0$ depending only on $\mathcal{S}$ such that if every protecting ball has radius less than λ_b and satisfies the relaxed ball properties, then the protecting balls satisfy the ball properties B1–B4.*

PROOF. Ball property B1(a) follows from relaxed ball property R1. Ball property B2 is a subset of relaxed ball property R2. If the radii of the protecting balls are sufficiently small, ball properties B3 and B4 hold because of the smoothness and differentiability of the manifolds.

Consider ball property B1(b). Let p and q be weighted vertices that are adjacent along a ridge γ with radius$(B_p) \geq$ radius(B_q). If R3(b)(i) holds, then $d_\gamma(q, \mathrm{seg}_\gamma(B_p)) \geq r_{\min}/12$. As γ is smooth, for a small enough λ_b, $d(q, B_p)$ is not much less than $d_\gamma(q, \mathrm{seg}_\gamma(B_p))$, so $d(p, q) >$ radius$(B_p) \geq$ radius(B_q), implying that $p \notin B_q$ and $q \notin B_p$. If R3(b)(ii) holds, then radius(B_p) = radius(B_q) and one of $d_\gamma(q, \mathrm{seg}_\gamma(B_p))$ or $d_\gamma(p, \mathrm{seg}_\gamma(B_q))$ is at least $r_{\min}/12$. Again, for a small enough λ_b, $d(q, B_p)$ and $d(p, B_q)$ are not much less than $d_\gamma(q, \mathrm{seg}_\gamma(B_p))$ and $d_\gamma(p, \mathrm{seg}_\gamma(B_q))$, respectively. Thus, $d(p, q) >$ radius(B_p) = radius(B_q), implying that $p \notin B_q$ and $q \notin B_p$. If R3(b)(iii) holds, then $d_\gamma(p, \mathrm{seg}_\gamma(B_q)) \geq$ radius$(B_q)/3 = \frac{4}{15}$radius(B_p). For a small enough λ_b, $d(p, B_q)$ is not much less than $d_\gamma(p, \mathrm{seg}_\gamma(B_q))$, so $d(p, q) >$ radius(B_q) + radius$(B_q)/4$ = radius(B_p), implying that $p \notin B_q$ and $q \notin B_p$.

Ball property B1(c) follows from relaxed ball property R3(a) by the following reasoning. Let p and q be two weighted vertices that are adjacent along a ridge, with $\omega_p \geq \omega_q$. Let Π be the plane that includes the circle $\mathrm{Bd}\, B_p \cap \mathrm{Bd}\, B_q$, and let x be a point on the circle. Let $\theta_p \geq \theta_q$ be the acute angles that px and qx make with Π, respectively, as illustrated in Figure 15.7; then $\angle pxq = \theta_p + \theta_q < 90° + \theta_q$. Relaxed ball property R3(a) states that $d(p, q) \leq \sqrt{\omega_p} + \frac{14}{15}\sqrt{\omega_q}$, which implies that $d(u, v) = d(q, v) + d(p, u) - d(p, q) = \sqrt{\omega_q} + \sqrt{\omega_p} - d(p, q) \geq \sqrt{\omega_q}/15$ and $d(v, w) \geq d(u, v)/2 \geq \sqrt{\omega_q}/30$. It follows that $d(q, w) = d(q, v) - d(v, w) \leq \frac{29}{30}\sqrt{\omega_q}$. As qwx is a right-angled triangle, $\theta_q \leq \arcsin(29/30) < 76°$ and $\angle pxq < 166°$. □

When the protecting ball radii are small enough, there is a gap between nonadjacent balls. The following definitions help make the statement more precise. For any subset $X \subset \mathbb{R}^3$, let ℓ_X be the minimum distance between two disjoint cells in $\{\xi \setminus X : \xi \in \mathcal{S}\}$. For example, $\ell_\emptyset$ is the minimum distance between two disjoint cells in $\mathcal{S}$; ℓ_{B_p} is the minimum distance between two cells that are disjoint after the protecting ball B_p is erased from every ridge and patch. In general, two cells in $\mathcal{S}$ that intersect might have disjoint counterparts in $\{\xi \setminus X : \xi \in \mathcal{S}\}$.

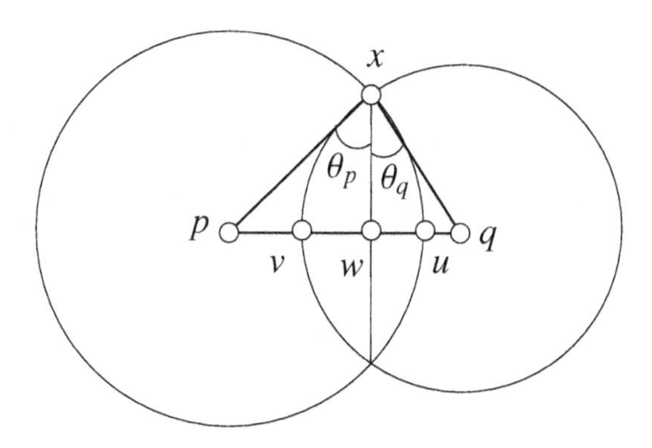

Figure 15.7: The angle $\angle pxq$ is less than 166°.

Proposition 15.5.

(i) *Suppose that p and q are positively weighted vertices, B_p is a corner-ball, q lies on a ridge not adjoining p, and* $\max\{\text{radius}(B_p), \text{radius}(B_q)\} \leq \ell_\emptyset/3$. *Then $d(B_p, B_q) \geq r_{\min}$.*

(ii) *Let V be the union of all the corner-balls. Suppose that $p \notin V$ and $q \notin V$ are positively weighted vertices that do not lie on the same ridge and* $\max\{\text{radius}(B_p), \text{radius}(B_q)\} \leq \ell_V/3$. *Then $d(B_p, B_q) \geq r_{\min}$.*

(iii) *Suppose that the protecting balls satisfy the relaxed ball properties. There exists a constant λ_n depending only on $\mathcal{S}$ such that for every ridge γ and every pair of positively weighted vertices $p, q \in \gamma$ that are not adjacent along γ, if* $\max\{\text{radius}(B_p), \text{radius}(B_q)\} \leq \lambda_n$, *then $d(B_p, B_q) \geq 0.07\, r_{\min}$.*

Proof. (i) As p is a vertex in $\mathcal{S}$, and q lies on a ridge not adjoining p, the distance $d(p, q)$ is at least $\ell_\emptyset$. Thus, $d(B_p, B_q) \geq \ell_\emptyset - \text{radius}(B_p) - \text{radius}(B_q) \geq \ell_\emptyset/3 \geq \text{radius}(B_p) \geq r_{\min}$.

(ii) As $p \notin V$ and $q \notin V$ do not lie on a common ridge, $d(p, q) \geq \ell_V$ and $d(B_p, B_q) \geq \ell_V - \text{radius}(B_p) - \text{radius}(B_q) \geq \ell_V/3 \geq \text{radius}(B_p) \geq r_{\min}$.

(iii) Let v be the weighted vertex adjacent to p along $\gamma(p, q)$. As p and q are not adjacent, $v \neq q$. By the following reasoning, $d(v, B_p) \geq r_{\min}/13$.

One of the relaxed ball properties R3(b) holds for B_p and B_v. If R3(b)(i) holds, then p is a corner-vertex and $d_\gamma(v, \text{seg}_\gamma(B_p)) \geq r_{\min}/12$. As γ is smooth, for a small enough λ_n, $d(v, B_p) \geq r_{\min}/13$ as claimed. If R3(b)(ii) holds, then either $d_\gamma(v, \text{seg}_\gamma(B_p)) \geq r_{\min}/12$, giving the same result, or $d_\gamma(p, \text{seg}_\gamma(B_v)) \geq r_{\min}/12$, giving $d(v, B_p) = d(p, B_v) \geq r_{\min}/13$ because the two balls have the same radius under property R3(b)(ii). If R3(b)(iii) holds and $\text{radius}(B_v) = \frac{5}{4}\text{radius}(B_p)$, then $d_\gamma(v, \text{seg}_\gamma(B_p)) \geq \text{radius}(B_p)/3$ and for a small enough λ_n, $d(v, B_p) \geq r_{\min}/13$. If R3(b)(iii) holds and $\text{radius}(B_p) = \frac{5}{4}\text{radius}(B_v)$, then $d_\gamma(p, \text{seg}_\gamma(B_v)) \geq \text{radius}(B_v)/3$ and for a small enough λ_n, $d(p, B_v) \geq \frac{17}{52}\text{radius}(B_v)$. Therefore, $d(v, B_p) = d(p, v) - \text{radius}(B_p) = d(p, B_v) + \text{radius}(B_v) - \text{radius}(B_p) \geq r_{\min}/13$.

Symmetrically, the weighted vertex w adjacent to q along $\gamma(p, q)$ is not in B_q. If λ_n is small enough, then $B_q \cap \gamma = \text{seg}_\gamma(B_q)$, and the fact that $w \notin B_q$ implies that $v \notin B_q$.

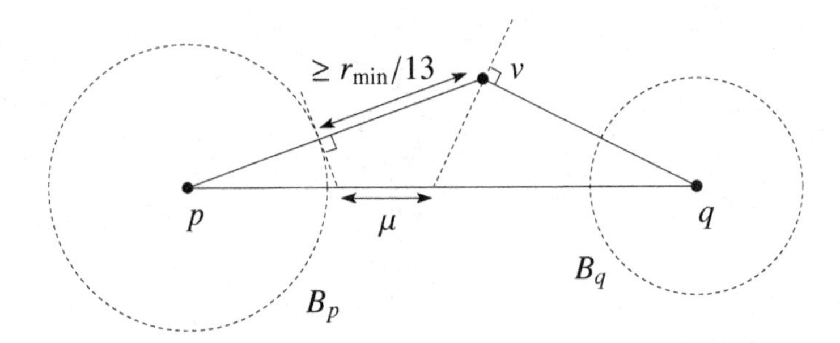

Figure 15.8: As the radii of the protecting balls and the distances between them shrink, $\angle pvq$ approaches $180°$ and μ approaches $d(v, B_p)$.

Suppose for the sake of contradiction that $d(B_p, B_q) < 0.07r_{\min}$. So B_p and B_q are contained in $B(a, 4r)$, where $B(a, r)$ denotes the larger of B_p and B_q. Since $r \leq \lambda_n$ by assumption, $\angle pvq$ approaches $180°$ as λ_n approaches zero. See Figure 15.8. As $\angle pvq$ approaches $180°$, μ approaches $d(v, B_p)$, so we can assume that $\mu > r_{\min}/14$ for a sufficiently small λ_n. But this contradicts the fact that $\mu \leq d(B_p, B_q) < 0.07\, r_{\min}$. Therefore, for a sufficiently small λ_n, $d(B_p, B_q) \geq 0.07\, r_{\min}$. □

15.3.2 The PSC Lemma

The intersection of a weighted Voronoi cell W_p and the domain $|\mathcal{S}|$ can be complicated. For a vertex $p \in S$ and a patch σ, the intersection $W_p \cap \sigma$ may have several connected components. Let σ_p be the union of the connected components that intersect at least one edge of W_p, in analogy to Σ_p in Definition 13.4. Figure 15.9 illustrates some of the possibilities. The PSC Lemma states that under certain conditions, σ_p has the generic topology (a ball), and so do the nonempty intersections of σ_p with the facets and edges of W_p. We introduce the following definition for convenience.

Definition 15.10 (λ-sized)**.** A set $S[\omega]$ of weighted points is *λ-sized* with respect to a patch σ if $\text{size}(\tau, \sigma) < \lambda$ for every triangle $\tau \in \text{Del}|^2_\sigma\, S[\omega]$.

Lemma 15.6 (PSC Lemma)**.** *Let $\mathcal{S}$ be a PSC, and let $S[\omega] \in |\mathcal{S}|$ be a set of weighted vertices. Suppose that the positively weighted vertices in $S[\omega]$ satisfy the ball properties, and that every unweighted vertex in $S[\omega]$ has a positive power distance to every other vertex in $S[\omega]$ (i.e. no protecting ball contains an unweighted vertex).*

(i) *Let $\gamma \in \mathcal{S}$ be a ridge. Let $p, q \in S \cap \gamma$ be two weighted vertices that are adjacent along γ.*

  (a) *The weighted Voronoi facet W_{pq} intersects $\gamma(p, q)$ exactly once.*

  (b) *If no two nonadjacent protecting balls intersect, then W_{pq} is the only weighted Voronoi facet in* Vor $S[\omega]$ *that intersects $\gamma(p, q)$.*

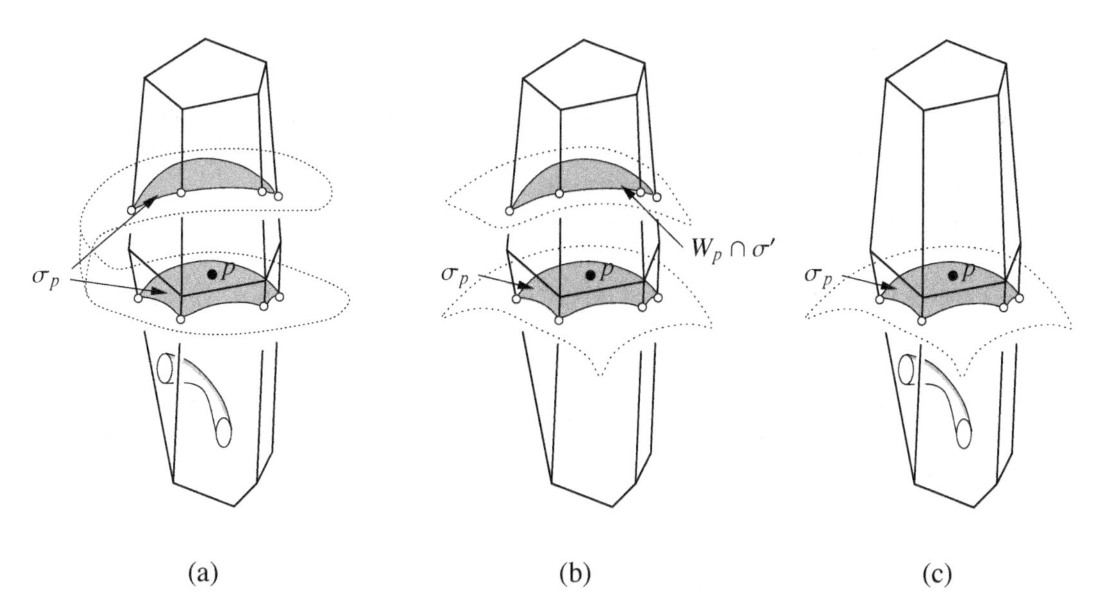

Figure 15.9: A weighted Voronoi cell W_p and its intersection with one or more patches. (a) The point set σ_p is not a topological disk. Note that W_p intersects σ or another patch in a tunnel that does not intersect any edge of W_p and, thus, is not included in σ_p. (b) The set σ_p is a topological disk and is equal to $W_p \cap \sigma$, but W_p intersects another patch σ'. (c) The set σ_p is a topological disk, but W_p intersects σ or another patch in a tunnel that does not intersect any edge of W_p.

(ii) *Let* $r_{\min} = \min\{\sqrt{\omega_p} : p[\omega_p] \in S[\omega]$ *and* $\omega_p > 0\}$ *be the minimum radius of a protecting ball. Let* $p \in S$ *be a vertex on a patch* σ, *and let* $W_p \in \text{Vor}\, S[\omega]$ *be its weighted Voronoi cell. Let* σ_p *be the union of the connected components of* $W_p \cap \sigma$ *that each intersect at least one edge of* W_p. *Suppose that every pair of positively weighted vertices connected by an edge in* $\text{Del}|_{\mathcal{S}}^2\, S[\omega]$ *are adjacent along some ridge in* $\mathcal{S}$. *Suppose also that every vertex in* $\text{Del}|_{\sigma}^2\, S[\omega]$ *lies on* σ.

Then there exists a constant $\lambda_* > 0$ *depending only on* $\mathcal{S}$ *such that for all* $\lambda \leq \min\{\lambda_*, r_{\min}\}$, *if* $S[\omega]$ *is* λ*-sized with respect to* σ *and* $d(p,q) \geq \lambda$ *for every pair of unweighted vertices* $p, q \in S$ *then the following statements hold.*

(a) *The set* σ_p *is a topological disk and it contains* p.

(b) *Every edge of* W_p *that intersects* σ *does so transversally at a single point.*

(c) *Every facet* g *of* W_p *that intersects* σ_p *does so in a topological interval* $g \cap \sigma_p = g \cap \sigma$, *and the intersection is transverse at every point in* $g \cap \sigma_p$.

Part (i)(a) of the PSC Lemma implies that the adjacent vertices p and q are connected by an edge in $\text{Del}|_{\gamma}\, S[\omega]$. Part (i)(b) ensures that if nonadjacent protecting balls are disjoint, then no other restricted Delaunay edge is generated by $\gamma(p,q)$. When part (ii) of the PSC Lemma applies to a patch σ, and part (i) applies to every pair of adjacent vertices on every ridge of σ, they imply that the union of the triangles incident to p in $\text{Del}|_{\sigma}\, S[\omega]$ is a topological disk. If this is true for every patch that contains p, then disk conditions D3 and D4 hold.

15.3.3 Proof of the PSC Lemma

Proof of the PSC Lemma, part (i). Let H_{pq} be the bisector of B_p and B_q with respect to the power distance. By ball property B1(b), $p \notin B_q$ and $q \notin B_p$, so H_{pq} lies between p and q. At least one of B_p and B_q is not a corner-ball, say, B_q.

Let z be a point in $\gamma(p,q) \cap H_{pq}$. Since $\gamma(p,q) \subset B_p \cup B_q$ by ball property B1(a) and $H_{pq} \cap (B_p \cup B_q) \subset B_p \cap B_q$, the point z lies in B_q, which implies that $\pi(z, B_q) \leq 0$. It suffices to show that z has positive power distance from the vertices in $S[\omega] \setminus \{p,q\}$ to show that W_{pq} intersects $\gamma(p,q)$ at z. By assumption, no unweighted vertex lies in $B_p \cup B_q$, so z has positive power distance to every unweighted vertex. As B_q is not a corner-ball, ball property B5 implies that z has positive power distance to every positively weighted vertex that does not lie on γ. It remains only to prove the same for every weighted vertex on γ. Suppose for the sake of contradiction that $\pi(z, v[\omega_v])$ is nonpositive for some weighted vertex $v \in \gamma$ other than p or q; then $z \in B_v$. By ball property B3(a), $\gamma(v,z) \subset B_v$. At least one weighted vertex other than v lies on $\gamma(v,z)$ (one of them being p or q), and the vertex adjacent to v violates ball property B1(b). This is a contradiction, so $\pi(z, v[\omega_v])$ is positive. Hence, W_{pq} intersects $\gamma(p,q)$ at z. To see that W_{pq} intersects $\gamma(p,q)$ exactly once, suppose for the sake of contradiction that W_{pq} intersects $\gamma(p,q)$ at two points y and z. Assume without loss of generality that $\text{radius}(B_p) \geq \text{radius}(B_q)$. By ball property B1(a), $\gamma(p,q) \subset B_p \cup B_q \subset B(p, 3\,\text{radius}(B_p))$. Applying ball property B3(b) to $B(p, 3\,\text{radius}(B_p)) \cap \gamma$, we obtain $\angle_a(\mathbf{h}_p, pq) > 90^\circ - \delta$ and $\angle_a(\mathbf{h}_p, yz) > 90^\circ - \delta$. This implies that $2\delta > \angle_a(pq, yz) = 90^\circ$, a contradiction. This proves (i)(a).

Consider (i)(b), for which we further assume that nonadjacent protecting balls are disjoint. As $\gamma(p,q) \subset B_p \cup B_q$ by ball property B1(a), every point in $\gamma(p,q)$ is at a nonpositive power distance from B_p or B_q. Let v be a vertex in $S[\omega] \setminus \{p,q\}$. If we can prove that $\gamma(p,q)$ does not intersect W_v, then W_{pq} is the only weighted Voronoi facet that intersects $\gamma(p,q)$. Suppose for the sake of contradiction that $\gamma(p,q)$ intersects W_v. We conduct a case analysis below and, in each case, show that $\gamma(p,q)$ intersects some weighted Voronoi cell W_u such that some point in $\gamma(p,q) \cap W_u$ is at a positive power distance from $u[\omega_u]$. This is a contradiction because every point in $\gamma(p,q)$ is at a nonpositive power distance from B_p or B_q. The point u may be v or a different point in S.

If v is unweighted or $v \in \gamma$, the same argument used in proving (i)(a) shows that every point in $\gamma(p,q)$ is at a positive power distance from $v[\omega_v]$, a contradiction.

Suppose that v is positively weighted and $v \notin \gamma$. As B_q is not a corner-ball by assumption, v and q do not lie on the same ridge, so B_v and B_q do not intersect by ball property B5. If B_v does not intersect B_p either, then B_v does not intersect $B_p \cup B_q$, implying that every point in $\gamma(p,q)$ is at a positive power distance from B_v, a contradiction. The remaining possibility is that v is positively weighted, $v \notin \gamma$, and B_v intersects B_p. By the assumption that nonadjacent protecting balls are disjoint, B_p must be a corner-ball (i.e. p is a vertex of γ), and both B_q and B_v are adjacent to B_p. Let H_{pv} be the bisector of B_p and B_v with respect to the power distance. By ball properties B1(b) and B2, $d(p, H_{pv}) \geq d(p,v)/2 \geq \text{radius}(B_p)/2$. Orient space so that the line L tangent to γ at p is horizontal, and the vertical plane containing L is perpendicular to H_{pv}. Let θ be the acute angle that H_{pv} makes with the horizontal.

Consider the case that $\theta \leq 28^\circ$. By ball properties B2 and B3(b), $\gamma(p,q)$ lies in the double cone with p as apex, the line L as axis, and angular aperture 2δ, where $\delta = 1.17^\circ$. The intersection between H_{pv} and this double cone is at distance at least $d(p, H_{pv})/\cos(90° - \theta - \delta) > \text{radius}(B_p)$ from p. Thus, $\gamma(p,q) \cap B_p$ cannot intersect H_{pv}, which implies that $\gamma(p,q) \cap B_q$ must intersect W_v as $\gamma(p,q)$ does by assumption. However, since B_v does not intersect B_q, every point in $\gamma(p,q) \cap B_q$ is at a positive power distance from B_v, a contradiction.

Consider the case that $\theta > 28°$. The subridge $\gamma(p,q)$ crosses H_{pv} before entering W_v, and $\gamma(p,q)$ must leave W_v before reaching q. Let W_u be the weighted Voronoi cell that $\gamma(p,q)$ enters after leaving W_v. By ball property B3(b), $\gamma(p,q)$ cannot turn by an angle greater than $28° - \delta$ to cross H_{pv} again. So $u \neq p$. If u and v are nonadjacent, B_u and B_v do not intersect by assumption. But then every point in $\gamma(p,q) \cap W_{uv}$ is at a positive power distance from B_u and B_v, a contradiction. If u and v are adjacent, then $u \notin \gamma$ and B_u is not adjacent to B_p. Thus, B_u intersects neither B_p nor B_q by the assumption that nonadjacent protecting balls are disjoint, but then every point in $\gamma(p,q)$ is at a positive power distance from B_u, a contradiction. □

The proof of part (ii) of the PSC Lemma requires several technical results. First we show that the ball properties imply that a restricted Delaunay triangle with a weighted vertex cannot have an angle arbitrarily close to 180°. We use this to derive a bound on the angle between the surface normal at a vertex on a patch and the normal to a restricted Delaunay triangle adjoining that vertex. That bound is 11δ where $\delta = 1.17°$.

Proposition 15.7. *Let σ be a patch in $\mathcal{S}$. Suppose that the following statements hold.*

(i) *The positively weighted vertices in $S[\omega]$ satisfy the ball properties, and no protecting ball contains an unweighted point.*

(ii) *If two positively weighted vertices are connected by an edge in $\text{Del}|_\sigma S[\omega]$, they are adjacent along some ridge in $\mathcal{S}$.*

For all $\lambda \leq r_{\min}$, if $S[\omega]$ is λ-sized with respect to σ and $d(p,q) \geq \lambda$ for every pair of unweighted vertices $p, q \in S$, then no triangle in $\text{Del}|_\sigma S[\omega]$ has both a positively weighted vertex and an angle greater than 166°.

Proof. Let $\tau = pqs$ be a triangle in $\text{Del}|_\sigma S[\omega]$ with a positively weighted vertex. At most one vertex of τ is a PSC vertex, because no two corner-balls are adjacent along a ridge. At least one vertex of τ is unweighted; otherwise, all three vertices would be adjacent along a single loop ridge, violating both ball property B3(b) and our assumption that no ridge is a loop. Let $\angle pqs$ be the largest angle of τ. Assume that $\angle pqs > 90°$; otherwise, there is nothing to prove.

If q has positive weight, assume without loss of generality that p is unweighted. Let H_{pq} be the weighted bisector plane of p and B_q with respect to the power distance. By assumption $p \notin B_q$, so H_{pq} lies between p and q and is further from q than from p. Hence, the distance from pq to the orthocenter of τ (which lies on H_{pq}) is at least $\frac{1}{2}d(p,q)\tan(\angle pqs - 90°) \geq \frac{1}{2}r_{\min}\tan(\angle pqs - 90°) \geq \frac{1}{2}\lambda\tan(\angle pqs - 90°)$. The fact that $S[\omega]$ is λ-sized with respect to σ implies that $\lambda \geq \text{size}(\tau, \sigma) \geq \frac{1}{2}\lambda\tan(\angle pqs - 90°)$; hence, $\angle pqs \leq 90° + \arctan 2 < 154°$.

Suppose that q is unweighted. If one of p or s is unweighted, say, p, then we can apply the argument in the previous paragraph to show that $\angle pqs < 154^\circ$ because $d(p,q) \geq \lambda$ by assumption. If both p and s are weighted, they are adjacent along some ridge. As q is not in either protecting ball B_p or B_s, $\angle pqs$ is maximized when q lies nearly on the circle $(\mathrm{Bd}\, B_p) \cap (\mathrm{Bd}\, B_s)$, in which case $\angle pqs < 166^\circ$ by ball property B1(c). □

Proposition 15.8. *Let σ be a patch in $\mathcal{S}$. Suppose that statements (i) and (ii) of Proposition 15.7 hold. Then there exists a constant $\lambda_* > 0$ depending only on $\mathcal{S}$ such that for all $\lambda \leq \min\{\lambda_*, r_{\min}\}$, if $S[\omega]$ is λ-sized with respect to σ and $d(p,q) \geq \lambda$ for every pair of unweighted vertices $p, q \in S$; then, for every triangle $\tau \in \mathrm{Del}|_\sigma S[\omega]$ whose vertices lie on σ and every vertex x of τ, $\angle_a(\mathbf{n}_x, W_\tau) < 11\delta$ where $\mathbf{n}_x$ is the normal to σ at x and W_τ is the weighted Voronoi edge dual to τ.*

Proof. Recall from Definition 15.1 that every ridge or patch $\xi \in \mathcal{S}$ is included in some manifold without boundary; let f_ξ denote the minimum local feature size over that manifold. We will show that the claim holds for

$$\lambda_* = 0.01 \min\left\{f_\xi : \xi \in \mathcal{S}_1 \cup \mathcal{S}_2\right\}.$$

Let $\tau = pqs$ be a triangle in $\mathrm{Del}|_\sigma S[\omega]$ whose vertices lie on σ. Let $\angle pqs$ be the largest angle of τ. As $S[\omega]$ is λ_*-sized, the orthoradius of τ is less than $\lambda_* \leq 0.01 f_\sigma$.

If every vertex of τ is unweighted, then $\angle_a(\mathbf{n}_q, W_\tau) < 2\delta$ by the Triangle Normal Lemma (Lemma 12.14). As τ's circumradius is less than $0.01 f_\sigma$, τ's edges have lengths less than $0.02 f_\sigma$, and by the Normal Variation Theorem (Theorem 12.8), $\angle(\mathbf{n}_p, \mathbf{n}_q) < \delta$ and $\angle(\mathbf{n}_s, \mathbf{n}_q) < \delta$. Therefore, $\angle_a(\mathbf{n}_p, W_\tau) < 3\delta$ and $\angle_a(\mathbf{n}_s, W_\tau) < 3\delta$, and the result follows.

Otherwise, τ has a weighted vertex. By Proposition 15.7, $\angle pqs < 166^\circ$, and as the proof of that proposition argues, τ has at least one unweighted vertex. Let C be the double cone whose apex is q, whose axis is $\mathrm{aff}\, \mathbf{n}_q$, and whose angular aperture is $180^\circ - 2\delta$. Let $\phi = \angle_a(\mathbf{n}_q, W_\tau)$. Let $H = \mathrm{aff}\, \tau$, a plane that intersects C at the apex q. We first show that $\phi < 9\delta$. If $\phi \leq \delta$, it follows immediately. Assume that $\phi > \delta$, so $H \cap C$ is a double wedge. If q is unweighted, let $r = 0.02 f_\sigma(q)$; otherwise, let $r = \mathrm{radius}(B_q)$. We claim that $B(q,r) \cap \sigma$ intersects the double cone C only at q. If q is unweighted, the claim follows from the Edge Normal Lemma (Lemma 12.12). If q is weighted, the claim follows from ball property B4(c). It follows from this claim that p and s are not in C. By elementary trigonometry, the angular aperture θ of $H \cap C$ satisfies

$$\tan\frac{\theta}{2} = \frac{\sqrt{\sin(\phi - \delta)\sin(\phi + \delta)}}{\sin\delta}.$$

If $\phi \geq 9\delta$, this equation yields $\theta > 167^\circ$, so the angular aperture of the open double wedge $H \setminus C$ is less than 26°. As $\angle pqs \geq 60^\circ$, the points p and s do not lie in the same wedge of $H \setminus C$. It follows that $\angle pqs$ is greater than the angular aperture of $H \cap C$, i.e. greater than 167°. But this is impossible as $\angle pqs < 166^\circ$. Hence, $\phi < 9\delta$.

Let p be the vertex of τ with the maximum weight. The triangle τ lies in $B(p, 4\,\mathrm{radius}(B_p))$ by the assumption that $\mathrm{size}(\tau,\sigma) \leq r_{\min} \leq \mathrm{radius}(B_p)$. By ball property B4(b), $\angle(\mathbf{n}_q, \mathbf{n}_p) < \delta$ and $\angle(\mathbf{n}_s, \mathbf{n}_p) < \delta$. Therefore, $\angle_a(\mathbf{n}_p, W_\tau) \leq \angle(\mathbf{n}_q, \mathbf{n}_p) + \angle_a(\mathbf{n}_q, W_\tau) < 10\delta$ and $\angle_a(\mathbf{n}_s, W_\tau) \leq \angle(\mathbf{n}_s, \mathbf{n}_p) + \angle_a(\mathbf{n}_p, W_\tau) < 11\delta$. □

Proof of PSC Lemma, part (ii). Let Σ be a smooth 2-manifold without boundary that includes the patch σ. Consider the following facts.

- By assumption, every vertex in $\mathrm{Del}|^2_\sigma\, S[\omega]$ lies on σ.

- By Proposition 15.8, there exists a constant $\lambda_* > 0$ depending only on $\mathcal{S}$ such that for all $\lambda \leq \min\{\lambda_*, r_{\min}\}$, if $S[\omega]$ is λ-sized with respect to σ and $d(p,q) \geq \lambda$ for every pair of unweighted vertices $p, q \in S$; then, for every triangle $\tau \in \mathrm{Del}|^2_\sigma\, S[\omega]$ and every vertex p of τ, $\angle_a(\mathbf{n}_p, W_\tau) < 11\delta$.

- Let W_{pqs} be an edge of a facet W_{pq} of a weighted Voronoi cell W_p, and suppose that W_{pqs} intersects σ_p; thus, $pqs \in \mathrm{Del}|^2_\sigma\, S[\omega]$. By the following reasoning, $\angle_a(\mathbf{n}_p, W_{pq}) < \delta$ when λ_* is sufficiently small.

  If p and q are unweighted, the length of pq is at most twice the orthoradius of pqs, which is at most $2\lambda_*$ because $S[\omega]$ is λ_*-sized. Therefore, by Proposition 13.3, $\angle_a(\mathbf{n}_p, W_{pq})$ approaches zero as λ_* does.

  If p is positively weighted and q is unweighted, let o be the orthocenter of pqs. As pqs is $r_{\min}$-sized, $r^2_{\min} > \pi(o, p[\omega_p]) = d(p,o)^2 - \mathrm{radius}(B_p)^2 \geq d(p, W_{pq})^2 - \mathrm{radius}(B_p)^2$, so $d(p, W_{pq}) < r_{\min} + \mathrm{radius}(B_p)$. Symmetrically, $d(q, W_{pq}) < r_{\min}$ as q is unweighted. Thus, $d(p,q) < 2r_{\min} + \mathrm{radius}(B_p) < 3\,\mathrm{radius}(B_p)$. Then, ball property B4(c) implies that $\angle_a(\mathbf{n}_p, W_{pq}) < \delta$.

  If both p and q are positively weighted, by assumption p and q are adjacent along a ridge of σ. Either $q \in B(p, 2\,\mathrm{radius}(B_p))$ or $p \in B(q, 2\,\mathrm{radius}(B_q))$, depending on which of B_p or B_q is bigger. Again, ball property B4(c) implies that $\angle_a(\mathbf{n}_p, W_{pq}) < \delta$.

- For an unweighted vertex p, the Feature Ball Lemma (Lemma 12.6) and the Normal Variation Theorem (Theorem 12.8) imply that for a fixed positive constant c, if λ_* is sufficiently small, then $B(p, c\lambda_*) \cap \Sigma$ is a topological disk in which the surface normal varies little. For a weighted vertex p, the same properties hold for $B(p, 4\,\mathrm{radius}(B_p))$ by ball properties B4(a) and B4(b).

These four facts are weighted analogs of those used in Chapter 13 to reason about how a Voronoi cell can intersect a surface Σ. Here, Σ is a manifold without boundary that includes the patch σ. If $W_p \cap \Sigma = W_p \cap \sigma$, then for a sufficiently small λ_*, the fact that $S[\omega]$ is λ_*-sized implies that the subset $\Sigma_p \subseteq W_p \cap \Sigma$ is 0.09-small (recall Definition 13.4), and the PSC Lemma (ii) follows immediately from the Small Intersection Theorem (Theorem 13.6).

Unfortunately, it is possible that $W_p \cap \Sigma \supset W_p \cap \sigma$ and $\Sigma_p \supset \sigma_p$. Nevertheless, the proofs of Proposition 13.3 through Lemma 13.12 can be adapted straightforwardly to work with Σ_p replaced by σ_p. For a vertex p in the interior of σ, the adaptation is relatively simple, because ball property B5, Proposition 15.5(iii) and the PSC Lemma (i) imply that W_p does not intersect $\mathrm{Bd}\,\sigma$ for a sufficiently small λ_*. A vertex p on the boundary of σ requires a little more work: we observe that when λ_* is small enough, $W_p \cap \mathrm{Bd}\,\sigma$ is a topological interval by ball property B5, Proposition 15.5(iii) and the PSC Lemma (i); hence, the closure of one of the connected components of $(W_p \cap \Sigma) \setminus \mathrm{Bd}\,\sigma$ is in σ_p, which also contains p. By revising the proofs of Proposition 13.3 through Lemma 13.12 accordingly, the Small Intersection Theorem (Theorem 13.6) becomes the PSC Lemma (ii). □

15.4 A proof of termination

Here we show that the algorithm DelPSC always terminates. We begin by showing that after a corner-ball has been refined to a sufficiently small size, it cannot contain the center of any other ball. Recall the definition of ℓ_X before Proposition 15.5.

Proposition 15.9. *Suppose that the protecting balls satisfy the relaxed ball properties. There exists a constant $\lambda_c > 0$ depending only on $\mathcal{S}$ such that for every corner-ball B_p with radius less than λ_c and every non-corner-ball B_q, $q \notin B_p$.*

Proof. Let γ be the ridge that contains q. Choose $\lambda_c > 0$ small enough that $\lambda_c \leq \ell_\emptyset$ and for each vertex x of γ, $B(x, \lambda) \cap \gamma$ is a single topological interval for every $\lambda < \lambda_c$.

If p does not lie on γ, then $d(p, q) \geq \ell_\emptyset$, implying that $q \notin B_p$ as radius$(B_p) < \lambda_c \leq \ell_\emptyset$.

Otherwise, p is a vertex of γ. Suppose for the sake of contradiction that $q \in B_p$. Then $q \in B_p \cap \gamma = \text{seg}_\gamma(B_p)$, which implies that the weighted vertex v adjacent to p along γ also lies on $\text{seg}_\gamma(B_p)$. This contradicts relaxed ball property R3(b)(i), which states that $d_\gamma(v, \text{seg}_\gamma(B_p)) \geq r_{\min}/12$. Therefore, $q \notin B_p$. □

Next we show that, under the right conditions, every vertex of $\text{Del}|_\sigma S[\omega]$ lies on σ.

Proposition 15.10. *Let V be the union of all the corner-balls, and let $U \supset V$ be the union of all the protecting balls. Suppose the following statements hold.*

- *No protecting ball contains an unweighted vertex.*
- *For every ridge γ and every pair of weighted vertices p and q that are adjacent along γ, $\gamma(p, q) \subset B_p \cup B_q$.*
- *No corner-ball contains the center of a non-corner-ball.*
- *For every patch σ, no edge in $\text{Del}|^2_\sigma S[\omega]$ connects two weighted vertices that are not adjacent along some ridge.*
- *If two protecting balls are not centered on a common ridge, they are disjoint.*

Then the following statements hold.

(i) *For every patch σ, if $S[\omega]$ is $(\sqrt{8}\ell_\emptyset/3)$-sized with respect to σ, then σ contains every positively weighted vertex $p \in (\text{Del}|^2_\sigma S[\omega]) \cap \mathcal{S}_0$ for which $\sqrt{\omega_p} \leq \ell_\emptyset/3$.*

(ii) *For every patch σ, if $S[\omega]$ is $(\sqrt{8}\ell_V/3)$-sized with respect to σ, then σ contains every positively weighted vertex $p \in (\text{Del}|^2_\sigma S[\omega]) \setminus \mathcal{S}_0$ for which $\sqrt{\omega_p} \leq \ell_V/3$.*

(iii) *For every patch σ, if $S[\omega]$ is ℓ_U-sized with respect to σ, then σ contains every unweighted vertex of $\text{Del}|^2_\sigma S[\omega]$.*

Proof. Suppose that there is a patch σ and a vertex $p \in \text{Del}|^2_\sigma S[\omega]$ such that $p \notin \sigma$. Let pqs be any triangle in $\text{Del}|^2_\sigma S[\omega]$ that adjoins p; i.e. $W_{pqs} \cap \sigma \neq \emptyset$.

If p is a vertex in $\mathcal{S}_0$, then $d(p,\sigma) \geq \ell_\emptyset$ as $p \notin \sigma$. If $\sqrt{\omega_p} \leq \ell_\emptyset/3$, the power distance from p to any point in $W_{pqs} \cap \sigma$ is at least $8\ell_\emptyset^2/9$; i.e. size$(pqs,\sigma) \geq \sqrt{8}\ell_\emptyset/3$. Conversely, such a vertex p in $\mathcal{S}$ does not exist if size$(pqs,\sigma) < \sqrt{8}\ell_\emptyset/3$. This proves (i).

If p is an unweighted vertex, the power distance from p to any point in $W_{pqs} \cap \sigma$ is positive, which implies that $W_{pqs} \cap \sigma$ does not intersect U, so $d(p, W_{pqs} \cap \sigma) \geq \ell_U$. It follows that size$(pqs,\sigma) \geq \ell_U$. Conversely, such an unweighted vertex p does not exist if size$(pqs,\sigma) < \ell_U$. This proves (iii).

Otherwise, p is a weighted vertex but not a vertex in $\mathcal{S}$. By assumption, nonadjacent weighted vertices are not connected in $\mathrm{Del}|_\sigma^2\, S[\omega]$, so either q or s is unweighted. No protecting ball contains an unweighted vertex by assumption, so the unweighted vertex or vertices of pqs have positive distance from every point in $W_{pqs} \cap \sigma$, which implies that $W_{pqs} \cap \sigma$ does not intersect V. Also, $p \notin V$ by our assumption that no corner-ball contains the center of a non-corner-ball. Thus, $d(p, W_{pqs} \cap \sigma) \geq \ell_V$. If $\sqrt{\omega_p} \leq \ell_V/3$, the power distance from p to any point in $W_{pqs} \cap \sigma$ is at least $8\ell_V^2/9$; i.e. size$(pqs,\sigma) \geq \sqrt{8}\ell_V/3$. Conversely, such a weighted vertex p does not exist if size$(pqs,\sigma) < \sqrt{8}\ell_V/3$. This proves (ii). □

Proposition 15.11. *Let x be an unweighted vertex inserted by* DELPSC.

(i) *No protecting ball contains x.*

(ii) *Let $r_{\min}$ be the minimum radius among the protecting balls before x is inserted. Among the sets of protecting balls constructed before x is inserted, pick the set such that the union $\bar{U}$ of protecting balls in it minimizes $\ell_{\bar{U}}$. The power distance from x to every prior vertex in $S[\omega]$ is at least* $(\min\{0.035r_{\min}, \lambda_{\mathrm{ref}}, \lambda_{\bar{U}}\})^2$.

PROOF. We prove the proposition by induction on the chronological order of the insertions of unweighted vertices.

Step 2(c)(i) of DELPSC inserts a vertex at the center x of a surface Delaunay ball only if $s_{\max} \geq 0.035\, r_{\min}$. Step 2(f) inserts x only if $s_{\max} > \lambda_{\mathrm{ref}}$.

Step 2(d)(i) inserts a vertex if there exists a vertex $p \in S$, a patch σ, and a vertex q of $\mathrm{Umb}_\sigma(p)$ such that $q \notin \sigma$, and $s_{\max} \geq$ radius(B_q) or q is unweighted. If q has positive weight, then $s_{\max} \geq$ radius$(B_q) \geq r_{\min}$. If q is unweighted, recall that when execution reaches Step 2(d), the preconditions of Proposition 15.10 are satisfied. Proposition 15.10(iii) implies that $S[\omega]$ is not ℓ_U-sized with respect to σ, where U is the union of all the protecting balls at the time when x is inserted, and thus, $s_{\max} \geq \ell_U \geq \ell_{\bar{U}}$.

Step 2(e)(i) inserts x only if $s_{\max}$ is at least $r_{\min}$ or the minimum distance among the existing unweighted vertices. The latter is at least $\min\{0.035r_{\min}, \lambda_{\mathrm{ref}}, \lambda_{\bar{U}}\}$ by induction.

We conclude that x is at a power distance of at least $(\min\{0.035\, r_{\min}, \lambda_{\mathrm{ref}}, \ell_{\bar{U}}\})^2$ from every prior vertex in $S[\omega]$. Hence, x is not contained in any protecting ball. □

Proposition 15.12. DELPSC$(\mathcal{S}, \lambda_{\mathrm{prot}}, \lambda_{\mathrm{ref}})$ *generates no protecting ball with a radius smaller than some positive constant that depends only on $\mathcal{S}$ and λ_{prot}.*

PROOF. PROTECTPSC and REFINEBALL maintain the relaxed ball properties. After the initial call to PROTECTPSC, new protecting balls are created only by calls to REFINEBALL in Steps 2(a), 2(b), 2(c)(ii), 2(d)(ii), and 2(e)(ii) of DELPSC. After DELPSC has called REFINEBALL any finite number of times, the minimum protecting ball radius $r_{\min}$ is positive. By Proposition 15.11(i), no unweighted vertex is inserted into a protecting ball. Steps 1(e) and 2(b) of REFINEBALL ensure that no unweighted vertex is contained in any protecting ball constructed by REFINEBALL.

If Step 2(a) of DELPSC is invoked, then a corner-ball contains the center of a non-corner-ball. By Proposition 15.9, the radius of the former ball is at least a constant $\lambda_c > 0$. Step 2(a) refines the larger of the two balls, whose radius is at least λ_c.

If Step 2(b) is invoked, then two protecting balls B_p and B_q intersect, but p and q do not lie on a common ridge. If one of the balls is a corner-ball, then $\max\{\text{radius}(B_p), \text{radius}(B_q)\} \geq d(p,q)/2 \geq \ell_\emptyset/2$. Otherwise, observe that neither p nor q lies in any corner-ball, because Step 2(a) was not invoked instead. Proposition 15.5(ii) implies that $\max\{\text{radius}(B_p), \text{radius}(B_q)\} \geq \ell_V/3$. In either case, Step 2(b) refines the larger ball, whose radius is at least $\min\{\ell_\emptyset/2, \ell_V/3\}$.

Consider the invocation of Step 2(c)(ii). Let B_q be the protecting ball refined in this step, which means that q is connected to some nonadjacent weighted vertex s by some restricted Delaunay triangle τ, and $\text{radius}(B_q) \geq \text{radius}(B_s)$. Suppose that q and s lie on the same ridge. If $\text{radius}(B_q)$ is less than the constant λ_n in Proposition 15.5(iii), then $d(B_q, B_s) \geq 0.07\, r_{\min}$; hence, the orthoradius of τ is at least $0.035\, r_{\min}$ and $s_{\max} \geq 0.035\, r_{\min}$. But in that case, Step 2(c)(i) is invoked instead of 2(c)(ii). Suppose that q and s do not lie on a common ridge. If either of B_q or B_s is a corner-ball and $\text{radius}(B_q) < \ell_\emptyset/3$, Proposition 15.5(i) applies. If both are non-corner-balls and $\text{radius}(B_q) < \ell_V/3$, Proposition 15.5(ii) applies. In either case, $d(B_q, B_s) \geq r_{\min}$, the circumradius of τ is at least $0.5\, r_{\min}$, $s_{\max} \geq 0.5\, r_{\min}$, and Step 2(c)(i) is invoked instead of 2(c)(ii). In summary, if a corner-ball is refined in Step 2(c)(ii), its radius is at least $\min\{\lambda_n, \ell_\emptyset/3\}$; if a non-corner-ball is refined in Step 2(c)(ii), its radius is at least $\min\{\lambda_n, \ell_\emptyset/3, \ell_V/3\}$.

When execution reaches Step 2(d), the preconditions of Proposition 15.10 are satisfied. Suppose Step 2(d)(ii) refines a corner-ball B_q; then $q \notin \sigma$. If $\text{radius}(B_q) \leq \ell_\emptyset/3$, by Proposition 15.10(i), $S[\omega]$ is not $(\ell_\emptyset/3)$-sized with respect to σ, so $\text{radius}(B_q) \leq s_{\max}$ and Step 2(d)(i) is invoked instead of 2(d)(ii). It follows that $\text{radius}(B_q) > \ell_\emptyset/3$ for every corner-ball B_q refined by Step 2(d)(ii). Likewise, Proposition 15.10(ii) implies that $\text{radius}(B_q) > \ell_V/3$ for every non-corner-ball B_q refined by Step 2(d)(ii).

Consider the invocation of Step 2(e)(ii) to shrink the largest ball B_m in $\mathcal{B}$. Suppose that $\text{radius}(B_m) < \min\{\lambda_b, \lambda_*\}$, where λ_b and λ_* are the constants in Proposition 15.4 and the PSC Lemma (ii) (Lemma 15.6(ii)), respectively. By Proposition 15.4 and Step 2(b) of DELPSC, $\mathcal{B}$ satisfies the ball properties. By Proposition 15.11 and the removal of unweighted vertices inside protecting balls by REFINEBALL, every unweighted vertex in $S[\omega]$ has a positive power distance from every other vertex in $S[\omega]$. Since D1 and D2 are not violated, every pair of positively weighted vertices connected by an edge in $\text{Del}|^2_{\mathcal{S}}\, S[\omega]$ is adjacent along some ridge in $\mathcal{S}$, and every vertex in $\text{Del}|^2_{\sigma}\, S[\omega]$ lies in σ for every patch σ. Therefore, the preconditions of the PSC Lemma (ii) are satisfied. Step 2(e)(ii) is invoked only if $s_{\max} < \min\{r_{\min}, d_{\min}\}$, where $d_{\min}$ is the minimum distance among the unweighted vertices. It means that there exists $\lambda \leq \min\{r_{\min}, d_{\min}\}$ such that $S[\omega]$ is λ-sized and $d_{\min} \geq \lambda$. But then PSC Lemma (ii) is applicable because $r_{\min} \leq \text{radius}(B_m) < \lambda_*$, and it implies that D3 and D4 cannot be violated in the first place. In other words, $\text{radius}(B_m) \geq \min\{\lambda_b, \lambda_*\}$ in order that B_m is shrunk in Step 2(e)(ii) due to the violation of D3 or D4.

In summary, if a corner-ball is refined, its radius is at least $\min\{\lambda_b, \lambda_c, \lambda_n, \lambda_*, \ell_\emptyset/3\}$, which implies that the refinement of corner-balls must stop eventually. Henceforth, let V and ℓ_V refer to the final values of V and ℓ_V after the last corner-ball is refined, and observe that ℓ_V attains its smallest value then. REFINEBALL can still refine non-corner-balls thereafter. If REFINEBALL is directly (not recursively) invoked on a non-corner-ball B in Step 2(a), 2(b), 2(c)(ii), or 2(d)(ii), then $\text{radius}(B) \geq \min\{\lambda_b, \lambda_c, \lambda_n, \lambda_*, \ell_\emptyset/3, \ell_V/3\}$, so REFINEBALL is invoked directly on non-corner-balls finitely many times. This implies that the recursive refinement of non-corner-balls by REFINEBALL must also stop eventually. □

Proposition 15.13. DELPSC *terminates, with no unweighted vertex in a protecting ball.*

PROOF. By Proposition 15.12, DELPSC constructs only finitely many sets of protecting balls. Proposition 15.11 implies that every unweighted vertex that DELPSC inserts is at a power distance of at least some positive constant from every prior vertex in $S[\omega]$. Moreover, Steps 1(e) and 2(b) of REFINEBALL ensure that no unweighted vertex lies in any protecting ball constructed by REFINEBALL. This implies that no unweighted vertex lies in a protecting ball, and that there is a constant lower bound on the Euclidean distance between every pair of vertices in S throughout the execution of DELPSC. Hence, the insertion of unweighted vertices by DELPSC must stop eventually. □

15.5 Manifold patch triangulations and homeomorphism

Here we show that $\text{DELPSC}(\mathcal{S}, \lambda_{\text{prot}}, \lambda_{\text{ref}})$ returns a mesh $\mathcal{T} = \text{Del}|^2_{\mathcal{S}}\, S[\omega]$ that satisfies the following topological guarantees.

G1. For each patch $\sigma \in \mathcal{S}_2$, $\left|\text{Del}|^2_\sigma\, S[\omega]\right|$ is a 2-manifold with boundary, every vertex in $\text{Del}|^2_\sigma\, S[\omega]$ lies on σ, the boundary of $\left|\text{Del}|^2_\sigma\, S[\omega]\right|$ is homeomorphic to $\text{Bd}\,\sigma$, and every vertex in $\text{Del}|^1_{\text{Bd}\,\sigma}\, S[\omega]$ lies on $\text{Bd}\,\sigma$.

G2. Let λ_b and λ_n be the constants in Propositions 15.4 and 15.5(iii), respectively. If $\lambda_{\text{prot}} < \min\{\lambda_b, \lambda_n\}$ and λ_{ref} is sufficiently small, then $\mathcal{T}$ is a triangulation of $\mathcal{S}$ (recall Definition 15.3). Furthermore, there is a homeomorphism h from $|\mathcal{S}|$ to $|\mathcal{T}|$ such that for every $\xi \in \mathcal{S}_i$ with dimension $i \in [0, 2]$, h is a homeomorphism from ξ to $\left|\text{Del}|^i_\xi\, S[\omega]\right|$, every vertex in $\text{Del}|^i_\xi\, S[\omega]$ lies on ξ, and the boundary of $\left|\text{Del}|^i_\xi\, S[\omega]\right|$ is $\left|\text{Del}|^{i-1}_{\text{Bd}\,\xi}\, S[\omega]\right|$.

Proposition 15.14. *For all* $\lambda_{\text{prot}}, \lambda_{\text{ref}} > 0$, *the mesh returned by* $\text{DELPSC}(\mathcal{S}, \lambda_{\text{prot}}, \lambda_{\text{ref}})$ *satisfies topological guarantee G1.*

PROOF. Let σ be a patch. DELPSC terminates only when all the disk conditions hold, ensuring that every vertex in $\text{Del}|^2_\sigma\, S[\omega]$ lies on σ and the triangles adjoining each vertex form a topological disk. It follows from a standard result in piecewise linear topology that $|\mathrm{Del}|^2_\sigma\, S[\omega]|$ is a 2-manifold, possibly with boundary.

Upon termination, disk condition D4 holds, ensuring that the vertices in $\mathrm{Del}|^2_\sigma\, S[\omega]$ on the boundary of $|\mathrm{Del}|^2_\sigma\, S[\omega]|$ are exactly the vertices lying on $\mathrm{Bd}\,\sigma$ and that each of them is connected by boundary edges to two other boundary vertices. Disk condition D1 ensures that edges in $\mathrm{Del}|^2_\sigma\, S[\omega]$ connect no two boundary vertices that are not adjacent along $\mathrm{Bd}\,\sigma$. By the pigeonhole principle, every boundary vertex is connected to the two boundary vertices that are adjacent along $\mathrm{Bd}\,\sigma$. Therefore, the boundary of $|\mathrm{Del}|^2_\sigma\, S[\omega]|$ is homeomorphic to $\mathrm{Bd}\,\sigma$. □

The proof of guarantee G2 uses an extension of the topological ball property (Definition 13.3) and the Topological Ball Theorem (Theorem 13.1). The *extended topological ball properties* (Definition 15.12) govern the relationship between a PSC $\mathcal{S}$ and the restricted Voronoi diagram $\mathrm{Vor}|_{|\mathcal{S}|}\, S[\omega]$. In the service of generality, we define these properties in terms of a type of topological complex called a *regular CW-complex*. The Extended Topological Ball Theorem (Theorem 15.15) uses a subdivision of $\mathcal{S}$ into restricted Voronoi cells that are subsets of PSC cells. The subdivision we will use to prove guarantee G2 is $\mathcal{R} = \{\xi \cap g : \xi \in \mathcal{S} \text{ and } g \in \mathrm{Vor}\, S[\omega]\}$, which is not necessarily a regular CW-complex but becomes one after sufficient refinement.

Definition 15.11 (regular CW-complex). A *regular CW-complex* $\mathcal{R}$ is a set of closed topological balls whose interiors are pairwise disjoint and whose boundaries are unions of other closed topological balls in $\mathcal{R}$.

Definition 15.12 (extended topological ball properties). Let $\mathcal{S}$ be a PSC embedded in $\mathbb{R}^d$. A weighted vertex set $S[\omega] \subset |\mathcal{S}|$ has the *extended topological ball properties* (extended TBP) for $\mathcal{S}$ if there exists a regular CW-complex $\mathcal{R}$ such that $|\mathcal{R}| = |\mathcal{S}|$ and each weighted Voronoi face $g \in \mathrm{Vor}\, S[\omega]$ that intersects $|\mathcal{S}|$ satisfies the following conditions.

C1. There is a subcomplex $\mathcal{R}_g \subseteq \mathcal{R}$ whose underlying space $|\mathcal{R}_g|$ is the restricted Voronoi face $g \cap |\mathcal{S}|$.

C2. There is a unique, closed, topological ball $b_g \in \mathcal{R}_g$ that is a face of every closed ball in $\mathcal{R}_g$ that intersects $\mathrm{Int}\, g$.

C3. $b_g \cap \mathrm{Bd}\, g$ is a topological sphere of dimension one less than b_g.

C4. For each j-ball $b \in \mathcal{R}_g \setminus \{b_g\}$ that intersects $\mathrm{Int}\, g$, $b \cap \mathrm{Bd}\, g$ is a $(j-1)$-ball.

Figure 15.10 shows two examples of a Voronoi facet g that satisfies all four conditions.

Theorem 15.15 (Extended Topological Ball Theorem). *If a weighted vertex set* $S[\omega] \subset |\mathcal{S}|$ *has the extended TBP for* $\mathcal{S}$*, then the underlying space of* $\mathrm{Del}|_{\mathcal{S}}\, S[\omega]$ *is homeomorphic to* $|\mathcal{S}|$.

One can show (see Exercise 6) that the following two properties imply that $\mathcal{R} = \{\xi \cap g : \xi \in \mathcal{S} \text{ and } g \in \mathrm{Vor}\, S[\omega]\}$ satisfies the extended TBP.

P1. Every nonempty intersection of a k-face in $\mathrm{Vor}\, S[\omega]$ and an i-cell in $\mathcal{S}_i$ is a closed $(k+i-3)$-ball.

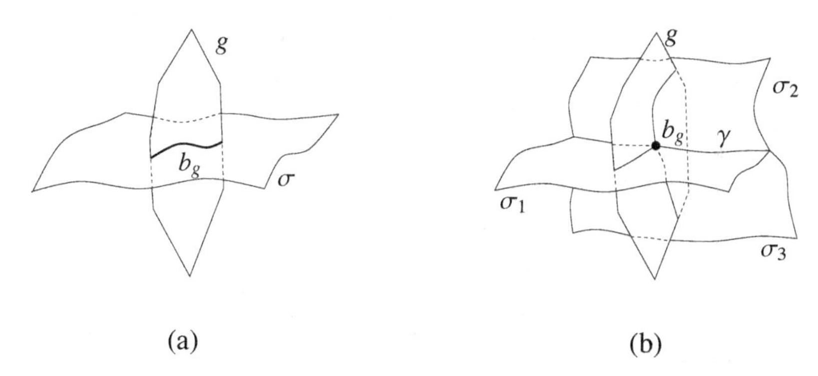

Figure 15.10: (a) A Voronoi facet g and a patch σ whose intersection is a topological interval (1-ball) b_g. The intersection of b_g with Bd g is two points, a 0-sphere. (b) A Voronoi facet g and a ridge γ whose intersection is a single point b_g. The intersection $b_g \cap \mathrm{Bd}\, g$ is the empty set, which by convention is a (-1)-sphere. For each patch σ_1, σ_2, and σ_3, b_g is a face of the topological interval $g \cap \sigma_i$.

P2. For every face $g \in \mathrm{Vor}\, S[\omega]$ that intersects $|\mathcal{S}|$, there is a unique cell in $\mathcal{S}$ that is a face of every cell in $\mathcal{S}$ that intersects g.

Propositions 15.16 and 15.17 below prove that P1 and P2 hold for the vertices that DELPSC returns. Recall that for a patch σ and a vertex $p \in S \cap \sigma$, σ_p is the union of the connected components of $W_p \cap \sigma$ that each intersect at least one weighted Voronoi edge of W_p. Analogously, for a ridge γ and a vertex $p \in S \cap \gamma$, define $\gamma_p = W_p \cap \gamma$. Observe that each connected component of γ_p must intersect some Voronoi facet of W_p because γ has two vertices.

Proposition 15.16. *Let $S[\omega]$ be the set of vertices in the mesh that* DELPSC($\mathcal{S}, \lambda_{\text{prot}}, \lambda_{\text{ref}}$) *returns. Let W_p be the weighted Voronoi cell of a vertex $p \in S$. Let λ_b and λ_n be the constants in Propositions 15.4 and 15.5(iii), respectively. If $\lambda_{\text{prot}} < \min\{\lambda_b, \lambda_n\}$ and $\lambda_{\text{ref}} \leq \lambda_{\text{prot}}$ is sufficiently small, then*

- *PSC Lemma (i)(a,b) and (ii)(a,b,c) hold.*
- *For every ridge or patch $\xi \in \mathcal{S}$ that intersects W_p, $p \in \xi$ and $\xi_p = W_p \cap \xi$.*

PROOF. All the ball properties B1–B5 are satisfied: Proposition 15.4 ensures that the protecting balls generated by PROTECTPSC satisfy the ball properties B1–B4 because $\lambda_{\text{prot}} < \lambda_b$; ball property B5 is ensured by DELPSC. Also, DELPSC ensures that no unweighted vertex lies in any protecting ball. As $\lambda_{\text{prot}} < \lambda_n$, Proposition 15.5(iii) is applicable. By B5 and Proposition 15.5(iii), no two nonadjacent protecting balls intersect. As a result, the preconditions of the PSC Lemma (i)(a,b) are satisfied, so PSC Lemma (i)(a,b) hold.

The disk condition D1 and D2 are guaranteed by DELPSC upon termination. They further imply that the preconditions of the PSC Lemma (ii) are also satisfied. By Proposition 15.12, the final minimum protecting ball radius $r_{\min}$ depends on $\mathcal{S}$ and λ_{prot} only. Let $\bar{U}$ be the set of protecting balls generated so far that minimizes $\ell_{\bar{U}}$. Recall the constant λ_* in the PSC Lemma (ii). As λ_{ref} can be made arbitrarily small, we can assume that

$\lambda_{\text{ref}} \leq \min\{\lambda_*, 0.035 r_{\min}, \ell_{\bar{U}}\}$. By Proposition 15.11(ii), the minimum distance among the unweighted vertices in S is at least λ_{ref}. On the other hand, DELPSC ensures that $S[\omega]$ is λ_{ref}-sized. Therefore, PSC Lemma (ii)(a,b,c) hold.

Assume for the sake of contradiction that $p \notin \xi$ or $\xi_p \subset W_p \cap \xi$. In the former case, let C_p be any connected component of $W_p \cap \xi$; in the latter case, let C_p be any connected component of $(W_p \cap \xi) \setminus \xi_p$. In either case, $p \notin C_p$.

If ξ is a ridge, C_p is a subridge of ξ that contains no vertex in S and intersects a facet of W_p at each of its two endpoints. Therefore, ξ intersects more than one weighted Voronoi facet between two adjacent positively weighted vertices in $\xi \cap S$, or the same facet twice, contradicting the PSC Lemma (i).

If ξ is a patch, let $q \in S \cap \xi$ be an arbitrary vertex on ξ. By the PSC Lemma (ii)(a), there is a component $C_q \subseteq \xi_q$ that contains q. Imagine following a path on ξ from q to $\operatorname{Int} C_p$, chosen so the path does not intersect any Voronoi edge. This path traverses two or more Voronoi cells in $\operatorname{Vor} S[\omega]$; specifically, it traverses two or more connected components of restricted Voronoi cells, starting with the component C_q of ξ_q and ending with C_p. Let $C_v \subset W_v$ and $C_w \subset W_w$ be two successive components visited along the path. We will show that if $v \in C_v \subseteq \xi_v$, then $w \in C_w \subseteq \xi_w$; thus by induction along the path, $p \in C_p \subseteq \xi_p$, so the result follows.

By the PSC Lemma (ii)(a,c), the set ξ_v is a topological disk that intersects the facet W_{vw} transversally in a topological interval. Hence $C_v = \xi_v$ intersects at least one edge of W_{vw}, and the edge vw is in the restricted Delaunay subcomplex $\operatorname{Del}|^2_{\xi} S[\omega]$ and is an edge of some triangle in $\operatorname{Umb}_{\xi}(v)$. Because DELPSC terminated, the disk condition D2 is satisfied, so $w \in \xi$. Because C_w adjoins C_v along the interval $C_v \cap W_{vw}$, it also intersects an edge of W_{vw}, so $C_w \subseteq \xi_w$. By the PSC Lemma (ii)(a), ξ_w is also a topological disk, so $w \in C_w = \xi_w$. □

Proposition 15.17. *Let $S[\omega]$ be the set of vertices in the mesh that* DELPSC($\mathcal{S}, \lambda_{\text{prot}}, \lambda_{\text{ref}}$) *returns. Let λ_b and λ_n be the constants in Propositions 15.4 and 15.5(iii), respectively. If $\lambda_{\text{prot}} < \min\{\lambda_b, \lambda_n\}$ and $\lambda_{\text{ref}} \leq \lambda_{\text{prot}}$ is sufficiently small, then* $\operatorname{Vor} S[\omega]$ *satisfies the properties P1 and P2.*

PROOF. The preconditions of the proposition allow us to assume that Proposition 15.16 and PSC Lemma (i)(a,b) and (ii)(a,b,c) hold.

Let g be a face of a weighted Voronoi cell W_p and let $\xi \in \mathcal{S}$ be a cell that intersects g. If ξ is a vertex, then $g = W_p$ and the property P1 holds. Suppose that ξ is a ridge or patch. By Proposition 15.16, $p \in \xi$ and $W_p \cap \xi = \xi_p$, and the property P1 follows from the PSC Lemma.

Consider the property P2. Let g be a face in $\operatorname{Vor} S[\omega]$ that intersects $|\mathcal{S}|$. If g is a Voronoi vertex, the property P2 holds trivially. We consider the other three cases individually: g is a Voronoi cell, a Voronoi facet, or a Voronoi edge.

- g is a Voronoi cell W_p. Let ξ be the lowest-dimensional cell in $\mathcal{S}$ that contains p; thus p lies in the interior of ξ, and every cell in $\mathcal{S}$ that contains p has ξ for a face. We claim that every cell in $\mathcal{S}$ that intersects W_p has ξ for a face; thus, property P2 holds. If not, there is a cell $\xi' \in \mathcal{S}$ intersecting W_p such that $\xi \not\subset \xi'$, which implies that $p \notin \xi'$, contradicting Proposition 15.16.

- g is a Voronoi facet W_{pq}. Let $\xi \in \mathcal{S}$ be a cell of least dimension that intersects W_{pq}. Suppose for the sake of contradiction that some other cell $\xi' \not\supset \xi$ intersects W_{pq}. By Proposition 15.16, both p and q lie on both ξ and ξ'. It follows that p and q each lie on a proper face of ξ; otherwise, they could not lie on both ξ and ξ' without ξ being a face of ξ'. Therefore, p and q have positive weights.

  There is a patch σ that has ξ for a face (possibly $\sigma = \xi$), and thus, $\mathrm{Del}|^2_\sigma\, S[\omega]$ contains the edge pq. As DELPSC terminated, disk condition D1 holds, so p and q must be adjacent along some ridge γ of σ. Every ridge has at least one vertex of S in its interior, so one of p or q lies in the interior of γ. Therefore, γ is a face of both ξ and ξ'. By the PSC Lemma (i), γ intersects W_{pq}. As ξ is a cell of least dimension that intersects W_{pq}, $\xi = \gamma$, and thus, ξ is a face of ξ', a contradiction.

  Hence, ξ is a face of every cell in $\mathcal{S}$ that intersects W_{pq}.

- g is a Voronoi edge W_{pqs}. By the PSC Lemma (ii)(b), W_{pqs} intersects some patch σ transversally at a single point. Moreover, W_{pqs} cannot intersect another patch σ'; if it did, then by Proposition 15.16 both σ and σ' contain p, q, and s, which implies that all three vertices have positive weights. All three vertices are connected by edges in $\mathrm{Del}|^2_\sigma\, S[\omega]$, and at least two of them are nonadjacent along the boundary of a patch, so they violate disk condition D1, contradicting the fact that DELPSC terminated.

□

The following theorem summarizes the topology of any mesh DELPSC returns.

Theorem 15.18. *Let $\mathcal{S}$ be a PSC embedded in $\mathbb{R}^3$. For all $\lambda_{\text{prot}}, \lambda_{\text{ref}} > 0$, DELPSC$(\mathcal{S}, \lambda_{\text{prot}}, \lambda_{\text{ref}})$ returns a triangulation of $\mathcal{S}$ that has the topological guarantee G1 and, if λ_{prot} is small enough, the triangulation also has the topological guarantee G2 for sufficiently small $\lambda_{\text{ref}} \le \lambda_{\text{prot}}$.*

PROOF. By Proposition 15.13, DELPSC always terminates and returns a mesh. By Proposition 15.14, the output mesh has guarantee G1 for all $\lambda_{\text{prot}}, \lambda_{\text{ref}} > 0$. The preconditions of the theorem allow us to assume that the PSC Lemma (i)(a,b) and (ii)(a,b,c), and P1 and P2 hold.

P1 and P2 imply that $S[\omega]$ has the extended topological ball properties for $\mathcal{S}$ (Exercise 6), and thus, the output mesh has an underlying space homeomorphic to $|\mathcal{S}|$ by the Extended Topological Ball Theorem (Theorem 15.15). The homeomorphism constructed in the proof of that theorem is also a homeomorphism between ξ and $\left|\mathrm{Del}|^i_\xi\, S[\omega]\right|$ for each cell $\xi \in \mathcal{S}_i$; thus, $\mathrm{Del}|^2_{\mathcal{S}}\, S[\omega]$ is a triangulation of $\mathcal{S}$.

For a vertex v, $\mathrm{Del}|^0_v\, S[\omega]$ is the vertex v itself. The PSC Lemma (i) ensures that for every ridge γ, the vertices of $\mathrm{Del}|^1_\gamma\, S[\omega]$ lie on γ. The disk condition D2 ensures that for every patch σ, the vertices of $\mathrm{Del}|^2_\sigma\, S[\omega]$ lie on σ.

By the PSC Lemma (i), for each ridge γ, $\mathrm{Del}|^1_\gamma\, S[\omega]$ is the polygonal curve that connects each pair of vertices that are adjacent along γ. Hence, the boundary of $\left|\mathrm{Del}|^1_\gamma\, S[\omega]\right|$ is equal to $\mathrm{Del}|^0_{\mathrm{Bd}\,\gamma}\, S[\omega]$, namely, the two vertices of γ.

Consider a patch σ. By the property G1, the boundary of $\left|\mathrm{Del}|^2_\sigma\, S[\omega]\right|$ is homeomorphic to $\mathrm{Bd}\,\sigma$, and every vertex of $\mathrm{Del}|^1_{\mathrm{Bd}\,\sigma}\, S[\omega]$ lies on $\mathrm{Bd}\,\sigma$. The disk condition D4 implies that the boundary of $|\mathrm{Del}|^2_\sigma\, S[\omega]|$ is the union of the edges connecting the vertices adjacent along $\mathrm{Bd}\,\sigma$. By the PSC Lemma (i), $\mathrm{Del}|^1_{\mathrm{Bd}\,\sigma}\, S[\omega]$ is the complex whose edges connect the vertices adjacent along $\mathrm{Bd}\,\sigma$. Therefore, the boundary of $|\mathrm{Del}|^2_\sigma\, S[\omega]|$ is $|\mathrm{Del}|^1_{\mathrm{Bd}\,\sigma}\, S[\omega]|$, and the topological guarantee G2 holds. □

15.6 Extensions: polygonal surfaces, quality, and tetrahedra

In Section 14.4, we observe that Delaunay refinement algorithms often work well when a polygonal surface is substituted where a smooth surface is expected, so long as the polygonal surface is "close enough" to smooth, in the sense of having dihedral angles close to 180°. Although DELPSC can take a polygonal surface as its input and generate a mesh that conforms exactly to the surface, it is sometimes desirable to treat some edges of the input surface as if they were smooth transitions between polygonal faces, rather than treat them as ridges. The faces sharing such edges are united into single patches. One application is *remeshing*, in which we replace a surface mesh with another one that has finer, coarser, or higher-quality triangles.

A user could explicitly specify which edges should be treated as "smooth," or the mesh generator might try to infer it automatically. An input edge that is not incident to exactly two polygons clearly must be treated as a ridge. An edge that adjoins exactly two polygons should be treated as a ridge if the angle at which the polygons meet deviates too far from 180°; the threshold is best chosen by trial and error. The greater the threshold, the more likely that DELPSC will fail to correctly mesh the polygonal surface because of its nonsmoothness. Figure 15.11 depicts meshes obtained by applying this approach to six models.

Once DELPSC generates a surface mesh, Delaunay refinement algorithms similar to DELSURF or DELTETSURF can further refine the mesh to improve the quality of the surface triangles or the tetrahedra they enclose. As in Chapter 9, the algorithm can attack only triangles and tetrahedra with large orthoradius-edge ratios; some simplices with large circumradius-edge ratios may survive. The tetrahedral refinement algorithm also differs from DELTETSURF in that if inserting a vertex at the circumcenter of a poor-quality tetrahedron will destroy some surface triangle, the algorithm simply opts not to insert the new vertex. This approach has the flaw that tetrahedra with large radius-edge ratios sometimes survive near the boundary; we hope that further research will alleviate this flaw. Figure 15.12 depicts three volume meshes generated this way.

15.7 Notes and exercises

The notion of a piecewise smooth complex has a precedent in the idea of a *stratification* of a semialgebraic point set, which is essentially the simplest PSC whose underlying space is the point set. See Gomes [104] for an introduction to the topic.

After the invention of Delaunay mesh generation algorithms for smooth closed surfaces, it was natural to explore nonsmooth surfaces. Boissonnat and Oudot [31] show that their algorithm for smooth surfaces [30] can be extended to a special class of piecewise smooth

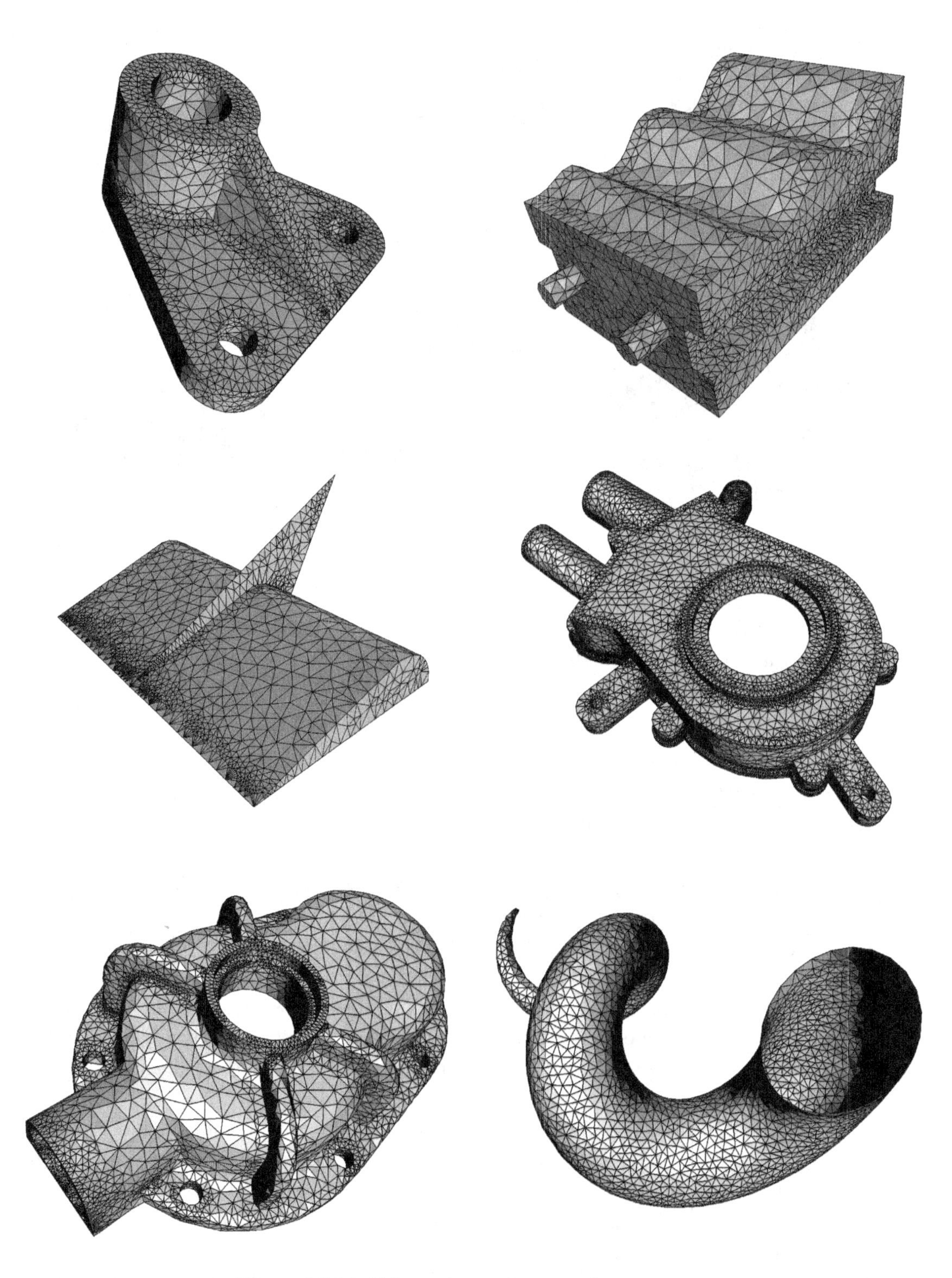

Figure 15.11: Triangular surface meshes.

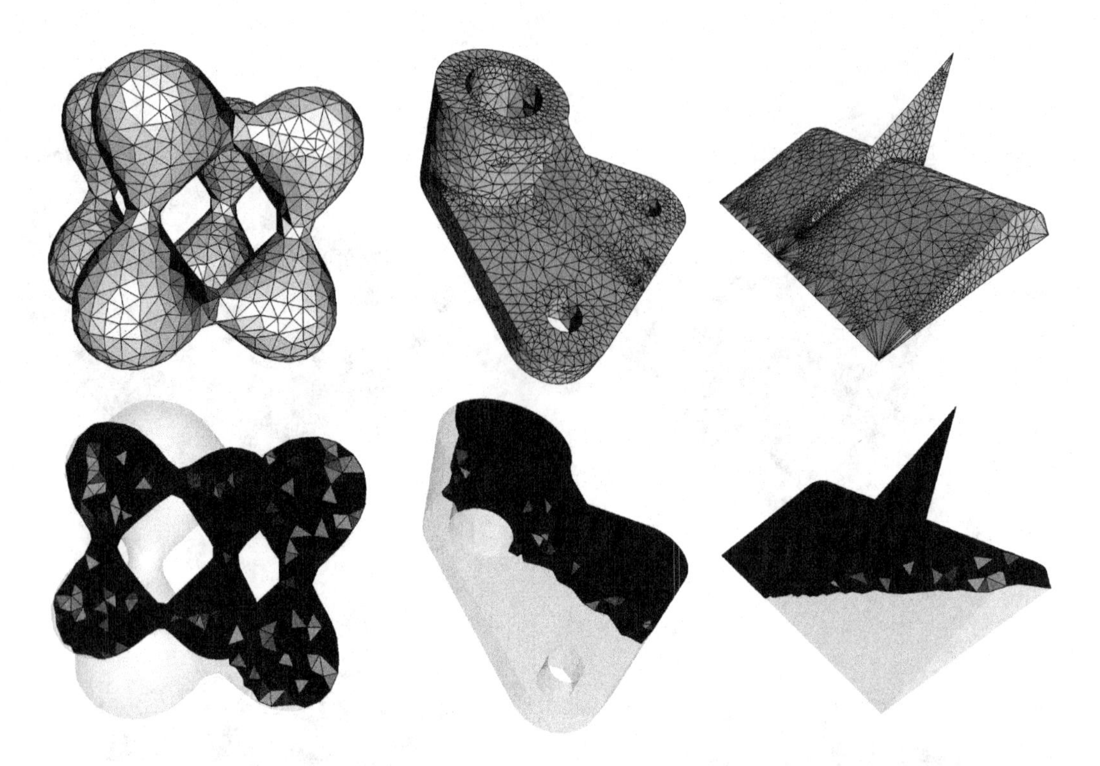

Figure 15.12: Tetrahedral volume meshes. From left to right, a smooth 2-manifold without boundary, a PSC whose underlying space is a 2-manifold, and a PSC whose underlying space is not a manifold and has small dihedral angles.

surfaces called Lipschitz surfaces. Their algorithm is not reliable for domains with small angles—the analysis requires that wherever two smooth surfaces meet along a ridge or at a vertex, their tangent planes locally subtend angles close to 180°. Dey and Ray [79] obtain a similar result for a polygonal surface that approximates a smooth surface.

The first guaranteed algorithm for piecewise smooth complexes with small angles appears in an article by Cheng, Dey, and Ramos [51], who introduce the idea to use weighted vertices as protecting balls in a weighted Delaunay triangulation, as this chapter uses them. Their algorithm accommodates PSCs with arbitrary domain angles, and it guarantees that the final mesh is a triangulation of the input PSC and that the two are related by homeomorphism. The proof of homeomorphism uses an extension of the topological ball property to accommodate PSCs. The paper by Edelsbrunner and Shah [92] that introduces the topological ball property also contains its extension stated in Theorem 15.15.

The algorithm of Cheng et al. [51] requires geometric computations that are very difficult to implement, such as feature size estimates. Cheng, Dey, and Levine [50] simplify some of these computations at the expense of weakening the topological guarantees. Like the DelPSC algorithm in this chapter, they guarantee that each input patch or ridge is represented by output simplices that form a manifold, and if the user-specified size parameter is small enough, the output mesh is a triangulation of the PSC, offering homeomorphism as well. In practice, the correct input topology is usually recovered easily. Dey and Levine [77] eliminate the feature size computations by adopting a strategy that refines the protecting balls (like RefineBall). They further simplify the computations with the idea of aiding balls, discussed in Section 15.2.2. The algorithm DelPSC described in this chapter combines ideas from both papers [50, 77], adds further improvements, and corrects some errors in the article by Dey and Levine. The mesh images in this chapter are generated by Dey and Levine's software.

Exercises

1. (a) Consider a smooth surface patch σ with a smooth boundary $\mathrm{Bd}\,\sigma$. Let M and M' be the medial axes of σ and $\mathrm{Bd}\,\sigma$, respectively. Define

$$\mathrm{lfs}(x) = \min\{d(x, M), \max\{d(x, \mathrm{Bd}\,\sigma), d(x, M')\}\}.$$

   Prove that the function lfs is positive and 1-Lipschitz everywhere on σ.

   (b) Let x and y be two points on a smooth surface patch with a smooth boundary. Prove that if $d(x, y) \leq \varepsilon\,\mathrm{lfs}(x)$ for some $\varepsilon < 1$, then $\angle(\mathbf{n}_x, \mathbf{n}_y) \leq 2\varepsilon$.

2. Consider enlarging the class of domains to permit patches that are piecewise smooth 2-manifolds with boundaries. Modify DelPSC to handle these domains.

3. In the relaxed ball property R3(b), we switch from the distance measure $d(\cdot, \cdot)$ in $\mathbb{R}^3$ to the distance measure $d_\gamma(\cdot, \cdot)$ along γ. Explain whether Cover can preserve R3(b) with $d_\gamma(\cdot, \cdot)$ replaced by $d(\cdot, \cdot)$ and perhaps slightly different constants.

4. Fill in the details left out from the proof of the PSC Lemma (ii).

5. Let $\mathcal{S}$ be a PSC. Suppose that an initial set of protecting balls is given that satisfies the ball properties.

   (a) Can you simplify DelPSC to mesh $\mathcal{S}$ without having to change the given set of protecting balls?

   (b) If you believe that (a) cannot be done, explain why. Suppose that you are allowed to modify the initial set of protecting balls, and then run a simplified DelPSC that meshes $\mathcal{S}$ without having to change the protecting balls again. How would you modify the initial set of protecting balls? Describe your simplified DelPSC.

6. Prove that the properties P1 and P2 imply the extended topological ball properties.

7. The Extended Topological Ball Theorem (Theorem 15.15) was proved by Edelsbrunner and Shah [92]. Prove it yourself.

8. Define extended topological ball properties (Definition 15.12) with CW-complexes (see Hatcher [110] for a definition) instead of regular CW-complexes. Can we extend Theorem 15.15 to accommodate this new definition of extended TBP? We claim that if this is achieved, the guarantees of DelPSC apply to the wider class of PSCs in Definition 15.2, which accommodates ridges and vertices floating in patch interiors. What are the modifications needed to prove this claim?

9. Extend DelPSC to generate tetrahedral volume meshes. Begin by formally defining PSCs that can include three-dimensional cells.

Bibliography

[1] Pankaj K. Agarwal, Lars Arge, and Ke Yi. I/O-efficient construction of constrained Delaunay triangulations. Unpublished manuscript, 2005. Most of this paper appears in the Proceedings of the Thirteenth European Symposium on Algorithms, pages 355–366, October 2005, but the published version omits the analysis of the number of structural changes performed by randomized incremental segment insertion.

[2] Pierre Alliez, David Cohen-Steiner, Mariette Yvinec, and Mathieu Desbrun. Variational tetrahedral meshing. *ACM Transactions on Graphics*, 24(3):617–625, 2005. Special issue on Proceedings of SIGGRAPH 2005.

[3] Nina Amenta and Marshall Bern. Surface reconstruction by Voronoi filtering. *Discrete & Computational Geometry*, 22(4):481–504, June 1999.

[4] Nina Amenta, Marshall Bern, and David Eppstein. Optimal point placement for mesh smoothing. In *Proceedings of the Eighth Annual Symposium on Discrete Algorithms*, pages 528–537, New Orleans, Louisiana, January 1997. Association for Computing Machinery.

[5] Nina Amenta, Marshall Bern, and David Eppstein. The crust and the β-skeleton: Combinatorial curve reconstruction. *Graphical Models and Image Processing*, 60(2):125–135, March 1998.

[6] Nina Amenta, Sunghee Choi, Tamal Krishna Dey, and Naveen Leekha. A simple algorithm for homeomorphic surface reconstruction. *International Journal of Computational Geometry and Applications*, 12(1–2):125–141, 2002.

[7] Nina Amenta, Sunghee Choi, and Günter Rote. Incremental constructions con BRIO. In *Proceedings of the Nineteenth Annual Symposium on Computational Geometry*, pages 211–219, San Diego, California, June 2003. Association for Computing Machinery.

[8] Nina Amenta and Tamal Krishna Dey. Normal variation with adaptive feature size. http://www.cse.ohio-state.edu/~tamaldey/paper/norvar/norvar.pdf, 2007.

[9] Franz Aurenhammer. Power diagrams: Properties, algorithms, and applications. *SIAM Journal on Computing*, 16(1):78–96, February 1987.

[10] Franz Aurenhammer. Voronoi diagrams—A survey of a fundamental geometric data structure. *ACM Computing Surveys*, 23(3):345–405, September 1991.

[11] Franz Aurenhammer and Rolf Klein. Voronoi diagrams. In Jörg-Rüdiger Sack and Jorge Urrutia, editors, *Handbook of Computational Geometry*, pages 201–290. Elsevier Science Publishing, Amsterdam, The Netherlands, 2000.

[12] Ivo Babuška and Abdul Kadir Aziz. On the angle condition in the finite element method. *SIAM Journal on Numerical Analysis*, 13(2):214–226, April 1976.

[13] Brenda S. Baker, Eric Grosse, and Conor S. Rafferty. Nonobtuse triangulation of polygons. *Discrete & Computational Geometry*, 3(2):147–168, December 1988.

[14] Randolph E. Bank and L. Ridgway Scott. On the conditioning of finite element equations with highly refined meshes. *SIAM Journal on Numerical Analysis*, 26(6):1383–1394, December 1989.

[15] Jernej Barbič and Gary L. Miller. A quadratic running time example for Ruppert's refinement algorithm. Technical Report 12-925, Computer Science Department, University of Southern California, Los Angeles, California, 2012.

[16] Marshall Bern and David Eppstein. Mesh generation and optimal triangulation. In Ding-Zhu Du and Frank Hwang, editors, *Computing in Euclidean Geometry*, volume 1 of *Lecture Notes Series on Computing*, pages 23–90. World Scientific, Singapore, 1992.

[17] Marshall Bern, David Eppstein, and Jeff Erickson. Flipping cubical meshes. *Engineering with Computers*, 18(3):173–187, October 2002.

[18] Marshall Bern, David Eppstein, and John R. Gilbert. Provably good mesh generation. In *31st Annual Symposium on Foundations of Computer Science*, pages 231–241. IEEE Computer Society Press, 1990.

[19] John Desmond Bernal and John Leslie Finney. Random close-packed hard-sphere model. II. Geometry of random packing of hard spheres. *Discussions of the Faraday Society*, 43:62–69, 1967.

[20] Ted D. Blacker and Ray J. Meyers. Seams and wedges in Plastering: A 3-D hexahedral mesh generation algorithm. *Engineering with Computers*, 9:83–93, 1993.

[21] Ted D. Blacker and Michael B. Stephenson. Paving: A new approach to automated quadrilateral mesh generation. *International Journal for Numerical Methods in Engineering*, 32(4):811–847, September 1991.

[22] Daniel K. Blandford, Guy E. Blelloch, David E. Cardoze, and Clemens Kadow. Compact representations of simplicial meshes in two and three dimensions. *International Journal of Computational Geometry and Applications*, 15(1):3–24, February 2005.

[23] Guy E. Blelloch, Hal Burch, Karl Crary, Robert Harper, Gary L. Miller, and Noel J. Walkington. Persistent triangulations. *Journal of Functional Programming*, 11(5):441–466, September 2001.

[24] Jules Bloomenthal. Polygonization of implicit surfaces. *Computer Aided Geometric Design*, 5(4):341–355, November 1988.

[25] Harry Blum. A transformation for extracting new descriptors of shape. In Weiant Wathen-Dunn, editor, *Models for the Perception of Speech and Visual Form*, pages 362–380. MIT Press, Cambridge, Massachusetts, 1967.

[26] Jean-Daniel Boissonnat and Frédéric Cazals. Natural coordinates of points on a surface. *Computational Geometry: Theory and Applications*, 19:155–173, July 2001.

[27] Jean-Daniel Boissonnat, David Cohen-Steiner, Bernard Mourrain, Günter Rote, and Gert Vegter. Meshing of surfaces. In Jean-Daniel Boissonnat and Monique Teillaud, editors, *Effective Computational Geometry for Curves and Surfaces*, Mathematics and Visualization series, chapter 5, pages 181–229. Springer, Berlin, 2006.

[28] Jean-Daniel Boissonnat, David Cohen-Steiner, and Gert Vegter. Isotopic implicit surface meshing. *Discrete & Computational Geometry*, 39:138–157, 2008.

[29] Jean-Daniel Boissonnat and Steve Oudot. Provably good surface sampling and approximation. In *Symposium on Geometry Processing*, pages 9–18. Eurographics Association, June 2003.

[30] Jean-Daniel Boissonnat and Steve Oudot. Provably good sampling and meshing of surfaces. *Graphical Models*, 67(5):405–451, September 2005.

[31] Jean-Daniel Boissonnat and Steve Oudot. Provably good sampling and meshing of Lipschitz surfaces. In *Proceedings of the Twenty-Second Annual Symposium on Computational Geometry*, pages 337–346, Sedona, Arizona, June 2006.

[32] Charles Boivin and Carl Ollivier-Gooch. Guaranteed-quality triangular mesh generation for domains with curved boundaries. *International Journal for Numerical Methods in Engineering*, 55(10):1185–1213, 20 August 2002.

[33] Adrian Bowyer. Computing Dirichlet tessellations. *The Computer Journal*, 24(2):162–166, 1981.

[34] Kevin Q. Brown. Voronoi diagrams from convex hulls. *Information Processing Letters*, 9(5):223–228, December 1979.

[35] Kevin Buchin and Wolfgang Mulzer. Linear-time Delaunay triangulations simplified. In *Proceedings of the 25th European Workshop on Computational Geometry*, pages 235–238, Brussels, Belgium, March 2009.

[36] Scott A. Canann, S. N. Muthukrishnan, and R. K. Phillips. Topological refinement procedures for triangular finite element meshes. *Engineering with Computers*, 12(3 & 4):243–255, 1996.

[37] Scott A. Canann, Michael Stephenson, and Ted Blacker. Optismoothing: An optimization-driven approach to mesh smoothing. *Finite Elements in Analysis and Design*, 13:185–190, 1993.

[38] James C. Cavendish. Automatic triangulation of arbitrary planar domains for the finite element method. *International Journal for Numerical Methods in Engineering*, 8(4):679–696, 1974.

[39] James C. Cavendish, David A. Field, and William H. Frey. An approach to automatic three-dimensional finite element mesh generation. *International Journal for Numerical Methods in Engineering*, 21(2):329–347, February 1985.

[40] Timothy M. Chan and Mihai Pătraşcu. Transdichotomous results in computational geometry, I: Point location in sublogarithmic time. *SIAM Journal on Computing*, 39(2):703–729, 2009.

[41] Donald R. Chand and Sham S. Kapur. An algorithm for convex polytopes. *Journal of the Association for Computing Machinery*, 17(1):78–86, January 1970.

[42] Frédéric Chazal and André Lieutier. Weak feature size and persistent homology: Computing homology of solids in $\mathbb{R}^n$ from noisy data samples. In *Proceedings of the 21st Annual Symposium Computational Geometry*, pages 255–262, Pisa, Italy, June 2005.

[43] Frédéric Chazal and André Lieutier. The λ-medial axis. *Graphical Models*, 67(4):304–331, July 2005.

[44] Bernard Chazelle. Convex partitions of polyhedra: A lower bound and worst-case optimal algorithm. *SIAM Journal on Computing*, 13(3):488–507, August 1984.

[45] Bernard Chazelle and Leonidas Palios. Triangulating a nonconvex polytope. *Discrete & Computational Geometry*, 5(1):505–526, December 1990.

[46] Long Chen and Jin-chao Xu. Optimal Delaunay triangulations. *Journal of Computational Mathematics*, 22(2):299–308, 2004.

[47] Ho-Lun Cheng, Tamal Krishna Dey, Herbert Edelsbrunner, and John Sullivan. Dynamic skin triangulation. *Discrete & Computational Geometry*, 25(4):525–568, December 2001.

[48] Siu-Wing Cheng and Tamal Krishna Dey. Quality meshing with weighted Delaunay refinement. *SIAM Journal on Computing*, 33(1):69–93, 2003.

[49] Siu-Wing Cheng, Tamal Krishna Dey, Herbert Edelsbrunner, Michael A. Facello, and Shang-Hua Teng. Sliver exudation. *Journal of the Association for Computing Machinery*, 47(5):883–904, September 2000. Conference version appeared in *Proceedings of the Fifteenth Annual Symposium on Computational Geometry*, 1999.

[50] Siu-Wing Cheng, Tamal Krishna Dey, and Joshua A. Levine. A practical Delaunay meshing algorithm for a large class of domains. In *Proceedings of the 16th International Meshing Roundtable*, pages 477–494, Seattle, Washington, October 2007.

[51] Siu-Wing Cheng, Tamal Krishna Dey, and Edgar A. Ramos. Delaunay refinement for piecewise smooth complexes. *Discrete & Computational Geometry*, 43(1):121–166, 2010. Conference version appeared in *Proceedings of the Eighteenth Annual Symposium on Discrete Algorithms*, 2007.

[52] Siu-Wing Cheng, Tamal Krishna Dey, Edgar A. Ramos, and Tathagata Ray. Quality meshing for polyhedra with small angles. *International Journal of Computational Geometry and Applications*, 15(4):421–461, August 2005. Conference version appeared in *Proceedings of the Twentieth Annual Symposium on Computational Geometry*, 2004.

[53] Siu-Wing Cheng, Tamal Krishna Dey, Edgar A. Ramos, and Tathagata Ray. Sampling and meshing a surface with guaranteed topology and geometry. *SIAM Journal on Computing*, 37(4):1199–1227, 2007.

[54] Siu-Wing Cheng, Tamal Krishna Dey, Edgar A. Ramos, and Rephael Wenger. Anisotropic surface meshing. In *Proceedings of the Seventeenth Annual Symposium on Discrete Algorithms*, pages 202–211, Miami, Florida, January 2006.

[55] Siu-Wing Cheng, Tamal Krishna Dey, and Tathagata Ray. Weighted Delaunay refinement for polyhedra with small angles. In *Proceedings of the 14th International Meshing Roundtable*, pages 325–342, San Diego, California, September 2005.

[56] Siu-Wing Cheng and Sheung-Hung Poon. Graded conforming Delaunay tetrahedralization with bounded radius-edge ratio. In *Proceedings of the Fourteenth Annual Symposium on Discrete Algorithms*, pages 295–304, Baltimore, Maryland, January 2003. Society for Industrial and Applied Mathematics.

[57] Siu-Wing Cheng and Sheung-Hung Poon. Three-dimensional Delaunay mesh generation. *Discrete & Computational Geometry*, 36(3):419–456, October 2006.

[58] L. Paul Chew. Constrained Delaunay triangulations. *Algorithmica*, 4(1):97–108, 1989.

[59] L. Paul Chew. Guaranteed-quality triangular meshes. Technical Report TR-89-983, Department of Computer Science, Cornell University, Ithaca, New York, 1989.

[60] L. Paul Chew. Building Voronoi diagrams for convex polygons in linear expected time. Technical Report PCS-TR90-147, Department of Mathematics and Computer Science, Dartmouth College, Hanover, New Hampshire, 1990.

[61] L. Paul Chew. Guaranteed-quality mesh generation for curved surfaces. In *Proceedings of the Ninth Annual Symposium on Computational Geometry*, pages 274–280, San Diego, California, May 1993. Association for Computing Machinery.

[62] L. Paul Chew. Guaranteed-quality Delaunay meshing in 3D. In *Proceedings of the Thirteenth Annual Symposium on Computational Geometry*, pages 391–393, Nice, France, June 1997. Association for Computing Machinery.

[63] Philippe G. Ciarlet and Pierre-Arnaud Raviart. Maximum principle and uniform convergence for the finite element method. *Computer Methods in Applied Mechanics and Engineering*, 2:17–31, February 1973.

[64] Kenneth L. Clarkson and Peter W. Shor. Applications of random sampling in computational geometry, II. *Discrete & Computational Geometry*, 4(1):387–421, December 1989.

[65] David Cohen-Steiner, Éric Colin de Verdière, and Mariette Yvinec. Conforming Delaunay triangulations in 3D. *Computational Geometry: Theory and Applications*, 28(2–3):217–233, June 2004.

[66] Richard Courant, Kurt Friedrichs, and Hans Lewy. Über die Partiellen Differenzengleichungen der Mathematischen Physik. *Mathematische Annalen*, 100:32–74, August 1928.

[67] Ed F. D'Azevedo and R. Bruce Simpson. On optimal interpolation triangle incidences. *SIAM Journal on Scientific and Statistical Computing*, 10:1063–1075, 1989.

[68] Jesús De Loera, Jörg Rambau, and Francisco Santos. *Triangulations: Structures for Algorithms and Applications*. Springer, Berlin, 2010.

[69] Boris Nikolaevich Delaunay. Sur la sphère vide. *Izvestia Akademia Nauk SSSR, VII Seria, Otdelenie Matematicheskii i Estestvennyka Nauk*, 7:793–800, 1934.

[70] René Descartes. *Principia Philosophiae*. Ludovicus Elzevirius, Amsterdam, The Netherlands, 1644.

[71] Olivier Devillers. On deletion in Delaunay triangulations. *International Journal of Computational Geometry and Applications*, 12(3):193–205, June 2002.

[72] Olivier Devillers, Sylvain Pion, and Monique Teillaud. Walking in a triangulation. *International Journal on Foundations of Computer Science*, 13(2):181–199, April 2002.

[73] Tamal Krishna Dey. *Curve and surface reconstruction: Algorithms with mathematical analysis*. Cambridge University Press, New York, 2006.

[74] Tamal Krishna Dey, Chanderjit L. Bajaj, and Kokichi Sugihara. On good triangulations in three dimensions. *International Journal of Computational Geometry and Applications*, 2(1):75–95, 1992.

[75] Tamal Krishna Dey, Herbert Edelsbrunner, and Sumanta Guha. Computational topology. In Bernard Chazelle, Jacob E. Goodman, and Ricky Pollack, editors, *Advances in Discrete and Computational Geometry*. AMS, Providence, RI, 1998.

[76] Tamal Krishna Dey and Joshua A. Levine. Delaunay meshing of isosurfaces. *Visual Computer*, 24(6):411–422, June 2008.

[77] Tamal Krishna Dey and Joshua A. Levine. Delaunay meshing of piecewise smooth complexes without expensive predicates. *Algorithms*, 2(4):1327–1349, 2009.

[78] Tamal Krishna Dey, Joshua A. Levine, and Andrew G. Slatton. Localized Delaunay refinement for sampling and meshing. *Computer Graphics Forum*, 29:1723–1732, 2010.

[79] Tamal Krishna Dey and Tathagata Ray. Polygonal surface remeshing with Delaunay refinement. *Engineering with Computers*, 26(3):289–301, 2005.

[80] Tamal Krishna Dey and Andrew G. Slatton. Localized Delaunay refinement for volumes. *Computer Graphics Forum*, 30(5):1417–1426, August 2011.

[81] Peter Gustav Lejeune Dirichlet. Über die Reduktion der positiven quadratischen Formen mit drei unbestimmten ganzen Zahlen. *Journal für die Reine und Angewandte Mathematik*, 40:209–227, 1850.

[82] Manfredo P. do Carmo. *Differential Geometry of Curves and Surfaces.* Prentice Hall, Upper Saddle River, New Jersey, 1976.

[83] Rex A. Dwyer. A faster divide-and-conquer algorithm for constructing Delaunay triangulations. *Algorithmica*, 2(2):137–151, 1987.

[84] Rex A. Dwyer. Higher-dimensional Voronoi diagrams in linear expected time. *Discrete & Computational Geometry*, 6(4):343–367, 1991.

[85] Herbert Edelsbrunner. An acyclicity theorem for cell complexes in d dimension. *Combinatorica*, 10(3):251–260, September 1990.

[86] Herbert Edelsbrunner. Deformable smooth surface design. *Discrete & Computational Geometry*, 21(1):87–115, January 1999.

[87] Herbert Edelsbrunner and Damrong Guoy. An experimental study of sliver exudation. In *Proceedings of the 10th International Meshing Roundtable*, pages 307–316, Newport Beach, California, October 2001. Sandia National Laboratories.

[88] Herbert Edelsbrunner, Xiang-Yang Li, Gary Miller, Andreas Stathopoulos, Dafna Talmor, Shang-Hua Teng, Alper Üngör, and Noel J. Walkington. Smoothing and cleaning up slivers. In *Proceedings of the 32nd Annual Symposium on the Theory of Computing*, pages 273–278, Portland, Oregon, May 2000. Association for Computing Machinery.

[89] Herbert Edelsbrunner and Ernst Peter Mücke. Simulation of simplicity: A technique to cope with degenerate cases in geometric algorithms. *ACM Transactions on Graphics*, 9(1):66–104, 1990.

[90] Herbert Edelsbrunner and Raimund Seidel. Voronoi diagrams and arrangements. *Discrete & Computational Geometry*, 1:25–44, 1986.

[91] Herbert Edelsbrunner and Nimish R. Shah. Incremental topological flipping works for regular triangulations. *Algorithmica*, 15(3):223–241, March 1996.

[92] Herbert Edelsbrunner and Nimish R. Shah. Triangulating topological spaces. *International Journal of Computational Geometry and Applications*, 7(4):365–378, August 1997.

[93] Jeff Erickson. Nice point sets can have nasty Delaunay triangulations. *Discrete & Computational Geometry*, 30(1):109–132, July 2003.

[94] Hale Erten and Alper Üngör. Triangulations with locally optimal Steiner points. In *Symposium on Geometry Processing 2007*, pages 143–152, July 2007.

[95] David A. Field. Qualitative measures for initial meshes. *International Journal for Numerical Methods in Engineering*, 47:887–906, 2000.

[96] Steven Fortune. A sweepline algorithm for Voronoi diagrams. *Algorithmica*, 2(2):153–174, 1987.

[97] C. O. Frederick, Y. C. Wong, and F. W. Edge. Two-dimensional automatic mesh generation for structural analysis. *International Journal for Numerical Methods in Engineering*, 2:133–144, 1970.

[98] Lori A. Freitag, Mark Jones, and Paul E. Plassmann. An efficient parallel algorithm for mesh smoothing. In *Proceedings of the 4th International Meshing Roundtable*, pages 47–58, Albuquerque, New Mexico, October 1995. Sandia National Laboratories.

[99] Lori A. Freitag and Carl Ollivier-Gooch. Tetrahedral mesh improvement using swapping and smoothing. *International Journal for Numerical Methods in Engineering*, 40(21):3979–4002, November 1997.

[100] William H. Frey. Selective refinement: A new strategy for automatic node placement in graded triangular meshes. *International Journal for Numerical Methods in Engineering*, 24(11):2183–2200, November 1987.

[101] Isaac Fried. Condition of finite element matrices generated from nonuniform meshes. *AIAA Journal*, 10(2):219–221, February 1972.

[102] John Alan George. *Computer Implementation of the Finite Element Method.* PhD thesis, Stanford University, Stanford, California, March 1971. Technical report STAN-CS-71-208.

[103] Paul-Louis George and Houman Borouchaki. *Delaunay Triangulation and Meshing: Application to Finite Elements*. Hermès, Paris, 1998.

[104] Abel J. P. Gomes. A concise B-rep data structure for stratified subanalytic objects. In Leif Kobbelt, Peter Schröder, and Hugues Hoppe, editors, *Symposium on Geometry Processing*, pages 83–93, Aachen, Germany, June 2003. Eurographics Association.

[105] Nicolas Grislain and Jonathan Richard Shewchuk. The strange complexity of constrained Delaunay triangulation. In *Proceedings of the Fifteenth Canadian Conference on Computational Geometry*, pages 89–93, Halifax, Nova Scotia, Canada, August 2003.

[106] Leonidas J. Guibas and Jorge Stolfi. Primitives for the manipulation of general subdivisions and the computation of Voronoi diagrams. *ACM Transactions on Graphics*, 4(2):74–123, April 1985.

[107] Damrong Guoy. *Tetrahedral Mesh Improvement, Algorithms and Experiments*. PhD thesis, Department of Computer Science, University of Illinois at Urbana-Champaign, 2001.

[108] Hugo Hadwiger. *Vorlesungen über Inhalt, Oberfläche und Isoperimetrie*. Springer-Verlag, Berlin, 1957.

[109] Sariel Har-Peled and Alper Üngör. A time-optimal Delaunay refinement algorithm in two dimensions. In *Proceedings of the Twenty-First Annual Symposium on Computational Geometry*, pages 228–236, Pisa, Italy, June 2005. Association for Computing Machinery.

[110] Allen Hatcher. *Algebraic Topology*. Cambridge University Press, New York, 2002.

[111] François Hermeline. *Une Methode Automatique de Maillage en Dimension n*. PhD thesis, Université Pierre et Marie Curie, Paris, France, 1980.

[112] François Hermeline. Triangulation automatique d'un polyèdre en dimension *N*. *RAIRO Analyse Numérique*, 16(3):211–242, 1982.

[113] John G. Hocking and Gail S. Young. *Topology*. Dover, New York, 1961.

[114] Robert Elmer Horton. Rational study of rainfall data makes possible better estimates of water yield. *Engineering News-Record*, pages 211–213, 1917.

[115] Benoît Hudson, Gary L. Miller, and Todd Phillips. Sparse Voronoi refinement. In *Proceedings of the 15th International Meshing Roundtable*, pages 339–358, Birmingham, Alabama, September 2006. Springer.

[116] Antony Jameson, Timothy J. Baker, and Nigel P. Weatherill. Calculation of inviscid transonic flow over a complete aircraft. In *Proceedings of the 24th AIAA Aerospace Sciences Meeting*, Reno, Nevada, January 1986. AIAA paper 86-0103.

[117] Pierre Jamet. Estimations d'erreur pour des éléments finis droits presque dégénérés. *RAIRO Analyse Numérique*, 10(1):43–60, 1976.

[118] Barry Joe. Three-dimensional triangulations from local transformations. *SIAM Journal on Scientific and Statistical Computing*, 10:718–741, 1989.

[119] Barry Joe. Construction of k-dimensional Delaunay triangulations using local transformations. *SIAM Journal on Scientific Computing*, 14(6):1415–1436, November 1993.

[120] Barry Joe. Construction of three-dimensional improved-quality triangulations using local transformations. *SIAM Journal on Scientific Computing*, 16(6):1292–1307, November 1995.

[121] H. A. Kamel and K. Eisenstein. Automatic mesh generation in two- and three-dimensional inter-connected domains. In *Symposium on High Speed Computing of Elastic Structures*, Les Congrès et Colloques de l'Université de Liège, Liège, Belgium, 1970.

[122] Paul Kinney. CleanUp: Improving quadrilateral finite element meshes. In *Proceedings of the 6th International Meshing Roundtable*, pages 449–461, Park City, Utah, October 1997. Sandia National Laboratories.

[123] Rolf Klein and Andrzej Lingas. A note on generalizations of Chew's algorithm for the Voronoi diagram of a convex polygon. In *Proceedings of the Fifth Canadian Conference on Computational Geometry*, pages 370–374, Waterloo, Ontario, Canada, August 1993.

[124] Bryan Matthew Klingner and Jonathan Richard Shewchuk. Aggressive tetrahedral mesh improvement. In *Proceedings of the 16th International Meshing Roundtable*, pages 3–23, Seattle, Washington, October 2007. Springer.

[125] Michal Křížek. On the maximum angle condition for linear tetrahedral elements. *SIAM Journal on Numerical Analysis*, 29(2):513–520, April 1992.

[126] François Labelle. Sliver removal by lattice refinement. In *Proceedings of the 23rd Annual Symposium on Computational Geometry*, pages 347–356, Sedona, Arizona, June 2006.

[127] François Labelle and Jonathan Richard Shewchuk. Anisotropic Voronoi diagrams and guaranteed-quality anisotropic mesh generation. In *Proceedings of the Nineteenth Annual Symposium on Computational Geometry*, pages 191–200, San Diego, California, June 2003. Association for Computing Machinery.

[128] François Labelle and Jonathan Richard Shewchuk. Isosurface stuffing: Fast tetrahedral meshes with good dihedral angles. *ACM Transactions on Graphics*, 26(3):57.1–57.10, July 2007. Special issue on Proceedings of SIGGRAPH 2007.

[129] Timothy Lambert. The Delaunay triangulation maximizes the mean inradius. In *Proceedings of the Sixth Canadian Conference on Computational Geometry*, pages 201–206, Saskatoon, Saskatchewan, Canada, August 1994.

[130] Charles L. Lawson. Transforming triangulations. *Discrete Mathematics*, 3(4):365–372, 1972.

[131] Charles L. Lawson. Software for C^1 surface interpolation. In John R. Rice, editor, *Mathematical Software III*, pages 161–194. Academic Press, New York, 1977.

[132] Charles L. Lawson. Properties of n-dimensional triangulations. *Computer Aided Geometric Design*, 3(4):231–246, December 1986.

[133] Der-Tsai Lee and Arthur K. Lin. Generalized Delaunay triangulations for planar graphs. *Discrete & Computational Geometry*, 1:201–217, 1986.

[134] Der-Tsai Lee and Bruce J. Schachter. Two algorithms for constructing a Delaunay triangulation. *International Journal of Computer and Information Sciences*, 9(3):219–242, 1980.

[135] C. G. Lekkerkerker. *Geometry of Numbers*. Wolters-Noordhoff, Groningen, 1969.

[136] Xiang-Yang Li. *Sliver-free 3-dimensional Delaunay mesh generation*. PhD thesis, University of Illinois at Urbana-Champaign, 2000.

[137] Xiang-Yang Li and Shang-Hua Teng. Generating well-shaped Delaunay meshes in 3D. In *Proceedings of the Twelfth Annual Symposium on Discrete Algorithms*, pages 28–37, Washington, D.C., January 2001. Association for Computing Machinery.

[138] Long Lin and Chee Yap. Adaptive isotopic approximation of nonsingular curves: The parametrizability and nonlocal isotopy approach. In *Proceedings of the Twenty-Fifth Annual Symposium on Computational Geometry*, pages 351–360, Århus, Denmark, June 2009.

[139] S. H. Lo. A new mesh generation scheme for arbitrary planar domains. *International Journal for Numerical Methods in Engineering*, 21(8):1403–1426, August 1985.

[140] Rainald Löhner and Paresh Parikh. Generation of three-dimensional unstructured grids by the advancing-front method. *International Journal for Numerical Methods in Fluids*, 8(10):1135–1149, October 1988.

[141] William E. Lorensen and Harvey E. Cline. Marching cubes: A high resolution 3D surface construction algorithm. In *Computer Graphics (SIGGRAPH '87 Proceedings)*, pages 163–170, Anaheim, California, July 1987.

[142] David L. Marcum. Unstructured grid generation using automatic point insertion and local reconnection. In Joe F. Thompson, Bharat K. Soni, and Nigel P. Weatherill, editors, *Handbook of Grid Generation*, chapter 18, pages 18.1–18.31. CRC Press, Boca Raton, Florida, 1999.

[143] Georges Matheron. Examples of topological properties of skeletons. In Jean Serra, editor, *Image Analysis and Mathematical Morphology, Vol. 2*, pages 217–238. Academic Press, London, 1988.

[144] D. H. McLain. Two dimensional interpolation from random data. *The Computer Journal*, 19(2):178–181, May 1976.

[145] Peter McMullen. The maximum number of faces of a convex polytope. *Mathematika*, 17:179–184, 1970.

[146] Elefterios A. Melissaratos. L_p optimal d dimensional triangulations for piecewise linear interpolation: A new result on data dependent triangulations. Technical Report RUU-CS-93-13, Department of Computer Science, Utrecht University, Utrecht, The Netherlands, April 1993.

[147] Gary L. Miller, Steven E. Pav, and Noel J. Walkington. When and why Ruppert's algorithm works. *International Journal of Computational Geometry and Applications*, 15(1):25–54, February 2005.

[148] Gary L. Miller, Dafna Talmor, Shang-Hua Teng, and Noel J. Walkington. A Delaunay based numerical method for three dimensions: Generation, formulation, and partition. In *Proceedings of the Twenty-Seventh Annual Symposium on the Theory of Computing*, pages 683–692, Las Vegas, Nevada, May 1995.

[149] Gary L. Miller, Dafna Talmor, Shang-Hua Teng, Noel J. Walkington, and Han Wang. Control volume meshes using sphere packing: Generation, refinement and coarsening. In *Proceedings of the 5th International Meshing Roundtable*, pages 47–61, Pittsburgh, Pennsylvania, October 1996.

[150] Scott A. Mitchell. Cardinality bounds for triangulations with bounded minimum angle. In *Proceedings of the Sixth Canadian Conference on Computational Geometry*, pages 326–331, Saskatoon, Saskatchewan, Canada, August 1994.

[151] Scott A. Mitchell and Stephen A. Vavasis. Quality mesh generation in three dimensions. In *Proceedings of the Eighth Annual Symposium on Computational Geometry*, pages 212–221, 1992.

[152] Scott A. Mitchell and Stephen A. Vavasis. Quality mesh generation in higher dimensions. *SIAM Journal on Computing*, 29(4):1334–1370, 2000.

[153] Bernard Mourrain and Jean-Pierre Técourt. Computing the topology of real algebraic surfaces. In *MEGA 2005: Effective Methods in Algebraic Geometry*, Alghero, Italy, May 2005.

[154] Ketan Mulmuley. On levels in arrangements and Voronoi diagrams. *Discrete & Computational Geometry*, 6:307–338, 1991.

[155] James R. Munkres. *Topology*. Prentice Hall, Upper Saddle River, New Jersey, second edition, 2000.

[156] Michael Murphy, David M. Mount, and Carl W. Gable. A point-placement strategy for conforming Delaunay tetrahedralization. *International Journal of Computational Geometry and Applications*, 11(6):669–682, December 2001.

[157] Demian Nave, Nikos Chrisochoides, and L. Paul Chew. Guaranteed-quality parallel Delaunay refinement for restricted polyhedral domains. *Computational Geometry: Theory and Applications*, 28:191–215, 2004.

[158] Friedhelm Neugebauer and Ralf Diekmann. Improved mesh generation: Not simple but good. In *Proceedings of the 5th International Meshing Roundtable*, pages 257–270, Pittsburgh, Pennsylvania, October 1996. Sandia National Laboratories.

[159] Van Phai Nguyen. Automatic mesh generation with tetrahedral elements. *International Journal for Numerical Methods in Engineering*, 18:273–289, 1982.

[160] Paul Niggli. Die topologische Strukturanalyse I. *Zeitschrift für Kristallographie*, 65:391–415, 1927.

[161] J. Tinsley Oden and Leszek F. Demkowicz. Advances in adaptive improvements: A survey of adaptive finite element methods in computational mechanics. In *State-of-the-Art Surveys on Computational Mechanics*, pages 441–467. The American Society of Mechanical Engineers, 1989.

[162] Steve Oudot, Laurent Rineau, and Mariette Yvinec. Meshing volumes bounded by smooth surfaces. In *Proceedings of the 14th International Meshing Roundtable*, pages 203–219, San Diego, California, September 2005. Springer.

[163] V. N. Parthasarathy, C. M. Graichen, and A. F. Hathaway. A comparison of tetrahedron quality measures. *Finite Elements in Analysis and Design*, 15(3):255–261, January 1994.

[164] V. N. Parthasarathy and Srinivas Kodiyalam. A constrained optimization approach to finite element mesh smoothing. *Finite Elements in Analysis and Design*, 9(4):309–320, September 1991.

[165] Steven E. Pav and Noel J. Walkington. Robust three dimensional Delaunay refinement. In *Proceedings of the 13th International Meshing Roundtable*, pages 145–156, Williamsburg, Virginia, September 2004. Sandia National Laboratories.

[166] Steven E. Pav and Noel J. Walkington. Delaunay refinement by corner lopping. In *Proceedings of the 14th International Meshing Roundtable*, pages 165–181, San Diego, California, September 2005. Springer.

[167] Steven Elliot Pav. *Delaunay Refinement Algorithms*. PhD thesis, Department of Mathematical Sciences, Carnegie Mellon University, Pittsburgh, Pennsylvania, May 2003.

[168] Jaime Peraire, Joaquim Peiró, Luca Formaggia, Ken Morgan, and Olgierd C. Zienkiewicz. Finite element Euler computations in three dimensions. *International Journal for Numerical Methods in Engineering*, 26(10):2135–2159, October 1988.

[169] Jaime Peraire, Joaquim Peiró, and Ken Morgan. Advancing front grid generation. In Joe F. Thompson, Bharat K. Soni, and Nigel P. Weatherill, editors, *Handbook of Grid Generation*, chapter 17, pages 17.1–17.22. CRC Press, Boca Raton, Florida, 1999.

[170] Jaime Peraire, Mehdi Vahdati, Ken Morgan, and Olgierd C. Zienkiewicz. Adaptive remeshing for compressible flow computations. *Journal of Computational Physics*, 72(2):449–466, October 1987.

[171] Simon Plantinga and Gert Vegter. Isotopic approximation of implicit curves and surfaces. In *Symposium on Geometry Processing 2004*, pages 245–254, Nice, France, July 2004. Eurographics Association.

[172] P. L. Powar. Minimal roughness property of the Delaunay triangulation: A shorter approach. *Computer Aided Geometric Design*, 9(6):491–494, December 1992.

[173] V. T. Rajan. Optimality of the Delaunay triangulation in $\mathbb{R}^d$. In *Proceedings of the Seventh Annual Symposium on Computational Geometry*, pages 357–363, North Conway, New Hampshire, June 1991.

[174] Alexander Rand and Noel J. Walkington. Collars and intestines: Practical conformal Delaunay refinement. In *Proceedings of the 18th International Meshing Roundtable*, pages 481–497, Salt Lake City, Utah, October 2009. Springer.

[175] Shmuel Rippa. Minimal roughness property of the Delaunay triangulation. *Computer Aided Geometric Design*, 7(6):489–497, November 1990.

[176] Shmuel Rippa. Long and thin triangles can be good for linear interpolation. *SIAM Journal on Numerical Analysis*, 29(1):257–270, February 1992.

[177] Claude Ambrose Rogers. *Packing and Covering*. Cambridge University Press, Cambridge, United Kingdom, 1964.

[178] Jim Ruppert. A new and simple algorithm for quality 2-dimensional mesh generation. Technical Report UCB/CSD 92/694, Computer Science Division, University of California at Berkeley, Berkeley, California, 1992.

[179] Jim Ruppert. A new and simple algorithm for quality 2-dimensional mesh generation. In *Proceedings of the Fourth Annual Symposium on Discrete Algorithms*, pages 83–92, January 1993. Association for Computing Machinery.

[180] Jim Ruppert. A Delaunay refinement algorithm for quality 2-dimensional mesh generation. *Journal of Algorithms*, 18(3):548–585, May 1995.

[181] Jim Ruppert and Raimund Seidel. On the difficulty of triangulating three-dimensional nonconvex polyhedra. *Discrete & Computational Geometry*, 7(3):227–253, 1992.

[182] Edward A. Sadek. A scheme for the automatic generation of triangular finite elements. *International Journal for Numerical Methods in Engineering*, 15(12):1813–1822, December 1980.

[183] Hanan Samet. The quadtree and related hierarchical data structures. *Computing Surveys*, 16:188–260, 1984.

[184] Francisco Santos. A point configuration whose space of triangulations is disconnected. *Journal of the American Mathematical Society*, 13(3):611–637, 2000.

[185] Francisco Santos. Non-connected toric Hilbert schemes. *Mathematische Annalen*, 332(3):645–665, May 2005.

[186] Robert Schneiders. Quadrilateral and hexahedral element meshes. In Joe F. Thompson, Bharat K. Soni, and Nigel P. Weatherill, editors, *Handbook of Grid Generation*, chapter 21, pages 21.1–21.27. CRC Press, Boca Raton, Florida, 1999.

[187] E. Schönhardt. Über die Zerlegung von Dreieckspolyedern in Tetraeder. *Mathematische Annalen*, 98:309–312, 1928.

[188] William J. Schroeder and Mark S. Shephard. Geometry-based fully automatical mesh generation and the Delaunay triangulation. *International Journal for Numerical Methods in Engineering*, 26(11):2503–2515, November 1988.

[189] Hermann A. Schwarz. Sur une définition erronée de l'aire d'une surface courbe. In *Gesammelte Mathematische Abhandlungen II*, pages 309–311. Verlag von Julius Springer, Berlin, 1890.

[190] Raimund Seidel. Voronoi Diagrams in Higher Dimensions. Diplomarbeit, Institut für Informationsverarbeitung, Technische Universität Graz, Graz, Austria, 1982.

[191] Raimund Seidel. The upper bound theorem for polytopes: An easy proof of its asymptotic version. *Computational Geometry: Theory and Applications*, 5:115–116, 1985.

[192] Raimund Seidel. Constructing higher-dimensional convex hulls at logarithmic cost per face. In *Proceedings of the Eighteenth Annual Symposium on the Theory of Computing*, pages 404–413. Association for Computing Machinery, 1986.

[193] Raimund Seidel. Constrained Delaunay triangulations and Voronoi diagrams with obstacles. In H. S. Poingratz and W. Schinnerl, editors, *1978–1988 Ten Years IIG*, pages 178–191. Institute for Information Processing, Graz University of Technology, Graz, Austria, 1988.

[194] Raimund Seidel. Backwards analysis of randomized geometric algorithms. In János Pach, editor, *New Trends in Discrete and Computational Geometry*, volume 10 of *Algorithms and Combinatorics*, pages 37–67. Springer-Verlag, Berlin, 1993.

[195] Michael Ian Shamos and Dan Hoey. Closest-point problems. In *16th Annual Symposium on Foundations of Computer Science*, pages 151–162, Berkeley, California, October 1975. IEEE Computer Society Press.

[196] Jonathan Richard Shewchuk. Triangle: Engineering a 2D quality mesh generator and Delaunay triangulator. In Ming C. Lin and Dinesh Manocha, editors, *Applied Computational Geometry: Towards Geometric Engineering*, volume 1148 of *Lecture Notes in Computer Science*, pages 203–222. Springer-Verlag, Berlin, May 1996. From the First ACM Workshop on Applied Computational Geometry.

[197] Jonathan Richard Shewchuk. *Delaunay Refinement Mesh Generation*. PhD thesis, School of Computer Science, Carnegie Mellon University, Pittsburgh, Pennsylvania, May 1997. Available as Technical Report CMU-CS-97-137.

[198] Jonathan Richard Shewchuk. Tetrahedral mesh generation by Delaunay refinement. In *Proceedings of the Fourteenth Annual Symposium on Computational Geometry*, pages 86–95, Minneapolis, Minnesota, June 1998. Association for Computing Machinery.

[199] Jonathan Richard Shewchuk. Mesh generation for domains with small angles. In *Proceedings of the Sixteenth Annual Symposium on Computational Geometry*, pages 1–10, Hong Kong, June 2000. Association for Computing Machinery.

[200] Jonathan Richard Shewchuk. Sweep algorithms for constructing higher-dimensional constrained Delaunay triangulations. In *Proceedings of the Sixteenth Annual Symposium on Computational Geometry*, pages 350–359, Hong Kong, June 2000. Association for Computing Machinery.

[201] Jonathan Richard Shewchuk. Delaunay refinement algorithms for triangular mesh generation. *Computational Geometry: Theory and Applications*, 22(1–3):21–74, May 2002.

[202] Jonathan Richard Shewchuk. What is a good linear element? Interpolation, conditioning, and quality measures. In *Proceedings of the 11th International Meshing Roundtable*, pages 115–126, Ithaca, New York, September 2002. Sandia National Laboratories.

[203] Jonathan Richard Shewchuk. Updating and constructing constrained Delaunay and constrained regular triangulations by flips. In *Proceedings of the Nineteenth Annual Symposium on Computational Geometry*, pages 181–190, San Diego, California, June 2003. Association for Computing Machinery.

[204] Jonathan Richard Shewchuk. General-dimensional constrained Delaunay triangulations and constrained regular triangulations I: Combinatorial properties. *Discrete & Computational Geometry*, 39(1–3):580–637, March 2008.

[205] Jonathan Richard Shewchuk and Brielin C. Brown. Inserting a segment into a constrained Delaunay triangulation in expected linear time. Unpublished manuscript, 2012.

[206] Hang Si. Constrained Delaunay tetrahedral mesh generation and refinement. *Finite Elements in Analysis and Design*, 46:33–46, January–February 2010.

[207] John M. Snyder. Interval analysis for computer graphics. In *Computer Graphics (SIGGRAPH '92 Proceedings)*, pages 121–130, July 1992.

[208] Daniel A. Spielman, Shang-Hua Teng, and Alper Üngör. Parallel Delaunay refinement: Algorithms and analyses. *International Journal of Computational Geometry and Applications*, 17(1):1–30, February 2007.

[209] John Stillwell. *Classical Topology and Combinatorial Group Theory*. Springer-Verlag, New York, 1980.

[210] Peter Su and Robert L. Scot Drysdale. A comparison of sequential Delaunay triangulation algorithms. In *Proceedings of the Eleventh Annual Symposium on Computational Geometry*, pages 61–70, Vancouver, British Columbia, Canada, June 1995. Association for Computing Machinery.

[211] Garret Swart. Finding the convex hull facet by facet. *Journal of Algorithms*, 6(1):17–48, March 1985.

[212] John Lighton Synge. *The Hypercircle in Mathematical Physics*. Cambridge University Press, New York, 1957.

[213] Masaharu Tanemura, Tohru Ogawa, and Naofumi Ogita. A new algorithm for three-dimensional Voronoi tessellation. *Journal of Computational Physics*, 51(2):191–207, August 1983.

[214] Alfred H. Thiessen. Precipitation average for large area. *Monthly Weather Review*, 39:1982–1084, 1911.

[215] Joe F. Thompson. The National Grid Project. *Computer Systems in Engineering*, 3(1–4):393–399, 1992.

[216] Joe F. Thompson, Bharat K. Soni, and Nigel P. Weatherill, editors. *Handbook of Grid Generation*. CRC Press, Boca Raton, Florida, 1999.

[217] Joe F. Thompson and Nigel P. Weatherill. Aspects of numerical grid generation: Current science and art. In *Eleventh AIAA Applied Aerodynamics Conference*, pages 1029–1070, 1993. AIAA paper 93-3593-CP.

[218] Jane Tournois, Rahul Srinivasan, and Pierre Alliez. Perturbing slivers in 3D Delaunay meshes. In *Proceedings of the 18th International Meshing Roundtable*, pages 157–173, Salt Lake City, Utah, October 2009.

[219] Alper Üngör. Off-centers: A new type of Steiner points for computing size-optimal quality-guaranteed Delaunay triangulations. *Computational Geometry: Theory and Applications*, 42(2):109–118, February 2009.

[220] Georges Voronoi. Nouvelles applications des paramètres continus à la théorie des formes quadratiques. *Journal für die Reine und Angewandte Mathematik*, 133:97–178, 1908.

[221] Shayne Waldron. The error in linear interpolation at the vertices of a simplex. *SIAM Journal on Numerical Analysis*, 35(3):1191–1200, 1998.

[222] David F. Watson. Computing the n-dimensional Delaunay tessellation with application to Voronoi polytopes. *The Computer Journal*, 24(2):167–172, 1981.

[223] Jeffrey R. Weeks. *The Shape of Space*. Marcel Dekker Inc., New York, 1985.

[224] Eugene Wigner and Frederick Seitz. On the constitution of metallic sodium. *Physical Review*, 43:804–810, 1933.

[225] Franz-Erich Wolter. Cut locus and medial axis in global shape interrogation and representation. Design Laboratory Memorandum 92-2 and MIT Sea Grant Report, Department of Ocean Engineering, Massachusetts Institute of Technology, Cambridge, Massachusetts, 1992.

[226] Mark A. Yerry and Mark S. Shephard. A modified quadtree approach to finite element mesh generation. *IEEE Computer Graphics and Applications*, 3:39–46, January/February 1983.

[227] Mark A. Yerry and Mark S. Shephard. Automatic three-dimensional mesh generation by the modified-octree technique. *International Journal for Numerical Methods in Engineering*, 20(11):1965–1990, November 1984.

[228] Günter M. Ziegler. *Lectures on Polytopes*. Springer-Verlag, New York, revised first edition, 1995.

Index

Printed in the United States
by Baker & Taylor Publisher Services